Lecture Notes in Mathematics 2120

More information about this series at http://www.springer.com/series/304

Volker Schmidt

Editor

Stochastic Geometry, Spatial Statistics and Random Fields

Models and Algorithms

 Springer

Editor
Volker Schmidt
Ulm University
Institute of Stochastics
Ulm, Germany

ISBN 978-3-319-10063-0   ISBN 978-3-319-10064-7 (eBook)
DOI 10.1007/978-3-319-10064-7
Springer Cham Heidelberg New York Dordrecht London

Lecture Notes in Mathematics ISSN print edition: 0075-8434
ISSN electronic edition: 1617-9692

Library of Congress Control Number: 2014949195

Mathematics Subject Classification (2010): 60D05, 60G55, 60G60, 62H11, 62M40, 62P30, 62P10

Preface

This volume is an attempt to provide a graduate level introduction to various aspects of stochastic geometry, spatial statistics and random fields, with special emphasis placed on fundamental classes of models and algorithms as well as on their applications. This book has a strong focus on simulations and includes extensive code in *Matlab* and *R*, which are widely used in the mathematical community. It can be seen as a continuation of the recent volume 2068 of Lecture Notes in Mathematics, where other issues of stochastic geometry, spatial statistics and random fields were considered with a focus on asymptotic methods.

The present volume comprises selected contributions to the Summer Academy on Stochastic Analysis, Modelling and Simulation of Complex Structures (cf. http://www.uni-ulm.de/mawi/summer-academy-2011.html) which took place during September 11-24, 2011 at the Söllerhaus, an Alpine conference center of the University of Stuttgart and RWTH Aachen, in the village of Hirschegg (Kleinwalsertal). It was organized by the Institute of Stochastics of Ulm University. In contrast with previous summer schools on this subject (Sandbjerg 2000, Martina Franca 2004, Sandbjerg 2007, Hirschegg 2009), the focus of this summer school was on models and algorithms from stochastic geometry, spatial statistics and random fields which can be used for the analysis, modelling and simulation of geometrically complex microstructures. Examples of such microstructures are complex point patterns and networks which appear in advanced functional materials and whose geometrical structures are closely related to the (macroscopic) physical properties of the underlying technical and biological materials.

This summer school hosted 43 young participants from 8 countries (Australia, Czech Republic, Denmark, France, Germany, Russia, Switzerland, UK). Fourteen international experts gave lectures on various aspects of stochastic geometry, spatial statistics and random fields. In addition, students and young researchers were able to give their own short talks.

As stated above, this volume is focused on fundamental classes of models and algorithms from stochastic geometry, spatial statistics and random fields as well as on their applications to the analysis, modelling and simulation of geometrically complex objects like points patterns, tessellations, graphs, and trees. It reflects recent progress in these areas, with respect to both theory and applications.

Point processes play an especially important role in many parts of the book. Poisson processes, the most fundamental class of point processes, are considered in Chapter 1 as approximations of other (not necessarily Poisson) point processes. In Chapter 2, they are used as a basic reference model when comparing the clustering properties of various point processes and, in Chapters 3 and 5, they are used in order to construct some basic classes of random tessellations and Boolean random functions, respectively. Several further classes of point processes considered in this book can be seen as generalizations of Poisson processes and are derived in various ways from Poisson processes. For example, Gibbs processes (Chapters 1 and 2), Poisson cluster processes (Chapters 2, 4 and 13), Cox processes (Chapters 2, 4, 7 and 13), and hard-core processes (Chapters 2 and 4).

From another perspective, point processes can be seen as a special class of random closed sets, which are also fundamental objects in stochastic geometry. In addition to point processes, a number of other classes of random closed sets are considered in the book; in particular, germ-grain models (Chapters 2, 4 and 5), random fiber processes (Chapters 4 and 6), random surface processes (Chapter 5), random tessellations (Chapters 3 and 6), random spatial networks (Chapters 2 and 4) and isotropic convex bodies (Chapter 8). Random marked closed sets (Chapter 6) – in particular, random marked point processes (Chapter 13) – also play an important role in the book.

Another focus of the book is on various aspects of random fields. In Chapter 5, a class of random fields is considered which can be seen as a generalization of random closed sets and, in particular, of germ-grain models. In Chapter 6, random fields are used in order to construct random marked sets. Some basic ideas of principal component analysis for random fields are discussed in Chapter 9. Genetic models involving random fields are considered in Chapter 10. Chapter 11 deals with extrapolation techniques for two large classes of random fields: square-integrable stationary random fields and stable random fields. In addition, various simulation algorithms for random fields are discussed in Chapters 12 and 13, in particular for Gaussian Markov random fields, fractional Gaussian fields, spatial Lévy processes and random walks.

The book is organized as follows. The first four chapters deal with point processes, random tessellations and random spatial networks, with a number of different examples of their applications. Chapter 1 gives an introduction to Stein's method, a powerful technique for computing explicit error bounds for distributional approximation. The classical case of normal approximation is used as an initial motivation. Then, the main part of the chapter is devoted to presenting the key concepts of Stein's method in a much more general framework, where the approximating distribution and the space it lives on can be almost arbitrary. This is particularly appealing for distributional approximation of various point-process models considered in stochastic geometry and spatial statistics. Chapter 2 reviews examples, methods, and recent results concerning the comparison of clustering properties of point processes. The approach is founded on the observation that void probabilities and moment measures can be used as two complementary tools for capturing clustering phenomena in point processes. Various global and local functionals of random geometric models driven by point processes are considered which admit more or less explicit bounds involving void probabilities and moment measures. Directional convex ordering of point processes is also discussed. Such an ordering turns out to be an appropriate choice, combined with the notion of (positive or negative) association, when comparison to the Poisson point process is considered. Chapter 3 introduces various tessellation models and discusses their application as models for cellular materials. First, the notion of a random tessellation is introduced, along with the most well-known model types (Voronoi and Laguerre tessellations, hyperplane tessellations, STIT tessellations), and their basic geometric characteristics. Assuming that a cellular material is a realization of a suitable random tessellation model, characteristics of these models can be estimated from 3D images of the material. An explanation is given of how such estimates are obtained and how they can be used to fit tessellation models to the observed microstructure. In Chapter 4, three classes of stochastic morphology models are presented. These describe different microstructures of functional materials by means of methods from stochastic geometry, graph theory and time series analysis. The structures of these materials strongly differ from one another. In particular, the following are considered: organic solar cells, which are anisotropic composites of two materials; nonwoven gas-diffusion layers in proton exchange membrane fuel cells, which consist of a system of curved carbon fibers; and, graphite electrodes in Li-ion batteries, which are an isotropic two-phase system (i.e., consisting of a pore and a solid phase). The goal of this chapter is to give an overview of how models from stochastic geometry, graph theory and time series analysis can be applied to the stochastic modeling of functional materials and how these models can be used for material optimization with respect to functionality.

The three following chapters deal with Boolean random functions, random marked sets and space-time models in stochastic geometry. In Chapter 5, the notion of Boolean random functions is considered. These are generalizations of Boolean random closed sets. Their construction is based on the combination of a sequence of primary random functions using the operations of supremum or infimum. Their main properties are given in the case of scalar random functions built on Poisson point processes. Examples of applications to the modeling of rough surfaces are also given. In Chapter 6, random marked closed sets are investigated. Special models with integer Hausdorff dimension are presented based on tessellations and numerical solutions of stochastic differential equations. Statistical analysis is developed which involves the random-field model test and estimation of first and second order characteristics. Real data analyses from neuroscience (track modeling marked by spiking intensity) and materials research (grain microstructure with disorientations of faces) are presented. Dimension reduction of point processes with Gaussian random fields as covariates, a recent development, is generalized in three different ways. Chapter 7 deals with space-time models in stochastic geometry which are used in many applications. Most such models are space-time point processes. Other common models are based on growth models of random sets. This chapter aims to present more general models, where time is considered to be either discrete or continuous. In the discrete-time case the authors focus on state-space models and the use of Monte Carlo methods for the inference of model parameters. Two applications to real situations are presented: evaluation of a neurophysiological experiment and models of interacting discs. In the continuous-time case, the authors discuss space-time Lévy-driven Cox processes with different second-order structures.

The following four chapters are devoted to different issues of spatial statistics and random fields. Chapter 8 contains an introduction to rotational integral geometry. This is the key tool in local stereological procedures for estimating quantitative properties of spatial structures. In rotational integral geometry, the focus is on integrals of geometric functionals with respect to rotation invariant measures. Rotational integrals of intrinsic volumes are studied. The opposite problem of expressing intrinsic volumes as rotational integrals is also considered. An explanation is given of how intrinsic volumes can be expressed as integrals with respect to geometric functionals defined on lower dimensional linear subspaces. The rotational integral geometry of Minkowski tensors is briefly discussed as well as a principal rotational formula. These tools are then applied in local stereology leading to unbiased stereological estimators of mean intrinsic volumes for isotropic random sets. At the end of the chapter, emphasis is put on how these procedures can be implemented when automatic image analysis is available. Chapter 9 gives an introduction to the methods of functional data analysis. The authors present the basics from principal component analysis for functional data together with the functional analytic background as well as the data analytic counterpart. As prerequisites, they give an introduction into presentation techniques for functional data and some smoothing techniques. In Chapter 10, a challenging statistical problem in modern genetics is considered: how to identify the collection of factors responsible for increasing the risk of specified complex diseases. Enormous progress in the field of genetics has made possible the collection of very large genetic datasets for analysis by means of various complementary statistical tools. Thus, one has to operate with data of huge dimensions and this is one of the main difficulties in detection of genetic susceptibility to common diseases such as hypertension, myocardial infarction and others. In this chapter, the author concentrates on the multifactor dimensionality reduction method. Modifications and extensions are also discussed. Recent results on the central limit theorem related to this method are provided as well. In addition, the main features of logistic regression are discussed and simulated annealing for stochastic minimization of functions defined on a graph with forests as vertices is tackled. Chapter 11 introduces basic statistical methods for the extrapolation of stationary random fields. The problem of extrapolation (prediction) of random fields arises in geosciences, mining, oil exploration, hydrosciences, insurance, and many other fields. The techniques to solve this problem are fundamental tools in geostatistics that provide statistical inference for spatially referenced variables of interest. Examples of such quantities are the amount of rainfall, concentration of minerals and vegetation, soil texture, population density, economic wealth, storm insurance claim amounts, etc. For square integrable fields, kriging extrapolation techniques are considered. For (non–Gaussian) stable fields, which are known to be heavy-tailed, further extrapolation methods are described and their properties are discussed. Two of them can be seen as direct generalizations of kriging.

The book concludes with two chapters which deal with algorithms for Monte Carlo simulation of random fields. The generation of random spatial data on a computer is an important tool for understanding the behavior of spatial processes. Chapter 12 describes how to generate realizations of the main types of spatial processes, including Gaussian and Markov random fields, point processes (including the Poisson, compound Poisson, cluster, and Cox processes), spatial Wiener processes, and Lévy fields. Concrete *Matlab* code is also provided. The purpose of Chapter 13 is to exemplify construction of selected coupling-from-the-past algorithms, using simple examples and discussing code which can be run in the statistical scripting language *R*. The simple examples are the symmetric random walk with two reflecting boundaries; a very basic continuous state-space Markov chain; the Ising model with external field; and, a random walk with negative drift and a reflecting boundary at the origin. In parallel with this, a discussion is given of the relationship between coupling-from-the-past algorithms on the one hand, and uniform and geometric ergodicity on the other.

The authors of this book have tried to present many different methods developed in various fields of stochastic geometry, spatial statistics and random fields that merit communication to a broader audience. All chapters contain introductory sections which are easily accessible for non-specialists who want to become acquainted with modern techniques of stochastic geometry, spatial statistics and random fields. New results, which have been obtained only recently, are also presented. Each chapter provides a number of exercises which will help the reader to use the stochastic models and algorithms considered in this book autonomously.

Ulm   *Volker Schmidt*
December 2013

Acknowledgements

The generous financial support of the German Academic Exchange Service (DAAD) and of Ulm University is gratefully acknowledged. It made the organization of the Summer Academy on Stochastic Analysis, Modelling and Simulation of Complex Structures possible. This took place during September 11-24, 2011 and was the starting point for the preparation of the present book. A large part of the hard technical work of unifying the style and layout of the individual chapters was done by Gerd Gaiselmann and Ole Stenzel, assisted by Tim Brereton and Daniel Westhoff. The authors would like to thank them sincerely for their valuable help.

Contents

1 Stein's Method for Approximating Complex Distributions, with a View towards Point Processes 1
Dominic Schuhmacher
  1.1   Introduction 1
  1.2   Normal Approximation 2
       1.2.1   Two Important Lemmas 3
       1.2.2   Independent Random Variables 4
       1.2.3   Dependent Random Variables 7
       1.2.4   Kolmogorov Distance 9
  1.3   Approximation by General Distributions 10
       1.3.1   The Key Steps of Stein's Method 10
       1.3.2   The Generator Approach 12
  1.4   Point Processes 15
       1.4.1   Distances between Point Patterns 17
       1.4.2   Statistical Applications of Distances between Point Patterns 19
       1.4.3   Distances between Point Process Distributions 22
  1.5   Poisson Process Approximation of Point Process Distributions ... 24
       1.5.1   The Coupling Strategy 25
       1.5.2   Two Upper Bounds for Poisson Process Approximation . 28
  1.6   Gibbs Process Approximation of Point Process Distributions 30

2 Clustering Comparison of Point Processes, with Applications to Random Geometric Models 31
Bartłomiej Błaszczyszyn and Dhandapani Yogeshwaran
  2.1   Introduction 32
  2.2   Examples of Point Processes 33
       2.2.1   Elementary Models 33
       2.2.2   Cluster Point Processes — Replicating and Displacing Points 34
       2.2.3   Cox Point Processes— Mixing Poisson Distributions 37

      2.2.4   Gibbs and Hard-Core Point Processes.................. 39

      2.2.5   Determinantal and Permanental Point Process.......... 39

  2.3 Clustering Comparison Methods........................... 40

      2.3.1   Second-order Statistics 41

      2.3.2   Moment Measures..................................... 43

      2.3.3   Void Probabilities 45

      2.3.4   Positive and Negative Association..................... 48

      2.3.5   Directionally Convex Ordering 49

  2.4 Some Applications 56

      2.4.1   Non-trivial Phase Transition in Percolation Models 56

      2.4.2   U-Statistics of Point Processes........................ 64

      2.4.3   Random Geometric Complexes 67

      2.4.4   Coverage in the Germ-Grain Model 69

  2.5 Outlook 70

3 Random Tessellations and their Application to the Modelling of Cellular Materials 73

Claudia Redenbach and André Liebscher

  3.1 Introduction 73

  3.2 Random Tessellations..................................... 74

      3.2.1   Definitions 74

      3.2.2   Tessellation Models................................... 78

  3.3 Estimation of Geometric Characteristics of Open Foams from Image Data 83

      3.3.1   Characteristics based on Intrinsic Volumes 84

      3.3.2   Model-based Mean Value Analysis 85

      3.3.3   Cell Reconstruction................................... 87

  3.4 Stochastic Modelling of Cellular Materials 89

      3.4.1   Modelling of the Cell System 89

      3.4.2   Modelling of Open Foams 91

4 Stochastic 3D Models for the Micro-structure of Advanced Functional Materials... 95

Volker Schmidt, Gerd Gaiselmann and Ole Stenzel

  4.1 Introduction 95

  4.2 Point Process Models and Time Series....................... 99

      4.2.1   Point Processes 99

      4.2.2   Multivariate Time Series............................. 105

  4.3 Stochastic 3D Model for Organic Solar Cells 107

      4.3.1   Data and Functionality 107

      4.3.2   Data Preprocessing 109

      4.3.3   Description of Stochastic Model 110

      4.3.4   Model Fitting 115

      4.3.5   Model Validation 115

      4.3.6   Scenario Analysis 116

  4.4 Stochastic 3D Modeling of Non-woven GDL 117

4.4.1 Data and Functionality 119
4.4.2 Data Preprocessing 120
4.4.3 Description of Stochastic Model 120
4.4.4 Model Fitting 125
4.4.5 Model Validation 126
4.5 Stochastic 3D Model for Uncompressed Graphite Electrodes in
Li-Ion Batteries 127
4.5.1 Data and Functionality 127
4.5.2 Data Preprocessing 128
4.5.3 Description of Stochastic Model 129
4.5.4 Model Fitting 136
4.5.5 Model Validation 138
4.5.6 Further Numerical Results 140
4.6 Conclusion 140

5 Boolean Random Functions 143
Dominique Jeulin
5.1 Introduction 143
5.2 Some Preliminaries 144
5.2.1 Random Closed Sets 144
5.2.2 Upper Semi-Continuous Random Functions 145
5.2.3 Principle of Random Sets and of Random Function
Modeling 147
5.3 Boolean Random Functions 147
5.3.1 Construction of the BRF 147
5.3.2 BRF and Boolean Model of Random Sets 148
5.3.3 Choquet Capacity of the BRF 149
5.3.4 Supremum Stability and Infinite Divisibility 151
5.3.5 Characteristics of the Primary Functions 152
5.3.6 Some Stereological Aspects of BRF 154
5.3.7 BRF and Counting 154
5.3.8 Identification of a BRF Model 155
5.3.9 Tests of BRF 157
5.3.10 BRF and Random Tessellations 158
5.4 Multiscale Boolean Random Functions 159
5.5 Application of BRF to Modeling of Rough Surfaces 161
5.5.1 Simulation of the Evolution of Surfaces and of Stresses
during Shot Peening 162
5.5.2 Simulation of the Roughness Transfer on Steel Sheets ... 164
5.5.3 Modeling of Electro Discharge Textures (EDT) 165

6 Random Marked Sets and Dimension Reduction 171
Viktor Beneš, Jakub Staněk, Blažena Kratochvílová and Ondřej Šedivý
6.1 Preliminaries 172
6.1.1 Random Measures and Random Marked Closed Sets 172
6.1.2 Second-order Characteristics 174

6.2   Statistical Methods for RMCS 175
        6.2.1   Random-field Model Test 175
        6.2.2   Estimation of Characteristics 176
6.3   Modeling of RMCS; Simulation Results 177
        6.3.1   Tessellation Models 178
        6.3.2   Fibre Process Based on Diffusion 180
6.4   Real Data Analyses 184
        6.4.1   Random-field Model Test in a Neurophysiological
                 Experiment 184
        6.4.2   Second-order RMCS Analysis of Granular Materials 185
6.5   Dimension Reduction 187
6.6   Theoretical Results 188
        6.6.1   Investigation of $\mathcal{S}_1$ 189
        6.6.2   Investigation of $\mathcal{S}_2$ 191
6.7   Statistical Methods 196
        6.7.1   Estimation 196
        6.7.2   Statistical Testing 198
6.8   Simulation Studies 199
        6.8.1   Description of the Simulation 199
        6.8.2   Numerical Results 202

7   **Space-Time Models in Stochastic Geometry** 205
     Viktor Beneš, Michaela Prokešová, Kateřina Staňková Helisová and
     Markéta Zikmundová
     7.1   Discrete-Time Modelling 206
             7.1.1   State-space Models and Sequential Monte Carlo 206
             7.1.2   The PMMH Method 208
     7.2   Application: Firing Activity of Nerve Cells 209
             7.2.1   Space-Time Model and Particle Filter 209
             7.2.2   Model Checking 211
     7.3   Systems of Interacting Discs 212
             7.3.1   Background in Space 212
             7.3.2   Space-Time Model 214
             7.3.3   Model Checking 216
             7.3.4   Simulation Study 217
     7.4   Continuous-Time Modelling 218
             7.4.1   Space-Time Point Processes 218
             7.4.2   Space-Time Lévy-Driven Cox Point Processes 220
     7.5   Separability and Space-Time Point Processes 222
             7.5.1   Separability 222
             7.5.2   Spatial and Temporal Projection Processes 223
     7.6   Ambit Sets and Nonseparable Kernels 224
     7.7   Estimation Procedures 226
             7.7.1   Estimation of the Space-Time Intensity Function 226

7.7.2 Estimation by Means of the Space-Time K-Function 228
7.7.3 Estimation by Means of Projection Processes 229

8 Rotational Integral Geometry and Local Stereology - with a View to Image Analysis 233
Eva B. Vedel Jensen and Allan Rasmusson
8.1 Rotational Integral Geometry 233
8.1.1 Rotational Integrals of Intrinsic Volumes 234
8.1.2 Intrinsic Volumes as Rotational Integrals 237
8.1.3 Rotational Integral Geometry of Minkowski Tensors 241
8.1.4 A Principal Rotational Formula 245
8.2 Local Stereology 246
8.3 Variance Reduction Techniques 252
8.4 Computational Stereology 253

9 An Introduction to Functional Data Analysis 257
Ulrich Stadtmüller and Marta Zampiceni
9.1 Some Fundamental Tools 257
9.1.1 Basic Statistics 257
9.1.2 Smoothing Techniques 260
9.2 Principal Components Analysis (PCA) 270
9.2.1 PCA for Multivariate Data 270
9.2.2 PCA for Functional Data 272
9.3 Statistical Models for Functional Data 279
9.3.1 Linear Regression 279
9.3.2 Estimation, a Second Look 281
9.3.3 More General Regression Models 288

10 Some Statistical Methods in Genetics 293
Alexander Bulinski
10.1 Introduction 293
10.2 The MDR method and its Modifications 295
10.2.1 Implementation of the MDR Method 295
10.2.2 Prediction Algorithm 298
10.2.3 Estimated Prediction Error 299
10.2.4 Dimensionality Reduction 300
10.2.5 Central Limit Theorem 302
10.2.6 Modifications of the MDR Method 304
10.3 Logic Regression and Simulated Annealing 306
10.3.1 Logic Trees 307
10.3.2 Logic Regression 307
10.3.3 Operations on Logic Trees 309
10.3.4 Simulated Annealing 312
10.4 Models Involving Random Fields 315
10.5 Concluding Remarks 319

11   Extrapolation of Stationary Random Fields 321
Evgeny Spodarev, Elena Shmileva and Stefan Roth
    11.1   Introduction 321
    11.2   Basics of Random Fields 322
            11.2.1   Random Fields with Invariance Properties 322
            11.2.2   Elements of Correlation Theory for Square Integrable
                      Random Fields 326
            11.2.3   Stable Random Fields 334
            11.2.4   Dependence Measures for Stable Random Fields 341
            11.2.5   Examples of Stable Processes and Fields 343
    11.3   Extrapolation of Stationary Random Fields 345
            11.3.1   Kriging Methods for Square Integrable Random Fields .. 346
            11.3.2   Simple Kriging 346
            11.3.3   Ordinary Kriging 349
    11.4   Extrapolation of Stable Random Fields 352
            11.4.1   Least Scale Linear Predictor 352
            11.4.2   Covariation Orthogonal Predictor 354
            11.4.3   Maximization of Covariation 357
            11.4.4   The Case $\alpha \in (0,1]$ 359
            11.4.5   Numerical Examples 363
    11.5   Open problems 367

12   Spatial Process Simulation 369
Dirk P. Kroese and Zdravko I. Botev
    12.1   Introduction 369
    12.2   Gaussian Markov Random Fields 370
            12.2.1   Gaussian Property 370
            12.2.2   Generating Stationary Processes via Circulant Embedding 372
            12.2.3   Markov Property 378
    12.3   Point Processes 379
            12.3.1   Poisson Process 380
            12.3.2   Marked Point Processes 381
            12.3.3   Cluster Process 382
            12.3.4   Cox Process 384
            12.3.5   Point Process Densities 388
    12.4   Wiener Surfaces 392
            12.4.1   Wiener Process 393
            12.4.2   Fractional Brownian Motion 394
            12.4.3   Fractional Wiener Sheet in $\mathbb{R}^2$ 395
            12.4.4   Fractional Brownian Field 396
    12.5   Spatial Levy Processes 399
            12.5.1   Lévy Process 400
            12.5.2   Lévy Sheet in $\mathbb{R}^2$ 403

13 Introduction to Coupling-from-the-Past using *R* 405
Wilfrid S. Kendall
   13.1   Introduction 405
   13.2   Classic Coupling-from-the-Past 406
         13.2.1   Random Walk CFTP 407
         13.2.2   Read-once CFTP 413
         13.2.3   Small-set (Green-Murdoch) CFTP 417
         13.2.4   Image Analysis and Ising CFTP 422
         13.2.5   Other Remarks 427
   13.3   Dominated Coupling-from-the-Past 428
         13.3.1   Random Walk domCFTP 430
         13.3.2   Other Remarks 437
   13.4   Conclusion 437

References 441

Index 459

List of Contributors

Contributors (in alphabetical order)

- Viktor Beneš, Charles University, Prague, Czech Republic
- Bartłomiej Błaszczyszyn, Inria & École normale supérieure, Paris, France
- Zdravko I. Botev, University of New South Wales, Sydney, Australia
- Alexander Bulinski, Lomonosov Moscow State University, Moscow, Russia
- Gerd Gaiselmann, Ulm University, Ulm, Germany
- Dominique Jeulin, MINES Paris Tech, Fontainebleau, France
- Wilfrid S. Kendall, University of Warwick, Coventry, UK
- Blažena Kratochvilová, Palacky University, Ólomouc, Czech Republic
- Dirk P. Kroese, University of Queensland, Brisbane, Australia
- André Liebscher, University of Kaiserslautern, Kaiserslautern, Germany
- Michaela Prokešová, Charles University, Prague, Czech Republic
- Allan Rasmusson, Aarhus University, Aarhus, Denmark
- Claudia Redenbach, University of Kaiserslautern, Kaiserslautern, Germany
- Stefan Roth, Ulm University, Ulm, Germany
- Volker Schmidt, Ulm University, Ulm, Germany
- Dominic Schuhmacher, University of Göttingen, Göttingen, Germany
- Ondřej Šedivý, Charles University, Prague, Czech Republic
- Elena Shmileva, St. Petersburg State University, St. Petersburg, Russia
- Evgeny Spodarev, Ulm University, Ulm, Germany
- Ulrich Stadtmüller, Ulm University, Ulm, Germany
- Jakub Staněk, Charles University, Prague, Czech Republic
- Kateřiná Staňkova-Helisová, Czech Technical University, Prague, Czech Republic
- Ole Stenzel, Ulm University, Ulm, Germany
- Eva B. Vedel Jensen, Aarhus University, Aarhus, Denmark
- Dhandapani Yogeshwaran, Indian Statistical Institute, Bangalore, India
- Marta Zampiceni, Ulm University, Ulm, Germany
- Markéta Zikmundová, Charles University, Prague, Czech Republic

Acronyms

Sets

$\mathbb{C}$	complex numbers
$\mathbb{N}$	positive integer numbers
$\mathbb{Q}$	rational numbers
$\mathbb{R}$	real numbers
$\mathbb{Z}$	all integer numbers
G	class of all closed sets
$\mathcal{C}$	class of all compact sets
$\mathcal{K}_{conv}^{d}$	class of all convex bodies in $\mathbb{R}^d$
$\mathcal{E}_{k}^{d}$	family of affine k-planes in $\mathbb{R}^d$
$\mathcal{B}_0(\mathbb{R}^d)$	family of bounded Borel sets in $\mathbb{R}^d$

Probability theory

ξ, η	random variables
X, Y	random sets
$\mathbf{P}_\xi$	probability measure of ξ
$\mathbf{E}\xi$	expectation of ξ
$\mathbf{corr}(\xi, \eta)$	correlation of ξ and η
$\mathbf{cov}(\xi, \eta)$	covariance of ξ and η
$\mathbf{var}\,\xi$	variance of ξ
N	set of all locally finite simple point patterns
$\mathbf{1}$	indicator function
$\mathrm{Pois}(\lambda)$	Poisson distribution with parameter λ
$\mathrm{Exp}(\lambda)$	Exponential distribution with parameter λ
$\mathrm{Ber}(p)$	Bernoulli distribution with parameter p
$\mathrm{Binom}(n, p)$	Binomial distribution with parameters n and p

Other notations

E	energy
T	tessellation
diag	diagonal of a matrix
dist	distance function
conv	convex hull
supp	support
card	cardinality of a set
diam	diameter
SO	rotation group
$\Im z$	imaginary part of a complex number
$\Re z$	real part of a complex number
sgn	signum function
Lip	Lipschitz operator
span	linear hull
ν_d	Lebesgue measure
$V_0, \ldots, V_d$	intrinsic volumes
χ	Euler characteristic
K^r	r-neighbourhood of K
$\operatorname{Int} K$	interior of K
$K \mid L$	orthogonal projection of K onto L
$B_r(o)$	ball of radius r centred in the origin
$S\alpha S$	symmetric α-stable

Chapter 1
Stein's Method for Approximating Complex Distributions, with a View towards Point Processes

Dominic Schuhmacher

Abstract We give an introduction to Stein's method, a powerful technique for computing explicit error bounds for distributional approximation. The classical case of normal approximation is provided for initial motivation. Then the main part of this chapter is devoted to presenting the key concepts of Stein's method in a much more general framework, where the approximating distribution Q and the space S it lives on can be almost arbitrary. This is particularly appealing for distributional approximation in stochastic geometry and spatial statistics.

Rather than providing many concrete results, the emphasis of this chapter lies on conveying the techniques for developing Stein's method on new state spaces S and for new approximating distributions. These techniques are elaborated in detail for the case where S is a space of point patterns and Q is the distribution of a Poisson process or a more general Gibbs process. Questions on how to measure distances between probability distributions on complicated spaces are also addressed. It is convenient if S is equipped with a suitable metric. We present several ideas and examples about performing statistical analyses on metric spaces.

1.1 Introduction

A basic fact that any statistician will realize quite early in her/his career is that most probability distributions appearing in the wild are terribly complicated. This is true for distributions on $\mathbb{R}$, and even more so for distributions of random objects considered in stochastic geometry and spatial statistics.

The most fundamental step in any statistical analysis is the approximation of the complicated real life distribution with a comparatively simple familiar one that can be handled more or less easily. This typically happens on two different levels.

Dominic Schuhmacher
University of Göttingen, Institute for Mathematical Stochastics, 37077 Göttingen, Germany, e-mail: `dominic.schuhmacher@mathematik.uni-goettingen.de`

© Springer International Publishing Switzerland 2015

V. Schmidt (ed.), *Stochastic Geometry, Spatial Statistics and Random Fields*,
Lecture Notes in Mathematics 2120, DOI 10.1007/978-3-319-10064-7_1

First, when making modelling assumptions, we cast the real world problem into a simplified model $(P_\theta)_{\theta \in \Theta}$. Then, when doing inference about the posited model and the distributions involved are too complicated, we resort to inference based on simpler distributions that apply asymptotically.

For illustration consider data from a stress test of some brittle material under various combinations of covariates, such as type of material, structural features, stress applied, temperature, humidity, and so on. We might assume that the results of the stress tests of individual specimens are independent and that the fracture probability can be expressed as a logistic function of a linear combination of the covariates. This is usually not entirely correct, but it allows us to formulate an *approximative* logistic regression model for our data. We might then compute a likelihood ratio test in this model in order to show that one material is more stress resistant than the other using the asymptotic χ^2-distribution of the likelihood ratio statistic, which would be the second time we approximate.

Note that neither of these approximation steps is very well controlled from a quantitative point of view. For formulating the model, the statistician relies on his experience, intuition and common sense to ensure that not too much can go wrong. For asymptotic inference a limit theorem is quoted that typically applies as the sample size n goes to infinity, and in simpler cases there is maybe also a rule of thumb, usually based on simulation studies or heuristic arguments, to make plausible that the approximation is good enough for the finite sample size n available.

However, what is really of interest in almost all applied situations are explicit bounds on the error committed in performing these approximations. This is by no means an easy task, but the only way to ensure that the statistical inference performed is valid.

This chapter is about a powerful technique to obtain such explicit bounds for the comparison of two probability distributions, called Stein's method [375]. Stein's method is known by many probabilists as a tool for proving variants of the central limit theorem and giving good rates for the approximation. What is less commonly known is that it can also be used in much more general approximation settings, especially if the probability distributions involved live on more complicated spaces, as is often the case in stochastic geometry and spatial statistics.

We start out in Sect. 1.2 by giving a tour d'horizon of the basic ideas of Stein's method for normal approximation. In Sect. 1.3 we present a general recipe for Stein's method and a very successful variant called the *generator approach* in the abstract setting of approximation by arbitrary distributions. Sect. 1.4 to 1.6 elaborate on these ideas in the case of point process approximation. Sect. 1.4 also contains ideas and concrete examples about statistical analyses on metric spaces.

1.2 Normal Approximation

In order to start with a concrete situation that every reader can relate to, we give an overview of the central ideas used in the normal approximation case and apply them to obtain a rate for various versions of the central limit theorem (CLT). In passing we encounter many important concepts that are discussed in Sect. 1.3 on a more abstract level and in much greater generality.

1.2.1 Two Important Lemmas

The following two lemmas are crucial for the development of Stein's method for normal approximation.

Lemma 1.1. *Let* $\vartheta : \mathbb{R} \to \mathbb{R}$ *be a bounded measurable function. Then the differential equation*

$$\frac{d}{dz} g(z) - z g(z) = \vartheta(z) \tag{1.1}$$

(for Lebesgue almost every z) has as solution the function $g^{(\vartheta)} : \mathbb{R} \to \mathbb{R}$ *with*

$$g^{(\vartheta)}(z) = \exp(z^2/2) \int_{-\infty}^{z} \vartheta(x) \exp(-x^2/2)\, dx. \tag{1.2}$$

If $\int_{-\infty}^{\infty} \vartheta(x) \exp(-x^2/2)\, dx = 0$, *then* $g^{(\vartheta)}$ *is bounded and its derivative is essentially bounded.*

Proof. Multiplying equation (1.1) by $e^{-z^2/2}$ and integrating yields

$$\int_{-\infty}^{u} \vartheta(z) e^{-z^2/2}\, dz = \int_{-\infty}^{u} (g(z) e^{-z^2/2})'\, dz + c$$
$$= g(u) e^{-u^2/2} + c,$$

and hence the particular solution for $c = 0$ is

$$g^{(\vartheta)}(z) = e^{z^2/2} \int_{-\infty}^{z} \vartheta(x) e^{-x^2/2}\, dx \tag{1.3}$$

as claimed. The inequality in [147] about Mills's ratio of a standard normal random variable says that

$$\frac{1}{z+1/z} < \frac{1-\Phi(z)}{\varphi(z)} < \frac{1}{z} \tag{1.4}$$

for every $z > 0$. By the upper bound we obtain that $g^{(\vartheta)}(z) = O(1/|z|)$ as $z \to -\infty$, and if $\int_{-\infty}^{\infty} \vartheta(x) e^{-x^2/2}\, dx = 0$ also $g^{(\vartheta)}(z) = O(1/z)$ as $z \to \infty$. Thus $g^{(\vartheta)}$ and $[z \mapsto z g^{(\vartheta)}(z)]$ are bounded, because they are continuous. Furthermore we obtain by $\frac{d}{dz} g^{(\vartheta)}(z) = z g^{(\vartheta)}(z) + \vartheta(z)$ almost everywhere that the derivative of $g^{(\vartheta)}$ is essentially bounded, since both functions on the right hand side are bounded. $\qquad\square$

The above lemma is used for the proof of the following important characterization of the normal distribution, and also in a more general way for solving the Stein equation (1.6) further below.

Lemma 1.2 (Stein's lemma). *A random variable Z is standard normally distributed if and only if*

$$\mathbf{E}(Zg(Z)) = \mathbf{E}g'(Z) \tag{1.5}$$

for every bounded and absolutely continuous function $g \colon \mathbb{R} \to \mathbb{R}$ that satisfies $\mathbf{E}|g'(Z)| < \infty$, where $g' = \frac{d}{dz}g$ denotes the derivative of g.

Proof. If $Z \sim N(0,1)$ and g is as specified, then equation (1.5) holds by a short calculation using integration by parts for absolutely continuous functions. More precisely

$$\mathbf{E}g'(Z) = \frac{1}{\sqrt{2\pi}} \int_{-\infty}^{\infty} g'(z)e^{-z^2/2}\, dz$$

$$= \frac{1}{\sqrt{2\pi}} g(z)e^{-z^2/2}\big|_{-\infty}^{\infty} + \frac{1}{\sqrt{2\pi}} \int_{-\infty}^{\infty} zg(z)e^{-z^2/2}\, dz = \mathbf{E}(Zg(Z)),$$

using that g is bounded. Conversely, assume that $\mathbf{E}g'(Z) = E(Zg(Z))$ for every g as specified. We apply Lemma 1.1 in the special case

$$\vartheta_a(z) = \mathbf{1}\{z \le a\} - \Phi(a).$$

Note that $\int_{-\infty}^{\infty} \vartheta_a(z)\exp(-z^2/2)\, dz = 0$ for every a, so that the corresponding solution $g_a = g^{(\vartheta_a)}$ given by (1.2) has the required properties.

Hence by the prerequisite

$$0 = \mathbf{E}\big(g_a'(Z) - Zg_a(Z)\big) = \mathbf{P}(Z \le a) - \Phi(a)$$

for every $a \in \mathbb{R}$. Thus $Z \sim N(0,1)$. $\qquad\qquad\square$

1.2.2 Independent Random Variables

For $n \in \mathbb{N}$ assume that $Y_1 = Y_1^{(n)}, \ldots, Y_n = Y_n^{(n)}$ are i.i.d. random variables that are normalized so that $\mathbf{E}Y_1 = 0$, $\mathbf{E}Y_1^2 = 1/n$. Write $W = W_n = \sum_{i=1}^{n} Y_i$ and $Z \sim N(0,1)$. In order to show the classical CLT, we would like to establish

$$\mathbf{E}f(W_n) \longrightarrow \mathbf{E}f(Z) \quad \text{as } n \to \infty$$

for every $f \in \mathcal{F}$, where $\mathcal{F}$ is a rich enough class of functions $f : \mathbb{R} \to \mathbb{R}$. For the time being we assume that $\mathcal{F}$ is the class of bounded Lipschitz continuous functions, which we denote by $\mathcal{F}_{BL}^{(\infty)}$.

Stein's lemma suggests that it might be a good idea to check if $\mathbf{E}g'(W) - \mathbf{E}(Wg(W)) \approx 0$ in order to show that the distributions of W and Z are approximately equal, i.e. $W \approx_{\mathscr{D}} Z$. To turn this to our advantage we re-express for arbitrary $f \in \mathcal{F}_{BL}^{(\infty)}$

$$f(w) - \mathbf{E}f(Z) = g_f'(w) - wg_f(w), \tag{1.6}$$

which we may do by setting $\vartheta(w) = f(w) - \mathbf{E}f(Z)$ in Lemma 1.1 and then taking $g_f = g^{(\vartheta)}$. Equation (1.6) is usually called the *Stein equation*. Its solution g_f is not so easy to deal with. Fortunately it is enough to have a handle on certain derivatives of g_f. Note that g_f has a second derivative almost everywhere, because f is Lipschitz continuous and hence equation (1.1) yields that g_f' is Lipschitz continuous on every compact set. In fact we may show that $|g_f''|$ cannot get too large. Denote by $\|\cdot\|_\infty$ the L^∞-norm. The proof of the following lemma is based on the explicit form of the solution obtained in Lemma 1.1, but is quite technical. We therefore refer to the more general Lemma 2.4 in [68].

Lemma 1.3. *If* $f \in \mathcal{F}_{BL}^{(\infty)}$, *then it holds that*

$$\|g_f\|_\infty \le 2\|f'\|_\infty, \quad \|g_f'\|_\infty \le \sqrt{\frac{2}{\pi}}\|f'\|_\infty, \quad \text{and} \quad \|g_f''\|_\infty \le 2\|f'\|_\infty.$$

Plugging in the random variable W and taking expectations in the Stein equation (1.6), we have

$$\left|\mathbf{E}f(W) - \mathbf{E}f(Z)\right| = \left|\mathbf{E}g_f'(W) - \mathbf{E}\left(Wg_f(W)\right)\right|.$$

Writing $W^{(i)} = \sum_{j=1, j \ne i}^n Y_j = W - Y_i$, we obtain

$$
\begin{aligned}
\mathbf{E}\left(Wg_f(W)\right) &= \sum_{i=1}^n \mathbf{E}\left(Y_i g_f(W)\right) \\
&= n\,\mathbf{E}\left(Y_1 g_f(W^{(1)} + Y_1)\right) \\
&= n\left[\mathbf{E}\left(Y_1 g_f(W^{(1)})\right) + \mathbf{E}\left(Y_1^2 g_f'(W^{(1)})\right) + r\right] \\
&= \mathbf{E}g_f'(W^{(1)}) + nr
\end{aligned}
\tag{1.7}
$$

by Taylor approximation and the fact that Y_1 and $W^{(1)}$ are independent, where $|r| \le \frac{1}{2}\|g_f''\|_\infty \mathbf{E}|Y_1|^3$. Thus in total for every $f \in \mathcal{F}_{BL}^{(\infty)}$

$$
\begin{aligned}
\left|\mathbf{E}f(W) - \mathbf{E}f(Z)\right| &= \left|\mathbf{E}g_f'(W) - \mathbf{E}\left(Wg_f(W)\right)\right| \\
&\le \left|\mathbf{E}g_f'(W) - \mathbf{E}g_f'(W^{(1)})\right| + \frac{n}{2}\|g_f''\|_\infty \mathbf{E}|Y_1|^3 \\
&\le \|g_f''\|_\infty \mathbf{E}|Y_1| + \frac{n}{2}\|g_f''\|_\infty \mathbf{E}|Y_1|^3 \\
&\le \|f'\|\left(\frac{2}{\sqrt{n}} + n\mathbf{E}|Y_1|^3\right)
\end{aligned}
\tag{1.8}
$$

by Lemma 1.3 and $\mathbf{E}|Y_1| \le \sqrt{\mathbf{E}Y_1^2} = 1/\sqrt{n}$.

As a weak consequence we obtain immediately that $\sum_{i=1}^n Y_i \xrightarrow{\mathscr{D}} N(0,1)$ if $\mathbf{E}|Y_1|^3 = o(1/n)$ as $n \to \infty$. This implies for unnormalized i.i.d. random variables $X_1, X_2, \ldots$ with $\mathbf{E}X_1 = \mu$, $\mathbf{var}X_1^2 = \sigma^2$, and $\mathbf{E}|X_1|^3 < \infty$ that by setting $Y_i = (X_i - \mu)/(\sqrt{n}\sigma)$ we recover the CLT

$$\frac{1}{\sqrt{n}} \sum_{i=1}^{n} \frac{X_i - \mu}{\sigma} \xrightarrow{\mathscr{D}} N(0,1) \text{ as } n \to \infty.$$

Note that unlike in the classical CLT we require a third moment for Y_1. In fact we will see in Theorem 1.1 that this is not necessary. But even so, we are richly rewarded by the above approach, as we do not only have convergence for $n \to \infty$, but an explicit result about the goodness of the approximation for any finite n in terms of arbitrary functions $f \in \mathcal{F}_{BL}^{(\infty)}$.

It is convenient and customary to express such results in terms of probability metrics, i.e. metrics ρ on spaces of probability distributions, of the form

$$\rho(\mathbf{P}_W, \mathbf{P}_Z) = \sup_{f \in \mathcal{F}} |\mathbf{E}f(W) - \mathbf{E}f(Z)|,$$

where $\mathbf{P}_W, \mathbf{P}_Z$ denote the distributions of W and Z, respectively. So far we have used bounded Lipschitz continuous functions f and may therefore say something about the metric based on

$$\mathcal{F} = \mathcal{F}_{BW} = \{f \colon \mathbb{R} \to [0,1]; \ |f(x) - f(y)| \leq |x - y| \text{ for all } x, y \in \mathbb{R}\}.$$

Other common choices for the function class $\mathcal{F}$ include

$$\mathcal{F}_W = \{f \colon \mathbb{R} \to \mathbb{R}; \ |f(x) - f(y)| \leq |x - y| \text{ for all } x, y \in \mathbb{R}\},$$
$$\mathcal{F}_K = \{\mathbf{1}_{(-\infty, t]}; \ t \in \mathbb{R}\},$$
$$\mathcal{F}_{TV} = \{\mathbf{1}_A; \ A \subset \mathbb{R} \text{ measurable}\}.$$

We distinguish the metrics based on these function classes by subscripting ρ with the corresponding letters, and then refer to ρ_{TV}, ρ_K, ρ_W and ρ_{BW} as *total variation*, *Kolmogorov*, *Wasserstein* and *bounded Wasserstein metric*, respectively. Of these four metrics the bounded Wasserstein metric (also known as bounded Lipschitz metric) is the only one that exactly metrizes weak convergence, i.e. for which $\rho(\mathbf{P}_{X_n}, \mathbf{P}_X) \to 0$ is equivalent to X_n converging to X in distribution for arbitrary random variables; see [107], Theorem 11.3.3. The other metrics are in general stronger in the sense that convergence in the metric is sufficient but not necessary for convergence in distribution. The Kolmogorov metric achieves exact metrization for absolutely continuous limiting random variables.

So what we have proved until now implies

$$\rho_{BW}(\mathbf{P}_W, \mathbf{P}_Z) \leq \frac{2}{\sqrt{n}} + n\mathbf{E}|Y_1|^3.$$

With some more work one may replace ρ_{BW} by ρ_W at no additional cost. The following theorem shows what is possible for the CLT in the case of independent but not necessarily identically distributed random variables when the proof strategy presented above is refined.

Theorem 1.1 (Independent case). *Let $Y_1,\ldots,Y_n$ be independent random variables with $\mathbf{E}Y_i = 0$ and $\sum_{i=1}^n \mathbf{E}Y_i^2 = 1$. Then for $W = \sum_{i=1}^n Y_i$ it holds that*

$$\rho_W\left(\mathbf{P}_W, N(0,1)\right) \leq \sum_{i=1}^n \mathbf{E}|Y_i|^3$$

and

$$\rho_W\left(\mathbf{P}_W, N(0,1)\right) \leq 4\sum_{i=1}^n \mathbf{E}\min(Y_i^2, |Y_i|^3).$$

Proof. See [68], Theorems 3.1 and 3.2, and Corollary 4.2. $\square$

Note that the second result of Theorem 1.1 does not require the existence of third moments in order to be useful.

Exercise 1.1. Show under the conditions above that for every $\varepsilon > 0$,

$$\sum_{i=1}^n \mathbf{E}\min(Y_i^2, |Y_i|^3) \to 0$$

if and only if the Lindeberg condition holds, i.e.

$$\sum_{i=1}^n \mathbf{E}\left(Y_i^2 \mathbf{1}\{|Y_i| > \varepsilon\}\right) \longrightarrow 0.$$

1.2.3 Dependent Random Variables

A particular strength of Stein's method is that it is quite easily adaptable to situations where the approximated random variable W incorporates a substantial dependence structure. In such situations many other proof strategies break down completely.

For demonstration purposes we consider the simplest case of a stationary sequence $(Y_i)_{1\leq i\leq n}$ that is (strictly) locally dependent. To avoid boundary effects we identify the index set $\{1,\ldots,n-1,n\}$ with the cyclic group $\mathbb{Z}/n\mathbb{Z} = \{[1],\ldots,[n-1],[0]\}$, i.e. we "join the ends" of the index set. By local dependence we mean that there is a set $A \subset \mathbb{Z}/n\mathbb{Z}$ such that with $A_i = i+A = \{i+a;\ a \in A\}$ we have that Y_i and $\{Y_j;\ j \in A_i^c\}$ are independent. This implies $[0] \in A$. So far local dependence means no restriction on the joint distribution of the Y_is. However, for the upper bound in equation (1.10) to be small, it will be necessary that the cardinality of A is small too.

We still assume normalization of the Y_i, meaning here $\mathbf{E}Y_1 = 0$ and $\mathbf{var}W = 1$. With a little bit of extra effort we may now mimic the computations in formulae (1.7) and (1.8). Write $U_i = \sum_{j\in A_i} Y_j$ and $W^{(i)} = \sum_{j\in A_i^c} Y_j = W - U_i$, and note that by local dependence

$$\mathbf{E}(Y_iU_i) = \sum_{j\in A_i} \mathbf{E}(Y_iY_j) = \sum_{j=1}^n \mathbf{E}(Y_iY_j) = \mathbf{E}(Y_iW) = \frac{1}{n}$$

because of stationarity and $\mathbf{E}W^2 = 1$. Then

$$\begin{aligned}
\mathbf{E}\big(Wg_f(W)\big) &= n\,\mathbf{E}\big(Y_1 g_f(W^{(1)} + U_1)\big) \\
&= n\big[\mathbf{E}\big(Y_1 g_f(W^{(1)})\big) + \mathbf{E}\big(Y_1 U_1 g_f'(W^{(1)})\big) + r\big] \\
&= \mathbf{E}g_f'(W^{(1)}) + nr + n\,\mathbf{E}\big[(Y_1 U_1 - \mathbf{E}(Y_1 U_1))g_f'(W^{(1)})\big]
\end{aligned} \tag{1.9}$$

by Taylor approximation and the fact that Y_1 and $W^{(1)}$ are independent, where $|r| \leq \frac{1}{2}\|g_f''\|_\infty \mathbf{E}(|Y_1|U_1^2)$. Thus for every $f \in \mathcal{F}_{BL}^{(\infty)}$

$$\begin{aligned}
&\big|\mathbf{E}f(W) - \mathbf{E}f(Z)\big| \\
&\quad = \big|\mathbf{E}g_f'(W) - \mathbf{E}\big(Wg_f(W)\big)\big| \\
&\quad \leq \big|\mathbf{E}g_f'(W) - \mathbf{E}\big(g_f'(W^{(1)})\big)\big| + \frac{n}{2}\|g_f''\|_\infty \mathbf{E}(|Y_1|U_1^2) \\
&\qquad\qquad + n\big|\mathbf{E}\big[(Y_1 U_1 - \mathbf{E}(Y_1 U_1))g_f'(W^{(1)})\big]\big| \\
&\quad \leq \|g_f''\|_\infty \mathbf{E}|U_1| + \frac{n}{2}\|g_f''\|_\infty \mathbf{E}(|Y_1|U_1^2) + \|g_f'\|_\infty n\,\mathbf{E}\big|Y_1 U_1 - \mathbf{E}(Y_1 U_1)\big| \\
&\quad \leq \|f'\|_\infty\left(\frac{2m}{\sqrt{n}} + n\,\mathbf{E}(|Y_1|U_1^2) + \sqrt{\frac{2}{\pi}}\, n\,\mathbf{E}\big|Y_1 U_1 - \mathbf{E}(Y_1 U_1)\big|\right)
\end{aligned} \tag{1.10}$$

by Lemma 1.3, where $m = |A|$ denotes the cardinality of A.

Exercise 1.2. Let $I_1, \ldots, I_n$ be i.i.d. indicators with expectation p. Define $X_i = I_i I_{i+1}$ for $i = 1, 2, \ldots, n-1$ and $X_n = I_n I_1$. Let furthermore $S_n = \sum_{i=1}^n X_i$ and $W = (S_n - \mathbf{E}S_n)/\mathbf{sd}(S_n)$. Give an upper bound for $\rho_{BW}(\mathbf{P}_W, N(0,1))$ based on the above calculations.

With more careful arguments one may again obtain a better bound in a more general setting. The following result is for a stronger local dependence condition with nested dependence neighbourhoods that is often satisfied in applications. We use the symbol $\perp\!\!\!\perp$ to denote independence of collections of random variables.

Theorem 1.2 (Locally dependent case). *Let $Y_1, \ldots, Y_n$ be random variables with $\mathbf{E}Y_i = 0$, $\mathbf{E}Y_i^2 < \infty$, and $\mathbf{var}\sum_{i=1}^n Y_i = 1$. Suppose further that the following dependence condition holds: for every i there are sets $A_i \subset B_i \subset \{1, 2, \ldots, n\}$ such that $Y_i \perp\!\!\!\perp \{Y_j\,;\, j \in A_i^c\}$ and $\{Y_j\,;\, j \in A_i\} \perp\!\!\!\perp \{Y_k\,;\, k \in B_i^c\}$. Then for $W = \sum_{i=1}^n Y_i$ it holds that*

$$\rho_W\big(\mathbf{P}_W, N(0,1)\big)$$
$$\leq 2\sum_{i=1}^n\big(\mathbf{E}|Y_i U_i V_i| + |\mathbf{E}(Y_i U_i)|\,\mathbf{E}|V_i|\big) + \sum_{i=1}^n \mathbf{E}|Y_i U_i^2|,$$

where $U_i = \sum_{j \in A_i} Y_j$ and $V_i = \sum_{k \in B_i} Y_k$.

Proof. See [68], Theorem 4.13. $\square$

For independent random variables Y_i we may put $A_i = B_i = \{i\}$ in Theorem 1.2. Then $U_i = V_i = Y_i$ and we recover the first statement in Theorem 1.1 with a constant of 5 instead of 1. Note that $\mathbf{E}Y_i^2\mathbf{E}|Y_i| \le \mathbf{E}|Y_i|^3$ follows by two applications of Hölder's inequality or by a single application of the Fortuin–Kasteleyn–Ginibre (FKG) inequality (see e.g. Theorem 3.3 in Chap. 7 of [29]).

1.2.4 Kolmogorov Distance

Stein's method is known to often give the optimal rates and in many situations also very good constants. This however needs not always be the case and in particular for probability metrics based on "non-smooth" functions certain standard techniques for bounding the right-hand side of the Stein equation (1.6) may break down.

This is what happens for approximation in terms of the Kolmogorov metric, where a more complicated argument is needed and the optimal rate can only be obtained at the cost of a somewhat large constant. The big advantage of Stein's method to lend itself to various generalizations remains, especially when it comes to relaxing the independence condition.

The following result demonstrates the state of the art for independent random variables.

Theorem 1.3 (Chen and Shao). *Let $Y_1,\ldots,Y_n$ be independent random variables with $\mathbf{E}Y_i = 0$ and $\sum_{i=1}^{n}\mathbf{E}Y_i^2 = 1$. Then for $W = \sum_{i=1}^{n}Y_i$ it holds that*

$$\rho_K\bigl(\mathbf{P}_W,N(0,1)\bigr) = \sup_{x\in\mathbb{R}}\bigl|\mathbf{P}(W \le x) - \Phi(x)\bigr| \le 4.1\sum_{i=1}^{n}\mathbf{E}\min(Y_i^2,|Y_i|^3),$$

where $\Phi : \mathbb{R} \to [0,1]$ denotes the distribution function of the standard normal distribution.

Proof. See [69], Theorem 2.1. $\qquad\qquad\qquad\qquad\qquad\qquad\qquad\qquad\square$

To the best of our knowledge, the constant of $C = 4.1$ is also the smallest one that has ever been obtained by Stein's method in the classical Berry–Esseen bound with third moments [35, 113], i.e. in

$$\rho_K\bigl(\mathbf{P}_W,N(0,1)\bigr) \le C\sum_{i=1}^{n}\mathbf{E}|Y_i|^3.$$

In contrast, other methods were able to improve this constant quite substantially to $C = 0.56$, cf. [366, 398]. For identically distributed random variables a further improvement to $C = 0.4785$ is possible [398]. In [114] it has been shown that even in the i.i.d. case the constant cannot be better than 0.4097.

1.3 Approximation by General Distributions

The fundamental ideas employed in Sect. 1.2 for the normal approximation of a sum of random variables can be put to use in almost arbitrary generality. Suppose that on some measurable space S we would like to approximate a distribution P by a reasonably nice distribution Q in terms of a probability metric ρ. The present section describes abstractly the main ingredients of Stein's method to go about such an approximation.

1.3.1 The Key Steps of Stein's Method

Let P and Q be probability measures on a measurable space S. We assume that Q is a "distinguished" probability measure that is typically more tractable than the measure P. We will also refer to Q as the approximating measure. Let furthermore $\mathcal{F}$ be the class of "test functions" that we build our probability metric on, such as indicators, or Lipschitz functions if S is a metric space.

Our goal is to find an upper bound for

$$\rho(P,Q) = \sup_{f \in \mathcal{F}} \left| \int f \, dP - \int f \, dQ \right|.$$

By definition ρ is at least a pseudometric. Typically we want the class $\mathcal{F}$ to be rich enough so that it is a metric, i.e. that also $\rho(P,Q) = 0$ implies $P = Q$.

We first present an overview of the general procedure suggested by Stein's method for attaining this goal, before discussing the individual steps in more detail.

1. **Set up the Stein equation.** For $f \in \mathcal{F}$, write

$$f(x) - \int f \, dQ = \mathscr{A} g(x), \tag{1.11}$$

where $\mathscr{A}$ is an operator on a space of functions $g \colon S \to \mathbb{R}$ that characterizes the distribution Q in the sense that $Z_* \sim Q$ if and only if $\mathbf{E}\mathscr{A} g(Z_*) = 0$ for "all" functions g.

2. **Solve the Stein equation.** Given f, find g_f that satisfies (1.11).

3. **Bound the right-hand side.**
   (a) *Find bounds on differences/derivatives of g_f.* These depend on Q, but not on P. So these bounds will be useful for approximation by the measure Q in general.
   (b) *Bound* $\left| \int \mathscr{A} g(x) P(dx) \right|$. This is based on the results from (a). It is only here that more specific knowledge about P enters.

This recipe has been applied to many approximating distributions Q, such as the Poisson, compound Poisson, binomial, hypergeometric, negative binomial (including geometric), discrete Gibbs, normal, gamma (including χ^2 and exponential), beta, multivariate normal, Poisson process, compound Poisson process, Gibbs process, and Wiener process distributions.

The following are a few examples of successful Stein operators for various distributions Q:

$$\text{Pois}(\lambda) \qquad \mathscr{A}g(j) = \lambda g(j+1) - jg(j), \qquad\qquad j \in \mathbb{Z}_+$$

$$\text{NBinom}(r,p) \qquad \mathscr{A}g(j) = (r+j)(1-p)g(j+1) - jg(j), \qquad j \in \mathbb{Z}_+$$

$$N(\mu,\sigma^2) \qquad \mathscr{A}g(x) = \sigma^2 g'(x) - (x-\mu)g(x), \qquad\qquad x \in \mathbb{R}$$

$$\text{Gamma}(a,\lambda) \qquad \mathscr{A}g(x) = xg'(x) + (a - \lambda x)g(x), \qquad\qquad x \in \mathbb{R}_+$$

All these distributions have the characterizing property described in step 1 of the recipe. The negative binomial distribution is parametrized here such that $X \sim \text{NBinom}(r,p)$ is a random variable on $\mathbb{Z}_+$ with $\mathbf{P}(X = k) = \binom{k+r-1}{k} p^r (1-p)^k$ for every $k \in \mathbb{Z}_+$, i.e. if r is an integer, X describes the number of failures before the r-th success in a Bernoulli experiment with success probability p.

But how do we find a good Stein operator? Originally such operators were found by careful study of the approximating distribution Q. However it is much more appealing to have a general recipe also for obtaining a proposal for a Stein operator. In [376] it is suggested, if Q is a distribution on $\mathbb{R}$ with a density q that satisfies certain conditions and in particular has itself a derivative q', then we may use

$$\mathscr{A}g(x) = g'(x) + \frac{q'(x)}{q(x)}g(x) \tag{1.12}$$

as a promising operator. This has become known as the *density approach* and may to some degree be extended to more general distributions Q. For example if Q is a distribution on $\mathbb{Z}$ with probability mass function q we may consider analogously

$$\mathscr{A}g(j) = g(j+1) - g(j) + \frac{q(j) - q(j-1)}{q(j)}g(j) = g(j+1) - \frac{q(j-1)}{q(j)}g(j). \tag{1.13}$$

We sweep problems of division by zero under the rug here. The above proposal works directly if there exists a $j_0 \in \mathbb{Z}$ such that $q(j) > 0$ for every $j \geq j_0$ and we set $0/0 = 0$.

Exercise 1.3. Show that equivalent forms of the four Stein operators above may be obtained by (1.12) and (1.13). *Hint*: For $\text{NBinom}(r,p)$ plug in the transformed function $g(j) = (j+r-1)\tilde{g}(j)$ in (1.13); for $\text{Gamma}(a,\lambda)$ plug in $g(x) = x\tilde{g}(x)$ in (1.12).

For approximation problems on more complicated spaces the density approach is usually not applicable or at least less clear-cut. We therefore concentrate in what follows on a somewhat less automatic, but much more universal method, which may yield a useful Stein operator for virtually any approximating distribution Q. This technique has been dubbed the *generator approach* for reasons that will become obvious in a moment.

1.3.2 The Generator Approach

Barbour's generator approach [26] suggests to use as the operator $\mathscr{A}$ in step 1 of our recipe the infinitesimal generator $\mathscr{G}$ of a Markov process Z that has Q as its stationary initial distribution. We recall here just the basic definitions of continuous-time Markov process theory, which is all that will be needed later on, avoiding any discussion of technical problems. For more comprehensive treatments, see the standard reference [115] or the introduction [255].

It is generally assumed that S is a complete, separable metric space equipped with its Borel σ-algebra $\mathcal{B}(S)$.

Definition 1.1. Let $Z = \{Z(t),\, t \geq 0\}$ be an S-valued stochastic process.

(a) Z is called a *(time-homogeneous) Markov process* if

$$\mathbf{P}\big(Z(t+r) \in A \mid \{Z(s),\, s \leq t\}\big) = \mathbf{P}\big(Z(t+r) \in A \mid Z(t)\big) = \mathbf{P}\big(Z(r) \in A \mid Z(0)\big)$$

for all $t, r \geq 0$ and all $A \in \mathcal{B}(S)$. We refer to $\mathbf{P}_{Z(0)}$ as the *initial distribution* of the process.

(b) If Z is a Markov process, its *infinitesimal generator* is defined as the functional $\mathscr{G} : \mathrm{dom}(\mathscr{G}) \to \mathbb{R}^S$ given by

$$\mathscr{G}h(x) = \frac{d}{dt}\Big[\mathbf{E}\big(h(Z(t)) \mid Z(0) = x\big)\Big]_{t=0} = \lim_{t \to 0} \frac{1}{t}\Big(\mathbf{E}\big(h(Z(t)) \mid Z(0) = x\big) - h(x)\Big),$$

where $\mathrm{dom}(\mathscr{G})$ consists of all $h : S \to \mathbb{R}$ for which the limit above exists.

(c) If Z is a Markov process, a *stationary distribution* Q is defined as any probability distribution on S for which

$$\int_S \mathbf{P}\big(Z(r) \in A \mid Z(0) = x\big)\, Q(dx) = Q(A)$$

for all $r \geq 0$ and all $A \in \mathcal{B}(S)$.

Intuitively for a Markov process the distribution of its future path depends just on the present state, regardless of any of its past or of the time on the clock. Hence, technicalities aside, it is just a continuous-time analogue of a discrete-time Markov chain. The generator describes the stochastic evolution of the Markov process in an "infinitesimal time step". The stationary distribution is such that, if chosen as initial distribution of the process, it is maintained for all times. Under very general conditions the stationary distribution Q may be characterized by

$$\int_S \mathscr{G}h(x)\, Q(dx) = 0 \qquad (1.14)$$

for "enough" functions $h\colon S \to \mathbb{R}$; see Theorem 9.2 in Chap. 4 of [115]. Intuitively this makes good sense, as it means that under the stationary distribution there is no stochastic evolution. Note that (1.14) is exactly the property required for our Stein operator $\mathscr{A}$ with respect to the approximating distribution Q. The reason why the functions are now called h instead of g will become clear in the examples.

Example 1.1 (Normal approximation). Let Z be an Ornstein–Uhlenbeck process, more precisely the solution of the stochastic differential equation

$$Z(t) = -Z(t)\,dt + \sqrt{2}\,dB(t),$$

which can be roughly thought of as a Brownian motion $\{B(t),\ t \geq 0\}$ with a pull towards the origin. It is well known that $N(0,1)$ is the stationary distribution of this process. Its generator $\mathscr{G}$ is given by

$$\mathscr{G}h(x) = h''(x) - xh'(x),$$

which is exactly our usual Stein operator for the normal distribution, but with g replaced by h'.

Example 1.2 (Pois(λ)-approximation). Let Z be an immigration-death process with immigration rate $\lambda > 0$ and unit-per-capita death rate. This is a pure-jump Markov process on $\mathbb{Z}_+$ whose evolution may be described as follows. Given $Z(t) = j$ ("there are j individuals at time t"), the process stays in the state j for an $\mathrm{Exp}(\lambda + j)$-distributed time, after which it jumps to $j+1$ with probability $\lambda/(\lambda + j)$ ("an immigration occurs") or it jumps to $j-1$ with probability $j/(\lambda + j)$ ("a death occurs"). This process has Pois(λ) as its stationary distribution; see Sect. 6.11 in [153]. Its generator is

$$\mathscr{G}h(j) = \lambda\big[h(j+1) - h(j)\big] + j\big[h(j-1) - h(j)\big].$$

If we write $g(j) = h(j) - h(j-1)$ for $j \geq 1$, we have exactly our Stein operator for the Poisson distribution from Sect. 1.3.1, i.e. $\mathscr{A}g(j) = \lambda g(j+1) - jg(j)$.

Like in these examples, the generator approach often yields an interesting candidate for the operator $\mathscr{A}$ needed for step 1 of our recipe. If we use this operator, we also have a default form for the solution of the corresponding Stein equation, as required for step 2. Denote by $Z_x = \{Z_x(t),\ t \geq 0\}$ an arbitrary Markov process with generator $\mathscr{G}$ that is started in state x, meaning $Z(0) = x$ almost surely.

Proposition 1.1. *Choose $\mathscr{A} = \mathscr{G}$ as the Stein operator and let the function $f\colon S \to \mathbb{R}$ be such that*

$$\vartheta\colon \mathbb{R}_+ \to \mathbb{R},\, t \mapsto \mathbf{E}f(Z_x(t))$$

is continuous at 0 for every x. Suppose that the function $h_f\colon S \to \mathbb{R}$ given by

$$h_f(x) = -\int_0^\infty \left[\mathbf{E}f(Z_x(t)) - \mathbf{E}f(Z_*)\right] dt,$$

where $Z_* \sim Q$, is well-defined. Then h_f solves the Stein equation

$$f(x) - \int f \, dQ = \mathscr{A}h(x). \tag{1.15}$$

Proof. Writing $\mu_x^{(s)}$ for the distribution of $Z_x(s)$, we can compute the expectation of the random variable $Y = \mathbf{E}(f(Z(t)) \mid Z(0) = \cdot) \circ Z_x(s)$ by the Markov property as

$$\begin{aligned}
\mathbf{E}Y &= \int_S \mathbf{E}(f(Z(t)) \mid Z(0) = z) \, \mu_x^{(s)}(dz) \\
&= \int_S \mathbf{E}(f(Z_x(s+t)) \mid Z_x(s) = z) \, \mu_x^{(s)}(dz) \\
&= \mathbf{E}f(Z_x(s+t)).
\end{aligned}$$

Hence

$$\begin{aligned}
\frac{1}{s}\left(\mathbf{E}h_f(Z_x(s)) - h_f(x)\right) &= \frac{1}{s}\left(-\int_0^\infty \left[\mathbf{E}f(Z_x(s+t)) - \mathbf{E}f(Z_*)\right] dt \right. \\
&\qquad\qquad \left. + \int_0^\infty \left[\mathbf{E}f(Z_x(t)) - \mathbf{E}f(Z_*)\right] dt \right) \\
&= \frac{1}{s}\int_0^s \left[\mathbf{E}f(Z_x(v)) - \mathbf{E}f(Z_*)\right] dv \\
&= f(x) - \mathbf{E}f(Z_*) + \frac{1}{s}\int_0^s \left[\mathbf{E}f(Z_x(v)) - f(x)\right] dv,
\end{aligned}$$

substituting $v = s + t$ for the second equality. Note that the well-definedness of h_f guarantees that all of the above integrals exist. Letting $s \to 0$ on both sides yields

$$\mathscr{G}h(x) = f(x) - \mathbf{E}f(Z_*)$$

as required. $\qquad\square$

Whether the function h_f is well-defined depends only on the speed of the convergence $\mathbf{E}f(Z_x(t)) \to \mathbf{E}f(Z_*)$ as $t \to \infty$. If the space S is not too complicated a promising strategy both for proving well-definedness of h_f and for bounding difference terms of h_f (following step 3(a) of the recipe) is by constructing a "rapid" coupling (Z_x, Z_y) of two Markov processes Z_x and Z_y with generator $\mathscr{G}$ started in states x and y, respectively. This *coupling strategy* is demonstrated in detail in Sect. 1.5.1.

From here on many further ramifications of Stein's method will depend on more detailed knowledge about the space S and the approximating distribution Q. For the rest of this chapter we therefore focus on the special case where $S = \bar{\mathcal{N}}$ is the set of finite point patterns on a compact subset of $\mathbb{R}^d$, and Q is typically a Poisson process distribution.

Step 3 of our general recipe is only demonstrated in this setting, because it offers a nice structure and has been rather well studied in the literature. Nevertheless it should be stressed that there is no fundamental obstacle in sight to developing Stein's method also for obtaining distance bounds between distributions of more complicated random geometric objects such as line processes, random tesselations, or germ-grain models.

1.4 Point Processes

From now on we consider for S the space $(\bar{\mathcal{N}}, \bar{\mathfrak{N}})$ of finite counting measures ("point patterns") on a compact metric space (W, ρ_0), where $\bar{\mathfrak{N}}$ is the canonical σ-algebra generated by the sets $\{\psi \in \bar{\mathcal{N}} : \psi(A) = n\}$ for all $n \in \mathbb{Z}_+$ and $A \in \mathcal{B}$. Here $\mathcal{B}$ denotes the family of Borel sets on W. In what follows we always take $W \subset \mathbb{R}^d$ with non-empty interior, and use Lebesgue measure whenever a reference measure is required. All the results hold in the more general setting and for an arbitrary reference measure, with the proviso that for some purposes this measure has to be diffuse.

Unlike in many other treatments on spatial statistics, we allow for the possibility of multipoints at a single location in W. We may write any element $\psi \in \bar{\mathcal{N}}$ in the form $\psi = \sum_{i=1}^{n} \delta_{x_i}$ where $n \in \mathbb{Z}_+$ and $x_1, \ldots, x_n \in W$ may contain repeats. If $x_1, \ldots, x_n$ are pairwise different we speak of a *simple* point pattern and tacitly identify ψ with the set $\{x_1, \ldots, x_n\}$. Denote by $\mathcal{N}$ the subset of $\bar{\mathcal{N}}$ of simple point patterns. Working with the space $\bar{\mathcal{N}}$ has the major technical advantage that by the compactness of W the vague and the weak topology commonly considered on spaces of measures coincide and $\bar{\mathcal{N}}$ is complete in this topology, which we simply denote by $\mathcal{T}$.

A *point process* Ψ is just a random element of $\bar{\mathcal{N}}$. We call Ψ *simple* if $\mathbf{P}(\Psi \in \mathcal{N}) = 1$. For a simple point process we ignore the null set $\bar{\mathcal{N}} \setminus \mathcal{N}$ as long as only distributional properties are concerned. Thus we use notation like $\{X_1, \ldots, X_N\}$ for the random set of its points.

We briefly summarize some basic concepts about point processes and present some important models. Note the much more detailed treatment of these materials in the previous volume [373] of these lecture notes, see e.g. Chap. 3 of [373]. We start with a concept that is an analogue to the expectation of a real-valued random variable.

Definition 1.2. Let Ψ be a point process on W. The measure $\Lambda = \mathbf{E}\Psi$ on W that is given by $\Lambda(A) = \mathbf{E}\Psi(A)$ for every $A \in \mathcal{B}$ is called the *expectation measure* of Ψ. If it is finite, we say *the expectation measure exists*. If it exists and has a density $\lambda : W \to \mathbb{R}_+$, we call λ the *intensity (function)* of Ψ.

The Poisson process is arguably the most important point process model, although not necessarily the most interesting one. Its high degree of independence allows for many explicit computations and makes it an ideal null model against which we may test if a given point pattern has any dependence structure. It is furthermore the basic building block from which we may construct by one or the other technique a large number of much more complicated point-process models; see [21], Sect. 3.1.4. One such technique is by defining a density with respect to a Poisson process, leading to the important class of Gibbs processes.

Definition 1.3. (a) Let Λ be a finite measure on W. A *Poisson process* $\Pi = \Pi_\Lambda$ on W with expectation measure Λ is characterized by the following two properties.

1. $\Pi_\Lambda(A) \sim \mathrm{Pois}(\Lambda(A))$ for every $A \in \mathcal{B}$;
2. $\Pi_\Lambda(A_1), \Pi_\Lambda(A_2), \ldots, \Pi_\Lambda(A_k)$ are independent for disjoint $A_1, \ldots, A_k \in \mathcal{B}$.

We denote the distribution of Π_Λ by π_Λ. If Λ has a density λ, we may also write Π_λ and π_λ. In particular π_1 denotes the distribution of the Poisson process whose expectation measure is Lebesgue measure ν_d. Such a process can be obtained by scattering a $\mathrm{Pois}(\nu_d(W))$ number of points uniformly and independently over W. It is therefore simple. We often use π_1 as a reference measure on $\mathcal{N}$.

(b) A point process Ψ on W is called a *Gibbs process* if it has a density $u : \mathcal{N} \to \mathbb{R}_+$ with respect to π_1 that is *hereditary* in the sense that $u(\psi) = 0$ implies $u(\varphi) = 0$ whenever $\psi \subset \varphi$.

(c) A Gibbs process is called a *pairwise interaction process* if its density is of the form

$$u(\psi) = \alpha \left(\prod_{x \in \psi} \vartheta_1(x) \right) \left(\prod_{\{x,y\} \subset \psi} \vartheta_2(x,y) \right)$$

for $\vartheta_1 : W \to \mathbb{R}_+$ and symmetric $\vartheta_2 : W \times W \to \mathbb{R}_+$, where $\alpha > 0$ is a normalizing constant. We refer to ϑ_1 as the activity function and to ϑ_2 as the (pairwise) interaction function.

Sect. 3.1.5 in [21] discusses various concrete Gibbs process models in detail, many of which are pairwise interaction processes. One of the most elementary models is the stationary *Strauss process*, where

$$\vartheta_1(x) = \beta, \qquad \vartheta_2(x,y) = \begin{cases} \gamma & \text{if } \|x - y\| \leq R, \\ 1 & \text{otherwise,} \end{cases}$$

with parameters $\beta, R > 0$ and $\gamma \in [0,1]$. See Figs. 3.20 and 3.21 in [21] for some realizations. For this as for most other processes the normalizing constant α cannot be calculated explicitly.

For Gibbs processes we can define a very appealing descriptive function.

Definition 1.4. Let Ψ be a Gibbs process on W with density u. We call the function $\beta^* : W \times \mathcal{N} \to \mathbb{R}_+$, which is given by

$$\beta^*(x; \psi) = \frac{u(\psi + \delta_x)}{u(\psi)},$$

the *conditional intensity (function)* of Ψ, where we put $0/0 = 0$.

In fact the conditional intensity characterizes the Gibbs process distribution, as it is possible to recover from β^* the density u recursively at larger and larger point patterns ψ. Intuitively $\beta^*(x;\psi)$ may be interpreted as the intensity at x ("expected number of points in an infinitesimal ball around x") given Ψ is equal to ψ everywhere else.

The conditional intensity β^* satisfies the *Georgii–Nguyen–Zessin equation*,

$$\mathbf{E}\left(\int_W h(x,\Psi-\delta_x)\,\Psi(dx)\right) = \int_W \mathbf{E}\left(h(x,\Psi)\beta^*(x;\Psi)\right)\,dx \qquad (1.16)$$

for every measurable $h\colon W\times\mathcal{N}\to\mathbb{R}_+$. This equation may be thought of as the Swiss Army knife of point process statistics. We make vital use of it in the sequel.

For the Strauss process above we obtain $\beta^*(x;\psi)=\beta\gamma^{\psi(B_R(x))}$, where $B_R(x)$ denotes the closed ball in $\mathbb{R}^d$ with centre at $x\in\mathbb{R}^d$ and radius $R>0$. Note that for any Gibbs process the unpleasant normalizing constant α in the density u cancels, and the conditional intensity β^* is then usually known explicitly.

1.4.1 Distances between Point Patterns

In order to study distributional approximation for point processes, we need a metric between probability measures on $\mathcal{N}$. We may always use the total variation metric on *any* measurable space S, but especially on complicated spaces it may be very strong and sometimes too strong to be useful. Often a better choice, if there is a reasonable metric structure on S, is the bounded Wasserstein metric. It will give a smaller approximation error while exactly metrizing convergence in distribution, and it captures the underlying topological structure of the state space.

This raises the question what might be an appropriate metric on $\mathcal{N}$. Among various possibilities proposed in the literature, we concentrate here on the metric introduced in [357]. On the one hand this metric has the very desirable property that it metrizes the topology $\mathcal{T}$, and on the other hand it reflects well human intuition about how similar or distinct two point patterns are. The second point is especially important for statistical applications, two of which are presented in Sect. 1.4.2 below.

We denote by Σ_n the set of permutations of $\{1,2,\dots,n\}$ and choose for $c>0$ any upper bound on $\operatorname{diam}(W)$. For two point patterns $\psi,\varphi\in\bar{\mathcal{N}}$ with representations $\psi=\sum_{i=1}^m \delta_{x_i}$ and $\varphi=\sum_{i=1}^n \delta_{y_i}$, where $\max(1,m)\le n$, we set

$$\rho_1(\psi,\varphi)=\frac{1}{n}\left(\min_{\sigma\in\Sigma_n}\sum_{i=1}^m \rho_0(x_i,y_{\sigma(i)})+c(n-m)\right). \qquad (1.17)$$

This is readily extended to a metric on $\bar{\mathcal{N}}$ by symmetrizing and setting $\rho_1(\emptyset,\emptyset)$ to zero.

In words, the ρ_1-distance is obtained by pairing the points of the smaller point pattern with an equally sized subpattern of the larger point pattern in such a way that the average pairing distance in terms of ρ_0 is minimal. The actual ρ_1-distance is then the average over all pairing distances including a penalty of c for each unpaired point. See Fig. 1.1 for an evident and a less evident example of such pairings in the case where ρ_0 is the Euclidean metric.

Exercise 1.4. Prove the following statement that is nicely illustrated in the right-hand panel of Fig. 1.1. Suppose that ψ and φ are point patterns in $W \subset \mathbb{R}^2$ that are jointly in general position, i.e. there is no straight line in $\mathbb{R}^2$ passing through more than two points at once. Furthermore take ρ_0 to be the Euclidean metric. Then the lines connecting the points of ψ to the points of φ in an optimal ρ_1-pairing do not cross.

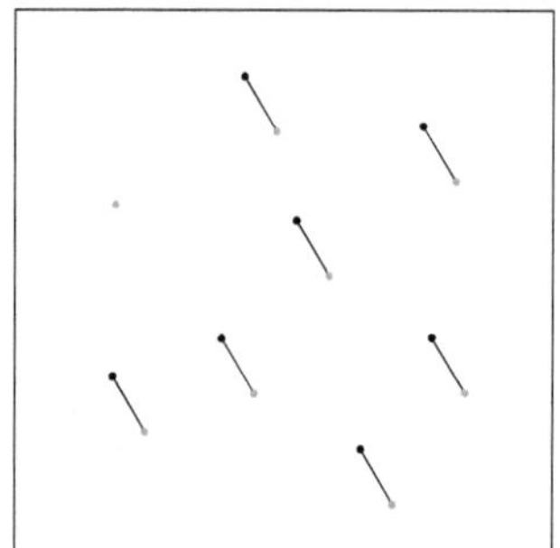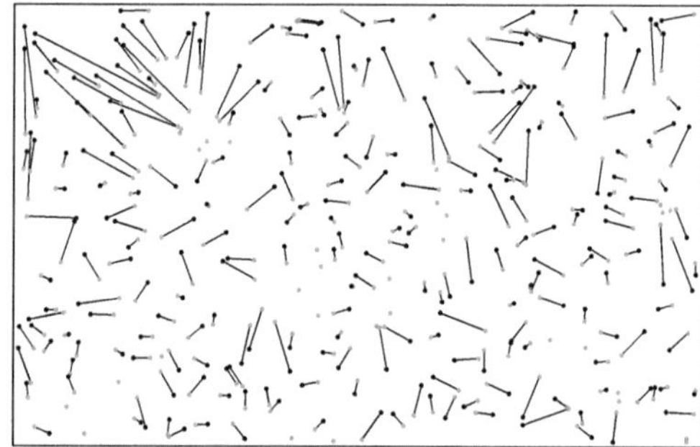

Fig. 1.1 The point pairing underlying the ρ_1-metric. *Left:* Two small point patterns with 7 and 8 points; *right:* two larger point patterns with 250 and 280 points.

An important feature is the fact that an optimal pairing of two point patterns can be found efficiently. First we can get rid of the second summand in (1.17) by adding an element to the space W that is at distance c from all the locations in W. Let $\widetilde{W} = W \cup \{x_\infty\}$, where x_∞ is an arbitrary element not contained in W, and extend ρ_0 by setting $\rho_0(x, x_\infty) = c$ for every $x \in W$. Note that the extended ρ_0 is still a metric. We then replace the smaller point pattern ψ in (1.17) by $\tilde{\psi} = \psi + (n-m)\delta_{x_\infty}$, so that it has the same number of points as φ and the $n-m$ new points can only be paired at distance c. Any pairing of the points of ψ with a subpattern of φ corresponds to a full pairing between $\tilde{\psi}$ and φ in the sense that the total costs of the pairings are the same. Therefore also $\rho_1(\psi, \varphi) = \rho_1(\tilde{\psi}, \varphi)$.

The problem of finding an optimal pairing between $\tilde{\psi}$ and φ is the classical *assignment problem* from operations research: Given n workers, n tasks and a matrix $C = (c_{ij})_{1 \le i, j \le n}$, where $c_{ij} \ge 0$ is the cost incurred for assigning worker i to task j, find a 1-1 assignment $\sigma \in \Sigma_n$ (the set of permutations defined above) of workers to tasks that minimizes the total cost $\sum_{i=1}^{n} c_{i,\sigma(i)}$. A first algorithm that solves this problem in polynomial time was already given in 1865 in a posthumously published article by Carl Gustav Jacobi [182], where the problem arose in the context of determining the order of a system of differential equations. However, when faced with the assignment problem in the economical context, people where unaware of the previous work, and a solution in polynomial time was rediscovered in [238] and called the *Hungarian method* to acknowledge the contribution of the Hungarian mathematicians Kőnig and Egerváry. The Hungarian method was later improved to a time complexity of $O(n^3)$ independently in [110] and [396]. A presentation of this method is beyond the scope of these lecture notes, but detailed descriptions can be found in many books on linear programming or combinatorial optimization, see e.g. [309], Sect. 11.2.

The readers may experiment for themselves with point pairings and ρ_1-distance computations by using the function `pppdist` in the package `spatstat` [19] contributed to the statistical computing environment R [323]. In fact not only the order of complexity, but also real computation times are reasonably low. E.g. finding the pairing in the right-hand panel of Fig. 1.1 required slightly less than a second on a reasonably modern laptop.

1.4.2 Statistical Applications of Distances between Point Patterns

Modern statistics often studies data that is most fruitfully thought of as elements in a more complex space than just $\mathbb{R}$ or $\mathbb{R}^d$, including functional data, shape data, or structural data. The present lecture notes volume gives numerous examples. Analysing such data can be very challenging as there are many modelling choices and simplifications to be made, and a good compromise between the physical reality (of which our knowledge may be very limited) and the computational feasibility of the problem is not always easy to find.

In this section we take a rather minimalistic approach by saying that if our data space has a reasonable metric, i.e. one that reflects well an intuitive concept by which we would like to group or separate their elements in view of the statistical problem at hand, then there is a number of methods which we may try and which should do reasonably well.

In what follows we give two applications for the metric space $(\bar{\mathcal{N}}, \rho_1)$. Traditionally in point process statistics there is just one data point pattern. Several examples for this situation can be found e.g. in [21]. However, recent technological advances, such as in medical imaging or destructive and non-destructive sampling in materials science, produce an increasing amount of multi-observation point patterns. These include independent replicates of a single point process, groups of independent replicates stemming from several point processes, or realizations from a point process that depends on a covariate, possibly time.

Our first application uses simulated point patterns that could have been obtained as the result of a medical screening procedure (or of a sampling technique in materials science). The task is to distinguish between "healthy" and "pathological" samples. The second application uses image data from a real ant colony. The task is to learn something about the collective activity pattern in the colony.

1.4.2.1 Discrimination and Classification in Medical Testing

Suppose that we have point pattern data from two groups of patients. Such data may for example give the positions of cells of a certain type in tissue samples of some organ. For simplicity we just refer to one group as "healthy tissue" and the other group as "pathological tissue".

Fig. 1.2 gives a simulated example of such data on $W = [0,1]^2$, which we will study further. The healthy tissue in the first row has been generated from a stationary Poisson process with intensity $\lambda_h(x,y) = 30$, the pathological tissue in the second row from a Poisson process with intensity $\lambda_p(x,y) = 45e^{-1.5x}/(1 - e^{-1.5})$. Both processes have the same expected number of points in $W = [0,1]^2$. Note that it is not so easy to spot the difference between the two rows by eye.

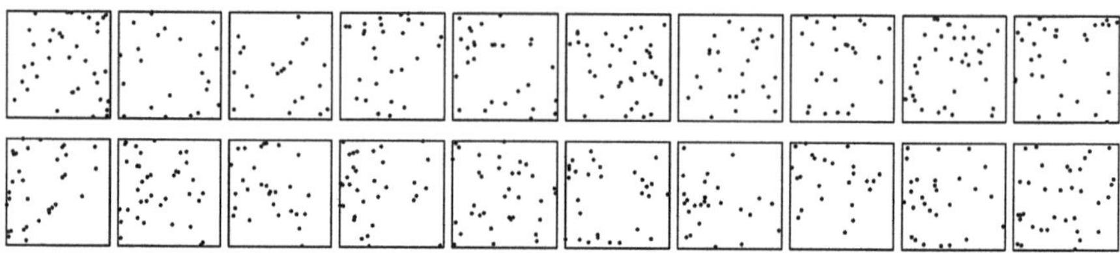

Fig. 1.2 Simulated data for cells in "healthy tissue" (top row) and in "pathological tissue" (bottom row). See text for simulation details.

We start with a permutation test for the discrimination of the pathological from the healthy tissue. There are several statistics that perform well for such a test. Denoting the healthy sample by $\psi_1, \ldots, \psi_m$ and the pathological sample by $\varphi_1, \ldots, \varphi_n$, we use

$$T = T(\psi_1, \ldots, \psi_m; \varphi_1, \ldots, \varphi_n)$$
$$= \frac{1}{mn} \sum_{i=1}^{m} \sum_{j=1}^{n} \rho_1(\psi_i, \varphi_j)$$
$$- \frac{1}{m(m-1)} \sum_{1 \le i < j \le m} \rho_1(\psi_i, \psi_j) - \frac{1}{n(n-1)} \sum_{1 \le i < j \le n} \rho_1(\varphi_i, \varphi_j),$$

which is in the same spirit as the statistic studied in [28] for two samples in $\mathbb{R}^d$. We base the ρ_1-metric now on the metric $\rho_0(x,y) = \min(\|x - y\|, 0.3)$, and the penalty c for extra points is correspondingly set to 0.3. This is rather arbitrary, but reflects our vague idea that in an optimal pairing with about 30 points in each pattern the pairing distances are typically still below 0.3.

Since we have only 10 patterns in each group, we can easily compute a deterministic p-value by finding the rank r of the observed value of T among the $\binom{20}{10}/2 = 92,378$ total values obtained by pooling the 20 patterns in Fig. 1.2, resplitting them into two groups of 10, and computing T based on the new split. Note that the distance matrix between the point patterns has to be computed only once.

For much larger samples of point patterns we may still proceed by taking a random (uniform) subsample of all possible splits. This still yields an exact p-value, but one that is random, as it depends on the concrete subsample picked. In any case the p-value is given by

$$\frac{\vartheta + 2 - r}{\vartheta + 1},$$

where ϑ is the number of other splits considered (in our case $92,377$). For the above data we obtain a p-value of 0.00296, so we have quite a clear discrimination of the healthy from the pathological group.

We next turn to the classification problem. We use the data shown in Fig. 1.2 as our *training data*, for which we know the true group membership. Suppose that we would like to assign new tissue samples reliably either to one or the other group. In a real medical context it might be the case that there is an invasive procedure that can decide for sure if the patient has a certain disease or not, and on which our knowledge of the group membership in the training data is based. Our goal might be to create a rapid test based solely on some imaging procedure.

Since we have a reasonable metric, a natural approach is k-nearest neighbour classification. This simply means that faced with a new point pattern we compute the ρ_1-metric to all of the training patterns. Then we determine the k closest patterns and do a majority vote among them to decide which group we assign it to. To avoid ties we only consider odd values of k.

Table 1.1 Misclassification rates for various choices of k.

	1	3	5	7	9	11	13	15
total rate	0.279	0.246	0.242	0.225	0.210	0.204	0.210	0.218
false positive rate	0.190	0.150	0.153	0.155	0.150	0.151	0.160	0.175
false negative rate	0.089	0.096	0.089	0.070	0.060	0.053	0.050	0.043

Table 1.1 gives the misclassification rates for various values of k based on 1000 test samples, each generated with probability $1/2$ from the healthy or pathological population. In practical applications, where we cannot afford a reasonably large test set, we may for example proceed by cross-validation in order to determine a good value of k.

1.4.2.2 Activity in an Ant Colony

In this example real data is considered. A whole colony of ants of the species *Temnothorax albipennis* has been placed in an artificial nest between two glass plates seperated by pieces of cardboard. The nest dimensions were $45 \times 30 \times 1$ mm, with an entrance so that the ants were able to leave and re-enter the nest at will. Photographs were taken at 5 minute intervals over the course of about 19 hours, from which the positions of the adult worker ants were converted to point pattern data.

Fig. 1.3 shows a short sequence of 1.5 hours from the data set. The experiment was carried out by Thomas Richardson at the Bristol Ant Lab, under the supervision of Ana Sendova-Franks. More details and various analyses conducted with this and similar data sets can be found in [331].

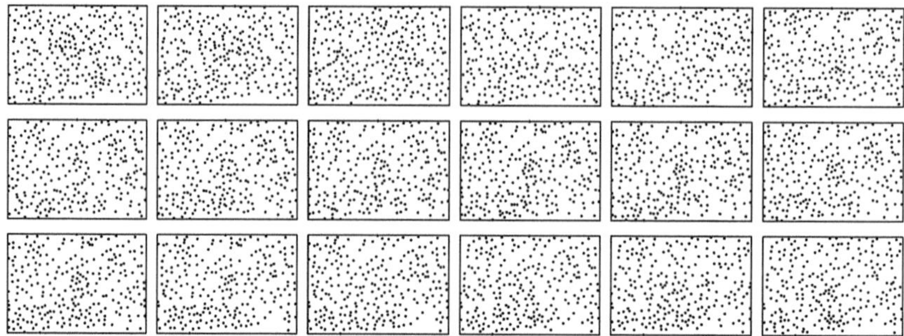

Fig. 1.3 A sequence of 18 point patterns of worker ants recorded at 5 minute intervals (by row).

Our main goal here is to demonstrate that the ρ_1-metric is an interesting tool for measuring overall activity in the ant colony. A natural measurement for this activity is the sum (or the average) of the velocities of the individual ants. We cannot track individual ants in our data set, because they have not been marked in the real world, but we may still compute the ρ_1-distance between subsequent pictures, hoping that this gives a reasonable approximation of this sum. For the ρ_0-metric we choose a sensibly cut off Euclidean metric again.

The plot of the ρ_1-distances as a function of the time in hours is given in Fig. 1.4. We see an interesting oscillating activity pattern that is known from other ant species, but is usually visible only with data recorded at much higher temporal (and sometimes also spatial) resolutions. Fig. 1.5 gives a slightly smoothed periodogram suggesting a period of about 50-70 minutes (corresponding to the dotted lines) for the main oscillation.

Note that a time gap of 5 minutes between subsequent point patterns is rather large, and we cannot hope that the point pairings found as part of the ρ_1-distance computation will allow us to track individual ants with any reliability. This, however, would be another interesting application of the ρ_1-distance in the case of unmarked ant data recorded at higher time resolutions.

1.4.3 Distances between Point Process Distributions

We define the *total variation metric* between the distributions of point processes Ψ and Φ in the analogous way as for distributions on $\mathbb{R}$, namely as

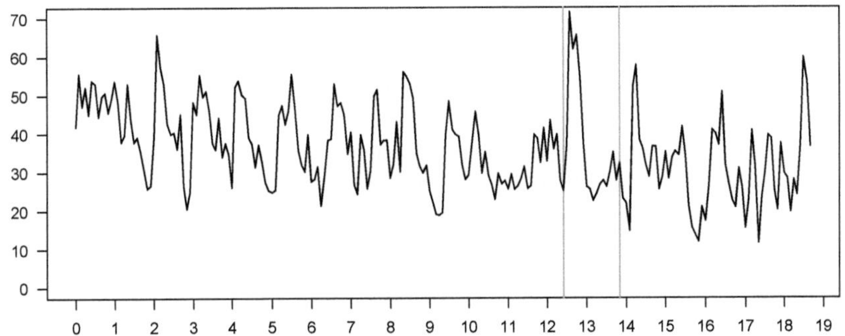

Fig. 1.4 Overall activity in the ant colony as determined by the ρ_1-distance between subsequent point patterns. The x-axis is in hours; the grey vertical lines mark the time window from which the sequence in Fig. 1.3 was taken.

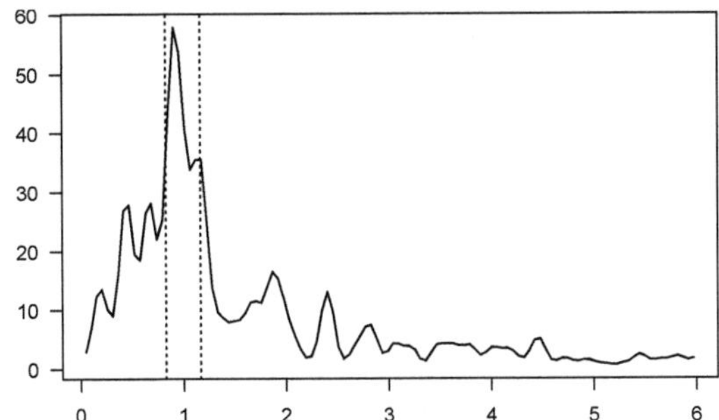

Fig. 1.5 Periodogram of the time series in Fig. 1.4 with vertical lines at 50 and 70 minutes.

$$\rho_{TV}(\mathbf{P}_\Psi, \mathbf{P}_\Phi) = \sup_{A \in \mathfrak{N}} |\mathbf{P}(\Psi \in A) - \mathbf{P}(\Phi \in A)|$$
$$= \sup_{f \in \mathcal{F}_*} |\mathbf{E}f(\Psi) - \mathbf{E}f(\Phi)|, \tag{1.18}$$

where $\mathcal{F}_*$ denotes the set of all measurable functions $f : W \to [0,1]$. The second equality follows by a standard extension argument.

We base the definition of the *Wasserstein metric* on the metric ρ_1 introduced in Sect. 1.4.1. For the simplicity of presentation we assume that $\rho_0 \leq 1$, which may always be arranged by scaling the problem, and set the penalty parameter $c = 1$. Then

$$\rho_W(\mathbf{P}_\Psi, \mathbf{P}_\Phi) = \sup_{f \in \bar{\mathcal{F}}_W} |\mathbf{E}f(\Psi) - \mathbf{E}f(\Phi)|$$
$$= \sup_{f \in \bar{\mathcal{F}}_{BW}} |\mathbf{E}f(\Psi) - \mathbf{E}f(\Phi)|,$$

where $\bar{\mathcal{F}}_W = \{f : \bar{\mathcal{N}} \to \mathbb{R} \,;\, |f(\psi) - f(\varphi)| \leq \rho_1(\psi, \varphi) \text{ for } \psi, \varphi \in \bar{\mathcal{N}}\}$ and $\bar{\mathcal{F}}_{BW} = \{f \in \bar{\mathcal{F}}_W \,;\, f(\bar{\mathcal{N}}) \subset [0,1]\}$. The second equality follows from the fact that $\rho_1 \leq c = 1$.

Since $\bar{\mathcal{F}}_{BW} \subset \mathcal{F}_*$, we have $\rho_W \leq \rho_{TV}$. Furthermore, since ρ_1 metrizes the topology $\mathcal{T}$ on $\bar{\mathcal{N}}$ and since ρ_W can be interpreted as the bounded Wasserstein metric with respect to ρ_1, it follows by Theorem 11.3.3 in [107] that ρ_W metrizes convergence in distribution of point processes.

An upper bound on the Wasserstein distance $\rho_W(\mathbf{P}_\Psi, \mathbf{P}_\Phi)$ gives us control over expressions of the form $\frac{1}{\ell}|\mathbf{E}f(\Psi) - \mathbf{E}f(\Phi)|$ for ρ_1-Lipschitz continuous functions f with an arbitrary Lipschitz constant ℓ. Such functions include the average nearest neighbour distance

$$f(\psi) = \frac{1}{n}\sum_{i=1}^{n}\min_{j\neq i}\rho_0(x_i,x_j)$$

for $\psi = \sum_{i=1}^{n}\delta_{x_i}$ and the average edge length of the minimum spanning tree

$$f(\psi) = \min_{\substack{(V,E)\text{ tree}\\V=\psi}}\frac{1}{|E|}\sum_{\{x,y\}\in E}\rho_0(x,y),$$

where the minimum is taken over all trees with vertex set $V = \psi$ and $|E|$ denotes the cardinality of the edge set E; see [357], Sect. 3, for further examples.

A bound on the total variation distance on the other hand gives us control over *all* expressions of the form $|\mathbf{E}f(\Psi) - \mathbf{E}f(\Phi)|$, and in particular over $|\mathbf{P}(\Psi \in A) - \mathbf{P}(\Phi \in A)|$. This, however, comes at the price that the total variation distance may be very large, and sometimes too large to be useful. For example if we approximate a point process Ψ on a grid $G \subset W$ by a continuous point process Φ, we always have $\rho_{TV}(\mathbf{P}_\Psi, \mathbf{P}_\Phi) \geq \mathbf{P}(\Phi \neq \emptyset)$, no matter how fine the grid gets.

Exercise 1.5. Show the following more general result. Suppose that Ψ and Φ are point processes, and there is a set $G \in \mathcal{B}$ such that $\mathbf{P}(\Psi(G^c) > 0) = P(\Phi(G) > 0) = 0$. Then $\rho_{TV}(\mathbf{P}_\Psi, \mathbf{P}_\Phi) = \max(\mathbf{P}(\Psi \neq \emptyset), \mathbf{P}(\Phi \neq \emptyset))$.

1.5 Poisson Process Approximation of Point Process Distributions

In this section we derive some concrete upper bounds using the general procedures presented in Sect. 1.3 for the case $\mathsf{S} = \bar{\mathcal{N}}$ and $Q = \pi = \pi_\Lambda$, i.e. the default random element Z_* is $\Pi = \Pi_\Lambda$. Note however that the emphasis remains on presenting techniques rather than results. More detailed upper bounds for Poisson process approximation can be found in [27, 70, 355, 357, 421].

Write ρ for either of the metrics ρ_{TV} or ρ_W between point process distributions. We are interested then in bounding $\rho(\mathbf{P}_\Psi, \pi_\Lambda)$ for some finite measure Λ. Denote the total mass of any finite measure by absolute value bars, thus $|\Lambda| = \Lambda(W)$ and likewise $|\psi| = \psi(W)$. For setting up the Stein equation (1.15) via the generator approach we choose

$$\mathscr{A}h(\psi) = \int_W \left[h(\psi + \delta_x) - h(\psi) \right] \Lambda(dx) + \int_W \left[h(\psi - \delta_x) - h(\psi) \right] \psi(dx), \quad (1.19)$$

which is the generator of a spatial immigration-death process on W with immigration measure Λ and unit per-capita death rate. This is a pure-jump Markov process on $\bar{\mathcal{N}}$ that is a natural generalization of the immigration-death process used for approximation by a Poisson random variable on $\mathbb{Z}_+$ in Sect. 1.3.2. Such a process evolves as follows. Given it is at $\psi \in \bar{\mathcal{N}}$ it stays there for an $\mathrm{Exp}(|\Lambda| + |\psi|)$-distributed time, after which a point is added to it with probability $|\Lambda|/(|\Lambda| + |\psi|)$ or removed with probability $|\psi|/(|\Lambda| + |\psi|)$. If added, the position of the point follows the distribution $\Lambda/|\Lambda|$; if removed, the point is chosen uniformly at random from all the points in ψ.

If the measure Λ has a density λ, we obtain immediately by the Georgii–Nguyen–Zessin equation and the fact that for a π_λ-process the conditional intensity is $\beta^*(\cdot\,; \psi) = \lambda$ (see (1.16), Example 3.9 in [21]) that $\mathbf{E}\mathscr{A}h(\Pi_\Lambda) = 0$. By (1.14) this is equivalent to π_λ being the stationary distribution of the spatial immigration–death process above.

1.5.1 The Coupling Strategy

We are now ready to present the coupling strategy announced at the end of Sect. 1.3. Its key idea is to find a general method for coupling two copies of the "characterizing" Markov process started at different states in such a way that they will meet "as soon as possible".

For the spatial immigration-death process such couplings are quite natural by virtue of the following lemma; see [421], Proposition 3.5, for a proof.

Lemma 1.4. *The immigration-death process* $Z_\psi = \{Z_\psi(t),\ t \geq 0\}$ *with immigration measure* Λ *and unit per-capita death rate started at* $\psi = \sum_{i=1}^n \delta_{x_i}$ *can be decomposed as*

$$Z_\psi(t) \overset{\mathscr{D}}{=} Z_\emptyset(t) + D_\psi(t), \quad (1.20)$$

where $Z_\emptyset = \{Z_\emptyset(t),\ t \geq 0\}$ *is an immigration-death process started at the empty point pattern, and* $D_\psi = \{D_\psi(t),\ t \geq 0\}$ *is a pure-death process started at* ψ *that is independent of* $Z_\emptyset$. *Moreover, it can be shown that* $Z_\emptyset(t)$ *is a Poisson process with expectation measure* $(1 - e^{-t})\Lambda$ *and that* D_ψ *may be represented as* $D_\psi(t) = \sum_{i=1}^n \mathbf{1}\{U_i > t\}\delta_{x_i}$ *for independent standard exponentially distributed random variables* $U_1, \ldots, U_n$.

Suppose that $\psi = \sum_{i=1}^n \delta_{x_i}$, $\varphi = \sum_{j=1}^m \delta_{y_j}$, $\chi = \sum_{k=1}^l \delta_{z_k} \in \bar{\mathcal{N}}$, where φ and χ are mutually singular, i.e. $\{y_1, \ldots, y_m\} \cap \{z_1, \ldots, z_l\} = \emptyset$. The goal is to couple immigration-death processes started at $\psi + \varphi$ and $\psi + \chi$, respectively. We base the construction on the process Z_ψ. Let $V_1, \ldots, V_m, W_1, \ldots, W_l$ be i.i.d. $\mathrm{Exp}(1)$-distributed random variables that are independent of Z_ψ. We may then define $Z_{\psi+\varphi}(t) = Z_\psi(t) + \sum_{j=1}^m \mathbf{1}\{V_j > t\}\delta_{y_j}$ and $Z_{\psi+\chi}(t) = Z_\psi(t) + \sum_{k=1}^l \mathbf{1}\{W_k > t\}\delta_{z_k}$.

By Lemma 1.4 it can be seen that the two processes $Z_{\psi+\varphi}$ and $Z_{\psi+\chi}$ have the right distributions.

It is now easy to compute the distribution of their *coupling time*

$$\tau_{\psi+\varphi,\psi+\chi} = \inf\{t > 0;\, Z_{\psi+\varphi}(t) = Z_{\psi+\chi}(t)\}.$$

Note that once the processes $Z_{\psi+\varphi}$ and $Z_{\psi+\chi}$ have met, they will be equal for all times.

Lemma 1.5. *The coupling time $\tau_{\psi+\varphi,\psi+\chi}$ is independent of Z_ψ and has the same distribution as $\sum_{i=1}^{m+l} E_i$, where $E_1,\ldots,E_{m+l}$ are independent random variables with $E_i \sim \mathrm{Exp}(i)$. Its distribution function is given by $F(t) = (1 - e^{-t})^{m+l}$ for $t \geq 0$.*

Proof. We have $Z_{\psi+\varphi}(t) = Z_{\psi+\chi}(t)$ if and only if by time t all the points of φ and all the points of χ have died. Thus, writing $D_{\varphi+\chi}(t) = \sum_{j=1}^{m} \mathbf{1}\{V_j > t\}\,\delta_{y_j} + \sum_{k=1}^{l} \mathbf{1}\{W_k > t\}\,\delta_{z_k}$, we have

$$\tau_{\psi+\varphi,\psi+\chi} = \inf\{t > 0;\, D_{\varphi+\chi}(t) = 0\}.$$

By construction the lifetimes V_j and W_k are all independent of Z_ψ, so $\tau_{\psi+\varphi,\psi+\chi}$ is independent of Z_ψ as well. Since $D_{\varphi+\chi}$ is a pure-jump Markov process, the inter-event times are independent and the time from the jump to i points until the jump to $i-1$ points is just the minimum of i standard exponentially distributed random variables and hence $\mathrm{Exp}(i)$-distributed. This yields the equality in distribution to $\sum_{i=1}^{m+l} E_i$. The distribution function is obtained as

$$F(t) = \mathbf{P}(\tau_{\psi+\varphi,\psi+\chi} \leq t) = \mathbf{P}(V_j \leq t \text{ for all } j,\, W_k \leq t \text{ for all } k)$$

$$= \left(\prod_{j=1}^{m} \mathbf{P}(V_j \leq t)\right)\left(\prod_{k=1}^{l} \mathbf{P}(W_k \leq t)\right) = (1 - e^{-t})^{m+l},$$

which completes the proof. $\qquad\qquad\qquad\qquad\qquad\qquad\qquad\qquad\qquad\quad\square$

We may now invoke Proposition 1.1 to find the solution to the Stein equation (1.15).

Proposition 1.2. *Let $\mathscr{A}$ be given by (1.19) and let $f\colon \bar{\mathcal{N}} \to \mathbb{R}$ be bounded and measurable. Then the function $h_f\colon \bar{\mathcal{N}} \to \mathbb{R}$ given by*

$$h_f(\psi) = -\int_0^\infty \left[\mathbf{E}f(Z_\psi(t)) - \mathbf{E}f(\Pi)\right] dt$$

is well-defined and solves the Stein equation (1.15), i.e.

$$f(\psi) - \mathbf{E}f(\Pi) = \mathscr{A}h(\psi).$$

Proof. In order to show the well-definedness of h_f, consider the coupling construction presented above for the immigration-death processes Z_ψ and Z_π, where the latter is started with distribution $\pi = \pi_\Lambda$, so that $\mathbf{P}_{Z_\pi(t)} = \pi$ for every $t \geq 0$. Denote the corresponding coupling time by $\tau_{\psi,\pi}$. Writing $|\psi|$ for the total number of points in ψ, we have by Lemma 1.5 that $\mathbf{E}\tau_{\psi,\varphi} \leq |\psi| + |\varphi|$ for any $\varphi \in \bar{\mathcal{N}}$. Thus

$$\mathbf{E}\tau_{\psi,\pi} = \int_{\bar{\mathcal{N}}} \mathbf{E}\tau_{\psi,\varphi}\,\pi(d\varphi) \leq \mathbf{E}(|\psi| + |\Pi|) = |\psi| + |\Lambda| < \infty.$$

Hence

$$\int_0^\infty \left|\mathbf{E}f(Z_\psi(t)) - \mathbf{E}f(\Pi)\right| dt \leq \int_0^\infty \mathbf{E}\left|f(Z_\psi(t)) - f(Z_\pi(t))\right| \mathbf{1}\{\tau_{\psi,\pi} > t\}\,dt$$

$$\leq 2\|f\|_\infty \int_0^\infty \mathbf{P}(\tau_{\psi,\pi} > t)\,dt$$

$$= 2\|f\|_\infty \mathbf{E}(\tau_{\psi,\pi}) < \infty.$$

It remains to show that $\vartheta \colon \mathbb{R}_+ \to \mathbb{R}, t \mapsto \mathbf{E}f(Z_\psi(t))$ is continuous at 0 for every ψ. Since Z_ψ is a pure-jump Markov process, we have $\mathbf{P}(Z_\psi(t) \neq \psi) = O(t)$ as $t \to 0$. Thus

$$\mathbf{E}f(Z_\psi(t)) = f(\psi)\,\mathbf{P}(Z_\psi(t) = \psi) + \mathbf{E}\big(f(Z_\psi(t)) \mid Z_\psi(t) \neq \psi\big)\,\mathbf{P}(Z_\psi(t) \neq \psi)$$

$$= f(\psi) + O(t)$$

because f is bounded. Hence ϑ is continuous at 0 and Proposition 1.1 yields the statement. $\qquad\square$

Next we proceed to step 3(a) of our general recipe stated in Sect. 1.3.1. We settle for considering only the first differences of h_f for functions f from the classes underlying the metrics ρ_{TV} and ρ_W presented in Sect. 1.4.3.

Lemma 1.6. *It holds that*

$$\Delta h_f = \sup_{\psi \in \bar{\mathcal{N}}, x \in W} |h_f(\psi + \delta_x) - h_f(\psi)| \leq \begin{cases} 1 & \text{if } f \in \mathcal{F}_*, \\[2mm] \min\left(1, \frac{\log(|\Lambda|) + \gamma + e^{-|\Lambda|}/|\Lambda|}{|\Lambda|}\right) & \text{if } f \in \bar{\mathcal{F}}_W, \end{cases}$$

where $\gamma \approx 0.57722$ is the Euler–Mascheroni constant.

Proof. Using the coupling introduced after Lemma 1.4 with $\varphi = \emptyset$ and $\chi = \delta_x$, we obtain for any $f \in \mathcal{F}_*$ that

$$\left|h_f(\psi + \delta_x) - h_f(\psi)\right|$$

$$= \left|\int_0^\infty \left[\mathbf{E}f(Z_{\psi + \delta_x}(t)) - \mathbf{E}f(\Pi)\right] dt - \int_0^\infty \left[\mathbf{E}f(Z_\psi(t)) - \mathbf{E}f(\Pi)\right] dt\right|$$

$$= \left|\int_0^\infty \mathbf{E}\left[f(Z_\psi(t) + \delta_x) - f(Z_\psi(t))\right] \mathbf{1}\{\tau_{\psi,\psi+\delta_x} > t\}\,dt\right|$$

$$= \left|\int_0^\infty e^{-t}\,\mathbf{E}\left[f(Z_\psi(t) + \delta_x) - f(Z_\psi(t))\right] dt\right|$$

$$\leq \int_0^\infty e^{-t} = 1,$$

where the third equality follows from Lemma 1.5. If $f \in \bar{\mathcal{F}}_W$, we still have $|f(\psi) - f(\varphi)| \leq c = 1$ for all $\psi, \varphi \in \bar{\mathcal{N}}$, and thus obtain the same bound as above. However, we may proceed in a more subtle way and bound

$$\left| \mathbf{E}\left[f(Z_\psi(t) + \delta_x) - f(Z_\psi(t)) \right] \right| = \mathbf{E}\left(\frac{1}{|Z_\psi(t)| + 1} \right)$$
$$\leq \mathbf{E}\left(\frac{1}{|Z_\emptyset(t)| + 1} \right) = \frac{1 - \exp(-|\Lambda_t|)}{|\Lambda_t|},$$

by Lemma 1.4, where $\Lambda_t = (1 - e^{-t})\Lambda$. Hence with the substitution $s = |\Lambda|(1 - e^{-t})$ we obtain

$$\left| h_f(\psi + \delta_x) - h_f(\psi) \right| \leq \int_0^\infty e^{-t} \frac{1 - \exp(-|\Lambda_t|)}{|\Lambda_t|} \, dt = \frac{1}{|\Lambda|} \int_0^{|\Lambda|} \frac{1 - e^{-s}}{s} \, ds.$$

It may now be seen by standard formulae, see [2], Items 5.1.39 and 5.1.19, that the last integral is bounded by $\log(|\Lambda|) + \gamma + e^{-|\Lambda|}/|\Lambda|$. $\qquad\square$

1.5.2 Two Upper Bounds for Poisson Process Approximation

It remains to work out step 3(b) of our general recipe in Sect. 1.3.1. If we restrict ourselves to the approximation of Gibbs processes, this step is straightforward. We obtain the following theorem.

Theorem 1.4. *Let Ψ be a Gibbs process with conditional intensity β^*, and let Λ be a finite measure on W with density λ. Denote by ρ either ρ_{TV} or ρ_W. Then*

$$\rho(\mathbf{P}_\Psi, \pi_\Lambda) \leq c(\Lambda) \int_W \mathbf{E}|\beta^*(x; \Psi) - \lambda(x)| \, dx, \tag{1.21}$$

where

$$c(\Lambda) = \begin{cases} 1 & \text{if } \rho = \rho_{TV}; \\ \min\left(1, \frac{\log(|\Lambda|) + \gamma + e^{-|\Lambda|}/|\Lambda|}{|\Lambda|}\right) & \text{if } \rho = \rho_W. \end{cases}$$

Proof. By the Georgii–Nguyen–Zessin equation (1.16) we obtain that

$$\left| \mathbf{E}f(\Psi) - \mathbf{E}f(\Pi) \right|$$
$$= \left| \mathbf{E}\mathscr{A} h_f(\Psi) \right|$$
$$= \left| \mathbf{E} \int_W \left[h_f(\Psi + \delta_x) - h_f(\Psi) \right] \Lambda(dx) + \mathbf{E} \int_W \left[h_f(\Psi - \delta_x) - h_f(\Psi) \right] \Psi(dx) \right|$$
$$= \left| \mathbf{E} \int_W \left[h_f(\Psi + \delta_x) - h_f(\Psi) \right] (\lambda(x) - \beta^*(x; \Psi)) \, dx \right|.$$

This implies that

$$\left| \mathbf{E} f(\Psi) - \mathbf{E} f(\Pi) \right| \le \Delta h_f \int_W \mathbf{E} |\beta^*(x;\Psi) - \lambda(x)| \, dx.$$

Thus the claim follows by Lemma 1.6. $\qquad\square$

Note that in Theorem 1.4 we make no assumption about the relation between β^* and Λ. If Λ is the expectation measure of Ψ, the Georgii–Nguyen–Zessin equation (1.16) with $h(x,\psi) = \mathbf{1}\{x \in A\}$ yields that Λ has the density $\lambda(x) = \mathbf{E}\beta^*(x,\Psi)$. So the right-hand side of (1.21) describes the variability of $\beta^*(x,\Psi)$.

We consider the special case where Ψ is a pairwise interaction process; see Definition 1.3. For the sake of simplicity we restrict ourselves further to stationary inhibitory point processes.

Corollary 1.1. *Suppose that Ψ is a stationary, inhibitory pairwise interaction process, i.e. $\vartheta_1(x) = \beta$ is constant and $\vartheta_2(x,y) = \vartheta_2(x-y) \le 1$ depends only on the difference of the locations. Then, if $\Lambda = \beta v_d$, it holds that*

$$\rho(\mathbf{P}_\Psi, \pi_\Lambda) \le c(\Lambda)\beta\,\mathbf{E}|\Psi| \int_{\mathbb{R}^d} (1 - \vartheta_2(z))\, dz, \tag{1.22}$$

where $c(\Lambda)$ is the same quantity as in Theorem 1.4.

Proof. Denote again by β^* the conditional intensity of Ψ. For $\psi = \sum_{i=1}^n \delta_{y_i} \in \mathcal{N}$ we have

$$
\begin{aligned}
|\beta^*(x;\psi) - \beta| &= \beta \left| \prod_{i=1}^n \vartheta_2(x,y_i) - 1 \right| \\
&= \beta \left| \sum_{j=1}^n \left(\prod_{i=1}^j \vartheta_2(x,y_i) - \prod_{i=1}^{j-1} \vartheta_2(x,y_i) \right) \right| \\
&= \beta \left| \sum_{j=1}^n (\vartheta_2(x,y_j) - 1) \prod_{i=1}^{j-1} \vartheta_2(x,y_i) \right| \\
&\le \beta \int_W (1 - \vartheta_2(x,y))\, \psi(dy),
\end{aligned}
$$

since $\vartheta_2 \le 1$. It then follows by Theorem 1.4 that

$$
\begin{aligned}
\rho(\mathbf{P}_\Psi, \pi_\Lambda) &\le c(\Lambda) \int_W \mathbf{E}|\beta^*(x;\Psi) - \beta|\, dx \\
&\le c(\Lambda)\beta \int_W \mathbf{E}\left(\int_W (1 - \vartheta_2(x,y))\, \Psi(dy) \right) dx \\
&= c(\Lambda)\beta \int_W \int_W (1 - \vartheta_2(x,y))\, (\mathbf{E}\Psi)(dy)\, dx \\
&\le c(\Lambda)\beta\,\mathbf{E}|\Psi| \int_{\mathbb{R}^D} (1 - \vartheta_2(z))\, dz
\end{aligned}
$$

using Campbell's formula, i.e. $\mathbf{E}(\int g(y)\, \Psi(dy)) = \int g(y)\, (\mathbf{E}\Psi)(dy)$ and the transformation $z = x - y$. $\qquad\square$

In general, in the upper bound given in (1.22) we cannot compute $\mathbf{E}|\Psi|$ explicitly. However, we may always bound it rather crudely by β. Also there are good approximations to $\mathbf{E}|\Psi|$ if Ψ is still reasonably close to a Poisson process, see [18] and [385].

1.6 Gibbs Process Approximation of Point Process Distributions

The abstract procedures for developing Stein's method presented in the earlier sections are not limited to the very simple approximating distributions, although the fact that such a distribution has many nice properties is of enormous help.

To hint at the possibilities in more complicated settings, we present a result for the comparison of two pairwise interaction processes in the total variation metric that generalizes Corollary 1.1 considerably. For simplicity we concentrate on the situation where the approximating process Φ is in the "low activity, high temperature" regime. The latter terminology comes from statistical physics and means that the constant ε given in (1.24) is reasonably small.

Even in this simpler case computing an upper bound on the first difference term as we did in Lemma 1.6 becomes much more difficult, although it works along exactly the same lines. The main difference is that the immigration-death process used in Sect. 1.5.1 has to be replaced by a more general *birth*-death process, in which the arrival of new points depends on the current point pattern. Correspondingly there is no simple decomposition as in Lemma 1.4, so that the coupling has to be constructed in a more complicated way and the coupling time is much more difficult to deal with.

For a detailed proof of the following and more general results about Gibbs process approximation we refer to [356].

Theorem 1.5. *Suppose that Ψ and Φ are both stationary, inhibitory pairwise interaction processes with activity β, and with interaction functions $\vartheta_2(x,y) = \vartheta_2(x-y) \leq 1$ and $\tilde{\vartheta}_2(x,y) = \tilde{\vartheta}_2(x-y) \leq 1$, respectively. Then*

$$\rho_{TV}(\mathbf{P}_\Psi, \mathbf{P}_\Phi) \leq \frac{1+\varepsilon}{\varepsilon} \log\left(\frac{1}{1-\varepsilon}\right) \beta \, \mathbf{E}|\Psi| \int_{\mathbb{R}^D} |\vartheta_2(z) - \tilde{\vartheta}_2(z)| \, dz, \qquad (1.23)$$

where

$$\varepsilon = \beta \int_{\mathbb{R}^d} (1 - \tilde{\vartheta}_2(z)) \, dz. \qquad (1.24)$$

For $\varepsilon = 0$ the upper bound in (1.23) is to be interpreted in the limit sense. Note that $\lim_{\varepsilon \downarrow 0} \frac{1+\varepsilon}{\varepsilon} \log\left(\frac{1}{1-\varepsilon}\right) = 1$. So if Φ is a Poisson process with constant intensity β, we have $\tilde{\vartheta}_2 \equiv 1$ and Theorem 1.5 is reduced to Corollary 1.1.

Acknowledgements D.S. would like to thank Kaspar Stucki for proofreading and helpful suggestions. Many thanks to Thomas Richardson and Ana Sendova-Franks for interesting discussions and for permitting the use of the data set in Sect. 1.4.2.2.

Chapter 2
Clustering Comparison of Point Processes, with Applications to Random Geometric Models

Bartłomiej Błaszczyszyn and Dhandapani Yogeshwaran

Abstract In this chapter we review some examples, methods, and recent results involving comparison of clustering properties of point processes. Our approach is founded on some basic observations allowing us to consider void probabilities and moment measures as two complementary tools for capturing clustering phenomena in point processes. As might be expected, smaller values of these characteristics indicate less clustering. Also, various global and local functionals of random geometric models driven by point processes admit more or less explicit bounds involving void probabilities and moment measures, thus aiding the study of impact of clustering of the underlying point process. When stronger tools are needed, directional convex ordering of point processes happens to be an appropriate choice, as well as the notion of (positive or negative) association, when comparison to the Poisson point process is considered. We explain the relations between these tools and provide examples of point processes admitting them. Furthermore, we sketch some recent results obtained using the aforementioned comparison tools, regarding percolation and coverage properties of the germ-grain model, the SINR model, subgraph counts in random geometric graphs, and more generally, U-statistics of point processes. We also mention some results on Betti numbers for Čech and Vietoris-Rips random complexes generated by stationary point processes. A general observation is that many of the results derived previously for the Poisson point process generalise to some "sub-Poisson" processes, defined as those clustering less than the Poisson process in the sense of void probabilities and moment measures, negative association or dcx-ordering.

Bartłomiej Błaszczyszyn
Inria/ENS, 75214 Paris, France, e-mail: `Bartek.Blaszczyszyn@ens.fr`

Dhandapani Yogeshwaran
Statistics and Mathematics Unit, Indian Statistical Institute, Bangalore - 560098, India.
e-mail: `d.yogesh@isibang.ac.in`

© Springer International Publishing Switzerland 2015

V. Schmidt (ed.), *Stochastic Geometry, Spatial Statistics and Random Fields,*
Lecture Notes in Mathematics 2120, DOI 10.1007/978-3-319-10064-7_2

2.1 Introduction

On the one hand, various interesting methods have been developed for studying local and global functionals of geometric structures driven by Poisson or Bernoulli point processes (see [271, 314, 427]). On the other hand, as will be shown in the following section, there are many examples of interesting point processes that occur naturally in theory and applications. So, the obvious question arises how much of the theory developed for Poisson or Bernoulli point processes can be carried over to other classes of point processes.

Our approach to this question is based on the comparison of clustering properties of point processes. Roughly speaking, a set of points in $\mathbb{R}^d$ clusters if it lacks spatial homogeneity, i.e., one observes points forming groups which are well spaced out. Many interesting properties of random geometric models driven by point processes should depend on the "degree" of clustering. For example, it is natural to expect that concentrating points of a point process in well-spaced-out clusters should negatively impact connectivity of the corresponding random geometric (Gilbert) graph, and that spreading these clustered points "more homogeneously" in the space would result in a smaller critical radius for which the graph percolates. For many other functionals, using similar heuristic arguments one can conjecture whether increase or decrease of clustering will increase or decrease the value of the functional. However, to the best of our knowledge, there has been no systematic approach towards making these arguments rigorous.

The above observations suggest the following program. We aim at identifying a class or classes of point processes, which can be compared in the sense of clustering to a (say stationary) Poisson point process, and for which — by this comparison — some results known for the latter process can be extrapolated. In particular, there are point processes which in some sense cluster less (i.e. spread their points more homogeneously in the space) than the Poisson point process. We call them *sub-Poisson*. Furthermore, we hasten to explain that the usual strong stochastic order (i.e. coupling as a subset of the Poisson process) is in general not an appropriate tool in this context.

Various approaches to mathematical formalisation of clustering will form an important part of this chapter. By formalisation, we mean defining a partial order on the space of point processes such that being smaller with respect to the order indicates less clustering. The most simple approach consists in considering void probabilities and moment measures as two complementary tools for capturing clustering phenomena in point processes. As might be expected, smaller values of these characteristics indicate less clustering. When stronger tools are needed, directionally convex (dcx) ordering of point processes happens to be a good choice, as well as the notion of negative and positive association. Working with these tools, we first give some useful, generic inequalities regarding Laplace transforms of the compared point processes. In the case of dcx-ordering these inequalities can be generalised to dcx functions of shot-noise fields.

Having described the clustering comparison tools, we present several particular results obtained by using them. Then, in more detail, we study percolation in the germ-grain model (seen as a random geometric graph) and the SINR graph. In particular, we show how the classical results regarding the existence of a non-trivial phase transition extend to models based on additive functionals of various sub-Poisson point processes. Furthermore, we briefly discuss some applications of the comparison tools to U-statistics of point processes, counts of sub-graphs and simplices in random geometric graphs and simplicial complexes, respectively. We also mention results on Betti numbers of Čech and Vietoris-Rips random complexes generated by sub-Poisson point processes.

Let us conclude this short motivating introduction by citing an excerpt from a standard reference book on stochastic comparison methods by Müller and Stoyan [289]: "It is clear that there are processes of comparable variability. Examples of such processes are a stationary Poisson process and a cluster process of equal intensity or two hard-core Gibbs processes of equal intensity and different hard-core distances. It would be fine if these variability differences could be characterized by order relations ... [implying], for example, reasonable relationship[s] for second order characteristics such as the pair correlation function."; cf [289, page 253]. We believe that the results reported in this chapter present one of the first answers to this question, although "Still much work has to be done in the comparison theory for point processes." (ibid.)

The present chapter is organised as follows. In Sect. 2.2 we recall some examples of point processes. The idea is to provide as many as possible examples which admit various clustering comparison orders, described then in Sect. 2.3. An extensive overview of applications is presented in Sect. 2.4.

2.2 Examples of Point Processes

In this section we give some examples of point processes, where our goal is to present them in the context of modelling of clustering phenomena. Note that throughout this chapter, we consider point processes on the d-dimensional Euclidean space $\mathbb{R}^d$, $d \geq 1$, although much of the presented results have straightforward extensions to point processes on an arbitrary Polish space.

2.2.1 Elementary Models

A first example we probably think of when trying to come up with some spatially homogeneous model is a point process on a deterministic lattice.

Definition 2.1 (Lattice point process). By a lattice point process $\Phi_{\mathbb{L}}$ we mean a simple point process whose points are located on the vertices of some deterministic lattice $\mathbb{L}$. An important special case is the d-dimensional *cubic lattice* $\Delta \mathbb{Z}^d = \{\Delta(u_1,\ldots,u_d), u_i \in \mathbb{Z}\}$ of edge length $\Delta > 0$, where $\mathbb{Z}$ denotes the set of integers. Another specific (two-dimensional) model is the *hexagonal lattice* on the complex plane given by $\Delta\mathbb{H} = \{\Delta(u_1 + u_2 e^{i\pi/3}),\ u_1, u_2 \in \mathbb{Z}\}$. The stationary version $\Phi_{\mathbb{L}}^{\mathrm{st}}$ of a lattice point process $\Phi_{\mathbb{L}}$ can be constructed by randomly shifting the deterministic pattern through a vector U uniformly distributed in some fixed cell of the lattice $\mathbb{L}$, i.e. $\Phi_{\mathbb{L}}^{\mathrm{st}} = \Phi_{\mathbb{L}} + U$. Note that the intensity of the stationary lattice point process is equal to the inverse of the cell volume. In particular, the intensity $\lambda_{\Delta\mathbb{Z}^d}$ of $\Phi_{\Delta\mathbb{Z}^d}^{\mathrm{st}}$ is equal to $\lambda_{\Delta\mathbb{Z}^d} = 1/\Delta^d$, while that of $\Phi_{\Delta\mathbb{H}}^{\mathrm{st}}$ is equal to $\lambda_{\Delta\mathbb{H}} = 2/(\sqrt{3}/\Delta^2)$.

Lattice point processes are usually considered to model "perfect" or "ideal" structures, e.g. the hexagonal lattice on the complex plane is used to study perfect cellular communication networks. We will see however, without further modifications, they escape from the clustering comparison methods presented in Sect. 2.3.

When the "perfect structure" assumption cannot be retained and one needs a random pattern, then the Poisson point process usually comes as a natural first modelling assumption. We therefore recall the definition of the Poisson process for the convenience of the reader, see also the survey given in [21].

Definition 2.2 (Poisson point process). Let Λ be a (deterministic) locally finite measure on the Borel sets of $\mathbb{R}^d$. The random counting measure Π_Λ is called a *Poisson point process* with intensity measure Λ if for every $k = 1, 2, \ldots$ and all bounded, mutually disjoint Borel sets $B_1, \ldots, B_k$, the random variables $\Pi_\Lambda(B_1), \ldots, \Pi_\Lambda(B_k)$ are independent, with Poisson distribution $\mathrm{Pois}(\Lambda(B_1)), \ldots, \mathrm{Pois}(\Lambda(B_k))$, respectively. In the case when Λ has an integral representation $\Lambda(B) = \int_B \lambda(x)\,\mathrm{d}x$, where $\lambda : \mathbb{R}^d \to \mathbb{R}_+$ is some measurable function, we call λ the *intensity field* of the Poisson point process. In particular, if λ is a constant, we call Π_Λ a *stationary* Poisson point process and denote it by Π_λ.

The Poisson point process is a good model when one does not expect any "interactions" between points. This is related to the *complete randomness* property of Poisson processes, cf. [93, Theorem 2.2.III]. Furthermore, the stationary Poisson point process is commonly considered as a reference model in comparative studies of clustering phenomena.

2.2.2 Cluster Point Processes — Replicating and Displacing Points

We now present several operations on the points of a point process, which in conjunction with the two elementary models presented above allow us to construct various other interesting examples of point processes. We begin by recalling the following elementary operations.

Superposition of patterns of points consists of set-theoretic addition of these points. Superposition of (two or more) point processes $\Phi_1, \ldots, \Phi_n$, defined as random counting measures, consists of adding these measures $\Phi_1 + \cdots + \Phi_n$. Superposition of independent point processes is of special interest.

Thinning of a point process consists of suppressing some subset of its points. *Independent thinning* with retention function $p(x)$ defined on $\mathbb{R}^d$, $0 \leq p(x) \leq 1$, consists in suppressing points independently, given a realisation of the point process, with probability $1 - p(x)$, which might depend on the location x of the point to be suppressed or not.

Displacement consists in, possibly randomised, mapping of points of a point process to the same or another space. *Independent displacement* with displacement (probability) kernel $\mathcal{X}(x,\cdot)$ from $\mathbb{R}^d$ to $\mathbb{R}^{d'}$, $d' > 1$, consists of independently mapping each point x of a given realisation of the original point process to a new, random location in $\mathbb{R}^{d'}$ selected according to the kernel $\mathcal{X}(x,\cdot)$.

Remark 2.1. An interesting property of the class of Poisson point processes is that it is closed with respect to independent displacement, thinning and superposition, i.e. the result of these operations made on Poisson point processes is a Poisson point process, which is not granted in the case of an arbitrary (i.e. not independent) superposition or displacement, see e.g. [93, 94].

Now, we define a more sophisticated operation on point processes that will allow us to construct new classes of point processes with interesting clustering properties.

Definition 2.3 (Clustering perturbation of a point process). Let Φ be a point process on $\mathbb{R}^d$ and $\mathcal{N}(\cdot,\cdot)$, $\mathcal{X}(\cdot,\cdot)$ be two probability kernels from $\mathbb{R}^d$ to the set of non-negative integers $\mathbb{Z}_+$ and $\mathbb{R}^{d'}$, $d,d' \geq 1$, respectively. Consider the following subset of $\mathbb{R}^{d'}$. Let

$$\Phi^{\text{pert}} = \bigcup_{X \in \Phi} \bigcup_{i=1}^{N_X} \{Y_{iX}\}, \tag{2.1}$$

where, given Φ,

1. $(N_X)_{X \in \Phi}$ are independent, non-negative integer-valued random variables with (conditional) distribution $\mathbf{P}(N_X \in \cdot \,|\, \Phi) = \mathcal{N}(X,\cdot)$,
2. $Y_X = (Y_{iX}; i = 1,2,\ldots)$, $X \in \Phi$ are independent vectors of i.i.d. random elements of $\mathbb{R}^{d'}$, with Y_{iX} having the conditional distribution $\mathbf{P}(Y_{iX} \in \cdot \,|\, \Phi) = \mathcal{X}(X,\cdot)$.

Note that the inner sum in (2.1) is interpreted as $\emptyset$ when $N_X = 0$.

The random set Φ^{pert} given in (2.1) can be considered as a point process on $\mathbb{R}^{d'}$ provided it is a locally finite. In what follows, we will assume a stronger condition, namely that the itensity measure of Φ^{pert} is locally finite (Radon), i.e.

$$\int_{\mathbb{R}^d} n(x)\mathcal{X}(x,B)\,\alpha(\mathrm{d}x) < \infty, \tag{2.2}$$

for all bounded Borel sets $B \subset \mathbb{R}^{d'}$, where $\alpha(\cdot) = \mathbf{E}\Phi(\cdot)$ denotes the intensity measure of Φ and

$$n(x) = \sum_{k=1}^{\infty} k\mathcal{N}(x,\{k\}) \tag{2.3}$$

is the mean value of the distribution $\mathcal{N}(x,\cdot)$.

Thus, *clustering perturbation* of a given parent process Φ consists in independent replication and displacement of the points of Φ, with the number of replications of a given point $X \in \Phi$ having distribution $\mathcal{N}(X, \cdot)$ and the replicas' locations having distribution $\mathcal{X}(X, \cdot)$. The replicas of X form a *cluster*.

For obvious reasons, we call Φ^{pert} a *perturbation* of Φ driven by the *replication kernel* $\mathcal{N}$ and the *displacement kernel* $\mathcal{X}$. It is easy to see that the independent thinning and displacement operations described above are special cases of clustering perturbation. In what follows we present a few known examples of point processes arising as clustering perturbations of a lattice or Poisson point process. For simplicity we assume that replicas stay in the same state space, i.e. $d = d'$.

Example 2.1 (Binomial point process). A (finite) binomial point process has a fixed total number $n < \infty$ of points, which are independent and identically distributed according to some (probability) measure Λ on $\mathbb{R}^d$. It can be seen as a Poisson point process $\Pi_{n\Lambda}$ conditioned to have n points, cf. [93]. Note that this property might be seen as a clustering perturbation of a one-point process, with deterministic number n of point replicas and displacement distribution Λ.

Example 2.2 (Bernoulli lattice). The Bernoulli lattice arises as independent thinning of a lattice point process; i.e., each point of the lattice is retained (but not displaced) with some probability $p \in (0, 1)$ and suppressed otherwise.

Example 2.3 (Voronoi-perturbed lattices). These are perturbed lattices with displacement kernel $\mathcal{X}$, where the distribution $\mathcal{X}(x, \cdot)$ is supported on the Voronoi cell $\mathcal{V}(x)$ of vertex $x \in \mathbb{L}$ of the original (unperturbed) lattice $\mathbb{L}$. In other words, each replica of a given lattice point gets independently translated to some random location chosen in the Voronoi cell of the original lattice point. Note that one can also choose other bijective, lattice-translation invariant mappings of associating lattice cells to lattice points; e.g. associate a given cell of the square lattice on the plane $\mathbb{R}^2$ to its "south-west" corner.

By a *simple perturbed lattice* we mean the Voronoi-perturbed lattice whose points are uniformly translated in the corresponding cells, without being replicated. Interestingly enough, the Poisson point process Π_Λ with some intensity measure Λ can be constructed as a Voronoi-perturbed lattice. Indeed, it is enough to take the Poisson replication kernel $\mathcal{N}$ given by $\mathcal{N}(x, \cdot) = \mathrm{Pois}(\Lambda(\mathcal{V}(x)))$ and the displacement kernel $\mathcal{X}$ with $\mathcal{X}(x, \cdot) = \Lambda(\cdot \cap \mathcal{V}(x))/\Lambda(\mathcal{V}(x))$; cf. Exercise 2.1. Keeping the above displacement kernel and replacing the Poisson distribution in the replication kernel by some other distributions convexly smaller or larger than the Poisson distribution, one gets the following two particular classes of Voronoi-perturbed lattices, clustering their points less or more than the Poisson point process Π_Λ (in a sense that will be formalised in Sect. 2.3).

Sub-Poisson Voronoi-perturbed lattices are Voronoi-perturbed lattices such that $\mathcal{N}(x, \cdot)$ is convexly smaller than $\mathrm{Pois}(\Lambda(\mathcal{V}(x)))$. Examples of distributions convexly smaller than $\mathrm{Pois}(\lambda)$ are the hyper-geometric distributions $\mathrm{HGeo}(n, m, k)$, $m, k \leq n$, $km/n = \lambda$ and the binomial distributions. $\mathrm{Binom}(n, \lambda/n)$, $\lambda \leq n$, which can be ordered as follows:

$$\mathrm{HGeo}(n, m, \lambda n/m) \leq_{\mathrm{cx}} \mathrm{Binom}(m, \lambda/m) \leq_{\mathrm{cx}} \mathrm{Binom}(r, \lambda/r) \leq_{\mathrm{cx}} \mathrm{Pois}(\lambda), \quad (2.4)$$

for $\lambda \leq m \leq \min(r,n)$; cf. [413]. Recall that $\text{Binom}(n,p)$ has the probability mass function $p_{\text{Binom}(n,p)}(i) = \binom{n}{i}p^i(1-p)^{n-i}$ ($i = 0,\ldots,n$), whereas $\text{HGeo}(n,m,k)$ has the probability mass function $p_{\text{HGeo}(n,m,k)}(i) = \binom{m}{i}\binom{n-m}{k-i}/\binom{n}{k}$ ($\max(k-n+m,0) \leq i \leq m$); cf. Exercise 2.2.

Super-Poisson Voronoi-perturbed lattices are Voronoi-perturbed lattices with $\mathcal{N}(x,\cdot)$ convexly larger than $\text{Pois}(\Lambda(\mathcal{V}(x)))$. Examples of distributions convexly larger than $\text{Pois}(\lambda)$ are the negative binomial distribution $\text{NBionom}(r,p)$ with $rp/(1-p) = \lambda$ and the geometric distribution $\text{Geo}(p)$ distribution $1/p - 1 = \lambda$, which can be ordered in the following way:

$$\text{Pois}(\lambda) \leq_{\text{cx}} \text{NBionom}(r_2, \lambda/(r_2+\lambda)) \leq_{\text{cx}} \text{NBionom}(r_1, \lambda/(r_1+\lambda))$$
$$\leq_{\text{cx}} \text{Geo}(1/(1+\lambda)) \leq_{\text{cx}} \sum_j \lambda_j \, \text{Geo}(p_j) \tag{2.5}$$

with $r_1 \leq r_2$, $0 \leq \lambda_j \leq 1$, $\sum_j \lambda_j = 1$ and $\sum_j \lambda_j/p_j = \lambda + 1$, where the largest distribution in (2.5) is a mixture of geometric distributions having mean λ. Note that any mixture of Poisson distributions having mean λ is in cx-order larger than $\text{Pois}(\lambda)$. Furthermore, recall that the probability mass functions of $\text{Geo}(p)$ and $\text{NBionom}(r,p)$ are given by $p_{\text{Geo}(p)}(i) = p(1-p)^i$ and $p_{\text{NBionom}(r,p)}(i) = \binom{r+i-1}{i}p^i(1-p)^r$, respectively.

Example 2.4 (Generalised shot-noise Cox point processes). These are clustering perturbations of an arbitrary parent point process Φ, with replication kernel $\mathcal{N}$, where $\mathcal{N}(x,\cdot)$ is the Poisson distribution $\text{Pois}(n(x))$ and $n(x)$ is the mean value given in (2.3). Note that in this case, given Φ, the clusters (i.e. replicas of the given parent point) form independent Poisson point process $\Pi_{n(X)\mathcal{X}(X,\cdot)}$, $X \in \Phi$. This special class of Cox point processes (cf. Sect. 2.2.3) has been introduced in [285].

Example 2.5 (Poisson-Poisson cluster point processes). This is a special case of the generalised shot-noise Cox point processes, with the parent point process being Poisson, i.e. $\Phi = \Pi_\Lambda$ for some intensity measure Λ. A further special case is often discussed in the literature, where the displacement kernel $\mathcal{X}$ is such that $\mathcal{X}(X,\cdot)$ is the uniform distribution in the ball $B_r(X)$ of some given radius r. It is called the *Matérn cluster point process*. If $\mathcal{X}(X,\cdot)$ is symmetric Gaussian, then the resulting Poisson-Poisson cluster point process is called a *(modified) Thomas point process*.

Example 2.6 (Neyman-Scott point process). These point processes arise as a clustering perturbation of a Poisson parent point process Π_Λ, with arbitrary (not necessarily Poisson) replication kernel $\mathcal{N}$.

2.2.3 Cox Point Processes— Mixing Poisson Distributions

We now consider a rich class of point processes known also as doubly stochastic Poisson point process, which are often used to model patterns exhibiting more clustering than the Poisson point process.

Definition 2.4 (Cox point process).   Let $\mathfrak{L}$ be a random locally finite (non-null) measure on $\mathbb{R}^d$. A *Cox point process* $\mathrm{Cox}_{\mathfrak{L}}$ on $\mathbb{R}^d$ generated by $\mathfrak{L}$ is defined as point process having the property that $\mathrm{Cox}_{\mathfrak{L}}$ conditioned on $\mathfrak{L} = \Lambda$ is the Poisson point process Π_Λ. Note that $\mathfrak{L}$ is called the *random intensity measure* of $\mathrm{Cox}_{\mathfrak{L}}$. In case when the random measure $\mathfrak{L}$ has an integral representation $\mathfrak{L}(B) = \int_B \xi(x)\,\mathrm{d}x$, with $\{\xi(x), x \in \mathbb{R}^d\}$ being a random field, we call this field the *random intensity* field of the Cox process. In the special case that $\xi(x) = \xi\lambda(x)$ for all $x \in \mathbb{R}^d$, where ξ is a (non-negative) random variable and λ a (non-negative) deterministic function, the corresponding Cox point process is called a *mixed Poisson* point process.

Note that Cox processes may be seen as a result of an operation transforming some random measure $\mathfrak{L}$ into a point process $\mathrm{Cox}_{\mathfrak{L}}$, being a mixture of Poisson processes.

In Sect. 2.2.2, we have already seen that clustering perturbation of an arbitrary point process with Poisson replication kernel gives rise to Cox processes (cf. Example 2.4), where Poisson-Poisson cluster point processes are special cases with Poisson parent point process. This latter class of point processes can be naturally extended by replacing the Poisson parent process by a Lévy basis.

Definition 2.5 (Lévy basis). A collection of real-valued random variables $\{Z(B), B \in \mathcal{B}_0\}$, where $\mathcal{B}_0$ denotes the family of bounded Borel sets in $\mathbb{R}^d$, is said to be a *Lévy basis* if the $Z(B)$ are infinitely divisible random variables and for any sequence $\{B_n\}$, $n \geq 1$, of disjoint bounded Borel sets in $\mathbb{R}^d$, $Z(B_1), Z(B_2), \ldots$ are independent random variables (*complete independence property*), with $Z(\bigcup_n B_n) = \sum_n Z(B_n)$ almost surely provided that $\bigcup_n B_n$ is bounded.

In this chapter, we shall consider only non-negative Lévy bases. We immediately see that the Poisson point process is a special case of a Lévy basis. Many other concrete examples of Lévy bases can be obtained by "attaching" independent, infinitely divisible random variables ξ_i to a deterministic, locally finite sequence $\{x_i\}$ of (fixed) points in $\mathbb{R}^d$ and letting $Z(B) = \sum_i \xi_i \mathbf{1}(x_i \in B)$. In particular, clustering perturbations of a lattice, with infinitely divisible replication kernel and no displacement (i.e. $\mathcal{X}(x, \cdot) = \delta_x$, where δ_x is the Dirac measure at x) are Lévy bases. Recall that any degenerate (deterministic), Poisson, negative binomial, gamma as well as Gaussian, Cauchy, Student's distribution are examples of infinitely divisible distributions.

It is possible to define an integral of a measurable function with respect to a Lévy basis (even if the latter is not always a random measure; see [170] for details) and consequently consider the following classes of Cox point processes.

Example 2.7 (Lévy-driven Cox process). Consider a Cox point process $\mathrm{Cox}_{\mathfrak{L}}$ with random intensity field that is an integral shot-noise field of a Lévy basis, i.e. $\xi(y) = \int_{\mathbb{R}^d} k(x,y)Z(\mathrm{d}x)$, where Z is a Lévy basis and $k : \mathbb{R}^d \times \mathbb{R}^d \to \mathbb{R}_+$ is some non-negative function almost surely integrable with respect to $Z \otimes \mathrm{d}y$.

Example 2.8 (Log-Lévy-driven Cox process). These are Cox point processes with random intensity field given by $\xi(y) = \exp\left(\int_{\mathbb{R}^d} k(x,y)Z(\mathrm{d}x)\right)$, where Z and k satisfy the same conditions as above.

Both Lévy- and log-Lévy-driven Cox processes have been introduced in [170], where one can find many examples of these processes. We still mention another class of Cox point processes considered in [284].

Example 2.9 (Log-Gaussian Cox point process). Consider a Cox point process whose random intensity field is given by $\xi(y) = \exp(\eta(y))$ where $\{\eta(y)\}$ is a Gaussian random field.

2.2.4 Gibbs and Hard-Core Point Processes

Gibbs and hard-core point processes are two further classes of point processes, which should appear in the context of modelling of clustering phenomena.

Roughly speaking Gibbs point processes are point processes having a density with respect to the Poisson point process. In other words, we obtain a Gibbs point process, when we "filter" Poisson patterns of points, giving more chance to appear for some configurations and less chance (or completely suppressing) some others. A very simple example is a Poisson point process conditioned to obey some constraint regarding its points in some bounded Borel set (e.g. to have some given number of points there). Depending on the "filtering" condition we may naturally create point processes which cluster more or less than the Poisson point process.

Hard-core point processes are point process in which the points are separated from each other by some minimal distance, hence in some sense clustering is "forbidden by definition".

However, we will not give precise definitions, nor present particular examples from these classes of point processes, because, unfortunately, we do not have yet interesting enough comparison results for them, to be presented in the remaining part of this chapter.

2.2.5 Determinantal and Permanental Point Process

We briefly recall two special classes of point processes arising in random matrix theory, combinatorics, and physics. They are "known" to cluster their points, less or more, respectively, than the Poisson point process.

Definition 2.6 (Determinantal point process). A simple point process on $\mathbb{R}^d$ is said to be a *determinantal point process* with a kernel function $k : \mathbb{R}^d \times \mathbb{R}^d \to \mathbb{C}$ with respect to a Radon measure μ on $\mathbb{R}^d$ if the joint intensities $\rho^{(\ell)}$ of the factorial moment measures of the point process with respect to the product measure $\mu^{\otimes \ell}$ satisfy $\rho^{(\ell)}(x_1, \dots, x_\ell) = \det\left(k(x_i, x_j)\right)_{1 \le i,j \le \ell}$ for all ℓ, where $\left(a_{ij}\right)_{1 \le i,j \le \ell}$ stands for a matrix with entries a_{ij} and det denotes the determinant of the matrix.

Definition 2.7 (Permanental point process). Similar to the notion of a determinantal point process, one says that a simple point process is a *permanental point* *process* with a kernel function $k : \mathbb{R}^d \times \mathbb{R}^d \to \mathbb{C}$ with respect to a Radon measure μ on $\mathbb{R}^d$ if the joint intensities $\rho^{(\ell)}$ of the point process with respect to $\mu^{\otimes \ell}$ satisfy $\rho^{(\ell)}(x_1,\ldots,x_\ell) = \mathrm{per}\big(k(x_i,x_j)\big)_{1 \le i,j \le \ell}$ for all ℓ, where $\mathrm{per}(\,\cdot\,)$ stands for the permanent of a matrix. From [33, Proposition 35 and Remark 36], we know that each permanental point process is a Cox point process.

Naturally, the kernel function k needs to satisfy some additional assumptions for the existence of the point processes defined above. We refer to [34, Chap. 4] for a general framework which allows to study determinantal and permanental point processes, see also [33]. Regarding statistical aspects and simulation methods for determinantal point processes, see [246].

Here is an important example of a determinantal point process recently studied on the theoretical ground (cf. e.g. [144]) and considered in modelling applications (cf. [275]).

Example 2.10 (Ginibre point process). This is the determinantal point process on $\mathbb{R}^2$ with kernel function $k((x_1,x_2),(y_1,y_2)) = \exp[(x_1 y_1 + x_2 y_2) + i(x_2 y_1 - x_1 y_2)]$, $x_j, y_j \in \mathbb{R}$, $j = 1,2$, with respect to the measure $\mu(\mathrm{d}(x_1,x_2)) = \pi^{-1} \exp[-x_1^2 - x_2^2]\,\mathrm{d}x_1\,\mathrm{d}x_2$.

Exercise 2.1. Let Φ be a simple point process on $\mathbb{R}^d$. Consider its cluster perturbation Φ^{pert} defined in (2.1) with the Poisson replication kernel $\mathcal{N}(x,\cdot) = \mathrm{Pois}(\Lambda(\mathcal{V}(x)))$, where $\mathcal{V}(x)$ is the Voronoi cell of x in Φ, and the displacement kernel $\mathcal{X}(x,\cdot) = \Lambda(\cdot \cap \mathcal{V}(x))/\Lambda(\mathcal{V}(x))$, for some given deterministic Radon measure Λ on $\mathbb{R}^d$. Show that Φ^{pert} is Poisson with intensity measure Λ.

Exercise 2.2. Prove (2.4) and (2.5) by showing the logarithmic concavity of the ratio of the respective probability mass functions, which implies increasing convex order and, consequently, cx-order provided the distributions have the same means.

2.3 Clustering Comparison Methods

Let us begin with the following *informal* definitions. A set of points is spatially stationary if approximately the same numbers of points occur in any spherical region of a given volume. A set of points clusters if it lacks spatial stationarity; more precisely, if one observes points arranged in groups being well spaced out.

Looking at Fig. 2.1, it is intuitively obvious that (realisations of) some point processes cluster less than others. However, the mathematical formalisation of such a statement appears not so easy. In what follows, we present a few possible approaches. We begin with the usual statistical descriptors of spatial stationarity, then show how void probabilities and moment measures come into the picture, in particular in relation to another notion useful in this context: positive and negative association. Finally we deal with directionally convex ordering of point processes.

This kind of organisation roughly corresponds to presenting ordering methods from weaker to stronger ones; cf. Fig. 2.2. We also show how the different examples

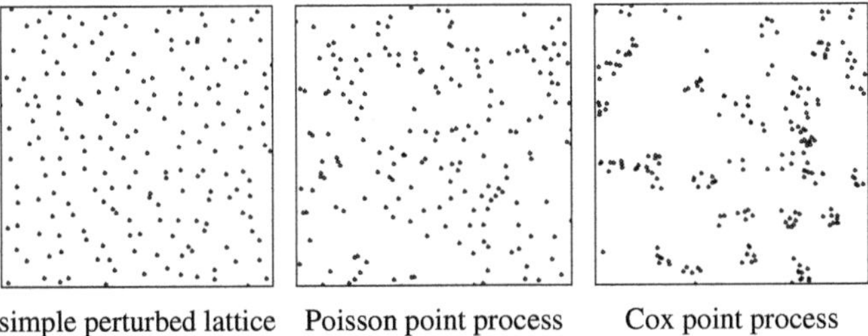

simple perturbed lattice   Poisson point process   Cox point process

Fig. 2.1 From left to right : patterns sampled from a simple perturbed lattice (cf. Example 2.3), Poisson point process and a doubly stochastic Poisson (Cox) point process, all with the same mean number of points per unit area.

presented in Sect. 2.2 admit these comparison methods, mentioning the respective results in their strongest versions. We recapitulate results regarding comparison to the Poisson process in Fig. 2.3.

2.3.1 Second-order Statistics

In this section we restrict ourselves to the stationary setting.

2.3.1.1 Ripley's K-Function

One of the most popular functions for the statistical analysis of spatial stationarity is *Ripley's K-function* $K : [0, \infty) \to [0, \infty)$ defined for stationary point processes (cf. [73]). Assume that Φ is a stationary point process on $\mathbb{R}^d$ with finite intensity $\lambda = \mathbf{E}\Phi([0, 1]^d)$. Then

$$K(r) = \frac{1}{\lambda v_d(B)} \mathbf{E} \sum_{X_i \in \Phi \cap B} (\Phi(B_{X_i}(r)) - 1),$$

where $v_d(B)$ denotes the Lebesgue measure of a bounded Borel set $B \subset \mathbb{R}^d$, assuming that $v_d(B) > 0$. Due to stationarity, the definition does not depend on the choice of B.

The value of $\lambda K(r)$ can be interpreted as the average number of "extra" points observed within the distance r from a randomly chosen (so-called typical) point. Campbell's formula from Palm theory of stationary point processes gives a precise meaning to this statement. Consequently, for a given intensity λ, the more one finds points of a point process located in clusters of radius r, the larger the value of $K(r)$ is, whence a first clustering comparison method follows. In other words, larger values $K(r)$ of Ripley's K-function indicate more clustering "at the cluster-radius scale" r. For the (stationary) Poisson process Π_λ on $\mathbb{R}^d$, which is often considered as a "reference model" for clustering, we have $K(r) = \kappa_d r^d$, where κ_d is the volume of the unit ball in $\mathbb{R}^d$. Note here that $K(r)$ describes clustering characteristics of the point process at the (cluster radius) scale r. Many point processes, which we tend to think that they cluster less or more than the Poisson point process, in fact are not comparable in the sense of Ripley's K-function (with the given inequality holding for all $r \geq 0$, neither, in consequence, in any stronger sense considered later in this section), as we can see in the following simple example.

The following result of D. Stoyan from 1983 can be considered as a precursor to our theory of clustering comparison. It says that the convex ordering of Ripley's K-functions implies ordering of variances of the numbers of observed points. We shall see in Remark 2.2 that variance bounds give us simple concentration inequalities for the distribution of the number of observed points. These inequalities help to control clustering. We will develop this idea further in Sections 2.3.2 and 2.3.3 showing that using moment measures and void probabilities one can obtain stronger, exponential concentration inequalities.

Proposition 2.1 ([383, Corollary 1]). *Consider two stationary, isotropic point processes Φ_1 and Φ_2 of the same intensity, with Ripley's functions K_1 and K_2, respectively. If $K_1 \leq_{dc} K_2$ i.e., $\int_0^\infty f(r)K_1(\mathrm{d}r) \leq \int_0^\infty f(r)K_2(\mathrm{d}r)$ for all decreasing convex functions $f : [0,\infty) \to [0,\infty)$ then $\mathbf{var}(\Phi_1(B)) \leq \mathbf{var}(\Phi_2(B))$ for all compact and convex sets B in $\mathbb{R}^d$.*

Exercise 2.3. For the stationary square lattice point process on the plane with intensity $\lambda = 1$ (cf. Definition 2.1), compare $K(r)$ and πr^2 for $r = 1, \sqrt{2}, 2$.

From Exercise 2.3, one should be able to see that though the square lattice is presumably more homogeneous (less clustering) than the Poisson point process of the same intensity, the differences of the values of their K-functions alternate between strictly positive and strictly negative. However, we shall see later that (cf. Example 2.18) this will not be the case for some perturbed lattices, including the simple perturbed ones and thus they cluster less than the Poisson point process in the sense of Ripley's K-function (and even in a much stronger sense). We will also discuss point processes clustering more than the Poisson processes in this sense.

2.3.1.2 Pair Correlation Function

Another useful characteristic for measuring clustering effects in stationary point processes is the *pair correlation function* $g : \mathbb{R}^d \times \mathbb{R}^d \to [0,\infty)$. It is related to the probability of finding a point at a given distance from another point and can be defined as

$$g(x,y) = \frac{\rho^{(2)}(x,y)}{\lambda^2},$$

where $\lambda = \mathbf{E}\Phi\left([0,1]^d\right)$ is the intensity of the point process and $\rho^{(2)}$ is its *joint second-order intensity*; i.e. the density (if it exists, with respect to the Lebesgue measure) of the second-order factorial moment measure $\alpha^{(2)}(\mathrm{d}(x,y))$ (cf. Sect. 2.3.2).

For stationary point processes the following relationship holds between the functions g and K:

$$K(r) = \int_{B_r(o)} g(o,y)\,dy,$$

which simplifies to

$$K(r) = d\kappa_d \int_0^r s^{d-1} g(o,s)\,ds,$$

in the case of isotropic processes; cf. [73, Eq. (4.97) and (4.98)].

Similarly as for Ripley's K-function, we can say that larger values $g(x,y)$ of the pair correlation function indicate more clustering "around" the vector $x - y$. For a Poisson point process Π_λ, we have that $g(x,y) \equiv 1$. Again, it is not immediate to find examples of point processes whose pair correlation functions are ordered for all values of x,y. Examples of such point processes will be provided in the following sections.

Exercise 2.4. Show that ordering of pair correlation functions implies ordering of Ripley's K-functions, i.e., for two stationary point processes Φ_1, Φ_2 with $\rho_1^{(2)}(x,y) \leq \rho_2^{(2)}(x,y)$ for almost all $(x,y) \in \mathbb{R}^{2d}$, it holds that $K_1(r) \leq K_2(r)$ for all $r \geq 0$.

Though Ripley's K-function and the pair-correlation function are very simple to compute, they define only a pre-odering of point processes, because their equality does not imply equality of the underlying point processes. We shall now present some possible definitions of partial ordering of point processes that capture clustering phenomena.

2.3.2 Moment Measures

Recall that the measure $\alpha^k : \mathcal{B}^{kd} \to [0,\infty]$ defined by

$$\alpha^k(B_1 \times \cdots \times B_k) = \mathbf{E} \prod_{i=1}^k \Phi(B_i)$$

for all (not necessarily disjoint) bounded Borel sets B_i ($i = 1,\ldots,k$) is called the *k-th order moment measure* of Φ. For simple point processes, the truncation of the measure α^k to the subset $\{(x_1,\ldots,x_k) \in (\mathbb{R}^d)^k : x_i \neq x_j, \text{ for } i \neq j\}$ is equal to the *k-th order factorial moment measure* $\alpha^{(k)}$. Note that $\alpha^{(k)}(B \times \cdots \times B)$ expresses the expected number of k-tuples of points of the point process in a given set B. Bearing this interpretation in mind we can say that in the class of point processes with some given intensity measure $\alpha = \alpha^1$, larger values $\alpha^k(B)$ and $\alpha^{(k)}(B)$ of the (factorial) moment measures α^k and $\alpha^{(k)}$, respectively, indicate point processes clustering more in $B \subset \mathbb{R}^d$. A first argument we can give to support the above statement is considered in Exercise 2.5 below.

Exercise 2.5. Show that comparability of $\alpha^{(2)}(B)$ for all bounded Borel sets B implies a corresponding inequality for the pair correlation functions and hence Ripley's K-functions. *Hint.* For stationary point processes of a given intensity, see [40] for details.

Remark 2.2. For a stronger justification of the relationship between moment measures and clustering, we can use concentration inequalities, which give upper bounds on the probability that the random counting measure Φ deviates from its intensity measure α. Smaller deviations can be interpreted as supportive for spatial stationarity. To be more specific, using Chebyshev's inequality we have

$$\mathbf{P}\left(|\Phi(B) - \alpha(B)| \geq a\right) \leq \left(\alpha^2(B) - (\alpha(B))^2\right)/a^2$$

for all bounded Borel sets B, and $a > 0$. Thus, for point processes with the same mean measure, second moments measures or Ripley's functions (via Proposition 2.1) allow to compare their clustering. Similarly, using Chernoff's bound, we get that

$$\mathbf{P}\left(\Phi(B) - \alpha(B) \geq a\right) \leq e^{-t(\alpha(B)+a)}\mathbf{E}e^{t\Phi(B)} = e^{-t(\alpha(B)+a)} \sum_{k=0}^{\infty} \frac{t^k}{k!}\alpha^k(B) \qquad (2.6)$$

for any $t, a > 0$. Both concentration inequalities give smaller upper bounds for the probability of the deviation from the mean (the upper deviation in the case of Chernoff's bound) for point processes clustering less in the sense of higher-order moment measures. We will come back to this idea in Propositions 2.2 and 2.4 below.

In Sect. 2.4 we will present results, in particular regarding percolation properties of point processes, for which it its enough to be able to compare factorial moment measures of point processes. We shall note casually that restricted to a "nice" class of point processes, the factorial moment measures uniquely determine the point process and hence the ordering defined via comparison of factorial moment measures is actually a partial order on this nice class of point processes.

We now concentrate on comparison to the Poisson point process. Recall that for a general Poisson point process Π_Λ we have $\alpha^{(k)}(\mathrm{d}(x_1,\ldots,x_k)) = \Lambda(\mathrm{d}x_1)\ldots\Lambda(\mathrm{d}x_k)$ for all $k \geq 1$, where $\Lambda = \alpha$ is the intensity measure Π_Λ. In this regard, we define the following class of point processes clustering less (or more) than the Poisson point process with the same intensity measure.

Definition 2.8 (α-weakly sub-Poisson point process). A point process Φ is said to be *weakly sub-Poisson* in the sense of moment measures (α-weakly sub-Poisson for short) if

$$\mathbf{E}\prod_{i=1}^{k}\Phi(B_i) \leq \prod_{i=1}^{k}\mathbf{E}\Phi(B_i), \qquad (2.7)$$

for all $k \geq 1$ and all mutually disjoint bounded Borel sets $B_1,\ldots,B_k \subset \mathbb{R}^d$. When the reversed inequality in (2.7) holds, we say that Φ is *weakly super-Poisson* in the sense of moment measures (α-weakly super-Poisson for short).

In other words, α-weakly sub-Poisson point processes have factorial moment measures $\alpha^{(k)}$ smaller than those of the Poisson point process with the same intensity measure. Similarly, α-weakly super-Poisson point processes have factorial moment measures larger than those of the Poisson point process with the same intensity measure. We also remark that the notion of sub- and super-Poisson distributions is used e.g. in quantum physics and denotes distributions for which the variance is smaller (respectively larger) than the mean. Our notion of α-weak sub- and super-Poissonianity is consistent with (and stronger than) this definition. In quantum optics, e.g. sub-Poisson patterns of photons appear in resonance fluorescence, where laser light gives Poisson statistics of photons, while the thermal light gives super-Poisson patterns of photons; cf. [415].

Exercise 2.6. Show that α-weakly sub- (super-) Poisson point processes have moment measures α^k smaller (larger) than those of the corresponding Poisson point process. *Hint.* Recall that the moment measures $\alpha^k : \mathcal{B}^{kd} \to [0,\infty]$ of a general point process can be expressed as non-negative combinations of products of its (lower-dimensional) factorial moment measures (cf. [93] Exercise 5.4.5, p. 143).

Here is an easy, but important consequence of the latter observation regarding Laplace transforms "in the negative domain", i.e. functionals $\mathcal{L}_\Phi(-f)$, where

$$\mathcal{L}_\Phi(f) = \mathbf{E}\exp\left(-\int_{\mathbb{R}^d} f(x)\,\Phi(\mathrm{d}x)\right),$$

for non-negative functions f on $\mathbb{R}^d$, which include as a special case the functional $\mathbf{E}e^{t\Phi(B)}$ appearing in the "upper" concentration inequality (2.6). By Taylor expansion of the exponential function at 0 and the well-known expression of the Laplace functional of the Poisson point process with intensity measure α which can be recognised in the right-hand side of (2.8), the following result is obtained.

Proposition 2.2. *Assume that Φ is a simple point process with locally bounded intensity measure α and consider $f \geq 0$. If Φ is α-weakly sub-Poisson, then*

$$\mathbf{E}\exp\left(\int_{\mathbb{R}^d} f(x)\,\Phi(\mathrm{d}x)\right) \leq \exp\left(\int_{\mathbb{R}^d}(e^{f(x)}-1)\,\alpha(\mathrm{d}x)\right). \qquad (2.8)$$

If Φ is α-weakly super-Poisson, then the reversed inequality is true.

The notion of weak sub(super)-Poissonianity is closely related to negative and positive association of point processes, as we shall see in Sect. 2.3.4 below.

2.3.3 Void Probabilities

The celebrated Rényi theorem says that the void probabilities $v(B) = \mathbf{P}(\Phi(B) = 0)$ of point processes, evaluated for all bounded Borel Sets B characterise the distribution of a simple point process. They also allow an easy comparison of clustering properties of point processes by the following interpretation: a point process having smaller void probabilities has less chance to create a particular hole (absence of points in a given region). Thus, in the class of point processes with some given intensity measure, larger void probabilities indicate point processes with stronger clustering.

Remark 2.3. Using void probabilities in the study of clustering is complementary to the comparison of moments considered in the previous section. An easy way to see this consists in using again Chernoff's bound to obtain the following "lower" concentration inequality (cf. Remark 2.2)

$$\mathbf{P}\left(\alpha(B) - \Phi(B) \geq a\right) \leq e^{t(\alpha(B)-a)}\mathbf{E}e^{-t\Phi(B)}, \tag{2.9}$$

which holds for any $t, a > 0$, and noting that

$$\mathbf{E}e^{-t\Phi(B)} = \sum_{k=0}^{\infty} e^{-tk}\mathbf{P}\left(\Phi(B) = k\right) = \mathbf{P}\left(\Phi'(B) = 0\right) = v'(B)$$

is the void probability of the point process Φ' obtained from Φ by independent thinning with retention probability $1 - e^{-t}$. It is not difficult to show that ordering of void probabilities of simple point processes is preserved by independent thinning (cf. [44]) and thus the bound in (2.9) is smaller for point processes less clustering in the latter sense. We will come back to this idea in Propositions 2.3 and 2.4. Finally, note that $\lim_{t\to\infty} \mathbf{E}e^{-t\Phi(B)} = v(B)$ and thus, in conjunction with what was said above, comparison of void probabilities is equivalent to the comparison of one-dimensional Laplace transforms of point processes for non-negative arguments.

In Sect. 2.4, we will present results, in particular regarding percolation properties, for which it is enough to be able to compare void probabilities of point processes. Again, because of Rényi's theorem, we have that ordering defined by void probabilities is a partial order on the space of simple point processes.

2.3.3.1 v-Weakly Sub(Super)-Poisson Point Processes

Recall that a Poisson point process Π_Λ can be characterised as having void probabilities of the form $v(B) = \exp(-\Lambda(B))$, with Λ being the intensity measure of Φ. In this regard, we define the following classes of point processes clustering less (or more) than the Poisson point process with the same intensity measure.

Definition 2.9 (v-weakly sub(super)-Poisson point process). A point process Φ is said to be *weakly sub-Poisson* in the sense of void probabilities (v-weakly sub-Poisson for short) if

$$\mathbf{P}\left(\Phi(B) = 0\right) \leq e^{-\mathbf{E}\Phi(B)} \tag{2.10}$$

for all Borel sets $B \subset \mathbb{R}^d$. When the reversed inequality in (2.10) holds, we say that Φ is *weakly super-Poisson* in the sense of void probabilities (v-weakly super-Poisson for short).

In other words, v-weakly sub-Poisson point processes have void probabilities smaller than those of the Poisson point process with the same intensity measure. Similarly, v-weakly super-Poisson point processes have void probabilities larger than those of the Poisson point process with the same intensity measure.

Example 2.11. It is easy to see by Jensen's inequality that all Cox point processes are v-weakly super-Poisson.

By using a coupling argument as in Remark 2.3, we can derive an analogous result as in Proposition 2.2 for v-weakly sub-Poisson point processes.

Proposition 2.3 ([44]). *Assume that Φ is a simple point process with locally bounded intensity measure α. Then Φ is v-weakly sub-Poisson if and only if (2.8) holds for all functions $f \leq 0$.*

2.3.3.2 Combining Void Probabilities and Moment Measures

We have already explained why the comparison of void probabilities and moment measures are in some sense complementary. Thus, it is natural to combine them, whence the following definition is obtained.

Definition 2.10 (Weakly sub- and super-Poisson point process). We say that Φ is *weakly sub-Poisson* if Φ is α-weakly sub-Poisson and v-weakly sub-Poisson. Weakly super-Poisson point processes are defined in the same way.

The following remark is immediately obtained from Propositions 2.2 and 2.3.

Remark 2.4. Assume that Φ is a simple point process with locally bounded intensity measure α. If Φ is weakly sub-Poisson then (2.8) holds for any f of constant sign ($f \geq 0$ or $f \leq 0$). If Φ is weakly super-Poisson, then in (2.8) the reversed inequality holds for such f.

Example 2.12. It has been shown in [45] that determinantal and permanental point process (with trace-class integral kernels) are weakly sub-Poisson and weakly super-Poisson, respectively.

Other examples (admitting even stronger comparison properties) will be given in Sect. 2.3.4 and 2.3.5.

As mentioned earlier, using the ordering of Laplace functionals of weakly sub-Poisson point processes, we can extend the concentration inequality for Poisson point processes to this class of point processes. In the discrete setting, a similar result is proved for negatively associated random variables in [106]. A more general concentration inequality for Lipschitz functions is known in the case of determinantal point processes ([313]).

Proposition 2.4. *Let Φ be a simple stationary point process with unit intensity which is weakly sub-Poisson, and let $B_n \subseteq \mathbb{R}^d$ be a Borel set of Lebesgue measure n. Then, for any $1/2 < a < 1$ there exists an integer $n(a) \geq 1$ such that for $n \geq n(a)$*

$$\mathbf{P}(|\Phi(B_n) - n| \geq n^a) \leq 2\exp\left(-n^{2a-1}/9\right).$$

Exercise 2.7. Prove Proposition 2.4. *Hint.* Use Markov's inequality, Propositions 2.2 and 2.3 along with the bounds for the Poisson case known from [314, Lemmas 1.2 and 1.4].

Note that the bounds we have suggested to use are the ones corresponding to the Poisson point process. For specific weakly sub-Poisson point processes, one expects an improvement on these bounds.

2.3.4 Positive and Negative Association

Denote covariance of (real-valued) random variables X, Y by $\mathbf{cov}(X,Y) = \mathbf{E}XY - \mathbf{E}X\mathbf{E}Y$.

Definition 2.11 ((Positive) association of point processes). A point process Φ is called *associated* if

$$\mathbf{cov}(f(\Phi(B_1),\ldots,\Phi(B_k)), g(\Phi(B_1),\ldots,\Phi(B_k))) \geq 0 \qquad (2.11)$$

for any finite collection of bounded Borel sets $B_1,\ldots,B_k \subset \mathbb{R}^d$ and $f,g : \mathbb{R}^k \to [0,1]$ (componentwise) increasing functions; cf. [61].

The property considered in (2.11) is also called *positive association*, or the *FKG property*. The theory for the opposite property is more tricky, cf. [312], but one can define it as follows.

Definition 2.12 (Negative association). A point process Φ is called *negatively associated* if

$$\mathbf{cov}(f(\Phi(B_1),\ldots,\Phi(B_k)), g(\Phi(B_{k+1}),\ldots,\Phi(B_l))) \leq 0$$

for any finite collection of bounded Borel sets $B_1,\ldots,B_l \subset \mathbb{R}^d$ such that $(B_1 \cup \cdots \cup B_k) \cap (B_{k+1} \cup \cdots \cup B_l) = \emptyset$ and f,g increasing functions.

Both definitions can be straightforwardly extended to arbitrary random measures, where one additionally assumes that f,g are continuous and increasing functions. Note that the notion of association or negative association of point processes does not induce any ordering on the space of point processes. Though association or negative association have not been studied from the point of view of stochastic ordering, it has been widely used to prove central limit theorems for random fields (see [57]).

The following result has been proved in [45]. It will be strengthened in the next section (see Proposition 2.12)

Proposition 2.5. *A negatively associated, simple point process with locally bounded intensity measure is weakly sub-Poisson. A (positively) associated point process with a diffuse locally bounded intensity measure is weakly super-Poisson.*

Exercise 2.8. Prove that a (positively) associated point process with a diffuse locally bounded intensity measure is α-weakly super-Poisson. Show a similar statement for negatively associated point processes as well.

Example 2.13. From [61, Th. 5.2], we know that any *Poisson cluster point process* is associated. This is a generalisation of the perturbation approach of a Poisson point process Φ considered in (2.1) having the form $\Phi^{\text{cluster}} = \sum_{X \in \Phi}(X + \Phi_X)$ with Φ_X being arbitrary i.i.d. (cluster) point processes. In particular, the Neyman-Scott point process (cf. Example 2.6) is associated. Other examples of associated point processes given in [61] are Cox point processes with random intensity measures being associated.

Example 2.14. Determinantal point processes are negatively associated (see [137, cf. Corollary 6.3]).

We also remark that there are negatively associated point processes, which are not weakly sub-Poisson. A counterexample given in [45] (which is not a simple point process, showing that this latter assumption cannot be relaxed in Proposition 2.5) exploits [210, Theorem 2], which says that a random vector having a *permutation distribution* (taking as values all $k!$ permutations of a given deterministic vector with equal probabilities) is negatively associated.

Exercise 2.9. Show that the binomial point process (cf. Example 2.1) and the simple perturbed lattice (cf. Example 2.3) are negatively associated.

2.3.5 Directionally Convex Ordering

2.3.5.1 Definitions and Basic Results

In this section, we present some basic results on directionally convex ordering of point processes that will allow us to see this order also as a tool to compare clustering of point processes.

A Borel-measurable function $f : \mathbb{R}^k \to \mathbb{R}$ is said to be *directionally convex* (dcx) if for any $x \in \mathbb{R}^k, \varepsilon, \delta > 0, i, j \in \{1, \dots, k\}$, we have that $\Delta_\varepsilon^i \Delta_\delta^j f(x) \geq 0$, where $\Delta_\varepsilon^i f(x) = f(x + \varepsilon e_i) - f(x)$ is the discrete differential operator, with $\{e_i\}_{1 \leq i \leq k}$ denoting the canonical basis vectors of $\mathbb{R}^k$. In the following, we abbreviate *increasing* and dcx by idcx and *decreasing* and dcx by ddcx (see [289, Chap. 3]). For random vectors X and Y of the same dimension, X *is said to be smaller than Y in dcx order* (denoted $X \leq_{\text{dcx}} Y$) if $\mathbf{E}f(X) \leq \mathbf{E}f(Y)$ for all f being dcx such that both expectations in the latter inequality are finite. Real-valued random fields are said to be dcx ordered if all finite-dimensional marginals are dcx ordered.

Definition 2.13 (dcx-order of point processes). Two point processes Φ_1 and Φ_2 are said to be dcx-ordered, i.e. $\Phi_1 \leq_{\text{dcx}} \Phi_2$, if for any $k \geq 1$ and bounded Borel sets $B_1, \dots, B_k$ in $\mathbb{R}^d$, it holds that $(\Phi_1(B_1), \dots, \Phi_1(B_k)) \leq_{\text{dcx}} (\Phi_2(B_1), \dots, \Phi_2(B_k))$.

The definition of comparability of point processes is similar for other orders, i.e. those defined by idcx, ddcx functions. It is enough to verify the above conditions for $B_1, \ldots, B_k$ mutually disjoint, cf. [40]. In order to avoid technical difficulties, we will consider only point processes whose intensity measures are locally finite. For such point processes, the dcx-order is a partial order.

Remark 2.5. It is easy to see that $\Phi_1 \leq_{dcx} \Phi_2$ *implies the equality of their intensity measures*, i.e: $\mathbf{E}\Phi_1(B) = \mathbf{E}\Phi_2(B) = \alpha(B)$ for any bounded Borel set $B \subset \mathbb{R}^d$ as both x and $-x$ are dcx functions.

We argue that, dcx-ordering is also useful in clustering comparison of point processes. Point processes larger in dcx-order cluster more, whereas point processes larger in idcx-order cluster more while having on average more points, and point processes larger in ddcx-order cluster more while having on average less points.

The two statements of the following result were proved in [40] and [45], respectively. They show that dcx-ordering is stronger than comparison of moments measures and void probabilities considered in the two previous sections.

Proposition 2.6. *Let Φ_1 and Φ_2 be two point process on $\mathbb{R}^d$. Denote their moment measures by α_j^k ($k \geq 1$) and their void probabilities by v_j, $j = 1, 2$, respectively.*

1. *If $\Phi_1 \leq_{idcx} \Phi_2$ then $\alpha_1^k(B) \leq \alpha_2^k(B)$ for all bounded Borel sets $B \subset (\mathbb{R}^d)^k$, provided that α_j^k is σ-finite for $k \geq 1$, $j = 1, 2$.*
2. *If $\Phi_1 \leq_{ddcx} \Phi_2$ then $v_1(B) \leq v_2(B)$ for all bounded Borel sets $B \subset \mathbb{R}^d$.*

Exercise 2.10. Show that $\prod_i (x_i \vee 0)$ is a dcx-function and $(1 - x) \vee 0$ is a convex function. Using these facts to prove the above proposition.

Note that the σ-finiteness condition considered in the first statement of Proposition 2.6 is missing in [40]; see [425, Proposition 4.2.4] for the correction. An important observation is that the operation of clustering perturbation introduced in Sect. 2.2.2 is dcx monotone with respect to the replication kernel in the following sense; cf. [45].

Proposition 2.7. *Consider a point process Φ with locally finite intensity measure α and its two perturbations Φ_j^{pert} ($j = 1, 2$) satisfying condition (2.2), and having the same displacement kernel $\mathcal{X}$ and possibly different replication kernels $\mathcal{N}_j$, $j = 1, 2$, respectively. If $\mathcal{N}_1(x, \cdot) \leq_{cx} \mathcal{N}_2(x, \cdot)$ (which means convex ordering of the conditional distributions of the number of replicas) for α-almost all $x \in \mathbb{R}^d$, then $\Phi_1^{pert} \leq_{dcx} \Phi_2^{pert}$.*

Thus clustering perturbations of a given point process provide many examples of point process comparable in dcx-order. Examples of convexly ordered replication kernels have been given in Example 2.3.

Another observation, proved in [40], says that the operations transforming some random measure $\mathfrak{L}$ into a Cox point process $Cox_{\mathfrak{L}}$ (cf. Definition 2.4) preserves the dcx-order.

Proposition 2.8. *Consider two random measures $\mathcal{L}_1$ and $\mathcal{L}_2$ on $\mathbb{R}^d$. If $\mathcal{L}_1 \leq_{\mathrm{dcx}\,(resp.\,\mathrm{idcx})}$ $\mathcal{L}_2$ then $Cox_{\mathcal{L}_1} \leq_{\mathrm{dcx}\,(resp.\,\mathrm{idcx})} Cox_{\mathcal{L}_2}$.*

The above result, combined with further results on comparison of shot-noise fields presented in Sect. 2.3.5.2 will allow us to compare many Cox point processes (cf. Example 2.15).

2.3.5.2 Comparison of Shot-Noise Fields

Many interesting quantities in stochastic geometry can be expressed by additive or extremal shot-noise fields. They are also used to construct more sophisticated point process models. For this reason, we state some results on dcx-ordering of shot-noise fields that are widely used in applications.

Definition 2.14 (Shot-noise fields). Let S be any (non-empty) index set. Given a point process Φ on $\mathbb{R}^d$ and a *response function* $h(x,y) : \mathbb{R}^d \times S \to (-\infty, \infty]$ which is measurable in the first variable, then the *(integral) shot-noise field* $\{V_\Phi(s), s \in S\}$ is defined as

$$V_\Phi(y) = \int_{\mathbb{R}^d} h(x,y)\,\Phi(\mathrm{d}x) = \sum_{X \in \Phi} h(X,y), \qquad (2.12)$$

and the *extremal shot-noise* field $\{U_\Phi(s), s \in S\}$ is defined as

$$U_\Phi(y) = \sup_{X \in \Phi} \{h(X,y)\}. \qquad (2.13)$$

As we shall see in Sect. 2.4.2 (and also in the proof of Proposition 2.11) it is not merely a formal generalisation to take S being an arbitrary set. Since the composition of a dcx-function with an increasing linear function is still dcx, linear combinations of $\Phi(B_1),\ldots,\Phi(B_n)$ for finitely many bounded Borel sets $B_1,\ldots,B_n \subseteq \mathbb{R}^d$ (i.e. $\sum_{i=1}^m c_i \Phi(B_i)$ for $c_i \geq 0$) preserve the dcx-order. An integral shot-noise field can be approximated by finite linear combinations of $\Phi(B)$'s and hence justifying continuity, one expects that integral shot-noise fields preserve dcx-order as well. This type of important results on dcx-ordering of point processes is stated below.

Proposition 2.9. *([40, Theorem 2.1]) Let Φ_1 and Φ_2 be arbitrary point processes on $\mathbb{R}^d$. Then, the following statements are true.*

1. *If $\Phi_1 \leq_{\mathrm{idcx}} \Phi_2$, then $\{V_{\Phi_1}(s), s \in S\} \leq_{\mathrm{idcx}} \{V_{\Phi_2}(s), s \in S\}$.*
2. *If $\Phi_1 \leq_{\mathrm{dcx}} \Phi_2$, then $\{V_{\Phi_1}(s), s \in S\} \leq_{\mathrm{dcx}} \{V_{\Phi_2}(s), s \in S\}$, provided that $\mathbf{E}V_{\Phi_i}(s) < \infty$, for all $s \in S$, $i = 1,2$.*

The results of Proposition 2.9, combined with those of Proposition 2.8 allow the comparison of many Cox processes.

Example 2.15 (Comparable Cox point processes). Let Z_1 and Z_2 be two Lévy-bases with mean measures α_1 and α_2, respectively. Note that $\alpha_i \leq_{\mathrm{dcx}} Z_i$ ($i = 1,2$). This can be easily proved using complete independence of Lévy bases and Jensen's inequality. In a sense, the mean measure α_i "spreads" (in the sense of dcx) the mass better than the corresponding completely independent random measure Z_i. Furthermore, consider the random fields ξ_1 and ξ_2 on $\mathbb{R}^d$ given by $\xi_i(y) = \int_{\mathbb{R}^d} k(x,y)Z_i(dx)$, $i = 1,2$ for some non-negative kernel k, and assume that these fields are a.s. locally Riemann integrable. Denote by Cox_{ξ_i} and $\mathrm{Cox}_{\exp(\xi_i)}$ the corresponding Lévy-driven and log-Lévy-driven Cox point process. The following inequalities hold.

1. If $Z_1 \leq_{\mathrm{dcx\,(resp.\,idcx)}} Z_2$, then $\mathrm{Cox}_{\xi_1} \leq_{\mathrm{dcx\,(resp.\,idcx)}} \mathrm{Cox}_{\xi_2}$ provided that, in case of dcx, $\mathbf{E}\int_B \xi_i(y)\,dy < \infty$ for all bounded Borel sets $B \subset \mathbb{R}^d$.
2. If $Z_1 \leq_{\mathrm{idcx}} Z_2$, then $\mathrm{Cox}_{\exp(\xi_1)} \leq_{\mathrm{idcx}} \mathrm{Cox}_{\exp(\xi_2)}$.

Suppose that $\{X_i(y)\}, i = 1,2$ are two Gaussian random fields on $\mathbb{R}^d$ and denote by $\mathrm{Cox}_{\exp(X_i)}$, $(i = 1,2)$ the corresponding log-Lévy-driven Cox point processes. Then the following is true.

3. If $\{X_1(y)\} \leq_{\mathrm{idcx}} \{X_2(y)\}$ (as random fields), then $\mathrm{Cox}_{\exp(X_1)} \leq_{\mathrm{idcx}} \mathrm{Cox}_{\exp(X_2)}$.

Note that the condition in the third statement is equivalent to $\mathbf{E}X_1(y) \leq \mathbf{E}X_2(y)$ for all $y \in \mathbb{R}^d$ and $\mathbf{cov}(X_1(y_1),X_1(y_2)) \leq \mathbf{cov}(X_2(y_1),X_2(y_2))$ for all $y_1,y_2 \in \mathbb{R}^d$. An example of a parametric dcx-ordered family of Gaussian random fields is given in [274].

Let Φ_1, Φ_2 be two point processes on $\mathbb{R}^d$ and denote by Cox_1, Cox_2 the generalised shot-noise Cox point processes (cf. Example 2.4) being clustering perturbations of Φ_1, Φ_2, respectively, with the same (Poisson) replication kernel $\mathcal{N}$ and with displacement distributions $\mathcal{X}(x,\cdot)$ having density $\mathcal{X}'(x,y)\,dy$ for all $x \in \mathbb{R}^d$. Then, the following result is true.

4. If $\Phi_1 \leq_{\mathrm{dcx\,(resp.\,idcx)}} \Phi_2$, then $\mathrm{Cox}_1 \leq_{\mathrm{dcx\,(resp.\,idcx)}} \mathrm{Cox}_2$ provided that, in case of dcx, $\int_{\mathbb{R}^d} \mathcal{X}'(x,y)\,\alpha(dx) < \infty$ for all $y \in \mathbb{R}^d$, where α is the (common) intensity measure of Φ_1 and Φ_2.

Proposition 2.9 allows us to compare extremal shot-noise fields using the following well-known representation $\mathbf{P}(U(y_i) \leq a_i, 1 \leq i \leq m) = \mathbf{E}e^{-\Sigma_i \hat{U}_i}$ where $\hat{U}_i = \sum_n -\log \mathbf{1}(h(X_n,y_i) \leq a_i)$ is an additive shot-noise field with response function taking values in $[0,\infty]$. Noting that $e^{-\Sigma_i x_i}$ is a dcx-function, we get the following result.

Proposition 2.10 ([40, Proposition 4.1]). *Let* $\Phi_1 \leq_{\mathrm{dcx}} \Phi_2$. *Then for any* $n \geq 1$ *and for all* $t_i \in \mathbb{R}, y_i \in S, 1 \leq i \leq n$, *it holds that*

$$\mathbf{P}(U_{\Phi_1}(y_i) \leq t_i, 1 \leq i \leq n) \leq \mathbf{P}(U_{\Phi_2}(y_i) \leq t_i, 1 \leq i \leq n).$$

An example of application of the above result is the comparison of *capacity functionals* of *germ-grain models*, whose definition we recall first, see also Chapter 5.

Definition 2.15 (Germ-grain model). Given (the distribution of) a random closed set Y and a point process Φ, a *germ-grain model* with the point process of *germs* Φ and the *typical grain* Y, is given by the random set $C(\Phi,Y) = \bigcup_{X_i \in \Phi}\{X_i + Y_i\}$, where $x + A = \{x + a : a \in A\}, a \in \mathbb{R}^d, A \subset \mathbb{R}^d$ and $\{Y_i\}$ is a sequence of i.i.d. random closed sets distributed as Y. We call Y a *fixed grain* if there exists a (deterministic) closed set $B \subseteq \mathbb{R}^d$ such that $Y = B$ a.s. In the case of spherical grains, i.e. $B = B_o(r)$, where

o is the origin of $\mathbb{R}^d$ and $r \geq 0$ a constant, we denote the corresponding germ-grain model by $C(\Phi, r)$.

A commonly made technical assumption about the distributions of Φ and Y is that for any compact set $K \subset \mathbb{R}^d$, the expected number of germs $X_i \in \Phi$ such that $(X_i + Y_i) \cap K \neq \emptyset$ is finite. This assumption, called "local finiteness of the germ-grain model" guarantees in particular that $C(\Phi, Y)$ is a random closed set in $\mathbb{R}^d$. The germ-grain models considered throughout this chapter will be assumed to have the local finiteness property.

Proposition 2.11 ([43, Propostion 3.4]). *Let $C(\Phi_j, Y)$, $j = 1, 2$ be two germ-grain models with point processes of germs Φ_j, $j = 1, 2$, respectively, and common distribution of the typical grain Y. Assume that Φ_1 and Φ_2 are simple and have locally finite moment measures. If $\Phi_1 \leq_{\mathrm{dcx}} \Phi_2$, then*

$$\mathbf{P}\left(C(\Phi_1, Y) \cap B = \emptyset\right) \leq \mathbf{P}\left(C(\Phi_2, Y) \cap B = \emptyset\right)$$

for all bounded Borel sets $B \subset \mathbb{R}^d$. Moreover, if Y is a fixed compact grain, then the same result holds, provided $v_1(B) \leq v_2(B)$ for all bounded Borel sets $B \subset \mathbb{R}^d$, where $v_i(B)$ denotes the void probabilities of Φ_i.

2.3.5.3 Sub- and Super-Poisson Point Processes

We now concentrate on dcx-comparison to the Poisson point process. To this end, we define the following classes of point processes.

Definition 2.16 (Sub- and super- Poisson point process). We call a point process *dcx sub-Poisson* (respectively *dcx super-Poisson*) if it is smaller (larger) in dcx-order than the Poisson point process (necessarily of the same mean measure). For simplicity, we will just refer to them as sub-Poisson or super-Poisson point process omitting the phrase dcx.

Proposition 2.12. *A negatively associated point processes Φ with convexly sub-Poisson one-dimensional marginal distributions, $\Phi(B) \leq_{cx} \mathrm{Pois}(\mathbf{E}\Phi(B))$ for all bounded Borel sets B, is sub-Poisson. An associated point processes with convexly super-Poisson one-dimensional marginal distributions is super-Poisson.*

Proof (sketch). This is a consequence of [75, Theorem 1], which says that a negatively associated random vector is supermodularly smaller than the random vector with the same marginal distributions and independent components. Similarly, an associated random vector is supermodularly larger than the random vector with the same marginal distributions and independent components. Since supermodular order is stronger than *dcx* order, this implies dcx ordering as well. Finally, a vector with independent coordinates and convexly sub-Poisson (super-Poisson) marginal distributions is *dcx* smaller (larger) than the vector of independent Poisson variables.

Example 2.16 (Super-Poisson Cox point process). Using Proposition 2.8 one can prove (cf. [40]) that Poisson-Poisson cluster point processes and, more generally, Lévy-driven Cox point processes are super-Poisson.

Also, since any mixture of Poisson distributions is cx larger than the Poisson distribution (with the same mean), we can prove that any mixed Poisson point process is super-Poisson.

Example 2.17 (Super-Poisson Neyman-Scott point process). By Proposition 2.7, any Neyman-Scott point process (cf. Example 2.6) with mean cluster size $n(x) = 1$ for all $x \in \mathbb{R}^d$ is super-Poisson. Indeed, for any $x \in \mathbb{R}^d$ and any replication kernel $\mathcal{N}$ satisfying $\sum_{k=1}^{\infty} k\mathcal{N}(x, \{k\}) = 1$, we have by Jensen's inequality that $\delta_1(\cdot) \leq_{cx} \mathcal{N}(x, \cdot)$, i.e. it is convexly larger than the Dirac measure on $\mathbb{Z}_+$ concentrated at 1. By the well-known displacement theorem for Poisson point processes, the clustering perturbation of the Poisson (parent) point process with this Dirac replication kernel is a Poisson point process. Using kernels of the form mentioned in (2.5) we can construct dcx-increasing super-Poisson point processes.

Example 2.18 (Sub- and super-Poisson perturbed lattices). Lattice clustering perturbations provide examples of both sub- and super-Poisson point process, cf. Example 2.3. Moreover, the initial lattice can be replaced by any fixed pattern of points, and the displacement kernel needs not to be supported by the Voronoi cell of the given point. Assuming Poisson replication kernels we still obtain (not necessarily stationary) Poisson point processes. Note, for example, that by (2.4) considering binomial replication kernels $\mathrm{Binom}(r, \lambda/r)$ for $r \in \mathbb{Z}^+$, $r \geq \lambda$ one can construct dcx-increasing families of sub-Poisson perturbed lattices converging to the Poisson point process Π_λ. Similarly, considering negative binomial replication kernels $\mathrm{NBionom}(r, \lambda/(r+\lambda))$ with $r \in \mathbb{Z}_+, r \geq 1$ one can construct dcx-decreasing families of super-Poisson perturbed lattices converging to Π_λ. The simple perturbed lattice (with $\mathrm{Binom}(1, 1)$, and necessarily $\lambda = 1$) is the smallest point process in dcx-order within the aforementioned sub-Poisson family.

Example 2.19 (Determinantal and permanental processes). We already mentioned in Example 2.12 that determinantal and permanental point processes are weakly sub- and super-Poisson point processes, respectively. Since determinantal point processes are negatively associated (Example 2.14) and have convexly sub-Poisson one-dimensional marginal distributions, cf [45, proof of Prop. 5.2], Proposition 2.12 gives us that determinantal point processes are *dcx* sub-Poisson. The statement for permanental processes can be strengthened to dcx-comparison to the Poisson point process with the same mean on *mutually disjoint, simultaneously observable compact subsets* of $\mathbb{R}^d$; see [45] for further details.

Exercise 2.11. Prove the statement of Example 2.11.

Exercise 2.12. Using Hadamard's inequality prove that determinantal point processes are α-weakly sub-Poisson.

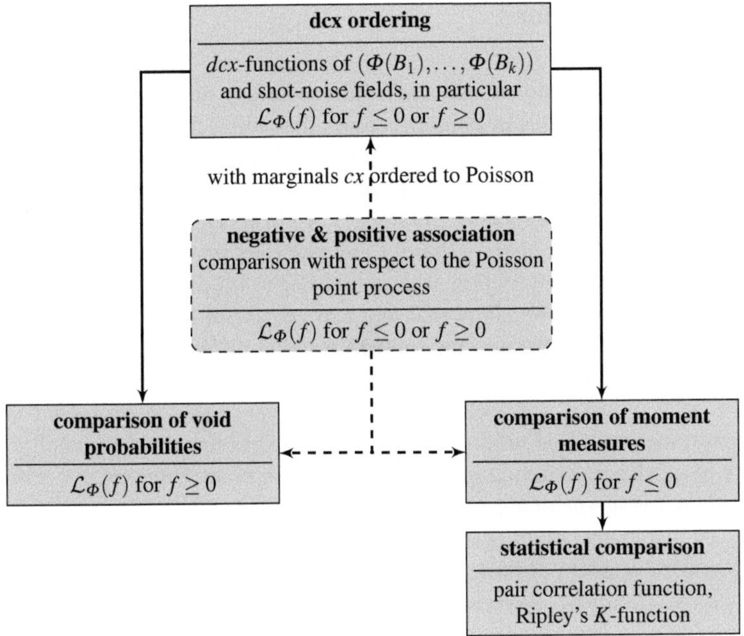

Fig. 2.2 Relationships between clustering comparison methods, and some characteristics that allow to compare them. Smaller in any type of comparison means that the point process clusters less. Recall, $\mathcal{L}_\Phi(f) := \mathbf{E}\exp\left[-\int_{\mathbb{R}^d} f(x)\,\Phi(\mathrm{d}x)\right]$.

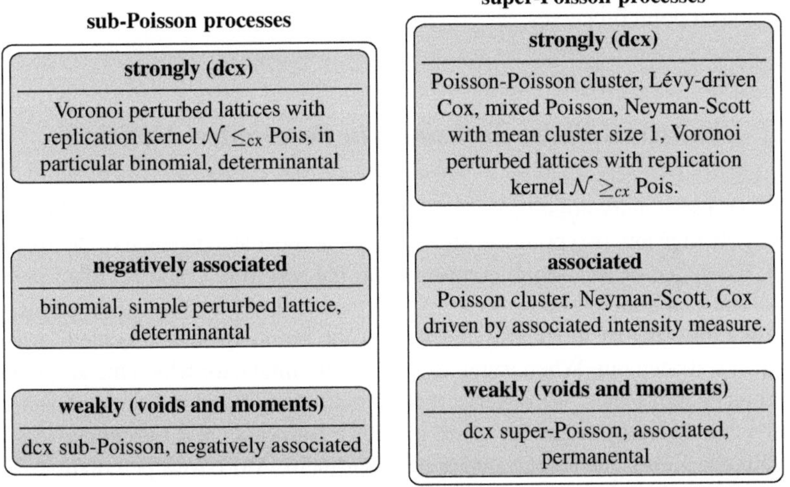

Fig. 2.3 Some point processes comparable to the Poisson point process using different comparison methods.

2.4 Some Applications

So far we introduced basic notions and results regarding ordering of point processes and we provided examples of point processes that admit these comparability properties. However, it remains to demonstrate the applicability of these methods to random geometric models which will be the goal of this section. Heuristically speaking, it is possible to easily conjecture the impact of clustering on various random geometric models, however there is hardly any rigorous treatment of these issues in the literature. The present section shall endeavour to fill this gap by using the tools of stochastic ordering. We show that one can get useful bounds for some quantities of interest on weakly sub-Poisson, sub-Poisson or negatively associated point processes and that in quite a few cases these bounds are as good as those for the Poisson point process. Various quantities of interest are often expressed in terms of moment measures and void probabilities. This explains the applicability of our notions of stochastic ordering of point processes in many contexts. As it might be expected, in this survey of applications, we shall emphasize breadth more than depth to indicate that many random geometric models fall within the purview of our methods. However, despite our best efforts, we would not be able to sketch all possible applications. Therefore, we briefly mention a couple of omissions. The notion of sub-Poissonianity has found usage in at least a couple of other models than those described below. In [173], connectivity of some approximations of minimal spanning forests is shown for weakly sub-Poisson point processes. Sans our jargon, in [92], the existence of the Lilypond growth model and its non-percolation under the additional assumption of absolutely continuous joint intensities is shown for weakly sub-Poisson point processes.

2.4.1 Non-trivial Phase Transition in Percolation Models

Consider a stationary point process Φ in $\mathbb{R}^d$. For a given "radius" $r \geq 0$, let us connect by an edge any two points of Φ which are at most at a distance of $2r$ from each other. Existence of an infinite component in the resulting graph is called *percolation* of the graph model based on Φ. As we have already mentioned in the previous section, clustering of Φ roughly means that the points of Φ are located in groups being well spaced out. When trying to find the minimal r for which the graph model based on Φ percolates, we observe that points belonging to the same cluster of Φ will be connected by edges for some smaller r but points in different clusters need a relatively high r for having edges between them. Moreover, percolation cannot be achieved without edges between some points of different clusters. It seems to be evident that spreading points from clusters of Φ "more homogeneously" in the space would result in a decrease of the radius r for which the percolation takes place. In other words, clustering in a point process Φ should increase the *critical radius* $r_c = r_c(\Phi)$ for the percolation of the graph model on Φ, also called Gilbert's disk graph or the germ-grain model with fixed spherical grains.

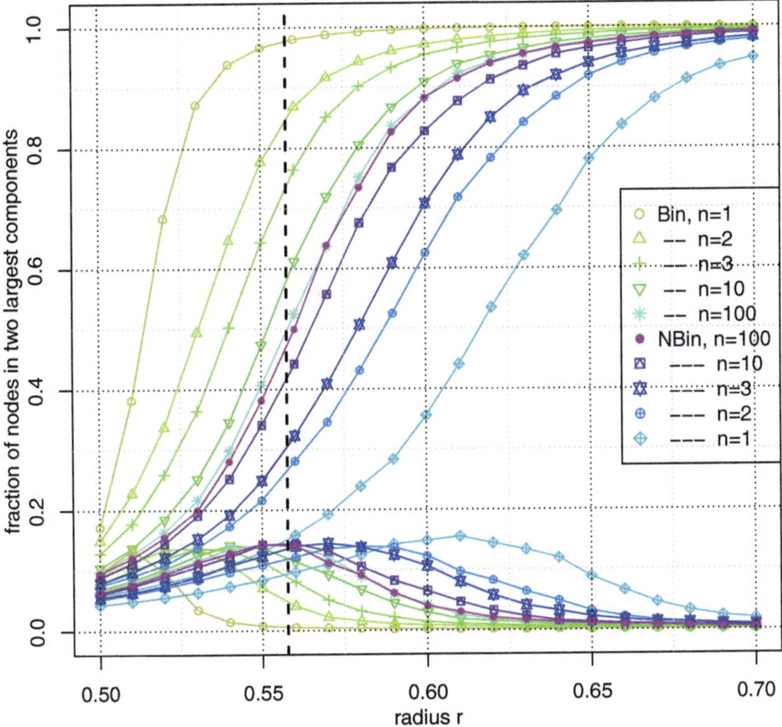

Fig. 2.4 Mean fractions of nodes in the two largest components of the sub- and super-Poisson germ-grain models $C(\Phi^{pert}_{\mathrm{Binom}(n,1/n)},r)$ and $C(\Phi^{pert}_{\mathrm{NBionom}(n,1/(1+n))},r)$, respectively, as functions of r. These families of underlying point processes converge to the Poisson point process Φ_λ with intensity $\lambda = 2/(\sqrt{3}) = 1.154701$ as n tends to ∞; cf. Example 2.18. The dashed vertical line corresponds to the radius $r = 0.5576495$ which is believed to be close to the critical radius $r_c(\Phi_\lambda)$.

We have shown in Sect. 2.3.5 that dcx-ordering of point processes can be used to compare their clustering properties. Hence, the above discussion tempts one to conjecture that r_c is increasing with respect to dcx-ordering of the underlying point processes; i.e. $\Phi_1 \leq_{\mathrm{dcx}} \Phi_2$ implies $r_c(\Phi_1) \leq r_c(\Phi_2)$. Some numerical evidences gathered in [44] (where we took Fig. 2.4 from) for a dcx-monotone family of perturbed lattice point processes, were supportive for this conjecture.

But it turns out that the conjecture is not true in full generality and a counterexample was also presented in [44]. It is a Poisson-Poisson cluster point process, which is known to be super-Poisson (cf. Example 2.16) whose critical radius is $r_c = 0$, hence smaller than that of the corresponding Poisson point process, for which r_c is known to be positive. In this Poisson-Poisson cluster point process, points concentrate on some carefully chosen larger-scale structure, which itself has good percolation properties. In this case, the points concentrate in clusters, however we cannot say that clusters are well spaced out. Hence, this example does not contradict our initial heuristic explanation of why an increase of clustering in a point process should increase the critical radius for the percolation. It reveals rather that dcx-ordering, while being able to compare the clustering tendency of point processes, is not capable of comparing macroscopic structures of clusters. Nevertheless, dcx-ordering, and some weaker tools introduced in Sect. 2.3, can be used to prove nontrivial phase transitions for point processes which cluster less than the Poisson process. In what follows, we will first present some intuitions leading to the above results and motivating our special focus on moment measures and void probabilities in the previous sections.

2.4.1.1 Intuitions — Some Non-Standard Critical Radii

Consider the radii $\underline{r}_c, \overline{r}_c$, which act as lower and upper bounds for the usual critical radius: i.e. $\underline{r}_c \leq r_c \leq \overline{r}_c$. We show that clustering acts differently on these bounding radii: It turns out that

$$\underline{r}_c(\Phi_2) \leq \underline{r}_c(\Phi_1) \leq r_c(\Phi_1) \leq \overline{r}_c(\Phi_1) \leq \overline{r}_c(\Phi_2)$$

for Φ_1 having smaller voids and moment measures than Φ_2. This sandwich inequality tells us that Φ_1 exhibits the usual phase transition $0 < r_c(\Phi_1) < \infty$, provided Φ_2 satisfies the stronger conditions $0 < \underline{r}_c(\Phi_2)$ and $\overline{r}_c(\Phi_2) < \infty$. Conjecturing that this holds if Φ_2 is a Poisson point process, one obtains the result on (uniformly) nontrivial phase transition for all weakly sub-Poisson processes Φ_1, which has been proved in [44] and will be presented in the subsequent sections in a slightly different way.

Let Φ be an arbitrary point process in $\mathbb{R}^d$. Let $W_m = [-m, m]^d$ and define $h_{m,k} : (\mathbb{R}^d)^k \to \{0, 1\}$ to be the indicator of the event that $x_1, \ldots, x_k \in (\Phi \cap W_m)^k, |x_1| \leq r, \inf_{x \in \partial W_m} |x - x_k| \leq r, \max_{i \in \{1,\ldots,k-1\}} |x_{i+1} - x_i| \leq r$, where ∂W_n denotes the boundary of set W_n. Let $N_{m,k}(\Phi, r) = \sum_{X_1,\ldots,X_k \in \mathbb{R}^d}^{\neq} h_{m,k}(X_1, \ldots, X_k)$ denote the number of distinct self-avoiding paths of length k from the origin $o \in \mathbb{R}^2$ to the boundary of the box W_m in the germ-grain model and $N_m(\Phi, r) = \sum_{k \geq 1} N_{m,k}(\Phi, r)$ to be the total number of distinct self-avoiding paths to the boundary of the box. We define the following "lower" critical radius

$$\underline{r}_c(\Phi) = \inf\{r : \liminf_m EN_m(\Phi, r) > 0\}.$$

Note that $r_c(\Phi) = \inf\{r : \lim_{m \to \infty} \mathbf{P}(N_m(\Phi, r) \geq 1) > 0\}$, with the limit existing because the events $\{N_m(\Phi, r) \geq 1\}$ form a decreasing sequence in m, and by Markov's inequality, we have that indeed $\underline{r}_c(\Phi) \leq r_c(\Phi)$ for a stationary point process Φ.

It is easy to see that $EN_m(\Phi, r)$ can be expressed in terms of moment measures, i.e. $EN_m(\Phi, r) = \sum_{k \geq 1} EN_{m,k}(\Phi, r)$ and

$$EN_{m,k}(\Phi, r) = \int_{(\mathbb{R}^d)^k} h_{m,k}(x_1, \ldots, x_k) \alpha_j^{(k)}(dx_1, \ldots, dx_k).$$

The following result is obtained in [45].

Proposition 2.13. *Let $C_j = C(\Phi_j, r)$, $j = 1, 2$ be two germ-grain models with simple point processes of germs Φ_j, $j = 1, 2$, and σ-finite k-th moment measures α_j^k for all $k \geq 1$ respectively. If $\alpha_1^{(k)}(\cdot) \leq \alpha_2^{(k)}(\cdot)$ for all $k \geq 1$, then $\underline{r}_c(\Phi_1) \geq \underline{r}_c(\Phi_2)$. In particular, for a stationary, α-weakly sub-Poisson point process Φ_1 with unit intensity, it holds that $\kappa_d \underline{r}_c(\Phi_1)^d \geq 1$ where κ_d is the volume of the unit ball in $\mathbb{R}^d$.*

In order to see void probabilities in action it is customary to use some discrete approximations of the continuum percolation model. For $r > 0, x \in \mathbb{R}^d$, define the following subsets of $\mathbb{R}^d$. Let $Q^r = (-\frac{1}{2r}, \frac{1}{2r}]^d$ and $Q^r(x) = x + Q^r$. We consider the following discrete graph parametrised by $n \in \mathbb{N}$. Let $\mathbb{L}_n^{*d} = (\mathbb{Z}_n^d, \mathbb{E}_n^{*d})$ be the usual close-packed lattice graph scaled down by the factor $1/n$. It holds that $\mathbb{Z}_n^d = \frac{1}{n}\mathbb{Z}^d$ for the set of vertices and $\mathbb{E}_n^{*d} = \{\langle z_i, z_j \rangle \in (\mathbb{Z}_n^d)^2 : Q^{\frac{n}{2}}(z_i) \cap Q^{\frac{n}{2}}(z_j) \neq \emptyset\}$ for the set of edges, where $\mathbb{Z}$ denotes the set of integers.

A *contour* in $\mathbb{L}_n^{*d}$ is a minimal collection of vertices such that any infinite path in $\mathbb{L}_n^{*d}$ from the origin has to contain one of these vertices (the minimality condition implies that the removal of any vertex from the collection will lead to the existence of an infinite path from the origin without any intersection with the remaining vertices in the collection). Let Γ_n be the set of all contours around the origin in $\mathbb{L}_n^{*d}$. For any subset of points $\gamma \subset \mathbb{R}^d$, in particular for paths $\gamma \in \Gamma_n$, we define $Q_\gamma = \bigcup_{z \in \gamma} Q^n(z)$.

With this notation, we can define the "upper" critical radius $\bar{r}_c(\Phi)$ by

$$\bar{r}_c(\Phi) = \inf\left\{ r > 0 : \text{for all } n \geq 1, \sum_{\gamma \in \Gamma_n} \mathbf{P}\left(C(\Phi, r) \cap Q_\gamma = \emptyset\right) < \infty \right\}. \qquad (2.14)$$

It might be seen as the critical radius corresponding to the phase transition when the discrete model $\mathbb{L}_n^{*d} = (\mathbb{Z}_n^d, \mathbb{E}_n^{*d})$, approximating $C(\Phi, r)$ with an arbitrary precision, starts percolating through the Peierls argument. As a consequence, $\bar{r}_c(\Phi) \geq r_c(\Phi)$ (see [43, Lemma 4.1]). The following ordering result follows immediately from the definitions.

Corollary 2.1. *Let $C_j = C(\Phi_j, r)$, $j = 1, 2$ be two germ-grain models with simple point processes of germs Φ_j, $j = 1, 2$. If Φ_1 has smaller voids probabilities then Φ_2, then $\bar{r}_c(\Phi_1) \leq \bar{r}_c(\Phi_2)$.*

Remark 2.6. The finiteness of $\bar{r}_c$ is not clear even for Poisson point process and hence Corollary 2.1 cannot be directly used to prove the finiteness of the critical radii of v-weakly sub-Poisson point processes. However, the approach based on void probabilities can be refined, as we shall see in what follows, to conclude the aforementioned property.

2.4.1.2 Percolation of Level-Sets of Shot-Noise Fields

Various percolation problems, including the classical continuum percolation model considered in the previous section, can be posed as percolation of some level sets of shot-noise fields. We say that a *level set percolates* if it has an unbounded connected subset. We now present a useful lemma that can be used in conjunction with other methods, in particular the famous *Peierls argument* (cf. [152, pp. 17–18]), to exhibit percolation of the level sets of shot-noise fields for an appropriate choice of parameters. We sketch these arguments in the simple case of the germ-grain model and the application to the SINR model in the subsequent sections. The proofs rely on coupling of a discrete model with the continuum model and showing percolation or non-percolation in the discrete model using the above bounds.

In what follows we will be interested in level-sets of shot-noise fields, i.e. sets of the form $\{y \in S : V_\Phi(y) \geq a\}$ or $\{y \in S : V_\Phi(y) \leq a\}$ for some $a \in \mathbb{R}$, where $\{V_\Phi(y), y \in S\}$ is a shot-noise field generated by some point process Φ with a non-negative response function h as introduced in Definition 2.14. For proving results on percolation of level-sets, we rely heavily on the bounds from the following lemma.

Lemma 2.1 ([44, Lemma 3.2]). *Let Φ be a stationary point process with positive and finite intensity λ. Then the following statements are true.*

1. If Φ is α-weakly sub-Poisson, then

$$\mathbf{P}\left(\min_{i \in \{1,\dots,m\}} V_\Phi(y_i) \geq a\right) \leq e^{-sma} \exp\left(\lambda \int_{\mathbb{R}^d} (e^{s\sum_{i=1}^m h(x,y_i)} - 1)\mathrm{d}x\right) \quad (2.15)$$

for any $y_1, \dots, y_m \in S$ and $s > 0$.
2. If Φ is ν-weakly sub-Poisson then,

$$\mathbf{P}\left(\max_{i \in \{1,\dots,m\}} V_\Phi(y_i) \leq a\right) \leq e^{sma} \exp\left(\lambda \int_{\mathbb{R}^d} (e^{-s\sum_{i=1}^m h(x,y_i)} - 1)\mathrm{d}x\right) \quad (2.16)$$

for any $y_1, \dots, y_m \in S$ and $s > 0$.

Proof. Observe that $\sum_{i=1}^m V_\Phi(y_i) = \sum_{X \in \Phi} \sum_{i=1}^m h(X, y_i)$ is itself a shot-noise field driven by the response function $\sum_{i=1}^m h(., y_i)$. Thus if (2.15) and (2.16) are true for $m = 1$, we can derive the general case as well. Using Chernoff's bound, we have that $\mathbf{P}(V_\Phi(y) \geq a) \leq e^{-sa} \mathbf{E} e^{sV_\Phi(y)}$ and $\mathbf{P}(V_\Phi(y) \leq a) \leq e^{sa} \mathbf{E} e^{-sV_\Phi(y)}$. The two statements now follow from Propositions 2.2 and 2.3 respectively. $\qquad\square$

2.4.1.3 k-Percolation in the Germ-Grain Model

By k-percolation in a germ-grain model, we understand percolation of the subset of the space covered by at least k grains of the germ-grain model; cf. Definition 2.15. Our aim is to show that for sub-Poisson point processes (i.e. point processes that are dcx-smaller than the Poisson point process) or negatively associated point processes, the critical connection radius r for k-percolation of the germ-grain model is non-degenerate, i.e., the model does not percolate for r too small and percolates for r sufficiently large. As will be seen in the proof given below, the finiteness of the critical radius r for $k = 1$ (i.e. the usual percolation) holds under a weaker assumption of ordering of void probabilities.

Given a point processes of germs Φ on $\mathbb{R}^d$, we define the coverage field $\{V_{\Phi,r}(x), x \in \mathbb{R}^d\}$ by $V_{\Phi,r}(x) = \sum_{X \in \Phi} \mathbf{1}(x \in B_r(X))$, where $B_r(x)$ denotes the Euclidean ball of radius r centred at x. The k-covered set is defined as the following level set. Let $C_k(\Phi,r) = \{x \in \mathbb{R}^d : V_{\Phi,r}(x) \geq k\}$. Note that $C_1(\Phi,r) = C(\Phi,r)$ is the usual germ-grain model. For $k \geq 1$, define the *critical radius for k-percolation* as

$$r_c^k(\Phi) = \inf\{r : \mathbf{P}(C_k(\Phi,r) \text{ percolates}) > 0\},$$

where, as before, percolation means existence of an unbounded connected subset. Clearly, $r_c^1(\Phi) = r_c(\Phi) \leq r_c^k(\Phi)$. As in various percolation models, the first question is whether $0 < r_c^k(\Phi) < \infty$? This is known for the Poisson point process ([141]) and not for many other point processes apart from that. The following result is a first step in that direction answering the question in affirmative for many point processes.

Proposition 2.14 ([44, Proposition. 3.4]). *Let Φ be a stationary point processes with intensity λ. For $k \geq 1, \lambda > 0$, there exist constants $c(\lambda)$ and $C(\lambda,k)$ (not depending on the distribution of Φ) such that $0 < c(\lambda) \leq r_c^1(\Phi)$ provided Φ is α-weakly sub-Poisson and $r_c^k(\Phi) \leq C(\lambda,k) < \infty$ provided Φ is v-weakly sub-Poisson. Consequently, for Φ being weakly sub-Poisson, combining both the above statements, it turns out that*

$$0 < c(\lambda) \leq r_c^1(\Phi) \leq r_c^k(\Phi) \leq C(\lambda,k) < \infty.$$

Remark 2.7. The above result not only shows non-triviality of the critical radius for stationary weakly sub-Poisson processes but also provides uniform bounds. Examples of particular point processes for which this non-triviality result holds are determinantal point processes with trace-class integral kernels (cf. Definition 2.6 and Example 2.12) and Voronoi-perturbed latices with convexly sub-Poisson replication kernels (cf. Example 2.3). For the case of zeros of Gaussian analytic functions which is not covered by us, a non-trivial critical radius for continuum percolation has been shown in [138], where uniqueness of infinite clusters for both zeros of Gaussian analytic functions and the Ginibre point process is also proved.

Sketch of Proof (of Proposition 2.14). A little more notation is required. For $r > 0, x \in \mathbb{R}^d$, define the following subsets of $\mathbb{R}^d$. Let $Q_r = (-r,r]^d$ and $Q_r(x) = x + Q_r$. Furthermore, we consider the following discrete graph. Let $\mathbb{T}^{*d}(r) = (r\mathbb{Z}^d, \mathbb{E}^{*d}(r))$ be a close-packed graph on the scaled-up lattice $r\mathbb{Z}^d$; the edge-set is given by $\mathbb{E}^{*d}(r) = \{\langle z_i, z_j \rangle \in (r\mathbb{Z}^d)^2 : Q_r(z_i) \cap Q_r(z_j) \neq \emptyset\}$. Recall that by site-percolation in a graph one means the existence of an infinite connected component in the random subgraph that remains after deletion of sites/vertices as per some random procedure. In order to prove the first statement, let Φ be α-weakly sub-Poisson and $r > 0$. Consider the close-packed lattice $\mathbb{T}^{*d}(3r)$. Define the response function $h_r(x,y) = \mathbf{1}(x \in Q_{3r/2}(y))$ and the corresponding shot-noise field $V_\Phi^r(.)$ on $\mathbb{T}^{*d}(3r)$. Note that if $C(\Phi,r)$ percolates then $\{z : V_\Phi^r(z) \geq 1\}$ percolates on $\mathbb{T}^{*d}(3r)$ as well. We shall now show that there exists a $r > 0$ such that the latter does not hold true. To prove this, we show that the expected number of paths from o of length n in the random subgraph tends to 0 as $n \to \infty$ and Markov's inequality gives that there is no infinite path (i.e. no percolation) in $\mathbb{T}^{*d}(3r) \cap \{z : V_\Phi^r(z) \geq 1\}$. There are $(3^d - 1)^n$ paths of length n from o in $\mathbb{T}^{*d}(2r)$ and the probability that a path $z_i \in r\mathbb{Z}^d, 1 \leq i \leq n$ is open can be bounded from above as follows. Using (2.15) and some further calculations, we get that

$$\mathbf{P}\left(\min_{i \in \{1,\dots,n\}} V_\Phi^r(z_i) \geq 1 \right) \leq (\exp\{-(s + (1 - e^s)\lambda(3r)^d)\})^n.$$

So, the expected number of paths from o of length n in $\mathbb{T}^{*d}(3r) \cap \{z : V_\Phi^r(z) \geq 1\}$ is at most $((3^d - 1)\exp\{-(s + (1 - e^s)\lambda(3r)^d)\})^n$. This term tends to 0 as $n \to \infty$ for r small enough and s large enough. Since the choice of r depends only on λ, we have shown that there exists a constant $c(\lambda) > 0$ such that $c(\lambda) \leq r_c^1(\Phi)$. For the upper bound, let Φ be ν-weakly sub-Poisson and consider the close-packed lattice $\mathbb{T}^{*d}(\frac{r}{\sqrt{d}})$. Define the response function $h_r(x,y) = \mathbf{1}(x \in Q_{\frac{r}{\sqrt{d}}}(y))$ and the corresponding additive shot-noise field $V_\Phi^r(\cdot)$. Note that if the random subgraph $\mathbb{T}^{*d}(\frac{r}{\sqrt{d}}) \cap \{z : V_\Phi^r(z) \geq k\}$ percolates, then $C(\Phi, r)$ also percolates. Then, the following exponential bound is obtained by using (2.16) and some more calculations:

$$\mathbf{P}(\max_{i \in \{1,\dots,n\}} V_\Phi^r(z_i) \leq k - 1) \leq (\exp\{-((1 - e^{-s})\lambda(\frac{r}{\sqrt{d}})^d - s(k-1))\})^n.$$

It now suffices to use the standard Peierls argument (cf. [152, pp. 17–18]) to complete the proof. $\qquad\square$

For $k = 1$; i.e., for the usual percolation in the germ-grain model, we can avoid the usage of the exponential estimates of Lemma 2.1 and work directly with void probabilities and factorial moment measures. This leads to improved bounds on the critical radius.

Proposition 2.15 ([44, Corollary 3.11]).

For a stationary weakly sub-Poisson point process Φ on $\mathbb{R}^d$, $d \geq 2$, it holds that

$$0 < \frac{1}{(\lambda \kappa_d)^{1/d}} \leq r_c(\Phi) \leq \sqrt{d}\left(\frac{\log(3^d - 2)}{\lambda} \right)^{1/d} < \infty,$$

where κ_d is the volume of the unit ball in $\mathbb{R}^d$. The lower and the upper bounds hold, respectively, for α-weakly and ν-weakly sub-Poisson processes.

Remark 2.8. Applying the above result to an α-weakly sub-Poisson point processes with unit intensity we observe that $r_c(\Phi) \geq \kappa_d^{-\frac{1}{d}} \to \infty$ as $d \to \infty$. This means that in high dimensions, it holds that $r_c(\Phi) \gg r_c(\mathbb{Z}^d) = \frac{1}{2}$, i.e. like the Poisson point process and even sub-Poisson point processes percolate much worse compared to the Euclidean lattice in high dimensions.

However, the question remains open whether the initial heuristic reasoning saying that more clustering worsens percolation (cf. p. 56) holds for sub-Poisson point processes. As we have seen and shall see below, sub-Poisson point processes are more tractable than super-Poisson point processes in many respects.

2.4.1.4 SINR Percolation

For a detailed background about this model of wireless communications, we refer to [42] and the references therein. Here we directly begin with a formal introduction to the model. We shall work only in $\mathbb{R}^2$ in this section.

The parameters of the model are the non-negative numbers P (signal power), N (environmental noise), γ, T (SINR threshold) and an attenuation function $\ell : \mathbb{R}_+^2 \to \mathbb{R}_+$ satisfying the following assumptions: $\ell(x,y) = l(|x-y|)$ for some continuous function $l : \mathbb{R}_+ \to \mathbb{R}_+$, strictly decreasing on its support, with $l(0) \geq TN/P$, $l(\cdot) \leq 1$, and $\int_0^\infty xl(x)\mathrm{d}x < \infty$. These are exactly the assumptions made in [105] and we refer to this article for a discussion on their validity.

Given a point processes Φ, the *interference* generated due to the point processes at a location x is defined as the following shot-noise field $\{I_\Phi(x), x \in \mathbb{R}^d\}$, where $I_\Phi(x) = \sum_{X \in \Phi \setminus \{x\}} l(|X-x|)$. Define the signal-to-noise ratio (SINR) as follows :

$$\mathrm{SINR}(x,y,\Phi,\gamma) = \frac{Pl(|x-y|)}{N + \gamma P I_{\Phi \setminus \{x\}}(y)}. \tag{2.17}$$

Let Φ_B and Φ_I be two point processes. Furthermore, let $P, N, T > 0$ and $\gamma \geq 0$. The SINR graph is defined as $\mathbb{T}(\Phi_B, \Phi_I, \gamma) = (\Phi_B, \mathbb{E}(\Phi_B, \Phi_I, \gamma))$ where

$$\mathbb{E}(\Phi_B, \Phi_I, \gamma) = \{\langle X, Y \rangle \in \Phi_B^2 : \min\{SINR(Y,X,\Phi_I,\gamma), SINR(X,Y,\Phi_I,\gamma)\} > T\}.$$

The SNR graph (i.e. the graph without interference, $\gamma = 0$) is defined as $\mathbb{T}(\Phi_B) = \mathbb{T}(\Phi_B, \emptyset, 0)$ and this is nothing but the germ-grain model $C(\Phi_B, r_l)$ with $2r_l = l^{-1}(\frac{TN}{P})$.

Recall that percolation in the above graphs is the existence of an infinite connected component in the graph-theoretic sense. Denote by $\lambda_c(r) = \lambda(r_c(\Phi_\lambda)/r)^2$ the *critical intensity* for percolation of the germ-grain model $C(\Pi_\lambda, r)$. There is much more dependency in this graph than in the germ-grain model where the edges depend only on the two corresponding vertices, but still we are able to suitably modify our techniques to obtain interesting results about non-trivial phase-transition in this model. More precisely, we are showing non-trivial percolation in SINR models with weakly sub-Poissonian set of transmitters and interferers and thereby considerably extending the results of [105]. In particular, the set of transmitters and interferers could be stationary determinantal point processes or sub-Poisson perturbed lattices and the following result still guarantees non-trivial phase transition in the model.

Proposition 2.16 ([44, Propositions 3.9 and 3.10]). *The following statements are true.*

1. *Let $\lambda > \lambda_c(r_l)$ and let Φ be a stationary α-weakly sub-Poisson point process with intensity μ for some $\mu > 0$. Then there exists $\gamma > 0$ such that $\mathbb{T}(\Pi_\lambda, \Phi, \gamma)$ percolates.*
2. *Let Φ be a stationary, γ-weakly sub-Poisson point processes and let Φ_I be a stationary α-weakly sub-Poisson point process with intensity μ for some $\mu > 0$.*

Furthermore, assume that $l(x) > 0$ for all $x \in \mathbb{R}_+$. Then there exist $P, \gamma > 0$ such that $\mathbb{T}(\Phi, \Phi_l, \gamma)$ percolates.

Exercise 2.13. Two related percolation models are the k-nearest neighbourhood graph (k-NNG) and the random connection model. Non-trivial phase transitions for percolation is shown for both models when defined on a Poisson point process in [126, Sect. 2.4 and 2.5]. We invite the reader to answer the challenging question of whether the methods of [126] combined with ours could be used to show non-trivial phase transition for weakly sub-Poisson point processes, too. *Hint.* The results of [92] on the Lilypond model could be useful to show non-percolation for 1-NNG.

2.4.2 U-Statistics of Point Processes

Denote by $\Phi^{(k)}(\mathrm{d}(x_1,\ldots,x_k)) = \Phi(\mathrm{d}x_1)\,(\Phi \setminus \delta_{x_1})\,(\mathrm{d}x_2)\ldots(\Phi \setminus \sum_{i=1}^{k-1}\delta_{x_i})\,(\mathrm{d}x_k)$ the (empirical) k-th order factorial moment measure of Φ. Note that this is a point process on $\mathbb{R}^{dk}$ with mean measure $\alpha^{(k)}$. In case when Φ is simple, $\Phi^{(k)}$ is simple too and corresponds to the point process of ordered k-tuples of distinct points of Φ.

In analogy with classical U-statistics, a U-statistic of a point process Φ can be defined as the functional $F(\Phi) = \sum_{X \in \Phi^{(k)}} f(X)$, for a non-negative symmetric function f ([330]). In case when Φ is infinite one often considers $F(\Phi \cap W)$, where $W \subset \mathbb{R}$ is a bounded Borel set. The reader is referred to [330] or [358, Sect. 2] for many interesting U-statistics of point processes of which two — subgraphs in a random geometric graph ([314, Chap. 3]) and simplices in a random geometric complex (see [427, Sect. 8.4.4]) — are described below.

2.4.2.1 Examples

Example 2.20 (Subgraphs counts in a random geometric graph). Let Φ be a finite point process and $r > 0$. The random geometric graph $\mathbb{T}(\Phi, r) = (\Phi, \mathbb{E}(\Phi, r))$ is defined through its vertices and edges, where the edge set is given by $\mathbb{E}(\Phi, r) = \{(X,Y) \in \Phi : |X - Y| \leq r\}$. For a connected subgraph Γ on k vertices, define $h : \mathbb{R}^{dk} \to \{0,1\}$ by $h_{\Gamma,r}(x) = \mathbf{1}(\mathbb{T}(x,r) \cong \Gamma)$, where $\cong$ stands for graph isomorphism. Now, the number of Γ-induced subgraphs in Φ is defined as

$$G_r(\Phi, \Gamma) = \frac{1}{k!} \sum_{X \in \Phi^{(k)}} h_{\Gamma,r}(X).$$

Clearly, $G_r(\Phi, \Gamma)$ is a U-statistic of Φ. In the special case that $|\Gamma| = 2$, then $G_r(\Phi, \Gamma)$ is the number of edges.

Example 2.21 (Simplices counts in a random geometric simplex). A non-empty family of finite subsets $\Delta(S)$ of a set S is an *abstract simplicial complex* if for every set $X \in \Delta(S)$ and every subset $Y \subset X$, we have that $Y \in \Delta(S)$. We shall from now on drop the adjective "abstract". The elements of $\Delta(S)$ are called *faces resp. simplices* of the simplicial complex and the dimension of a face X is $|X| - 1$. Given a finite point process Φ, one can define the following two simplicial complexes : The *Čech complex* $Ce(\Phi, r)$ is defined as the simplicial complex whose k-faces are $\{X_0, \ldots, X_k\} \subset \Phi$ such that $\cap_i B_r(X_i) \neq \emptyset$. The *Vietoris-Rips complex* $VR(\Phi, r)$ is defined as the simplicial complex whose k-faces are $\{X_0, \ldots, X_k\} \subset \Phi$ such that $B_r(X_i) \cap B_r(X_j) \neq \emptyset$ for all $0 \leq i \neq j \leq k$. The 1-skeleton (i.e. the subcomplex consisting of all 0-faces and 1-faces) of the two complexes are the same and it is nothing but the random geometric graph $\mathbb{T}(\Phi, r)$ of Example 2.20. Also, the Čech complex $Ce(\Phi, r)$ is homotopy equivalent to the germ-grain model $C(\Phi, r)$. The number of k-faces in the two simplicial complexes can be determined as follows:

$$S_k(Ce(\Phi, r)) = \frac{1}{k!} \sum_{X \in \Phi^{(k)}} \mathbf{1}(X \text{ is a } k\text{-face of } Ce(\Phi, r))$$

and similarly for $S_k(VR(\Phi, r))$. Clearly, both characteristics are examples of U-statistics of Φ.

Note that a U-statistic is an additive shot-noise of the point process $\Phi^{(k)}$ and this suggests the applicability of our theory to U-statistics. Speaking a bit more generally, consider a family of U-Statistics K and define an additive shot-noise field $\{F(\Phi)\}_{F \in K}$ indexed by K. Why do we consider such an abstraction? Here is an obvious example.

Example 2.22. Consider $K = \{F_B, F_B(\Phi) = \sum_{X \in \Phi^{(k)}} \mathbf{1}(f(X) \in B)\}_{B \in B_0(\mathbb{R}_+)}$ for some given non-negative symmetric function f defined on $\mathbb{R}^k$. Then the additive shot-noise field on $\Phi^{(k)}$, indexed by bounded Borel sets $B \in \mathbb{R}_+$, defined above is nothing but the random field characterising the following point process on $\mathbb{R}_+$ associated to the U-statistics of Φ:

$$\eta_{(f,\Phi)} = \{f(X) : X \in \Phi^{(k)}\}$$

in the sense that $F_B(\Phi) = \eta_{(f,\Phi)}(B)$. (Note that if $|f^{-1}([0,x])| < \infty$ for all $x \in \mathbb{R}_+$, then η_Φ is indeed locally finite and hence a point process and we always assume that f satisfies such a condition.) This point process has been studied in [358], in the special case when $\Phi = \Pi_\lambda$ is a stationary Poisson point process. It is shown that if $\lambda \to \infty$, then $\eta_{(f,\Phi)}$ tends to a Poisson point process with explicitly known intensity measure.

For any U-statistic F and for a bounded window $W \subset \mathbb{R}^d$, we have that

$$\mathbf{E}F(\Phi \cap W) = \frac{1}{k!} \int_W f(x_1, \ldots, x_k) \alpha^{(k)} d(x_1, \ldots, x_k).$$

Similarly, we can express higher moments of the shot-noise field $\{F(\Phi)\}_{F \in K}$ by those of Φ. With these observations in hand and using Proposition 2.9, we can state the following result.

Proposition 2.17. *Let Φ_1, Φ_2 be two point processes with respective factorial moment measures $\alpha_i^{(k)}$, $i = 1, 2$ and let W be a bounded Borel set in $\mathbb{R}^d$. Consider a family* K *of U-statistics. Then the following statements are true.*

1. *If $\alpha_1^{(j)}(\cdot) \leq \alpha_2^{(j)}(\cdot)$ for all $1 \leq j \leq k$, then*

$$\mathbf{E}(F_1(\Phi_1 \cap W)F_2(\Phi_1 \cap W)\ldots F_k(\Phi_1 \cap W))$$
$$\leq \mathbf{E}(F_1(\Phi_2 \cap W)F_2(\Phi_2 \cap W)\ldots F_k(\Phi_2 \cap W))$$

   for any k-tuple of U-statistics $F_1, \ldots, F_k \in$ K. In particular, for any given $F \in$ K based on a non-negative symmetric function f it holds that $\alpha_{\eta_{(f,\Phi_1)}}^{(k)}(\cdot) \leq \alpha_{\eta_{(f,\Phi_2)}}^{(k)}(\cdot)$, where $\alpha_{\eta_{(f,\Phi_i)}}^{(k)}$ is the k-th order factorial moment measure of $\eta_{(f,\Phi_i)}$.
2. *If $\Phi_1 \leq_{\mathrm{dcx}} \Phi_2$, then $\{F(\Phi_1 \cap W)\}_{F \in \mathsf{K}} \leq_{\mathrm{idcx}} \{F(\Phi_2 \cap W)\}_{F \in \mathsf{K}}$ and in particular $\eta_{(f,\Phi_1)} \leq_{\mathrm{idcx}} \eta_{(f,\Phi_2)}$.*

2.4.2.2 Some Properties of Random Geometric Graphs

The subgraph count $G_r(\cdot, \cdot)$ considered in Example 2.20 is only a particular example of a U-statistic but its detailed study in the case of the Poisson point process (see [314, Chap. 3]) was the motivation to derive results about subgraph counts of a random geometric graph over other point processes ([426]). Here, we explain some simple results about clique numbers, maximal degree and chromatic number that can be deduced as easy corollaries of ordering of subgraph counts known due to Proposition 2.17.

Slightly differing from [314], we consider the following asymptotic regime for a stationary point process Φ with unit intensity. We look at the properties of $\mathbb{T}(\Phi_n, r_n)$, $n \geq 1$, where $\Phi_n = \Phi \cap W_n$ with $W_n = [-\frac{n^{\frac{1}{d}}}{2}, \frac{n^{\frac{1}{d}}}{2}]^d$ and a radius regime r_n. To compare our results with those of [314], replace the r_n^d factor in our results by nr_n^d. Detailed asymptotics of $G_n(\Phi_n, \Gamma) = G_{r_n}(\Phi_n, \Gamma)$ for general stationary point processes have been studied in [426, Sect. 3].

Let $C_n = C_n(\Phi), \Delta_n = \Delta_n(\Phi), \mathcal{X}_n = \mathcal{X}_n(\Phi)$ denote the size of the largest clique, maximal vertex degree and chromatic number of $\mathbb{T}(\Phi_n, r_n)$, respectively. Heuristic arguments for these quantities say that they should increase with more clustering in the point process. We give a more formal statement of this heuristic at the end of this section.

Let Γ_k denote the complete graph on k vertices and $\Gamma_1', \ldots, \Gamma_m'$ be the maximum collection of non-isomorphic graphs on k vertices having maximum degree $k - 1$. Then for $k \geq 1$, we have the following two equalities and the graph-theoretic inequality that drive the result following them: $\{C_n < k\} = \{G_n(\Phi_n, \Gamma_k) = 0\}$, $\{\Delta_n < k - 1\} = \cap_{i=1}^{m}\{G_n(\Phi_n, \Gamma_i') = 0\}$, $C_n \leq \mathcal{X}_n \leq \Delta_n + 1$.

Corollary 2.2. *Let Φ be a stationary α-weakly sub-Poisson point process with unit intensity. If $nr_n^{d(k-1)} \rightarrow 0$, then*

$$\lim_{n\to\infty} \mathbf{P}(\mathcal{C}_n < k) = \lim_{n\to\infty} \mathbf{P}(\Delta_n < k-1) = \lim_{n\to\infty} \mathbf{P}(\mathcal{X}_n < k) = 1.$$

Sketch of Proof. To prove the result for $\mathcal{C}_n$, due to Markov's inequality and the fact that $\mathbf{E}G_n(\Phi_n,\Gamma) \le \mathbf{E}G_n(\Pi_{1_n},\Gamma)$ (see Proposition 2.17), it suffices to show that $\mathbf{E}G_n(\Pi_{1_n},\Gamma) \to 0$ for $nr_n^{d(k-1)} \to 0$ and any graph Γ on k vertices. This is already known from [314, Theorem 6.1]. The proof for Δ_n is similar and it also proves the result for $\mathcal{X}_n$. $\qquad\square$

Note that if $k = 2$, the results of Corollary 2.2 implies that $\lim_{n\to\infty} \mathbf{P}(\mathcal{C}_n = 1) = 1$ by using the trivial lower bound of $\mathcal{C}_n \ge 1$, and analogously for the other two quantities. To derive a similar result for $k \ge 2$, we need variance bounds for $G_n(\cdot,\cdot)$ to use the standard second moment method. These variance bounds for $G_n(\cdot,\cdot)$ are available in the case of negatively associated point processes (see [426, Sect. 3.4]).

Further, for point processes with $\alpha^{(k)}$ admitting a continuous density $\alpha^{(k)}(\mathrm{d}(x_1,\ldots,x_k)) = \rho^{(k)}(x_1,\ldots,x_k)\mathrm{d}x_1,\ldots,\mathrm{d}x_k$ in the neighbourhood of $(0,\ldots,0)$, such that $\rho^{(k)}(0,\ldots,0) = 0$ (for example, α-weakly sub-Poisson point processes such as the Ginibre point process, perturbed lattice, zeros of Gaussian analytic function et al.) we know from [426, Sect. 3.2] that $\frac{\mathbf{E}G_n(\Phi_n,\Gamma_k)}{nr_n^{d(k-1)}} \to 0$. Using this result, we can show that $\lim_{n\to\infty} \mathbf{P}(\mathcal{C}_n < k) \to 1$ even for $nr_n^{d(k-1)} \to \lambda > 0$. This is not true for the Poisson point process (see [314, Theorem 6.1]). Thus, we have the following inequality for point processes with $\rho^{(k)}(0,\ldots,0) = 0$:

$$\lim_{n\to\infty} \mathbf{P}(\mathcal{C}_n(\Pi_1) < k) \le \lim_{n\to\infty} \mathbf{P}(\mathcal{C}_n(\Phi) < k). \tag{2.18}$$

This inequality can be easily concluded from the fact that for radius regimes with $nr_n^{d(k-1)} \to \lambda \ge 0$, the expression on the right-hand side of (2.18) is equal to 1 while for the radius regime $nr_n^{d(k-1)} \to \infty$, the expression on the left-hand side of 2.18 is equal to 0 (see [314, Theorem 6.1]). In vague terms, we can rephrase the above inequality as that $\mathcal{C}_n(\Pi_1)$ is "stronger ordered" than $\mathcal{C}_n(\Phi)$ in the limit, i.e., the Poisson point process is likely to have a larger clique number than a point process with $\rho^{(k)}(0,\ldots,0) = 0$. Similar "strong ordering" results for Δ_n's and $\mathcal{X}_n$'s matching well with heuristics can also be derived.

2.4.3 Random Geometric Complexes

We have already noted in Example 2.21 that the number of k-faces in Čech and Vietoris-Rips complexes on point processes are U-statistics. In this section we will further describe the topological properties of these random geometric complexes. In the same manner as simplicial complexes are considered to be topological extensions of graphs, so are random geometric complexes to random geometric graphs. Random geometric graphs on Poisson or binomial point processes are a well-researched subject with many applications (see [314]). Motivated by research in topological data analysis ([64, 139]) and relying on results from random geometric graphs, random geometric complexes on Poisson or binomial point processes have been studied recently [214]. In [426], the investigation of the topology of random geometric complexes has been extended to a wider class of stationary point processes using tools from stochastic ordering of point processes. However, we shall content ourselves with just explaining one of the key phase-transition results given in [426].

One of the first steps towards the understanding of Čech and Vietoris-Rips complexes is to understand the behaviour of their *Betti numbers* $\beta_k(\cdot), k \geq 0$ as functions of r. Informally speaking, the k-th ($k \geq 1$) Betti number counts the number of $k+1$-dimensional holes in the appropriate Euclidean embedding of the simplicial complex. The 0-th Betti number is the number of connected components in the simplicial complex, β_1 is the number of two-dimensional or "circular" holes, β_2 is the number of three-dimensional voids, etc. If $\beta_0(\cdot) = 1$, then we say that the simplicial complex is connected. Unlike simplicial counts, Betti numbers are not U-statistics.

Regarding the dependence of the Betti number $\beta_k(r)$ of Čech complexes on r, for $k \geq 1$, unlike in earlier percolation models, there are *two phase-transitions* happening in this case: $\beta_k(r)$ goes from zero to positive (the complex "starts creating" $k+1$-dimensional holes) and, alternatively, it goes from positive to zero (the holes are "filled in"). If one were to think about the relative behaviours of the Betti numbers $\beta_k(\cdot)$, $k \geq 1$, of the Čech complexes on two point process Φ_1, Φ_2 where Φ_1 is "less clustered" than Φ_2, then it should be possible that the first threshold decreases with clustering and the second threshold increases with clustering. Indeed, depending on the strength of the result we require, weak sub-Poissonianity or negative association turn out to be the right notion to prove the above heuristic more rigorously.

Let Φ be a stationary weakly sub-Poisson point process with unit intensity and let $W_n = n^{1/d}[-\frac{1}{2}, \frac{1}{2}]^d$. Let $r_n \geq 0, n \geq 1$ be the corresponding radius regime with $\lim_n r_n \in [0, \infty]$. Based on whether $\lim_n r_n$ is $0, \infty$ or a constant between 0 and ∞, we shall get different scaling limits for the Betti numbers. Under certain technical assumptions, there is a function $f^k(\cdot)$ depending on the joint intensities of the point process ($f^k(r) \to 0$ as $r \to 0$ for point processes for which $\rho^{(k)}(0, \ldots, 0) = 0$ and otherwise $f^k(\cdot) \equiv 1$) such that the following statements hold for $k \geq 1$. 1. If

$$r_n^{d(k+1)} f^{k+2}(r_n) = o(n^{-1}) \text{ or } r_n^d = \omega(\log n),$$

then with high probability $\beta_k(Ce(\Phi \cap W_n, r_n)) = 0$. 2. Let Φ be negatively associated. If

$$r_n^{d(k+1)} f^{k+2}(r_n) = \omega(n^{-1}) \text{ and } r_n^d = O(1),$$

then with high probability $\beta_k(Ce(\Phi_n, r_n)) \neq 0$.

To see how the above statements vindicate the heuristic described before, consider the first threshold for the appearance of Betti numbers (with high probability). In the Poisson case ($f^k(\cdot) \equiv 1$) this threshold is exactly $r_n \to 0$ and $nr_n^{d(k+1)} \to \infty$, whereas for weakly sub-Poisson point processes with $\rho^{(k)}(0, \ldots, 0) = 0$ ($f^k(r) \to 0$

as $r \to 0$), this is at least $r_n \to 0$ and $nr_n^{d(k+1)} \to \infty$, i.e., the threshold is larger for these weakly sub-Poisson point processes. For specific point processes such as the Ginibre point process or the zeros of Gaussian analytic functions ($f^k(r) = r^{k(k-1)}$), this threshold is only at $r_n \to 0$ and $nr_n^{(k+1)(k+2)} \to \infty$ which is much larger than that of the Poisson point process.

Now if we consider the second threshold, when Betti numbers vanish, then for the Ginibre point process or the zeros of Gaussian analytic functions, with high probability $\beta_k(Ce(\Phi \cap W_n, r_n)) = 0$ for $r_n^d = \omega(\sqrt{\log n})$. This is of strictly smaller order compared to the Poisson case where $r_n^d = \omega(\log n)$. For other weakly sub-Poisson point processes, the above results imply that the order of the radius threshold for vanishing of Betti numbers cannot exceed $r_n^d = \omega(\log n)$, i.e. that of the Poisson point process. Hence, this second threshold is smaller for weakly sub-Poisson point processes. Negative association just assures positivity of Betti numbers for the intermediate regime of $r_n \to r \in (0, \infty)$. Now if we consider the second threshold, when Betti numbers vanish, then for the Ginibre point process or the zeros of Gaussian analytic functions, with high probability $\beta_k(Ce(\Phi \cap W_n, r_n)) = 0$ for $r_n^d = \omega(\sqrt{\log n})$. This is of strictly smaller order compared to the Poisson case where $r_n^d = \omega(\log n)$. For other weakly sub-Poisson point processes, the above results imply that the order of the radius threshold for vanishing of Betti numbers cannot exceed $r_n^d = \omega(\log n)$, i.e. that of the Poisson point process. Hence, this second threshold is smaller for weakly sub-Poisson point processes. Negative association just assures positivity of Betti numbers for the intermediate regime of $r_n \to r \in (0, \infty)$.

Barring the upper bound for the vanishing of Betti numbers, similar results hold true for the Vietoris-Rips complex, too. One can obtain asymptotics for Euler characteristic of the Čech complex using the Morse-theoretic point process approach. This and asymptotics for Morse critical points on weakly sub-Poisson or negatively associated point processes are also obtained in [426].

Furthermore, analogous to percolation in random geometric graphs, one can study percolation in random Čech complexes by defining the following graph. Any two k-faces of the Čech complex are said to be connected via an edge if both of them are contained in a $(k+1)$-face of the complex. In the case of $k = 0$, the graph obtained is the random geometric graph. The existence of non-trivial percolation radius for any $k \geq 1$ is guaranteed by Proposition 2.14 provided the point process Φ is weakly sub-Poisson. This question was raised in [214, Sect. 4] for the Poisson point process.

2.4.4 Coverage in the Germ-Grain Model

In previous sections of this chapter, we looked at percolation and connectivity aspects of the germ-grain model. Yet another aspect of the germ-grain model that has been well studied for the Poisson point process is coverage [163]. Here, one is concerned with the volume of the set $C_k(\Phi, r)$ defined in Sect. 2.4.1.3. The heuristic is again that the volume of the 1-covered region should decrease with clustering, but the situation would reverse for the k-covered region for large enough k. We are in a position to state a more formal statement once we introduce still another definition. We say that two discrete random variables X, Y are ordered in *uniformly convex variable order* (UCVO)($X \leq_{uv} Y$) if their respective density functions f, g satisfy the following conditions: $\mathrm{supp}(f) \subset \mathrm{supp}(g)$, $f(\cdot)/g(\cdot)$ is an unimodal function but their respective distribution functions are not ordered, i.e. $F(\cdot) \nleq G(\cdot)$ or vice-versa (see [413]) and where $\mathrm{supp}(\cdot)$ denotes the support of a function.

Proposition 2.18 ([45, Proposition 6.2]). *Let Φ_1 and Φ_2 be two simple, stationary point processes such that $\Phi_1(B_O(r)) \leq_{uv} \Phi_2(B_O(r))$ for $r \geq 0$. Then there exists $k_0 \geq 1$ such that for any bounded Borel set $W \subset \mathbb{R}^d$ it holds that*

$$\mathbf{E}v_d(C_k(\Phi_1, r) \cap W) \geq \mathbf{E}v_d(C_k(\Phi_2, r) \cap W) \qquad \text{for all } k : 1 \leq k \leq k_0,$$

and

$$\mathbf{E}v_d(C_k(\Phi_1, r) \cap W) \leq \mathbf{E}v_d(C_k(\Phi_2, r) \cap W) \qquad \text{for all } k > k_0,.$$

For Φ_1, we can take any of the sub-Poisson perturbed lattices presented in Example 2.3 or a determinantal point process (see Definition 2.6) and Φ_2 as a Poisson point process. We can also take Φ_1 to be a Poisson point process and Φ_2 to be any of the super-Poisson perturbed lattices presented in Example 2.3 or a permanental point process (see Definition 2.7).

Note that for a stationary point process Φ, we have that

$$\mathbf{E}v_d(C_k(\Phi, r) \cap W) = \int_W \mathbf{P}(\Phi(B_x(r)) \geq k)\mathrm{d}x = v_d(W)\mathbf{P}(\Phi(B_O(r)) \geq k).$$

It is easy to derive from the above equation that the expected 1-covered region of v-weakly sub-Poisson point processes is larger than that of the Poisson point process [40, Sect. 6.1].

The question of coverage also arises in the SINR model of Sect. 2.4.1.4. We shall not delve further into this question other than remarking that in a certain variant of the SINR model, it has been shown that the coverage and capacity which is defined as $\log(1 + SINR)$ increase with increase in dcx-ordering [425, Sect. 5.2.3].

2.5 Outlook

Let us mention some possible directions for future work. While several examples of point processes comparable to the Poisson point process were presented, a notable absentee from our list are Gibbs point processes, which should appear in the context of modelling of clustering phenomena. In particular, some Gibbs hard-core point processes are expected to be sub-Poisson. Bounds for the probability generating functionals with estimates for Ripley's K-function and the intensity and higher order correlation functions for some stationary locally stable Gibbs point process are given in [385]. Also, some geometric structures on specific Gibbs point processes have already been considered (see e.g. [84, 352]), and these processes perhaps could serve as new reference processes, replacing in this role the Poisson point process. As for today we are not aware of any such results.

The question of other useful orders for comparison of point processes also is worthy of investigation. In [425, Sect. 4.4], it has already been shown that related orders of supermodularity and componentwise convexity are not suitable orders whereas convexity could be useful. It might be interesting to study convex ordering of point processes.

Though we have presented applications to various geometric models, there are many other questions such as the ordering of critical radii for percolation, uniqueness of giant component in continuum percolation (see Remark 2.7), concentration inequalities for more general functionals (see Proposition 2.4), asymptotic analysis of other U-statistics along the lines of Sect. 2.4.3 and 2.4.2.2, etc., which are investigated.

Acknowledgements This chapter is based on research supported in part by Israel Science Foundation 853/10, AFOSR FA8655-11-1-3039 and FP7-ICT-318493-STREP and was written while DY was a post-doc at Technion, Israel.

Chapter 3
Random Tessellations and their Application to the Modelling of Cellular Materials

Claudia Redenbach and André Liebscher

Abstract This chapter introduces various tessellation models and discusses their application as models for cellular materials. First, the notion of a random tessellation, the most well-known model types (Voronoi and Laguerre tessellations, hyperplane tessellations, STIT tessellations), and their basic geometric characteristics are introduced. Assuming that a cellular material is a realisation of a suitable random tessellation model, these characteristics can be estimated from 3D images of the material. It is explained how estimates are obtained and how these characteristics can be used to fit tessellation models to the observed structure. All analysis and modelling steps are illustrated using the example of an open cell aluminium foam.

3.1 Introduction

The construction of vehicles, machines and buildings nowadays includes tasks such as energy saving, lightweight construction and heat insulation. Consequently, the materials used for these purposes are more and more optimised for the particular applications. An important element in this optimisation is the understanding of the influence of a material's microstructure on its macroscopic properties. The rapid development of image acquisition techniques and computing power allows for a virtual design of materials which provides a promising alternative to the construction and experimental testing of prototype materials. 3D images obtained for instance by micro computed tomography (μCT) capture the full spatial information on the microstructure geometry. Hence, geometric characteristics of a material can be es-

Claudia Redenbach
University of Kaiserslautern, Mathematics Department, 67653 Kaiserslautern, Germany e-mail: `redenbach@mathematik.uni-kl.de`

André Liebscher
University of Kaiserslautern, Mathematics Department, 67653 Kaiserslautern, Germany e-mail: `liebscher@mathematik.uni-kl.de`

© Springer International Publishing Switzerland 2015
V. Schmidt (ed.), *Stochastic Geometry, Spatial Statistics and Random Fields,*
Lecture Notes in Mathematics 2120, DOI 10.1007/978-3-319-10064-7_3

timated from the image data. Subsequently, these characteristics can be used to fit a model from stochastic geometry to the material. Changing the model parameters, realisations with altered microstructure can be generated. Prediction of macroscopic properties of these realisations then allows to study microstructure-property relations.

Random closed sets are classical models from stochastic geometry which can be used to describe the microstructure of materials (see [73, 267, 305]). The variety of these models makes them applicable for a wide range of of materials. Processes of lines or cylinders are used to model fibrous materials [340]. Systems of balls are used for porous materials as they appear in the beginning of the sinter process [243] or the system of grains and pores in concrete ([24] and [213], respectively). Boolean models with various kinds of grain shapes are used for example for cast iron (line segments, [73]), calcium ferrite (quadrangles, [361, Chap. XIII]) or the pore phase in carbonate rock (ellipsoids, [7]).

In this chapter we focus on cellular materials whose microstructure is characterised by a system of nearly polyhedral cells (see Fig. 3.1). In particular, we will describe the analysis and modelling of open foams. These materials consist of an open pore space whose cells are separated by a connected system of struts. Their microstructure will be modelled using random tessellations, i.e. space-filling collections of non-overlapping cells. Several random tessellation models have been introduced in the literature [73, 347]. Among these, Laguerre tessellations, a special type of weighted Voronoi tessellations, are of particular interest for the modelling of cellular materials. In the following, we will introduce several random tessellation models and their basic geometric characteristics. We will describe how these characteristics can be estimated from 3D images of open foams and how they can be used to fit tessellation models to the observed structure. All analysis and modelling steps are illustrated using the example of an open aluminium foam.

3.2 Random Tessellations

3.2.1 Definitions

Definition 3.1 (Point process). Let E be a locally compact space with countable basis and Borel σ-algebra $\mathcal{B}$. Denote by $\mathsf{N}(E)$ the set of all locally finite counting measures on E. Equip $\mathsf{N}(E)$ with the σ-algebra $\mathfrak{N}$ generated by the mappings $\varphi \mapsto \varphi(B)$ for $B \in \mathcal{B}$. Then a *point process* is a random variable Φ on a probability space $(\Omega, \mathcal{A}, \mathbf{P})$ taking values in the measurable space $(\mathsf{N}(E), \mathfrak{N})$. The support of Φ forms a locally finite collection of points of E which is identified with Φ.

If $E = \mathbb{R}^d \times M$, where M is a locally compact space with countable basis, we call Φ a marked point process with mark space M. In this case, a point $(x, m) \in \Phi$ is interpreted as a point $x \in \mathbb{R}^d$ to which a mark $m \in M$ is attached.

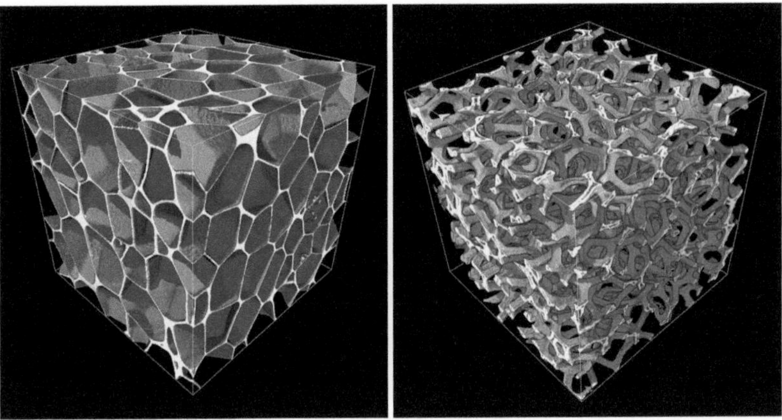

Fig. 3.1 Visualisations of μCT images of a closed polymer foam (left sample: ROHACELL WIND-F RC100, image: Fraunhofer ITWM, 600^3 voxels, voxel edge length 2.72 μm) and an open nickel-chromium foam (right sample: Recemat International, image: RJL Micro & Analytic, 500^3 voxels, voxel edge length 8.93 μm).

Definition 3.2 (Stationarity and isotropy). A point process on $\mathbb{R}^d$ is called *stationary* if its distribution is invariant under translations and *isotropic* if it is invariant under rotations.

Exercise 3.1. Give examples of point processes which are 1. stationary, but not isotropic, 2. isotropic, but not stationary, 3. neither stationary nor isotropic.

Definition 3.3 (Intensity measure). For any point process Φ on $\mathbb{R}^d$, its *intensity measure* $\Lambda : \mathcal{B} \to [0,\infty]$ is defined by

$$\Lambda(B) = \mathbf{E}\Phi(B), \quad B \in \mathcal{B}.$$

For stationary point processes we have $\Lambda = \lambda v_d$, where $\lambda = \mathbf{E}\Phi([0,1]^d)$ is the *intensity* of Φ and v_d denotes the d-dimensional Lebesgue measure.

Definition 3.4 (Stationary Poisson process). A stationary *Poisson process* Φ on $\mathbb{R}^d$ with intensity $\lambda > 0$ is characterised by the following properties:

1. The number of points $\Phi(B)$ contained in a bounded Borel set $B \in \mathcal{B}$ has a Poisson distribution with parameter $\lambda v_d(B)$.
2. For arbitrary $k \in \mathbb{N}$, the numbers of points in k disjoint Borel sets are independent random variables.

Definition 3.5 (Tessellation). A *tessellation* of $\mathbb{R}^d$ is a locally finite collection $T = \{C_i : i \in \mathbb{N}\}$ of compact convex sets C_i with interior points such that $\mathrm{int}(C_i) \cap \mathrm{int}(C_j) = \emptyset$ for $i \neq j$ and $\bigcup_{i \in \mathbb{N}} C_i = \mathbb{R}^d$. Locally finite means that $\#\{C \in T : C \cap B \neq \emptyset\} < \infty$ for all bounded $B \subset \mathbb{R}^d$. The sets $C_i \in T$ are the *cells* of the tessellation T.

By [347, Lemma 10.1.1], the cells are bounded d-dimensional polytopes. In the literature, also tessellation models with non-convex cells are investigated (for examples see [306]). Here, however, we will restrict attention to convex cells.

Definition 3.6 (k-faces). The *faces* of a convex polytope P are the intersections of P with its supporting hyperplanes [346, Sect. 2.4]. Let P be a d-dimensional polytope and $k \in \{0,\ldots,d-1\}$. A k-dimensional face of P is called a *k-face*. Then the 0-faces of P are the *vertices*, the 1-faces the *edges*, and the $(d-1)$-faces the *facets*. For convenience, the polytope P is considered as a d-face. Write $\Delta_k(P)$ for the set of k-faces of a polytope P and $\Delta_k(T) = \bigcup_{C \in T} \Delta_k(C)$ for the set of k-faces of all cells C of T. Furthermore, let

$$F(y) = \bigcap_{C \in T, y \in C} C, \quad y \in \mathbb{R}^d,$$

be the intersection of all cells of the tessellation containing the point y. Then $F(y)$ is a finite intersection of d-polytopes and, since it is non-empty, $F(y)$ is a k-dimensional polytope for some $k \in \{0,\ldots,d\}$. Therefore, we may introduce

$$\mathcal{F}_k(T) = \{F(y) : \dim F(y) = k, y \in \mathbb{R}^d\}, \quad k = 0,\ldots,d,$$

the set of *k-faces of the tessellation* T. A k-face $H \in \Delta_k(T)$ of a cell C of T is the union of all those k-faces $F \in \mathcal{F}_k(T)$ of the tessellation contained in H.

Definition 3.7 (Face-to-face and normal tessellation). A tessellation T is called *face-to-face* if the faces of the cells and the faces of the tessellation coincide, i.e. if $\Delta_k(T) = \mathcal{F}_k(T)$ for all $k = 0,\ldots,d$. For $k = 0$ and $k = d$ this is always true. In the case of face-to-face tessellations we will unify the notation by writing $\mathcal{F}_k(C)$ for the set of k-faces of a cell C of T. A tessellation T is called *normal* if it is face-to-face and every k-face of T is contained in the boundary of exactly $d - k + 1$ cells for $k = 0,\ldots,d-1$.

Exercise 3.2. Construct examples of tessellations which are 1. normal, 2. face-to-face, but not normal, 3. not face-to-face.

We write $\mathbb{T}$ for the set of all tessellations in $\mathbb{R}^d$. It is equipped with the σ-algebra $\mathcal{T}$ generated by the sets

$$\left\{ \{C_i\}_i : \left[\bigcup_i \partial C_i\right] \cap K \neq \emptyset \right\}, \quad K \subset \mathbb{R}^d \text{ compact.}$$

Definition 3.8 (Random tessellation). A *random tessellation* in $\mathbb{R}^d$ is a random variable X on a probability space $(\Omega, \mathcal{A}, \mathbf{P})$ with range $(\mathbb{T}, \mathcal{T})$. It is called *normal* and *face-to-face* if its realisations are almost surely normal and face-to-face, respectively.

The translation and the rotation of a tessellation $T \in \mathbb{T}$ are defined via

$$T + y = \{C + y : C \in T\}, \quad y \in \mathbb{R}^d, \text{ and}$$
$$\vartheta T = \{\vartheta C : C \in T\}, \quad \vartheta \in SO_d.$$

A random tessellation is called *stationary* if its distribution is invariant under translations and *isotropic* if it is invariant under rotations.

Stationary random tessellations contain infinitely many cells. For the definition of geometric tessellation characteristics, a finite subsample of the aggregate of cells is investigated. For this purpose, only cells with centroid in a given reference set are considered.

Definition 3.9 (Centroid of a compact set). Write $\mathcal{C}'$ for the system of compact non-empty sets in $\mathbb{R}^d$. Let $c : \mathcal{C}' \to \mathbb{R}^d$ be a measurable function such that

$$c(C+y) = y + c(C), \quad y \in \mathbb{R}^d, \quad C \in \mathcal{C}'. \tag{3.1}$$

The point $c(C)$ is called the *centroid* of the set $C \in \mathcal{C}'$.

Typical choices of centroids are the centre of gravity of the set C, the centre of its surrounding ball, or the "extreme" point of C with respect to a given direction. Let c_k denote a centroid function acting on the set of k-faces of a random tessellation X. Then we can define the point process Φ_k of centres of the k-faces of X as

$$\Phi_k(X) = \sum_{F \in \mathcal{F}_k(X)} \delta_{c_k(F)}.$$

Stationarity of the random tessellation X and property (3.1) imply stationarity of the point processes $\Phi_k(X)$. The intensity γ_k of Φ_k is given by the formula

$$\gamma_k = \mathbf{E}\left(\sum_{F \in \mathcal{F}_k(X)} \mathbf{1}_{[0,1]^d}(c_k(F)) \right), \quad k = 0, \ldots, d,$$

and can be interpreted as the mean number of k-faces per unit volume. The value of γ_k does not depend on the choice of the centroid function c_k [278, 347].

Further random measures induced by a random tessellation are the measures $M_k : \mathcal{B} \to [0, \infty]$ given by

$$M_k(B) = \sum_{F \in \mathcal{F}_k(X)} \mathcal{H}^k(F \cap B), \quad k = 0, \ldots, d, B \in \mathcal{B},$$

where $\mathcal{H}^k$ is the k-dimensional Hausdorff measure. In the stationary case, their intensities

$$\mu_k = \mathbf{E}\left(\sum_{C \in \mathcal{F}_k(X)} \mathcal{H}^k(C \cap [0,1]^d) \right), \quad k = 0, \ldots, d,$$

can be interpreted as the mean total k-content of the k-faces of the tessellation per unit volume.

The above characteristics carry some information on the aggregate of the tessellation's cells. In applications it is often interesting to investigate characteristics of single cells and their faces leading e.g. to the distributions of cell volumes or edge lengths. This is formalised using the typical k-face of the stationary tessellation X which is defined by means of Palm theory [347].

Definition 3.10 (Typical k-face). Let $\mathcal{P}_k$ denote the space of k-dimensional polytopes in $\mathbb{R}^d$. The *typical k-face* $\mathcal{C}_k$ of a random stationary tessellation is a $\mathcal{P}_k$-valued random variable such that for every bounded, measurable, and translation invariant function $f : \mathcal{P}_k \to \mathbb{R}$ and every Borel set $B \in \mathcal{B}$ with $0 < v_d(B) < \infty$ we have

$$\mathbf{E}f(\mathcal{C}_k) = \frac{1}{\gamma_k v_d(B)} \mathbf{E}\left(\sum_{\{C \in \mathcal{F}_k : c_k(C) \in B\}} f(C - c_k(C)) \right). \tag{3.2}$$

The typical k-face can be interpreted as a k-dimensional polytope picked at random from the system of k-faces of the tessellation.

Exercise 3.3. Show that the distribution of $\mathcal{C}_k$ is rotation invariant if X is isotropic and $c_k(\vartheta C) = \vartheta c_k(C)$ for all $\vartheta \in SO_d, C \in \mathcal{C}'$. Give examples of centroid functions with this property.

Besides by the characteristics γ_k and μ_k a random tessellation can be described by geometric characteristics of its typical k-faces:

N_{kl} - the expected number of l-faces adjacent to the typical k-face, $k,l \in \{0,1,2,3\}$,

    e. g. N_{13} denotes the expected number of cells neighbouring the typical edge,

L_1 - the expected length of the typical edge,

L_2 - the expected perimeter of the typical face,

A_2 - the expected area of the typical face,

B_3 - the expected mean width of the typical cell,

L_3 - the expected total edge length of the typical cell,

S_3 - the expected surface area of the typical cell,

V_3 - the expected volume of the typical cell.

These characteristics are related to each other. For example, for a planar normal random tessellation in $\mathbb{R}^2$ they can be expressed by $\mu_0 (= \gamma_0)$ and μ_1. In $\mathbb{R}^3$, γ_3, $\mu_0 (= \gamma_0)$, μ_1, and μ_2 are required [269].

3.2.2 Tessellation Models

Various models for random tessellations can be found in the literature. Two common construction principles are the division of space by hyperplanes and tessellations of the Voronoi type which are generated by a point process with respect to a given distance measure.

3.2.2.1 Tessellations Constructed by Hyperplanes

Definition 3.11 (Hyperplane tessellation). Let H be a locally finite set of hyperplanes in $\mathbb{R}^d$. Then the connected components of $\mathbb{R}^d \setminus \bigcup_{h \in H} h$ form a system of open subsets of $\mathbb{R}^d$. Their closures build the *hyperplane tessellation* induced by H.

If H is given by a Poisson hyperplane process, i.e., a Poisson process on the space of hyperplanes in $\mathbb{R}^d$, then the special case of a *Poisson hyperplane tessellation* is obtained. These tessellations are face-to-face but not normal. In the stationary case, this model depends on the intensity of the Poisson hyperplane process (which corresponds to μ_{d-1}) and the distribution $\mathcal{R}$ of normal directions of the hyperplanes. A visualisation of a three-dimensional isotropic Poisson hyperplane tessellation, where $\mathcal{R}$ is the uniform measure on S^2, is shown in Fig. 3.2. See [347, Sect. 10.3] for a summary of analytic results for this model.

Definition 3.12 (STIT tessellation). A random *STIT tessellation* in any bounded Borel set $W \subset \mathbb{R}^d$ with (finite) volume $v_d(W) > 0$ is obtained by the following spatio-temporal construction. First, a random life-time is assigned to W which is exponentially distributed with parameter related to the mean width of W. After this life time, a uniformly distributed random hyperplane with normal direction drawn from a distribution $\mathcal{R}$ is thrown onto W. This hyperplane splits W into two new cells W_+ and W_-. On these, the cell splitting process starts anew and develops independently in W_+ and W_-. This procedure is continued until a given time $t > 0$ is reached.

It can be shown that there exists a tessellation on the whole space $\mathbb{R}^d$ which extends the tessellation constructed by the algorithm described above [270]. In the stationary case, model parameters are the stopping time t of the construction process (which corresponds to μ_{d-1}) and the distribution $\mathcal{R}$ of normal directions of the hyperplanes.

The name STIT stems from the fact that these tessellations are STable with respect to ITeration [296]. Several first-order properties of this model in the planar and three-dimensional case were obtained in [297], [298], and [391]. Second-order theory for STIT tessellations was developed in [351].

Due to the separate splitting of their cells, STIT tessellations are not face-to-face. A visualisation of a three-dimensional isotropic STIT tessellation is shown in Fig. 3.2.

It has been shown that the interiors of the typical cells of a Poisson hyperplane tessellation and a STIT tessellation with the same parameters have the same distribution [296]. A comparison of second order properties and the arrangement of cells in these models has been discussed in [328] and [329], respectively.

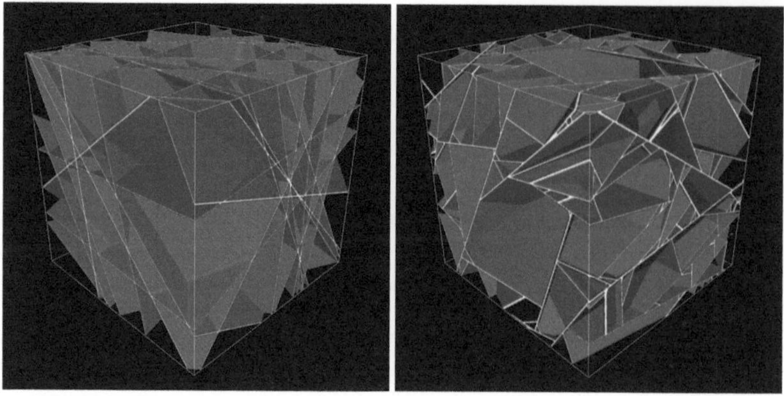

Fig. 3.2 Realisations of isotropic Poisson hyperplane (left) and STIT (right) tessellations in $\mathbb{R}^3$.

3.2.2.2 Tessellations of Voronoi Type

Definition 3.13 (Voronoi tessellation). Let φ be a locally finite subset of $\mathbb{R}^d$. The *Voronoi cell $C(x, \varphi)$* of $x \in \varphi$ is defined as

$$C(x, \varphi) = \{y \in \mathbb{R}^d : ||y - x|| \leq ||y - x'|| \quad \text{for all } x' \in \varphi.\}$$

The *Voronoi tessellation* of φ is the set $V(\varphi) = \{C(x, \varphi) : x \in \varphi\}$.

If Φ is a Poisson process in $\mathbb{R}^d$, the special case of a Poisson-Voronoi tessellation is obtained. Realisations of this model in $\mathbb{R}^2$ and $\mathbb{R}^3$ are shown in Fig. 3.3.

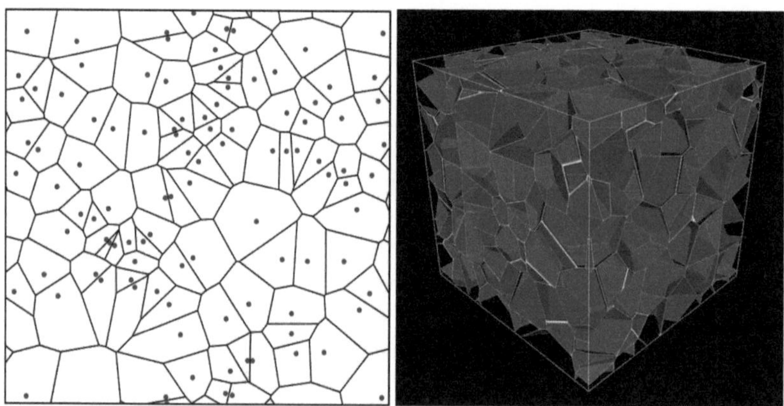

Fig. 3.3 Realisations of stationary Poisson-Voronoi tessellations in $\mathbb{R}^2$ (left) and $\mathbb{R}^3$ (right).

For a stationary Poisson-Voronoi tessellation in $\mathbb{R}^d$ with intensity λ, the densities μ_k are given by

$$\mu_k = \frac{2^{d-k+1}\pi^{\frac{d-k}{2}}}{d(d-k+1)!}\frac{\Gamma\left(\frac{d^2-kd+k+1}{2}\right)\Gamma\left(1+\frac{d}{2}\right)^{d-k+\frac{k}{d}}\Gamma\left(d-k+\frac{k}{d}\right)}{\Gamma\left(\frac{d^2-kd+k}{2}\right)\Gamma\left(\frac{d+1}{2}\right)^{d-k}\Gamma\left(\frac{k+1}{2}\right)}\lambda^{\frac{d-k}{d}}, \quad (3.3)$$

see [347, Theorem 10.2.4]. A summary of further analytical results for this model including formulae for the edge length distribution and contact distance distributions can be found in [306].

The range of cell structures which can be realised by Voronoi tessellations is limited as the boundary between cells is always located at equal distance between the generators. In order to produce more flexible models, several approaches of weighting Voronoi tessellations have been introduced. Some of these, e.g. the Johnson-Mehl tessellation [279], can contain non-convex cells. Here, we will concentrate on Laguerre tessellations which have convex polyhedral cells.

Definition 3.14 (Laguerre tessellation). Let φ be a locally finite subset of $\mathbb{R}^d \times \mathbb{R}_+$. The *Laguerre cell* of $(x,r) \in \varphi$ is defined as

$$L((x,r),\varphi) = \{y \in \mathbb{R}^d : ||y-x||^2 - r^2 \leq ||y-x'||^2 - r'^2 \quad \text{for all } (x',r') \in \varphi.\}$$

The *Laguerre tessellation* of φ is the set $L(\varphi) = \{L((x,r),\varphi) : (x,r) \in \varphi\}$. The 'distance' $\text{pow}((x,r),y) = ||y-x||^2 - r^2$ is called the *power* of y w.r.t. (x,r).

A generator $(x,r) \in \varphi$ can be interpreted as a ball with centre x and radius r. If all radii are equal, the special case of a Voronoi tessellation is obtained.

Exercise 3.4. Show that the section of a Laguerre tessellation on $\mathbb{R}^d$ with a k-dimensional plane $L \subset \mathbb{R}^d$ is a Laguerre tessellation in L. *Hint:* In this case, imaginary balls, i.e. balls with radius ir ($i = \sqrt{-1}$, $r \geq 0$) have to be allowed.

While each cell of a Voronoi tessellation contains its generating point this is no longer true for Laguerre tessellations (see Fig. 3.4, left). In fact, there may be points which do not generate a cell at all. However, if the system of generators consists of non-overlapping balls, then each ball is completely contained in its cell. Under certain regularity assumptions on the set φ, $L(\varphi)$ is a normal random tessellation. An important reason to choose this model is the fact that for $d \geq 3$ each normal tessellation of $\mathbb{R}^d$ is a Laguerre tessellation [14, 245].

In [245] integral formulae for the intensities μ_k of the Laguerre tessellation generated by a stationary, independently marked Poisson process Φ in $\mathbb{R}^d$ with intensity λ and mark distribution Q with finite d-th moment are given. Due to the lack of symmetries in the Laguerre tessellation these formulae are less explicit than the ones for the Poisson-Voronoi tessellation: For $m \in \mathbb{N}$ and $x_0,\ldots,x_m \in \mathbb{R}^m$ let $\Delta_m(x_0,\ldots,x_m)$ be the m-dimensional volume of the convex hull of $x_0,\ldots,x_m$ in $\mathbb{R}^m$. For $w_0,\ldots,w_m \geq 0$ define

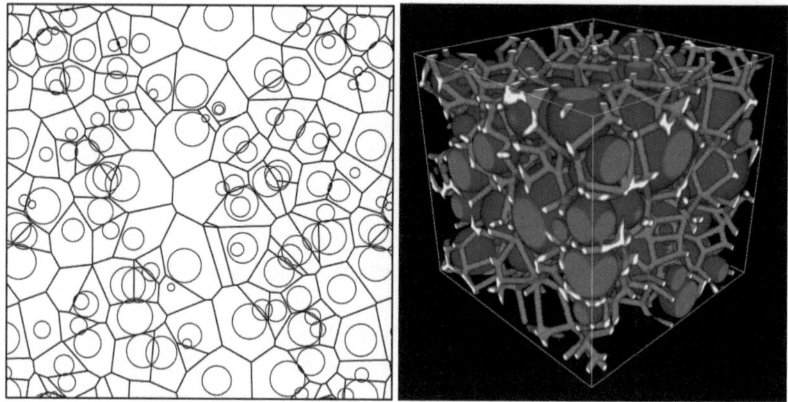

Fig. 3.4 Realisations of a stationary Poisson-Laguerre tessellation in $\mathbb{R}^2$ (left) and the edge system of a Laguerre tessellation generated by a ball packing in $\mathbb{R}^3$ (right) with some of the generating balls.

$$V_{m,k}(w_0,\ldots,w_m) = (m!)^{k+1} \int\limits_{S^{m-1}} \cdots \int\limits_{S^{m-1}} \Delta_m^{k+1}(w_0 u_0,\ldots,w_m u_m)\,\sigma(du_0)\ldots\sigma(du_m),$$

where σ is the surface measure on S^{d-1}. Furthermore, let

$$p(t) = \exp\left(-\lambda \kappa_d \int\limits_0^\infty ([t+r^2]_+)^{\frac{d}{2}} Q(dr)\right),$$

where $t_+ = \max\{t,0\}$ and κ_d is the volume of the d-dimensional unit ball. Then $p(t)$ is the probability that the power from the origin to each point of Φ exceeds t.

The intensities $\mu_k, 0 < k < d$, are given by the formula

$$\mu_k = \frac{\lambda^{m+1}}{4(m+1)!} c_{dm}\sigma_k \int\limits_0^\infty \cdots \int\limits_0^\infty \int\limits_{-\min\limits_i r_i^2}^\infty \prod_{i=0}^m (t+r_i^2)^{\frac{m-2}{2}} V_{m,k}\left((t+r_0^2)^{\frac{1}{2}},\ldots,(t+r_m^2)^{\frac{1}{2}}\right)$$

$$\times \int\limits_0^\infty p(s+t)s^{\frac{k-2}{2}}\,ds\,dt\,Q(dr_0)\ldots Q(dr_m),$$

$$(3.4)$$

where $m = d - k$, σ_k is the surface area of S^{k-1}, and $c_{dm} = \frac{\sigma_{d-m+1}\ldots\sigma_d}{\sigma_1\ldots\sigma_m}$. For $k = d$ we have $\mu_d = 1$, and for $k = 0$,

$$\mu_0 = \frac{\lambda^{d+1}}{2(d+1)!} \int\limits_0^\infty \cdots \int\limits_0^\infty \int\limits_{-\min_i r_i^2}^\infty \prod_{i=0}^d (t+r_i^2)^{\frac{d-2}{2}} V_{d,0}\left((t+r_0^2)^{\frac{1}{2}},\ldots,(t+r_d^2)^{\frac{1}{2}}\right) p(t)\,dt$$

$$\times Q(dr_0)\ldots Q(dr_d).$$

For the Poisson-Voronoi tessellation, obviously $\gamma_d = \lambda$. For Poisson-Laguerre tessellations no explicit formula for γ_d is known for general dimension. Nevertheless, for $d = 2$, the cell intensity can be computed via $\gamma_2 = \mu_0/2$. For $d = 3$, an explicit formula for γ_3 would be of particular interest as γ_3 is one of the four parameters determining the mean value characteristics of the tessellation, see Sect. 3.2.1.

For these and further results on Laguerre tessellations as well as analytic formulae for geometric characteristics of Poisson-Laguerre tessellations we refer to [245]. A detailed overview of results on Voronoi tessellations as well as their weighted generalisations can be found in [306].

3.3 Estimation of Geometric Characteristics of Open Foams from Image Data

Random tessellations are suitable models for a wide variety of structures. In the following we restrict attention to an application in materials science and discuss the analysis and modelling of open foams.

For our analyses we assume that the solid component of the foam is the parallel set of the edge system of a stationary random tessellation. In order to fit a tessellation model to a foam sample, we use geometric characteristics of the foam structure which are estimated from volume images of the material. Mean values of certain geometric characteristics can be computed from the densities of the intrinsic volumes which are estimated in a binary image of the foam structure. Furthermore, the reconstruction of the foam cells allows for the estimation of empirical distributions of the cell characteristics.

The techniques are illustrated by the example of an open aluminium foam. For the analysis of the material we use a μCT image with a size of $820 \times 820 \times 278$ voxels and a voxel size of $64.57\,\mu$m. Hence, the sample corresponds to $52.95\,\text{mm} \times 52.95\,\text{mm} \times 17.95\,\text{mm}$ of material. A visualisation of the tomographic image is shown in Fig. 3.5.

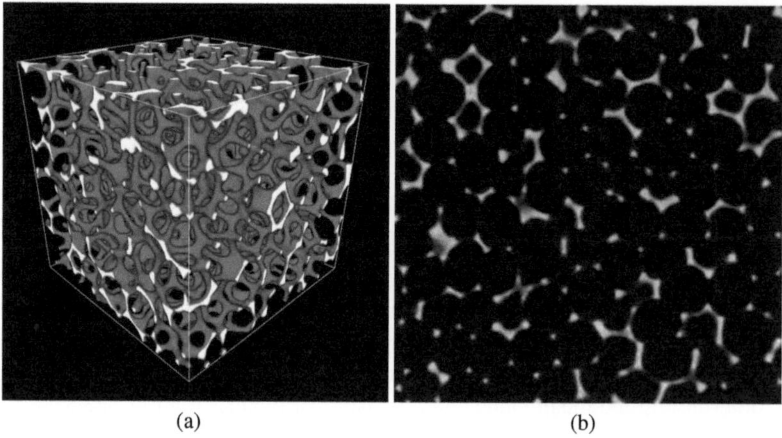

(a)   (b)

Fig. 3.5 3D visualisation (a) and a sectional image (b) of the open aluminium foam. Sample: M-Pore GmbH. Imaging: Fraunhofer IZFP.

3.3.1 Characteristics based on Intrinsic Volumes

Let $\mathcal{K}$ denote the system of compact and convex sets (convex bodies) in $\mathbb{R}^d$. For $K \in \mathcal{K}$, the intrinsic volumes $V_k(K)$, $k = 0,\ldots,d$, can be defined using the Steiner formula

$$V_d(K \oplus b(0,r)) = \sum_{k=0}^{d} \kappa_k V_{d-k}(K) r^k, \quad r \geq 0,$$

where κ_k is the volume of the k-dimensional unit ball. Recall that $\oplus$ denotes Minkowski addition, i.e., $A \oplus B = \{a + b : a \in A, b \in B\}$ for $A, B \in \mathcal{B}$. For $d = 3$ the intrinsic volumes are – up to constant factors – the volume $V = V_3$, the surface area $S = 2V_2$, the integral of mean curvature $M = \pi V_1$, and the Euler characteristic $\chi = V_0$, see e. g. [346, p. 210].

Exercise 3.5. Show that for $k = 0,\ldots,d$, the functional V_k is k-homogeneous in the sense that

$$V_k(rK) = r^k V_k(K), K \in \mathcal{K}, r > 0.$$

Exercise 3.6. Use the Steiner formula to compute the intrinsic volumes of a ball of radius $r > 0$ in $\mathbb{R}^d$ and a cube with edge length $a > 0$ in $\mathbb{R}^2$ and $\mathbb{R}^3$.

Important characteristics for stationary random closed sets Ξ are the densities of the intrinsic volumes. They are defined as the limits

$$V_{V,k}(\Xi) = \lim_{r \to \infty} \frac{\mathbb{E} V_k(\Xi \cap rW)}{V_d(rW)}, \qquad k = 0,\ldots,d, \tag{3.5}$$

where $W \in \mathcal{K}$ is a convex body with $V_d(W) > 0$ serving as an observation window. A sufficient condition for the existence of the limit is that Ξ is almost surely locally polyconvex, i.e., for $K \in \mathcal{K}$ the set $\Xi \cap K$ can almost surely be written as a finite union of convex bodies, and satisfies $\mathbf{E}2^{N(\Xi \cap [0,1]^d)} < \infty$ [347, Theorem 9.2.1]. Here $N(X)$ denotes the smallest number m such that $X = K_1 \cup \ldots \cup K_m$ with $K_1, \ldots, K_m \in \mathcal{K}$.

The intrinsic volumes and their densities can be estimated efficiently from binary volume images using discrete versions of the Crofton formula and the Euler-Poincaré formula as described in [305, 339].

For open foams further characteristics can be derived from the densities of the intrinsic volumes [341]. In particular,

p – the porosity $1 - V_V$,
L_V – the specific strut length (mean total strut length per unit volume)

$$L_V \approx \frac{M_V}{\pi(1 - V_V)}, \tag{3.6}$$

γ_0 – the intensity of nodes $\gamma_0 = -\chi_V$
$\bar{u}$ – the mean strut perimeter $\bar{u} = \frac{S_V}{L_V}$,
$\bar{\bar{b}}$ – the mean diameter of the struts $\bar{\bar{b}} = \frac{S_V}{\pi L_V}$, and
$\bar{a}$ – the mean cross section area of the struts $\bar{a} = \frac{V_V}{L_V}$.

These formulae are based on the assumption that the strut system is formed by a cylinder system obtained via dilation of the tessellation's edges. Therefore, an increasing strut thickness results in an increasing overlap of these cylinders at the nodes. Division by the porosity in (3.6) is an attempt to correct for this overlap.

For the open aluminium foam data, we obtained that $V_V = 14.624\%$, $S_V = 0.840$ $1/mm$, $M_V = 0.880\ 1/mm^2$, and $\chi_V = -0.220\ 1/mm^3$. This yields $p = 0.854$, $L_V = 0.328\ 1/mm^2$, $\gamma_0 = 0.220\ 1/mm^3$, $\bar{u} = 2.559$ mm, $\bar{\bar{b}} = 0.815$ mm, and $\bar{a} = 0.445$ mm^2.

3.3.2 Model-based Mean Value Analysis

Throughout the rest of this chapter we assume that $d = 3$ and the characteristics introduced in the previous section describe the random closed set X formed by the strut system of the foam. Assuming that this is the dilated edge system of a stationary random tessellation T allows to further characterise the foam using geometric characteristics of the typical cell of T. Hence, let T be a random tessellation which is stationary, normal and whose edge-skeleton $Z_1(T) = \bigcup_{F \in \mathcal{F}_1(T)} F$ is topologically equivalent to X in the sense that

$$\chi_V(Z_1(T)) = \chi_V(X)$$

and such that X can be modelled as the dilated system of the edges of T. The formulae given above imply that the node-intensity is $\gamma_0 = -\chi_V(Z_1(T)) = -\chi_V(X)$.

Now assume that the typical cell of the tessellation T is – up to a scaling factor c – distributed like the typical cell of a given tessellation model M. Then knowing c the characteristics of the typical cell can be derived immediately. The scaling factor is determined from the binary image of a realisation of X by using

$$\chi_V(X) = \frac{1}{c^3}\chi_V(M), \tag{3.7}$$

where the model structure M is normalised to $V_3(M) = 1$. Cell and node densities as well as the mean total facet area per unit volume are deduced from the corresponding quantities for the chosen model, i.e.

$$\gamma_3(T) = \frac{1}{c^3}\gamma_3(M),\ \gamma_0(T) = \frac{1}{c^3}\gamma_0(M),\ \mu_1(T) = \frac{1}{c^2}\mu_1(M),\ \text{and}\ \mu_2(T) = \frac{1}{c}\mu_2(M).$$

Similarly we get for the means of volume, surface area, mean width, edge length, and number of faces of the typical cell that

$$V_3(T) = c^3, \qquad S_3(T) = c^2 S_3(M), \qquad B_3(T) = cB_3(M),$$

$$L_3(T) = cL_3(M), \qquad N_{32}(T) = N_{32}(M).$$

In [341], the above quantities were determined for several tessellation models: the Poisson-Voronoi tessellation, a hardcore Voronoi tessellation generated by a Matérn hardcore process, and two Laguerre tessellation models generated by dense packings of balls with lognormal volume distribution, a packing fraction of 60%, and coefficients of variation of ball volumes of $cv = 0.2$ and $cv = 2.0$. The geometric characteristics of the typical cell of these models are given in Table 3.1. Sectional images of model realisations are shown in Fig. 3.6.

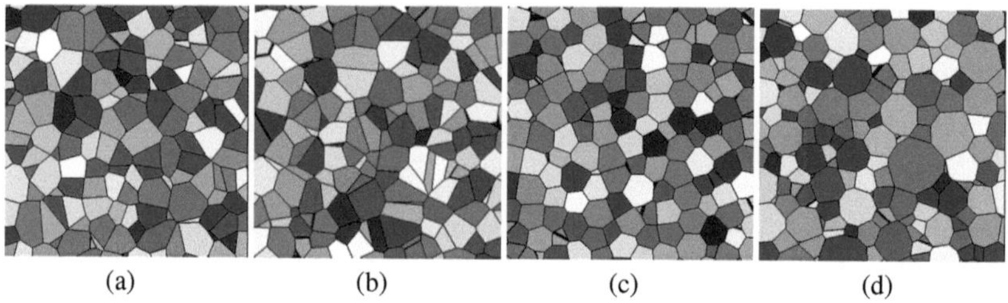

(a)   (b)   (c)   (d)

Fig. 3.6 Planar sections of three-dimensional random tessellation models with the same cell intensity: a Poisson-Voronoi tessellation (a), a hardcore Voronoi tessellation (b), and Laguerre tessellations of ball packings with low (c) and high (d) variation of ball volume.

The characteristics of the typical cell of the open aluminium foam data were estimated under all four model assumptions. The results given in Table 3.2 show that the influence of the model assumption on these parameters is relatively small. This is not surprising given the strong relations between the mean values in random

Table 3.1 Geometric characteristics of the typical cell of the Poisson Voronoi (PV), the Matérn hardcore Voronoi (HCV), and two Laguerre tessellations generated by ball packings (L1 with $cv = 0.2$ and L2 with $cv = 2.0$), normalised to $V_3 = 1$, $\gamma_3 = 1$.

	PV	HCV	L1	L2
γ_0	6.768	6.668	6.072	5.686
μ_1	5.832	5.728	5.436	5.064
μ_2	2.910	2.852	2.684	2.454
N_{32}	15.535	15.335	14.145	13.371
L_3	17.496	17.178	16.302	15.186
S_3	5.821	5.703	5.368	4.908
B_3	1.458	1.432	1.359	1.266
f_1	0.728	0.772	0.850	0.800

tessellations. In fact, the simulation study in [244] shows that the error due to the wrong model assumption is negligible compared to the error due to discretisation effects and insufficient resolution. As a consequence, higher order moments should also be considered when fitting tessellation models to real data.

Table 3.2 Geometric characteristics of the open aluminium foam data estimated under several model assumptions. The models are Poisson Voronoi (PV), the Matérn hardcore Voronoi (HCV), and two Laguerre tessellations generated by ball packings (L1 with $cv = 0.2$ and L2 with $cv = 2.0$)

	unit	PV	HCV	L1	L2
γ_3	mm^{-3}	0.033	0.033	0.036	0.039
V_3	mm^3	30.763	30.308	27.604	25.844
S_3	mm^2	57.150	55.438	49.029	42.902
B_3	mm	4.568	4.465	4.107	3.743
γ_2	mm^{-3}	0.252	0.253	0.256	0.259
A_2	mm^2	3.679	3.615	3.466	3.209
γ_1	mm^{-3}	0.440	0.440	0.440	0.4400
γ_0	mm^{-3}	0.220	0.220	0.220	0.220

3.3.3 Cell Reconstruction

The densities of the intrinsic volumes only obtain information on the mean values of certain geometric characteristics. The results of the previous section indicate that for the model fit they should be complemented by higher order moments or even distribution functions of the cell characteristics. For their estimation, the pore space of the foam has to be divided into cells whose edges coincide with the struts of the foam. Such a reconstruction of the foam cells can be obtained using the following image processing chain:

1. Binarisation yielding the strut system of the foam,
2. Euclidean distance transform on the pore space,
3. inversion, i.e. each voxel gets as new grey value the difference between the maximal grey value and its old value,
4. smoothing to remove superfluous local minima and avoid oversegmentation,
5. watershed transform.

Sectional images obtained during this procedure are shown in Fig. 3.7. To avoid oversegmentation, the h-minima transform, the height adaptive h-minima transform [372, Chap. 6] or a preflooded watershed transform [389] can be used. For details on the choice of the smoothing method and its parameters we refer to [341]. Although the resulting cells are obviously not convex, the deviation from convexity is often not very pronounced such that the cell system can be modelled by a tessellation consisting of convex polytopes.

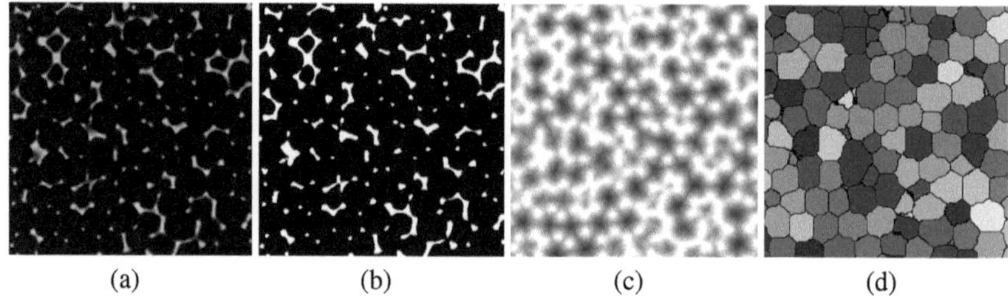

(a)   (b)   (c)   (d)

Fig. 3.7 Foam reconstruction: Planar sections of the original 3D image (a), the binarisation (b), the inverted distance image (c), and the reconstructed foam cells (d).

The definition of the typical cell in (3.2) suggests to estimate distributions of characteristics of the typical cell from the sample of cells with centroid in a given observation window.

In practice, neither the characteristics nor the centroid can be determined correctly if a cell hits the boundary of the image. Simply removing all boundary cells from the analysis induces a bias since large cells intersect the boundary with higher probability. Weighting with the reciprocal of the probability of being observed (Horvitz-Thompson procedure, see [20, 2.4]) is one way to correct for this bias. In our context that means weighting with the fraction of the volume of the whole sample and the volume of the sample reduced by the bounding box of the cell (Miles-Lantuejoul correction, see [361, p. 246]). If the sample is large enough, minus sampling is an alternative: Reduce the observation window such that all cells with centroid inside this sub-window are contained completely in the original observation window.

Exercise 3.7. Assume that the observation window is a cuboid. Compute the maximum window for minus sampling.

3.4 Stochastic Modelling of Cellular Materials

The assumption that the foam is the dilated edge system of a random tessellation suggests a two-step modelling procedure: First, a random Laguerre tessellation is fitted to the reconstructed cell system. In a second step, the system of edges of the tessellation is dilated to form the struts of the foam.

3.4.1 Modelling of the Cell System

For the first modelling step, we assume that the cell system of the foam is a realisation of a stationary and isotropic random tessellation. In practice, however, often anisotropies, measured e.g. by deviations of the mean cell diameters in the three coordinate directions, are observed. In this case, the anisotropies are removed by a suitable transformation of the cell system, e.g. the axes are scaled such that all three mean diameters are equal. Then, an isotropic model is fitted to the transformed microstructure. An appropriate transformation of this isotropic model finally yields a model for the original material.

Since the cells of real foams often show a high degree of regularity, the same should be true for their models. Tessellations generated by marked Poisson processes are typically too irregular for this application (see Fig. 3.6). A more suitable choice are tessellations generated by random systems of non-overlapping balls which can be simulated by random sequential adsorption (RSA, [179, p. 132]) or dense packing algorithms such as the force-biased algorithm [38] . Besides their regularity, these models have the advantage that each Laguerre cell completely contains its generating ball. Consequently, to a certain degree, the volume distribution of the balls may be used to control the volume distribution of the cells. A drawback of these models, however, is that no analytic formulae for their geometric characteristics exist, which means that they have to be studied by simulation. Fortunately, efficient algorithms for the generation of Laguerre tessellations are available [386].

The deviation of the models from the real foam data is measured using the relative distance measure

$$\rho(\hat{c},c) = \sqrt{\sum_{i=1}^{n} \left(\frac{c_i - \hat{c}_i}{\hat{c}_i} \right)^2}, \tag{3.8}$$

where the entries of $\hat{c} = (\hat{c}_1,\ldots,\hat{c}_n)$ and $c = (c_1,\ldots,c_n)$ are given by suitable geometric characteristics of the cells of the original foam and the model, respectively. In our case, these characteristics are the means and standard deviations of the volume v, the surface area s, the number of facets F_C, and the diameter (mean width) $\bar{b}$ of the cells. They have been chosen for the following reasons: For fixed cell intensity γ_3, the mean cell volume is given by $V_3 = \frac{1}{\gamma_3}$, the mean values of S_3 and B_3 are proportional to μ_2 and μ_1, respectively, and the mean value of F_C only depends on γ_0. Hence, all four characteristics determining the mean values of the tessellation are covered.

In practice, typically a parametric tessellation model is chosen by fixing the generating process and a parametric distribution for the ball volumes. Then the value of $\rho(\hat{c}, c)$ in (3.8) is minimised on the parameter space of the model by repeated simulation of the model. An approach avoiding the time-consuming simulation was presented in [327] for Laguerre tessellations of dense packings of balls with lognormal and gamma distributed volumes. These models are of particular interest for the modelling of foam structures due to their regularity. Furthermore, the volume distribution in cellular materials is often assumed to be a lognormal or a gamma distribution. Hence, model realisations for various packing fractions τ and coefficients of variation c of the volume distribution were generated. Subsequently, polynomials in c were fitted to the estimated geometric characteristics for each value of τ. Using these results, the minimisation of the value of $\rho(\hat{c}, c)$ in (3.8) reduces to the minimisation of a polynomial which allows for a quick and easy model fit.

Here, this technique is applied to the open aluminium foam. The best fit for the lognormal distribution was obtained with $\tau = 60\%$ and $c = 0.172$. For the gamma distribution the optimal parameters were $\tau = 60\%$ and $c = 0.169$. To obtain an isotropic structure, the foam cells were scaled by 0.92 along the z-axis. The geometric characteristics estimated from the cell reconstruction and the respective characteristics in the best-fit model are shown in Table 3.3.

Table 3.3 Estimated mean values and standard deviations of the cell characteristics of the aluminium foam and the best fit models for the lognormal and gamma distribution.

lognormal	scaled data		lognormal				gamma			
	mean	std	mean	deviation	std	deviation	mean	deviation	std	deviation
$v[mm^3]$	20.174	2.751	20.174	$\pm 0.0\%$	2.751	$\pm 0.0\%$	20.174	$\pm 0.0\%$	2.704	-1.7%
$s[mm^2]$	41.015	3.671	39.854	-2.8%	3.423	-6.8%	39.858	-2.8%	3.368	-8.9%
$d[mm]$	3.649	0.174	3.705	$+1.5\%$	0.179	$+2.9\%$	3.705	$+1.5\%$	0.176	$+1.1\%$
F_C	13.838	1.203	14.164	$+2.4\%$	1.298	$+7.9\%$	14.165	$+2.4\%$	1.288	$+7.1\%$

3.4.2 Modelling of Open Foams

The previous section explains how to fit a random tessellation model to the cell system of a foam. A model for the solid component of the foam can now be obtained via dilation of the edge system of this tessellation. A typical feature of real foams, however, is that their struts are significantly thicker near the vertices than at their centres. Therefore, we propose the use of locally adaptable dilations [91], where the size of the structuring element, a ball in our case, may vary locally. An illustration is shown in Fig. 3.8.

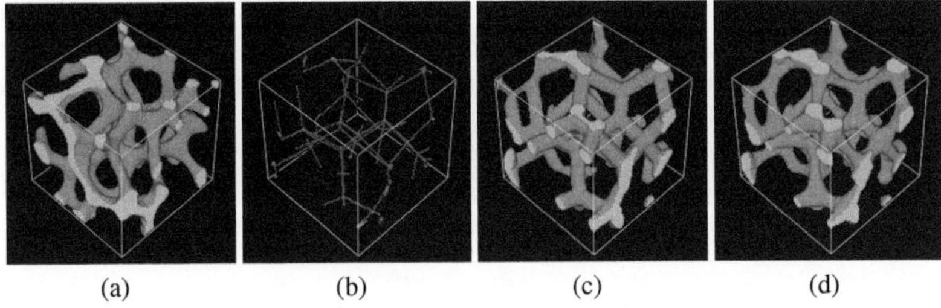

(a)   (b)   (c)   (d)

Fig. 3.8 Locally adaptable dilation: A part of a real foam (a), the edge system of a random tessellation (b), its dilation with a ball of constant size (c), and a locally adaptable dilation (d).

The size at a particular voxel is stored in a size map, a second image of the same size where the voxel grey values correspond to size information. To determine the size map, the thickness profiles of the struts are estimated from the image data as described in [253]. The first step is to compute the skeleton of the foam. By skeleton, we mean a one voxel thick subset of the solid component Y of the foam that has the same topology as Y and that is centred w. r. t. the Euclidean distance. It is obtained by iteratively removing points from Y that are not necessary to preserve the topology of Y. The skeleton is then decomposed into its topological components, namely curve segments (struts) and curve junctions (nodes).

Let us denote by Y_0 the particle process (i.e., a point process on $\mathcal{C}'$) forming the decomposed skeleton of Y. Each strut of the foam contributes a curve segment $Z \in Y_0$ whose length is denoted by ℓ. The local strut thickness at a point $x \in Z$ is defined as the radius of the largest ball with centre x inscribed in Y. It may be parametrised using the distance $\xi = \xi(x) \in [-\ell/2, \ell/2]$ of x to the strut centre yielding a function $p_Z(\xi)$. The spherical contact profile $P_Y(\xi, \ell)$ for the entire foam is obtained as the mean thickness at distance ξ of all struts with length ℓ. In practice, the local strut thickness is computed by a Euclidean distance transform [305] on the strut system.

The size map is computed by first normalising the spherical contact profile $P_Y(\xi, \ell)$ by the mid-span thickness $P_Y(0, \ell)$ and the strut length ℓ to yield a scale free representation of the strut thickness. Let us denote by $\tilde{\xi} = \xi/\ell$ the distance from the strut centre normalised by length ℓ. Then the strut thickness is modelled by a polynomial in $\tilde{\tilde{\xi}}$ which is fitted to the normalised thickness. Since the shape of the strut profiles is highly dependent on the strut length ℓ, the polynomial is fitted separately to the length classes $q_{i,j} = (x_i, x_j]$ formed by the successive deciles x_i and x_j of the strut length distribution. In [253], various choices for the polynomial were compared. The initial model was chosen as

$$\tilde{p}_{2468}(\tilde{\tilde{\xi}}) = a_4 \tilde{\xi}^8 + a_3 \tilde{\xi}^6 + a_2 \tilde{\xi}^4 + a_1 \tilde{\xi}^2 + 1, \quad a_1, \ldots, a_4 \in \mathbb{R},$$

where the indices denote the exponents of $\tilde{\tilde{\xi}}$ that are included. The odd exponents were not considered since the thickness profiles turned out to be symmetric about the strut centre. The intercept of one is caused by the normalisation of the data. Via all-subset regression of $\tilde{p}_{2468}$ guided by the MC_p-statistic [97, 130] the generally best fit polynomial model was found to be $\tilde{p}_{248}$. In general, the same holds for the aluminium foam considered here, albeit the profiles of struts smaller than x_4 were slightly better reproduced by $\tilde{p}_{268}$.

Table 3.4 Parameters of the polynomial model $\tilde{p}_{248}$ individually fitted to the length classes $q_{i,j}$.

	a_1	a_2	a_4
$q_{0,1}$	1.9463	4.4753	−152.5598
$q_{1,2}$	2.7510	3.5818	−127.8460
$q_{2,3}$	3.2397	3.6990	−116.8598
$q_{3,4}$	3.1954	4.7320	−96.8980
$q_{4,5}$	2.7230	1.1191	−156.3740
$q_{5,6}$	2.3683	12.7177	−158.7181
$q_{6,7}$	1.7256	18.6126	−221.9740
$q_{7,8}$	1.2100	2.8316	−29.4988
$q_{8,9}$	0.4753	24.4270	−218.3543
$q_{9,10}$	−0.7547	28.5711	−188.2299

The size map for the adaptable dilation of the edge system is now chosen according to the edge length classes given by $q_{i,j}$. To get a smooth transition from one strut to another, the maximal value of each polynomial (the thickness in the nodes) was fixed to the mean node radius $r_N = 0.36$ mm. This is achieved by multiplication of all polynomials with the factor $P_0 = r_N / \tilde{p}_{248}(0.5)$. A visualisation of the original foam along with its model is shown in Fig. 3.9.

Table 3.5 shows the average deviation of the intrinsic volumes computed over 25 realisations of the best fit lognormal and gamma model from the aluminium foam. Both models exhibit the same behaviour. They only differ by less than three per cent in V_V and S_V. However, for M_V and χ_V we observed a deviation of around 20 %. In case of M_V this could be explained by the different cross-sectional shape of the struts. The deltoidal shape of the aluminium foam exposes less curvature than the circular shape used in the model. The deviation in χ_V is caused by closed faces in the aluminium foam that increase the density of χ_V. These were not reproduced in the model.

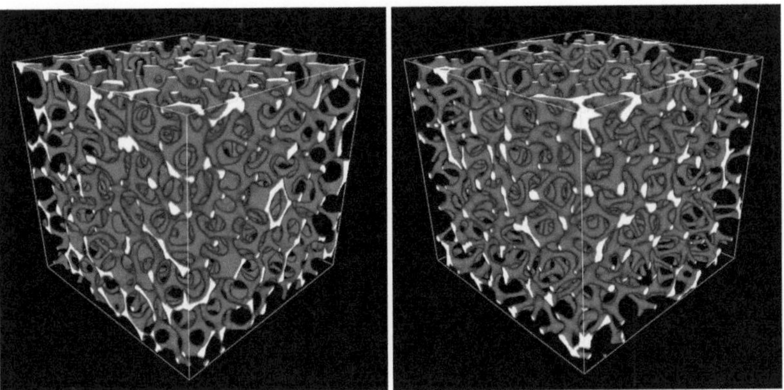

Fig. 3.9 Visualisation of the real foam and a realisation of the model. Visualised are 277^3 voxels.

Table 3.5 Deviation of the intrinsic volumes of the best fit models for the lognormal and gamma distribution from the aluminium foam averaged over 25 realisations.

	Data	Dev. lognormal	Dev. gamma
V_V	14.624	-1.41%	-1.56%
$S_V\,[mm^{-1}]$	0.840	$+2.94\%$	$+2.87\%$
$M_V\,[mm^{-2}]$	0.880	$+20.11\%$	$+20.21\%$
$\chi_V\,[mm^{-3}]$	-0.220	$+23.68\%$	$+23.54\%$

Chapter 4
Stochastic 3D Models for the Micro-structure of Advanced Functional Materials

Volker Schmidt, Gerd Gaiselmann and Ole Stenzel

Abstract Optimization of functional materials is a challenging task. Thereby, stochastic morphology models can provide helpful methods. Three classes of stochastic models are presented describing different micro-structures of functional materials by means of methods from stochastic geometry, graph theory and time series analysis. The structures of these materials strongly differ from each other, where we consider organic solar cells being an anisotropic composite of two materials, non-woven gas-diffusion layers in proton exchange membrane fuel cells consisting of a system of curved carbon fibers, and graphite electrodes in Li-ion batteries which are built up by an isotropic two-phase system (i.e., consisting of a pore and a solid phase). The goal is to give an overview how the stochastic modeling of functional materials can be organized and to provide an outlook how these models can be used for material optimization with respect to functionality.

4.1 Introduction

Often, the micro-structure of materials is closely related to their functionality making the study of morphology an important and growing research field. Thus, to produce materials with improved properties, the morphology of the material under consideration has to be optimized with regard to its functionality. In general, however,

Volker Schmidt
Institute of Stochastics, Ulm University, 89069 Ulm, Germany, e-mail: `volker.schmidt@ uni-ulm.de`

Gerd Gaiselmann
Institute of Stochastics, Ulm University, 89069 Ulm, Germany, e-mail: `gerd.gaiselmann@ uni-ulm.de`

Ole Stenzel
Institute of Stochastics, Ulm University, 89069 Ulm, Germany, e-mail: `ole.stenzel@ uni-ulm.de`

© Springer International Publishing Switzerland 2015

V. Schmidt (ed.), *Stochastic Geometry, Spatial Statistics and Random Fields,*
Lecture Notes in Mathematics 2120, DOI 10.1007/978-3-319-10064-7_4

a systematic understanding of the influence of the 3D micro-structure on functional properties is missing. Stochastic morphology models, fitted to experimental 3D (image) data of these materials, can help to elucidate the correlation between processing parameters, 3D micro-structure, and functional properties. Furthermore, stochastic simulation models can be applied to *virtual materials design*, that is, to detect micro-structures with improved functional properties. Such a design of virtual materials can be obtained by simulating a broad range of virtual structures according to the stochastic model (using different values for model parameters) and analyzing their functional properties using numerical (transport) calculations.

In this chapter, we present three stochastic morphology models for different functional materials: the 3D morphology of hybrid-organic solar cells (Fig. 4.1 (left)), the 3D micro-structure of non-woven gas-diffusion layers (GDL, used in fuel cells, Fig. 4.1 (center)), and the 3D micro-structure of graphite electrodes (used in Li-ion batteries, Fig. 4.1 (right)).

Fig. 4.1 Left: example of morphology of the active layer of polymer-ZnO solar cells (yellow phase: ZnO, transparent: P3HT), center: non-woven GDL (yellow phase: system of carbon fibers), right: graphite electrode (yellow phase: graphite, transparent: pore phase)

The general procedure for our stochastic simulation models is to first understand the functionality of the material and - if already identified - its correlation to the micro-structure, such that the most important aspects are reflected by the stochastic morphology model. Secondly, we aim to gain some information about the structure of the material. Therefore, we consider 3D high-resolution (tomographic) image data. A good quality of image acquisition plays an important role in stochastic modeling: what is not present in image data, cannot be modeled adequately. In the next step, we structurally segment the 3D image data of functional materials in order to simplify the fitting of model parameters. More precisely, we represent materials e.g. by unions of spheres (applied for the solar cells), by a system of 3D polygonal tracks (applied for GDL), or by a spatial 3D network (applied for the graphite electrode). Then, we develop a parametric stochastic model which is able to describe the micro-structure of a material sufficiently well. Next, these models are fitted to the experimentally measured 3D image data. Thus, the stochastic models, which are established in the three-dimensional Euclidean space, are discretized on a voxel grid in order to be compared to the image data by morphological characteristics. In short, the model is fitted by choosing the parameters such that the agreement between experimental data and synthetic (i.e., simulated) data with respect to morphological characters is maximized. Finally, we validate the stochastic model by comparing characteristics relevant for the functionality computed for experimental 3D image data, and for realizations sampled from the stochastic model.

In agreement with this general procedure, the outline for all three morphology examples (solar cells, GDL, graphite electrodes) is the same. Thus, in each section, we first describe *data and functionality*, then we deal with *data preprocessing*, with the description of the *stochastic model*, with *model fitting*, and finally with *model validation*.

In the following. we shortly present the stochastic models for three different functional materials (hybrid organic solar cells, non-woven GDL of proton exchange membrane fuel cells (PEMFC) and graphite electrodes in Li-ion batteries).

In Sect. 4.3, a spatial stochastic model is developed that describes the 3D morphology of anisotropic composite materials, being blends of two different solid phases with an uniaxial anisotropy. The model is flexible enough to describe morphologies with a high degree of structural complexity. It is based on a *multi-scale approach*, where the complexity of the morphology is split into two different length scales: a *macro-scale* model describing the main morphological aspects and a *micro-scale* model including morphological details omitted in the macro-scale model. The basic idea of the macro-scale model is to describe the main morphological aspects, i.e., the large clusters of the two phases, by a system of overlapping spheres, where the midpoints are modeled by a stack of 2D point processes with a suitably chosen correlation structure and where the radii are added such that they are positively correlated. The model is fitted to 3D image data describing the morphology of photoactive layers of hybrid organic solar cells consisting of poly(3-hexylthiophene) (P3HT) as electron-donor and ZnO as electron-acceptor. Such a solar cell consists of a blend of electron-donor and electron-acceptor materials. Here, roughly speaking, the two materials have been mixed by spin-coating (i.e., rotating the two materials on a plate). The model is fitted to several different solar cell morphologies fabricated with different processing conditions (varying spin coating velocities), see Fig. 4.1 (left) for an example of such a morphology. Polymer (hybrid) solar cells are a promising alternative to silicon-based solar cells since they offer the prospect of being cheap to produce and ecological. Despite enormous improvements, they still stuffer from relatively low efficiencies. Therefore, enormous efforts are undertaken to increase power conversion efficiency. One approach is to optimize solar cell morphology since there is a close relationship between morphology and efficiency, see [307]. The model is validated by comparing physically relevant characteristics of experimental and simulated data, like the efficiency of exciton quenching, which is important for the generation of charges. Finally, a scenario analysis is performed where 3D morphologies are simulated for different values of model parameters such that we mimic the results of the spin-coating process.

In Sect. 4.4 a stochastic 3D model is developed describing the micro-structure of non-woven GDL in PEMFC which consist of strongly curved and non-overlapping fibers, see Fig. 4.1 (center). Fuel cells are an attractive instrument for electrical power generation due to their high efficiency and environment-friendly emissions. Moreover, due to their weight, compactness and quick startup time they are ideal for the application in automotive industry. One of their components is the gas-diffusion layer (GDL) which is mainly responsible for the supply of the electrodes with gas and water storage / evacuation within the GDL. Stochastic models can help to quantitatively predict physical properties for a given 3D GDL micro-structure only by means of computer experiments. The model that we propose is based on ideas from stochastic geometry and multivariate time series analysis. It is constructed by a two-stage approach, in which we first introduce a system of overlapping fibers by means of a germ-grain model, where the germs form a 3D Poisson point process and the grains are given by random 3D polygonal tracks describing single fibers in terms of multivariate time series. Secondly, we transform the germ-grain model into a system of non-overlapping random fibers using an iterative procedure leaned on the so-called force-biased algorithm. This model is validated by comparing transport-relevant characteristics computed for experimental 3D synchrotron data, and for realizations sampled from the stochastic model.

In Sect. 4.5, a flexible stochastic 3D model for simulation of isotropic 2-phase materials is presented. The model is fitted to a 3D image displaying graphite electrodes used for in Li-ion batteries. Their applications are widely spread ranging from smart phones, laptops, electric cars, etc. These batteries are built up of an anode, a cathode and an electrolyte. During discharging, Li atoms release an electron at the anode and the so-formed Li-ions then diffuse from the anode to the cathode via the electrolyte. The electrons also move from the anode to the cathode. During charging, Li-ions move back to the anode and are embedded in the graphite phase. For more details on functionality of Li-ion batteries, we refer to [95]. It is well-known that the morphology of the graphite anode of Li-ion batteries is strongly related to the electrochemical performance of the battery, e.g. for a large capacity as well as a high performance, a high specific surface area of the graphite electrode as well as a large total volume are beneficial. The analysis of the micro-structure of graphite electrodes by our stochastic modeling approach helps to provide more inside into the relationship between micro-structure and functionality. The stochastic 3D model is based on a hybrid-approach where two well-established stochastic approaches are combined: spatial random networks and simulated annealing. The basic idea of this modeling approach can be summarized as follows. First, a spatial random network model is developed which describes the main structural features of the simulated micro-structure. Then, a realization drawn from the network model is discretized on a voxel grid, where voxels representing the edges of the network are colored 'white', and the remaining ones 'black'. Subsequently, the discretized network is combined with simulated annealing in a second step. Therefore, the set of white voxels is filled up with further white voxels around the edges of the network in order to get a suitable initial configuration for the simulated annealing algorithm. Thus, in the initial configuration, the white voxels tend to cluster along the edges of the network. Note that the initial white voxels that represent the edges of the network are not changed by simulated annealing and, therefore, they serve as a backbone for the simulated micro-structure. Thus, via the structural properties of the network, it is possible to govern properties of the resulting morphology.

4.2 Point Process Models and Time Series

The stochastic models developed in the present chapter are mostly based on tools from point processes and time series models. We therefore give a short introduction to the corresponding mathematical background in order to provide the reader with the basic concepts, notation and definitions used in this chapter, see also [21] and Chap. 2.

In the following, let $\mathbb{Q}^d$ be the family of all (bounded) half-open cuboids in $\mathbb{R}^d$, i.e.,

$$\mathbb{Q}^d = \left\{ B \subset \mathbb{R}^d : B = (a_1, b_1] \times \ldots \times (a_d, b_d], a_i, b_i \in \mathbb{R}, a_i \le b_i \text{ for all } i = 1, \ldots, d \right\}.$$

Furthermore, let $\mathcal{B}(\mathbb{R}^d)$ denote the Borel σ-algebra on $\mathbb{R}^d$ and let $\mathcal{B}_0(\mathbb{R}^d)$ be the family of all bounded Borel sets on $\mathbb{R}^d$.

To begin with, we introduce the notion of random counting measures, which are used to formally describe point processes.

Definition 4.1.

1. Let N be the family of all locally finite counting measures $\varphi : \mathcal{B}(\mathbb{R}^d) \to \{0, 1, \ldots, \} \cup \{\infty\}$, i.e., $\varphi(B) < \infty$ for all $B \in \mathbb{Q}^d$ and $\varphi\left(\bigcup_{n=1}^{\infty} B_n\right) = \sum_{n=1}^{\infty} \varphi(B_n)$ for pairwise disjoint $B_1, B_2 \ldots \in \mathcal{B}(\mathbb{R}^d)$.
2. Furthermore, let $\mathcal{N}$ be the smallest $\sigma-$algebra of subsets of N such that the mapping $\varphi \to \varphi(B)$, for $B \in \mathcal{B}(\mathbb{R}^d)$, is $(\mathcal{N}, \mathcal{B}(\mathbb{R}))$-measurable.
3. A *random counting measure* $N : \Omega \to$ N is a random variable over a certain probability space $(\Omega, \mathcal{A}, \mathbf{P})$, whose values are in the measurable space $(N, \mathcal{N})$. A random counting measure N can thus be interpreted as a set-indexed stochastic process $\{N_B, B \in \mathcal{B}(\mathbb{R}^d)\}$ such that $\{N_B(\omega), B \in \mathcal{B}(\mathbb{R}^d)\}$ is a locally finite counting measure for each $\omega \in \Omega$.

4.2.1 Point Processes

Definition 4.2. Let $S_1, S_2, \ldots : \Omega \to \mathbb{R}^d \cup \{\infty\}$ be a sequence of random vectors over some probability space $(\Omega, \mathcal{A}, \mathbf{P})$ such that

$$\#\{n : S_n \in B\} < \infty \quad \text{for all } B \in \mathcal{B}_0(\mathbb{R}^d).$$

Then the sequence $\{S_n, n \ge 1\}$ is called a *point process* in $\mathbb{R}^d$. The point process is called *simple* if $\mathbf{P}(S_i \ne S_j \text{ for all } i \ne j) = 1$.

Remark 4.1. Let $\{S_n, n \ge 1\}$ be a random point process in $\mathbb{R}^d$ and $\{N_B, B \in \mathcal{B}(\mathbb{R}^d)\}$ be defined by

$$N_B = \#\{n : S_n \in B\} \quad \text{for all } B \in \mathcal{B}(\mathbb{R}^d).$$

Then, $\{N_B, B \in \mathcal{B}(\mathbb{R}^d)\}$ is a random counting measure. Thus, each point process $\{S_n, n \geq 1\}$ can be assigned a corresponding random counting measure and an alternative approach is to consider point processes as random counting measures.

We are interested in the first moments of random point processes, which leads to the following definition of the so-called intensity measure.

Definition 4.3. Let $\{S_n, n \geq 1\}$ be a random point process with corresponding random counting measure $\{N_B, B \in \mathcal{B}(\mathbb{R}^d)\}$. The function $\mu : \mathcal{B}(\mathbb{R}^d) \to [0, \infty]$ with $\mu(B) = \mathbf{E}N_B$ for $B \in \mathcal{B}(\mathbb{R}^d)$ is called *intensity measure*. In the following, we assume that μ is locally finite, i.e., $\mu(B) < \infty$ for all $B \in \mathcal{B}_0(\mathbb{R}^d)$. If μ is absolutely continuous with respect to the d-dimensional Lebesgue measure ν_d, i.e., there exists a Borel-measurable function $\lambda : \mathbb{R}^d \to [0, \infty)$, such that

$$\mu(B) = \int_B \lambda(x)\mathrm{d}x$$

for all $B \in \mathcal{B}_0(\mathbb{R}^d)$, then the function $\lambda : \mathbb{R}^d \to [0, \infty)$ is called *intensity function*.

Definition 4.4. Let μ be a locally finite measure on $\mathbb{R}^d$ which is diffuse, i.e. $\mu(\{x\}) = 0$ for all $x \in \mathbb{R}^d$. A simple point process $\{S_n, n \geq 1\}$ is called a *Poisson process* with intensity measure μ if for the corresponding random counting measure $\{N_B, B \in \mathcal{B}(\mathbb{R}^d)\}$ it holds that

1. $N_{B_1}, \ldots, N_{B_n}$ are independent random variables for any pairwise disjoint $B_1, \ldots, B_n \in \mathcal{B}_0(\mathbb{R}^d)$, $n \geq 1$, and
2. N_B is Poisson-distributed with parameter $\mu(B)$ for each $B \in \mathcal{B}_0(\mathbb{R}^d)$.

The Poisson process $\{S_n, n \geq 1\}$ is called *stationary* if its intensity measure μ is proportional to the d-dimensional Lebesgue measure ν_d, i.e., there exists a constant $\lambda \in (0, \infty)$ (called *intensity*), such that $\mu(B) = \lambda \nu_d(B)$ for all $B \in \mathcal{B}_0(\mathbb{R}^d)$.

Realizations of a stationary and an non-stationary Poisson process are given in Fig. 4.2.

In the following, we define two general invariance properties of point processes, stationarity and isotropy. Stationarity means that the point process, more precisely, its finite-dimensional marginal distributions are invariant under affine translations, where isotropy refers to a corresponding invariance under rotations around the origin.

Definition 4.5. A point process $\{S_n, n \geq 1\}$ is called *stationary* if for the corresponding random counting measure $\{N_B, B \in \mathcal{B}(\mathbb{R}^d)\}$ it holds that

$$(N_{B_1}, \ldots, N_{B_n}) \stackrel{D}{=} (N_{B_1+x}, \ldots, N_{B_n+x}) \text{ for all } n \geq 1, B_1, \ldots, B_n \in \mathcal{B}(\mathbb{R}^d) \text{ and } x \in \mathbb{R}^d,$$

where $\stackrel{D}{=}$ means equality in distribution. The point process $\{S_n, n \geq 1\}$ is called *isotropic* if

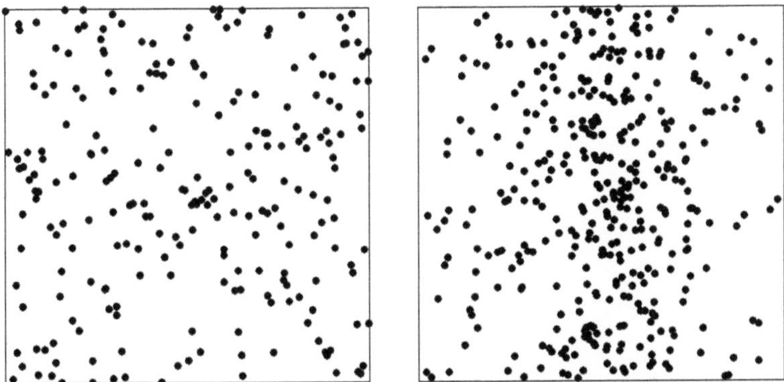

Fig. 4.2 Left: realization of stationary Poisson process, right: realization of non-stationary Poisson process with intensity function given by $\lambda(x_1,x_2) = 1000\exp(-5|x_1-0.5|)$ for $(x_1,x_2) \in \mathbb{R}^2$

$$(N_{B_1},\ldots,N_{B_n}) \overset{D}{=} \left(N_{\delta(B_1)},\ldots,N_{\delta(B_n)}\right) \text{ for all } B_1,\ldots,B_n \in \mathcal{B}(\mathbb{R}^d)$$

and for arbitrary rotations $\delta : \mathbb{R}^d \to \mathbb{R}^d$ around the origin.

Exercise 4.1. Let $\{S_n, n \geq 1\}$ be a stationary point process with random counting measure $\{N_B, B \in \mathcal{B}(\mathbb{R}^d)\}$. The *spherical contact distribution function* $H : [0,\infty) \to [0,1]$ of $\{S_n, n \geq 1\}$ is then defined by $H(r) = 1 - \mathbf{P}(N_{B(o,r)} = 0)$ for $r \geq 0$, where $B(x,r)$ denotes the d-dimensional sphere with center $x \in \mathbb{R}^d$ and radius $r \geq 0$. Show that

$$H(r) = 1 - \exp(-\lambda\,\kappa_d r^d)$$

if $\{S_n, n \geq 1\}$ is a stationary Poisson process with intensity λ and $\kappa_d = v_d(B(o,1))$ denotes the volume of the d-dimensional unit sphere.

Definition 4.6.

1. Let $\{S_n, n \geq 1\}$ be a stationary Poisson process with intensity $\lambda_0 \in (0,\infty)$. Furthermore, let $\{L_n, n \geq 1\}$ be a sequence of independent and identically distributed (iid) random variables with $L_1 \sim \text{Unif}[0,1]$ and let $\{L_n, n \geq 1\}$ be independent of $\{S_n, n \geq 1\}$.

2. Then, for a fixed $r_h \in (0,\infty)$, the point process $\left\{\widetilde{S}_n, n \geq 1\right\}$ with

$$\widetilde{S}_n = \begin{cases} S_n & \text{if } L_n < M_n, \\ \infty, & \text{else,} \end{cases} \tag{4.1}$$

where

$$M_n = \begin{cases} \min_{k \neq n: 0 < |S_k - S_n| \leq r_h} L_k, & \text{if } \#\{k \neq n : 0 < |S_k - S_n| \leq r_h\} > 0, \\ 1, & \text{else,} \end{cases}$$

is called a *Matérn hard-core process* with hard-core radius r_h.

A realization of a Matérn hard-core process is given in Fig. 4.3 (left).

Exercise 4.2. Show that the Matérn hard-core process introduced in Definition 4.6 is a stationary point process whose intensity λ is given by

$$\lambda = \frac{1 - \exp(\kappa_d \lambda_0 r_h^d)}{\kappa_d r_h^d},$$

where $\kappa_d = \nu_d(B(o,1))$ denotes the volume of the d-dimensional unit sphere.

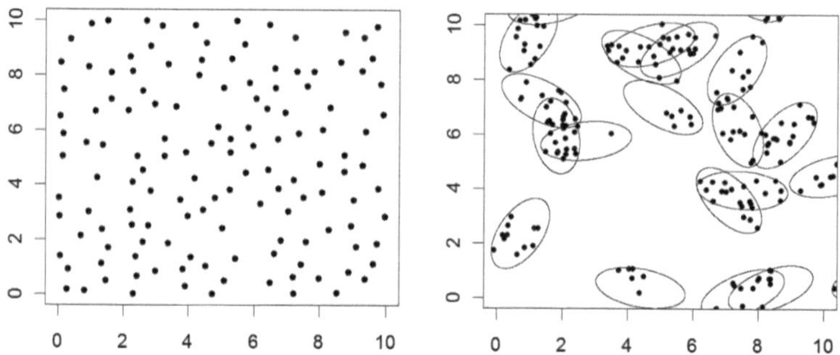

Fig. 4.3 Left: realization of Matérn-hardcore process; right: realization of elliptical Matérn cluster process

In the following, we introduce the notion of Cox processes, also known as doubly stochastic Poisson processes, which are a generalization of Poisson processes, where the intensity measure itself is a stochastic process.

Definition 4.7. Let $\Lambda = \{\Lambda_B, B \in \mathcal{B}(\mathbb{R}^d)\}$ be a random measure which is locally finite with probability 1. A point process $\{S_n, n \geq 1\}$ is called a *Cox process* if for the corresponding random counting measure $\{N_B, B \in \mathcal{B}(\mathbb{R}^d)\}$ it holds that

$$\mathbf{P}(N_{B_1} = k_1, \ldots, N_{B_n} = k_n) = \mathbf{E}\left(\prod_{i=1}^{n} \frac{\Lambda_{B_i}^{k_i}}{k_i!} \exp(-\Lambda_{B_i})\right)$$

for all $n \geq 1$ and pairwise disjoint $B_1, \ldots, B_n \in \mathcal{B}_0(\mathbb{R}^d)$.

Remark 4.2. Note that

$$\mathbf{P}(N_{B_1} = k_1, \ldots, N_{B_n} = k_n \,|\, \{\Lambda\} = \mu) = \prod_{i=1}^{n} \frac{\mu(B_i)^{k_i}}{k_i!} \exp(-\mu(B_i)) \qquad (4.2)$$

for all $n \geq 1$ and pairwise disjoint $B_1, \ldots, B_n \in \mathcal{B}_0(\mathbb{R}^d)$. Thus, given a realization μ of the intensity random measure $\{\Lambda_B, B \in \mathcal{B}(\mathbb{R}^d)\}$, the Cox process has the same distribution as a Poisson process with intensity measure μ. In other words, a Cox process can be interpreted as a mixture of Poisson processes since

$$\mathbf{P}(N_{B_1} = k_1, \ldots, N_{B_n} = k_n) = \int \mathbf{P}(N_{B_1} = k_1, \ldots, N_{B_n} = k_n \,|\, \{\Lambda\} = \mu)\, P_\Lambda(\mathrm{d}\mu)$$

$$= \int \prod_{i=1}^{n} \frac{\mu(B_i)^{k_i}}{k_i!} \exp(-\mu(B_i))\, P_\Lambda(\mathrm{d}\mu)$$

for all $n \geq 1$ and pairwise disjoint $B_1, \ldots, B_n \in \mathcal{B}_0(\mathbb{R}^d)$, where $\mathbf{P}_\Lambda$ denotes the distribution of the random measure Λ.

Equation (4.2) also suggests a simulation approach in some bounded observation window $W \in \mathcal{B}_0(\mathbb{R}^d)$: In a first stage, a random intensity measure Λ is generated in W, say $\mu(\cdot \cap W)$ and then, in a second stage, a Poisson process is simulated according to the intensity measure $\mu(\cdot \cap W)$.

The following theorem relates stationarity of Cox processes to stationarity of their random intensity measures.

Theorem 4.1. *Let $\{S_n, n \geq 1\}$ be a Cox process. Then, $\{S_n, n \geq 1\}$ is stationary if and only if its random intensity measure $\{\Lambda_B, B \in \mathcal{B}(\mathbb{R}^d)\}$ is stationary, i.e.,*

$$(\Lambda_{B_1}, \ldots, \Lambda_{B_n}) \overset{D}{=} (\Lambda_{B_1+x}, \ldots, \Lambda_{B_n+x})$$

for all $n \geq 1$, $B_1, \ldots, B_n \in \mathcal{B}_0(\mathbb{R}^d)$ and $x \in \mathbb{R}^d$.

Exercise 4.3. Provide a proof of Theorem 4.1.

In the following, we introduce an example of doubly stochastic processes, the class of modulated Matérn hard-core point process, which exhibits repulsion of points for *small* distances while clustering of points for *medium* distances. In Fig. 4.4 the principle idea of the modulated Matérn hard-core point process is displayed.

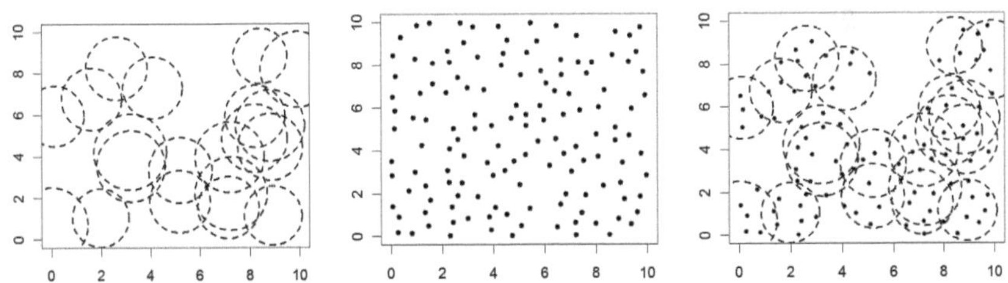

Fig. 4.4 Construction of modulated Matérn hard-core point process (displayed in 2D). First: realization of random sphere system Ξ (left); second: realization of Matérn hard-core process $\{S_n^{(1)}, n \geq 1\}$ (center); third: only points inside sphere system are considered, i.e. $\{S_n^{(1)}, n \geq 1\} \cap \Xi$ (right)

Definition 4.8.

1. Let $\{S_n, n \geq 1\}$ be a stationary Poisson process with intensity $\lambda_0 \in (0, \infty)$.
2. Let Ξ be the random sphere system that evolves by assigning each point S_n the sphere $B(S_n, R_n)$ with iid random radii $R_n \sim F$ following some distribution function $F : [0, \infty) \rightarrow [0, 1]$, i.e.,

$$\Xi = \bigcup_{i=1}^{\infty} B(S_n, R_n),$$

and let $\{R_n, n \geq 1\}$ be independent of $\{S_n, n \geq 1\}$.
3. Let $\left\{ S_n^{(1)}, n \geq 1 \right\}$ be a Matérn hard-core process with intensity λ_1 and hard-core radius r_h, which is independent of $\{R_n, n \geq 1\}$ and $\{S_n, n \geq 1\}$.
4. Then the point process $\left\{ S_n^{(2)}, n \geq 1 \right\}$ given by $\left\{ S_n^{(2)}, n \geq 1 \right\} = \left\{ S_n^{(1)}, n \geq 1 \right\} \cap \Xi$ is called a *modulated Matérn hard-core process*.

Remark 4.3. The modulated Matérn hard-core process introduced in Definition 4.8, is referred to as being *modulated* since a Matérn hard-core process is generated according to the (random) intensity function $\left\{ \lambda_x^{(2)}, x \in \mathbb{R}^d \right\}$ given by

$$\lambda_x^{(2)} = \begin{cases} \lambda_1 \mathbf{E}(v_d(\Xi) \cap [0, 1]^d), & \text{if } x \in \Xi, \\ 0, & \text{if } x \notin \Xi. \end{cases}$$

Moreover, the definition of a 2D elliptical Matérn cluster process is given.

Definition 4.9.

1. Let $\{S_n, n \geq 1\}$ be a stationary Poisson process with intensity $\lambda_0 \in (0, \infty)$.
2. Let $\{\psi_n, n \geq 1\}$ be a sequence of independent and identically distributed (iid) angles with $\psi_1 \sim \text{Unif}[0, \pi)$, independent of $\{S_n, n \geq 1\}$, and let $\left\{ E_{a,b}(o, \psi_n), n \geq 1 \right\}$ be a sequence of ellipses with semi-axes a, b with $a > b > 0$ with random rotation ψ_n around the origin o.
3. Let $\left\{ S_k^{(1)}, k \geq 1 \right\}$, $\left\{ S_k^{(2)}, k \geq 1 \right\}$, ... be a sequence of independent and identically distributed Cox processes with random intensity measures $\left\{ \Lambda_B^{(n)}, B \in \mathcal{B}(\mathbb{R}^d) \right\}$, $n \geq 1$, independent of $\{S_n, n \geq 1\}$ with

$$\Lambda_B^{(n)} = \lambda_1 v_2(B \cap E_{a,b}(o, \psi_n)) \quad \text{for } n \geq 1,$$

where $\lambda_1 \in (0, \infty)$ is some positive intensity.
4. Then the point process $\left\{ \widetilde{S}_n, n \geq 1 \right\} = \bigcup_{n=1}^{\infty} \left(\left\{ S_k^{(n)}, k \geq 1 \right\} + S_n \right)$ is called a *2D elliptical Matérn cluster process*.

Remark 4.4. The point process introduced in Definition 4.9 can be interpreted as follows: Each point process $\left\{ S_k^{(n)}, k \geq 1 \right\}$, $n \geq 1$, can be interpreted as a Poisson process whose points a released in an ellipse with random orientation and deterministic semi-axes a,b. Since each point process $\left\{S_k^{(n)}, k \geq 1\right\}$, $n \geq 1$, is then translated by S_n, the stationary Poisson process $\{S_n, n \geq 1\}$ serves as cluster centers (i.e. centers of the ellipses). Consequently, the intensity λ_0 is the intensity of clusters.

A realization of a 2D elliptical Matérn cluster process is given in Fig. 4.3 (right).

Next, we recall the notion of marked point processes, see also [169]. The idea is to assign each point S_n of some point process $\{S_n, n \geq 1\}$ an additional information L_n, called mark.

Definition 4.10. Let $\{S_n, n \geq 1\}$ be a point process as described above. Furthermore, let $\{L_n, n \geq 1\}$ be a sequence of random variables $L_n : \Omega \to \mathsf{L}$, where $(\Omega, \mathcal{A}, \mathbf{P})$ is the same probability space as for the point process, and $(\mathsf{L}, \mathcal{L})$ be an arbitrary measurable space. Then the sequence $\{(S_n, L_n), n \geq 1\}$ is called a *marked point process*.

4.2.2 Multivariate Time Series

Time series techniques deliver powerful and well-established methods for the analysis and modeling of data measured typically at successive points in time in equidistant time intervals. In particular, the widespread application of time series analysis can be found in the fields of financial mathematics, weather forecast, etc.

In this chapter, ideas from time series analysis are used to describe 3D polygonal tracks. They are incorporated into stochastic 3D models for fiber-based materials.

4.2.2.1 Basic Definitions

A time series is understood to be a realization (or trajectory) of a (time-)discrete stochastic process. In general, stochastic processes can be defined as follows.

Definition 4.11. Let $(\Omega, \mathcal{A}, \mathbf{P})$ be an arbitrary probability space and let $\mathcal{T}$ and E be arbitrary sets called *index set* and *image space*, respectively. Moreover, $(E, \mathcal{E})$ is a measurable space called *state space*, where $\mathcal{E}$ is an arbitrary σ-algebra of E. Then, a *stochastic process* is a family $\{Y_t, i \in \mathcal{T}\}$ of random variables $Y_t : \Omega \to E$ defined on the probability space $(\Omega, \mathcal{A}, \mathbf{P})$. If the index set $\mathcal{T}$ of a stochastic process countable, e.g., $\mathcal{T} = \mathbb{Z}$ or $\mathcal{T} = \mathbb{N}$, then $\{Y_t, i \in \mathcal{T}\}$ is called a *discrete* stochastic process.

Recall that a random counting measure $\{N(B), B \in \mathcal{B}(\mathbb{R}^d)\}$ is also a stochastic process where $\mathcal{T} = \mathcal{B}(\mathbb{R}^d)$ and $E = \{0, 1, \ldots\}$. Furthermore, for each $\omega \in \Omega$ the function $\{Y_t(\omega), t \in \mathcal{T}\}$ is called *trajectory* of $\{Y_t\}$. A stochastic process $\{Y_t\}$ is uniquely determined by its finite-dimensional distributions $\mathbf{P}(Y_{t_1} \in B_1, \ldots, Y_{t_n} \in B_n)$ for all $n \geq 1, t_1, \ldots, t_n \geq 0$, and $B_1, \ldots, B_n \in \mathcal{E}$. In the following, we always consider discrete stochastic processes with state space $(\mathbb{R}^d, \mathcal{B}(\mathbb{R}^d))$, where we speak about an *univariate* process if $d = 1$ and a *multivariate* process if $d > 1$.

Stochastic processes are characterized by their moments. Since Y_t is a random vector for each $t \in \mathcal{T}$, the moments of the process are given in dependence of t. Of particular interest in terms of stochastic processes are their mean-value, variance, and covariance functions.

Definition 4.12. Let us consider a stochastic process $\{Y_t, t \in \mathcal{T}\}$ with $\mathcal{T} \subset \mathbb{R}$.

1. The *mean-value function* $\mu : \mathcal{T} \to \mathbb{R}^d$ of the process $\{Y_t, t \in \mathcal{T}\}$ is defined by $\mu(t) = \mathbf{E} Y_t$ for all $t \in \mathcal{T}$.
2. The *variance function* $\sigma : \mathcal{T} \to \mathbb{R}^d$ of the process $\{Y_t, t \in \mathcal{T}\}$ is defined by $\sigma(t) = \mathbf{var} Y_t$ for all $t \in \mathcal{T}$.
3. The *covariance function* $c : \mathcal{T} \times \mathcal{T} \to \mathbb{R}^{d \times d}$ of the process $\{Y_t, t \in \mathcal{T}\}$ is defined by $c(t,s) = \mathbf{E}\left((Y_t - \mathbf{E}Y_t)(Y_s - \mathbf{E}Y_s)^\top\right)$ for all $t,s \in \mathcal{T}$.

Definition 4.13. A stochastic process $\{Y_t, t \in \mathcal{T}\}$ with $\mathcal{T} \subset \mathbb{R}$ is said to be *weakly stationary* if it holds that $\mu(t) = \mu$ for all $t \in \mathcal{T}$ and $c(t,s) = c(t+h, s+h)$ (and thus $\sigma(t) = \sigma$) for all $t, s \in \mathcal{T}$ and $h \in \mathbb{R}$ such that $t + h \in \mathcal{T}$ and $s + h \in \mathcal{T}$. In this case, μ and σ are called *process mean* and *process variance*, respectively.

After having introduced some basic properties of stochastic processes, we state the notion of a time series.

Definition 4.14. Let $\{Y_t, t \in \mathcal{T}\}$ be a discrete stochastic process. Then, any finite trajectory $\{y_t, t \in \mathcal{T}'\}$ where $\mathcal{T}' \subset \mathcal{T}$ and $|\mathcal{T}'| < \infty$, is called a time series.

Typical examples of time series are the daily stock prices of a stock corporation given for the last two years, or the daily mean temperature measured at the same location over the last 5 years, etc.

The main goal of time series analysis is the prediction of future values of the time series based on the values from the past. A promising approach of forecasting future values is to fit a stochastic process to the observed data and compute predictions based on this model. In the following, we introduce a specific class of discrete stochastic processes (i.e., time series models).

4.2.2.2 Vectorial Autoregressive Processes

In this section, we focus on a discrete stochastic process $\{Y_i\}$, where Y_i is a linear combination of its foregoing variables and a random error term.

Definition 4.15. The *vectorial autoregressive* (VAR) *process* $\{Y_i, i \in \mathbb{Z}\}$ of order $q \geq 0$ and dimension $d \geq 1$ is given by

$$Y_i = \eta + A_1 Y_{i-1} + \ldots + A_q Y_{i-q} + \varepsilon_i \qquad \text{for each } i \in \mathbb{Z}, \tag{4.3}$$

where $\eta \in \mathbb{R}^d$ denotes the intercept vector of the process, and $A_1, \ldots, A_q \in \mathbb{R}^{d \times d}$ are the coefficient matrices. The errors $\{\varepsilon_i, i \in \mathbb{Z}\}$ are assumed to form a sequence of d-dimensional random vectors which are independent and identically distributed with vanishing mean vector $\mathbf{E}\varepsilon_i = o$ and some (non-singular) covariance matrix $\Sigma = \mathbf{E}(\varepsilon_i \varepsilon_i^\top)$. If $\{\varepsilon_i, i \in \mathbb{Z}\}$ is a Gaussian process where $\varepsilon_i \sim N(o, \Sigma)$ is multivariate normal distributed with zero mean and covariance matrix Σ, the $\{Y_i, i \in \mathbb{Z}\}$ is called a *Gaussian* VAR *process*.

Remark 4.5. Let us deonte the d-dimensional unit matrix by I_d. If it holds that $\det\left(I_d - (A_1 z + \ldots + A_q z^q)^{-1}\right) \neq 0$ for all $z \in \mathbb{R}^d$ with $||z|| < 1$, then it follows that $\{Y_i, i \in \mathbb{Z}\}$ is weakly stationary [258, p. 25].

Exercise 4.4. Show that alternatively to formula (4.3) of Definition 4.15, a stationary VAR process of order $q \geq 0$ and dimension $d \geq 1$ can be represented by its *mean-adjusted form*, i.e.,

$$Y_i - \mu = A_1 (Y_{i-1} - \mu) + \ldots + A_q (Y_{i-q} - \mu) + \varepsilon_i \qquad \text{for each } i \in \mathbb{Z}, \qquad (4.4)$$

where $\mu = \mathbf{E}Y_i \in \mathbb{R}^d$ is the *process mean* which can be expressed by means of the intercept vector η of the process, i.e., $\mu = (I_d - A_1 - \ldots - A_q)^{-1} \eta$.

4.3 Stochastic 3D Model for Organic Solar Cells

In the following, we present a stochastic model to describe the 3D morphology of the photoactive layer of hybrid P3HT-ZnO solar cells. The aim is to construct a parameterized stochastic model that can be fitted to a variety of experimental 3D morphologies, which have been processed with different processing conditions, see Fig. 4.5. Note that the structure exhibits an anisotropy in $z-$direction, see Fig. 4.6. This anisotropy is to be captured by the stochastic model.

Organic and hybrid solar cells are a promising alternative to classical silicon solar cells as they offer the prospect of being cheap in production and ecological. However, despite enormous improvements, they still suffer from relatively low efficiencies. Up to now, the most efficient organic solar cells reach power conversion efficiencies of about 10-12%. One approach to optimize efficiency is by optimizing their morphology. To better understand the relation between morphology and efficiency, we present a spatial stochastic simulation model for the morphology of organic solar cells described in [378].

4.3.1 Data and Functionality

In organic solar cells, for a good efficiency, the electron-donor and electron-acceptor materials need to be finely mixed and continuous percolation pathways need to exist

Fig. 4.5 Photoactive layers of P3HT-ZnO solar cells fabricated for different spin coating velocities (resulting in different layer thicknesses). ZnO appears yellow, P3HT transparent. Left: spin coating velocity $\omega = 5000$ rpm (57 nm layer thickness), center: $\omega = 1500$ rpm (100 nm layer thickness), right: $\omega = 1000$ rpm (167 nm layer thickness)

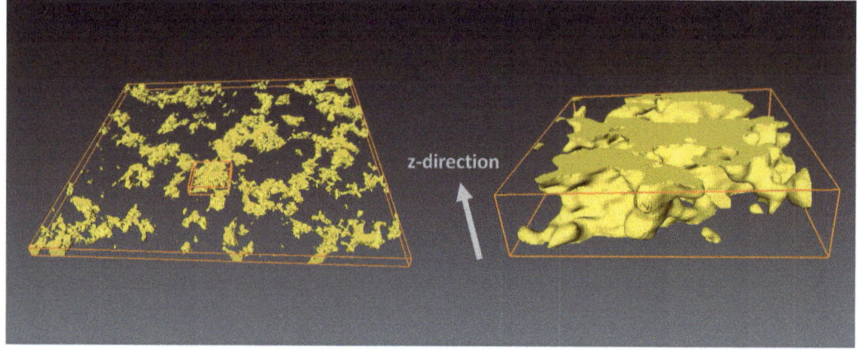

Fig. 4.6 Left: example of solar cell morphology; right: enlarged cutout from the left image

to transport charges towards the electrodes. Upon exposure to light, photons are absorbed in the polymer phase and so-called excitons, i.e., photoexcited electron-hole pairs, evolve. Excitons are neutral quasiparticles which diffuse inside the polymer phase within a limited lifetime; see [365]. If an exciton reaches the interface to the ZnO phase, it is split up into a free electron (negative charge) in the ZnO and a hole (positive charge) in the polymer phase. This process is commonly referred to as quenching. Provided that the electrons in the ZnO phase and the holes in the polymer phase reach the electrodes at the top and bottom of the photoactive layer, respectively, current is generated. A schematic illustration of the morphology of photoactive layers in hybrid polymer-ZnO solar cells is shown in Fig. 4.7, where the electrodes are supposed to be parallel to the x-y-plane. For further information about polymer solar cells and the physical processes therein we refer e.g. to [47].

The stochastic model, presented in this section, is based on a rather complex random sphere system with spheres on different length scales. Thus, we call the model multi-scale sphere model (MSM). The model is fitted to high-resolution three-dimensional data, gained by electron tomography (ET), describing the morphology of photoactive layers consisting of poly(3-hexylthiophene)-ZnO solar cells. In particular, we fit the parameters of the stochastic model to three photoactive layers produced with different production parameters, called 57, 100, and 167 nm films.

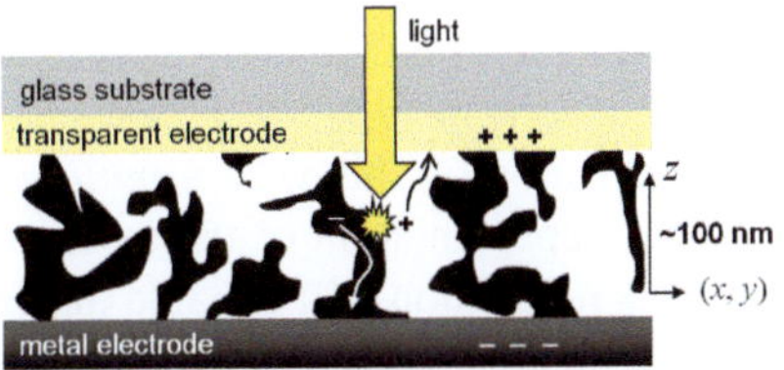

Fig. 4.7 Schematic layout of a polymer-ZnO thin film solar cell, showing the percolation of photogenerated holes $(+)$ and electrons $(-)$

For each of the three different layer thicknesses, the 3D ET images are given as stacks of 2D grayscale images (being parallel to the x-y-plane, say), which are numbered according to their location in z-direction. The sizes of these images in the x-y-plane are 934×911 voxels for the 57 nm film, and 942×911 voxels for the 100 nm film and the 167 nm film. Each voxel represents a cube with side length of 0.71 nm. Binarization has been executed by local thresholding, to compensate for varying brightnesses, see [392].

4.3.2 Data Preprocessing

First, the complexity of the solar cell morphology data - to which the model will be fitted - is split into two different length scales, the macro- and the micro-scale (Fig. 4.8a). The macro-scale, which is obtained by morphological smoothing, contains the main structural features of the ZnO phase. The micro-scale consists of all details that were omitted on the macro-scale.

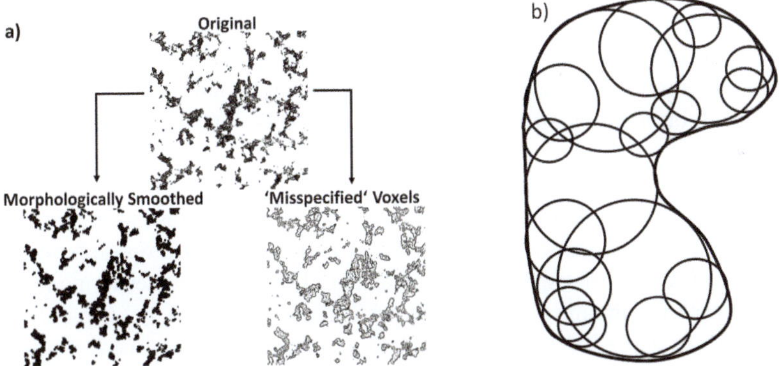

Fig. 4.8 a) Original image split up into structural components at two different length scales (macro-scale and micro-scale), b) schematic 2D representation of a ZnO domain by union of circles

The smoothed ZnO phase (Fig. 4.8a) is then represented by a union of overlapping spheres, such that the agreement is 99 %, see Fig. 4.8b) for a cartoon example in 2D. For more information on this *sphere-putting* algorithm, the reader is referred to [392]. The sphere representation of the macro-scale allows to interpret the smoothed ZnO phase as a realization of a suitable marked point process model, where the marks are the radii and the points the corresponding sphere centers.

4.3.3 Description of Stochastic Model

A stochastic simulation model is developed for each scale individually, i.e., for the macro- and the micro-scale. By this partitioning of complexity, a very flexible model is obtained. The composition of the models for the macro- and the micro-scale is the final 3D simulation model, which we call a multi-scale sphere model (MSM).

4.3.3.1 Simulation Model for the Macro-Scale

The basic idea is to represent the ZnO domains on the macro-scale as a system of overlapping spheres. Therefore, a sphere-putting algorithm is applied to the experimental image data [392]. It represents the morphologically smoothed ZnO domains in an efficient way by unions of overlapping spheres as schematically illustrated in Fig. 4.8b.

This representation allows interpreting the ZnO domains as a realization of a marked point process, where the points are the sphere centers and the marks the corresponding radii. For more information on marked point processes, see [179]. After a thorough analysis of the point patterns for the 57, 100, and 167 nm films, a suitable point process model has been developed which is parameterized and sufficiently flexible to represent all three film thicknesses, i.e., the model type is the same for all three film thicknesses with varying parameters indicating different morphological structures. In particular, it turned out that the point patterns in subsequent z-slices exhibit a strong similarity, which is due to the anisotropy of the solar cell morphologies.

Since the solar cell morphologies exhibit an uniaxial anisotropy in z-direction, the stochastic model should also be anisotropic. In more detail, to include an uniaxial anisotropy into the morphology, say, in z-direction, we propose a multi-layer approach consisting of sequences of correlated 2D point processes to model the 3D point pattern of midpoints. In particular, the 2D point processes, being parallel to the x-y-plane, are described by elliptical Matérn cluster processes, see Definition 4.9. To model the 3D point patterns of midpoints, a Markov chain with stationary initial distribution is constructed, which consists of highly correlated Matérn cluster processes. It can be seen as a stationary point process in 3D.

Point Process Model

The aim is to simulate a 3D point process $\{\{\widetilde{S}_n^{(z)}\}, z \geq 1\}$ of correlated 2D point processes $\{\widetilde{S}_n^{(z)}\}$, where $\{\widetilde{S}_n^{(z)}\}$ is a 2D point process in the plane (or 'slice') $\mathbb{R} \times \mathbb{R} \times \{z\}$. Such a correlation is obtained by constructing a Markov chain of 2D point processes, i.e. given the point process $\{\widetilde{S}_n^{(z)}\}$, the point process $\{\widetilde{S}_n^{(z+1)}\}$ is obtained as a stochastic modification of $\{\widetilde{S}_n^{(z)}\}$. For an overview of the construction of the point process, see Fig. 4.9. As 2D point processes we consider elliptical Matérn cluster processes, where the cluster points are scattered in ellipses of uniformly distributed orientation around their cluster centers, see Definition 4.9.

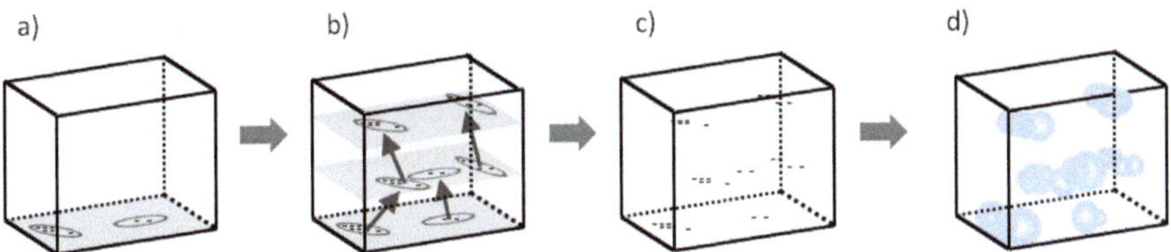

Fig. 4.9 a) First 2D slice of points, b) displacement of clusters, including spatial birth and death, c) resulting 3D point pattern, d) 3D morphology by union of overlapping spheres

Initial 2D Point Process

In the following, we introduce the notation required for the 'initial' point process $\{\widetilde{S}_n^{(0)}\}$.

1. Let $\left\{ S_n^{(0)}, n \geq 1 \right\}$ be a stationary Poisson process with intensity $\lambda_0 \in (0, \infty)$.

2. Let $\left\{ \psi_n^{(z)}, n \geq 1 \right\}$, $z \geq 0$, be sequences of independent and identically distributed angles with $\psi_1^{(0)} \sim \text{Unif}[0, \pi)$ and let $\left\{ E_{a,b}(o, \psi_n^{(z)}), n \geq 1 \right\}$, $z \geq 0$, be sequences of ellipses with semi-axes a, b with $a > b > 0$ with random rotation $\psi_n^{(z)}$ around the origin o.

3. Let $\left\{ \left\{ S_k^{(1),(z)}, k \geq 1 \right\}, \left\{ S_k^{(2),(z)}, k \geq 1 \right\}, \dots \right\}$, $z \geq 0$, be sequences of independent and identically distributed Cox processes with random intensity measures $\left\{ \Lambda_B^{(n),(z)}, B \in \mathcal{B} \right\}$, $n \geq 1, z \geq 1$, where

$$\Lambda_B^{(1),(0)} = \lambda_1 \nu_2(B \cap E_{a,b}(o, \psi_1^{(0)})),$$

and $\lambda_1 \in (0, \infty)$ is some positive intensity.

With these definitions, the 'initial' point process $\{\widetilde{S}_n^{(0)}\}$ in the plane $\mathbb{R} \times \mathbb{R} \times \{0\}$ can be introduced as follows:

$$\left\{ \widetilde{S}_n^{(0)}, n \geq 1 \right\} = \bigcup_{n=1}^{\infty} \left(\left\{ S_k^{(n),(0)}, k \geq 1 \right\} + S_n^{(0)} \right).$$

This initial 2D (elliptical) Matérn cluster process $\{\widetilde{S}_n^{(0)}\}$ in $\mathbb{R}^2$ can be described by a vector of four parameters: $(\lambda_0, \lambda_1, a, b)$, where λ_0 is the intensity of the stationary Poisson point process $\{S_n^{(0)}, n \geq 1\}$ of cluster centers, a and b with $a > b > 0$ are the semi-axes of (random) ellipses $\mathsf{E}_{a,b}(o, \psi_n^{(0)}) \subset \mathbb{R}^2$, and λ_1 is the intensity of the stationary Poisson processes $\{S_k^{(n),(0)}, k \geq 1\}$ of cluster members within the random ellipses.

Transitions of the Markov Chain

To obtain a (vertical) correlation structure between consecutive 2D point processes, we model a 3D point process by a Markov chain of 2D Matérn cluster processes. In particular, the transition from $\left\{\widetilde{S}_n^{(z)}, n \geq 1\right\}$ to $\left\{\widetilde{S}_n^{(z+1)}, n \geq 1\right\}$ is organized such that small displacements of clusters can be included, as well as 'births' and 'deaths' of clusters. In other words, we consider a certain class of spatial birth-and-death processes with random displacement of points, see e.g. [287].

In the following, we introduce some notation that is required for the construction of the Markov chain.

1. Let $\{B_n^{(z)}, n \geq 1\}$, $z \geq 1$, be sequences of independent stationary Poisson point processes in $\mathbb{R}^2$ with intensity λ_0' such that $0 < \lambda_0' < \lambda_0$. The point process $\{B_n^{(z)}, n \geq 1\}$, $z \geq 1$ models the cluster centers of new clusters ('birth').

2. Let $\{\delta_n^{(z)}, n \geq 1\}$, $z \geq 1$, be sequences of independent and identically distributed Bernoulli random variables, which are independent of $\{B_n^{(z)}\}$, where $\mathbf{P}(\delta_n^{(z)} = 1) = p$ for some $p \in (0,1)$. Note that $\{\delta_n^{(z)}\}$ will be used in order to model 'deaths' of complete clusters.

3. Let $\{D_n^{(z)}\}$, $z \geq 1$, be sequences of random displacement vectors with values in $\mathbb{R}^2$, which are independent of $\{B_n^{(z)}\}$ and $\{\delta_n^{(z)}\}$. We assume that the random vectors $D_1^{(z)}, D_2^{(z)}, \ldots$ are uniformly distributed in the set $B(o, r'') \setminus B(o, r')$, where r' and r'' denote the size of minimum and maximum displacement, respectively; $0 < r' < r''$.

Then, a Markov chain $\{\{\widetilde{S}_n^{(z)}\}, z \geq 0\}$ of Matérn cluster processes can be constructed as follows. For $z = 0$, let $\{\widetilde{S}_n^{(0)}\}$ be a 2D elliptical Matérn cluster process as introduced above, i.e.,

$$\left\{\widetilde{S}_n^{(0)}, n \geq 1\right\} = \bigcup_{n=1}^{\infty} \left(\left\{S_k^{(n),(0)}, k \geq 1\right\} + S_n^{(0)}\right).$$

The subsequent point process $\left\{\widetilde{S}_n^{(1)}, n \geq 1\right\}$ is given by

$$\left\{\widetilde{S}_n^{(1)}, n \geq 1\right\} = \bigcup_{n=1, \delta_n^{(1)}=1}^{\infty} \left(\left\{S_k^{(n),(0)}, k \geq 1\right\} + S_n^{(0)} + D_n^{(1)}\right)$$

$$\cup \left(\bigcup_{n=1}^{\infty} \left(\left\{S_k^{(n),(1)}, k \geq 1\right\} + B_n^{(1)}\right)\right).$$

This transition from $\left\{\widetilde{S}_n^{(0)}, n \geq 1\right\}$ to $\left\{\widetilde{S}_n^{(1)}, n \geq 1\right\}$ can be described as follows: All clusters from $\left\{\widetilde{S}_n^{(0)}, n \geq 1\right\}$ with index n with $\delta_n^{(1)} = 0$ are deleted. The surviving clusters are translated by the random displacement vector $D_n^{(1)}$. In addition, some new clusters are born. The cluster centers of these new clusters follow the Poisson process $\{B_n^{(1)}, n \geq 1\}$. The clusters members have the same distribution as in $\left\{\widetilde{S}_n^{(0)}, n \geq 1\right\}$.

The subsequent point processes $\left\{\widetilde{S}_n^{(z)}, n \geq 1\right\}$, $z \geq 2$ follow analogously. Since we want to obtain a stationary point process, it is necessary that all 2D point processes have the same distribution. Note that all clusters $\left\{S_k^{(n),(z)}, k \geq 1\right\}$, $n \geq 1, z \geq 0$ are independent and identically distributed. Thus, it must be assured that the intensity of clusters is equal for all $z \geq 0$. The intensity of clusters of $\left\{\widetilde{S}_n^{(1)}, n \geq 1\right\}$ is given by $\lambda_0 p + \lambda_0'$ (intensity of surviving clusters plus intensity of new clusters). Thus, we require that the 'birth rate' λ_0' and the 'survival probability' p satisfy

$$\lambda_0 p + \lambda_0' = \lambda_0, \tag{4.5}$$

where λ_0 is the intensity of the Poisson process $\{S_n^{(0)}\}$ of cluster centers.

Exercise 4.5. Show that the Markov chain $\left\{\widetilde{S}_n^{(z)}, z \geq 0\right\}$ is stationary if and only if (4.5) holds.

The Markov chain $\{\{\widetilde{S}_n^{(z)}\}, z \geq 0\}$ of Matérn cluster processes introduced above can be seen as point process in 3D. Note that this point process is stationary in 2D, but not (yet) stationary in 3D since points can only occur in the slices $\mathbb{R} \times \mathbb{R} \times \{z\}$. To obtain a stationary 3D point process, we shift each point in z-direction uniformly in the interval $[-0.5, 0.5)$, i.e., as final point process we use

$$\{\{S_n'^{(z)}\}, z \geq 1\} = \{\{\widetilde{S}_n^{(z)} + (0, 0, U_n^{(z)})\}, z \geq 1\},$$

where $\{U_n^{(z)}, n, z \geq 1\}$ is a sequence of iid random variables with $U_n^{(z)} \sim \mathrm{Unif}(-0.5, 0.5]$. The point process $\{\{S_n^{(z)}\}, z \geq 1\}$ has points scattered not only in thin slices $\mathbb{R} \times \mathbb{R} \times \{z\}$, but with continuous z-component. The point process $\{\{S_n'^{(z)}\}, z \geq 1\}$ is stationary in 3D. It possesses seven (free) parameters: $\lambda_0, \lambda_1, a, b$ of the 'initial' point process $\{\{\widetilde{S}_n^{(0)}\}, z \geq 1\}$, and p, r', r'' describing the transitions from step to step, whereas the 'birth intensity' λ_0' of (new) cluster centers is given by $\lambda_0' = \lambda_0(1 - p)$ (due to (4.5)).

Modeling of Radii

Our aim is to model the radii of spheres such that they are possitively correlated in space. Therefore, the radii of spheres are modeled by a Gamma distribution, where the radii of spheres are not modeled just by independent marking, but by the following moving-average procedure, see Fig. 4.10.

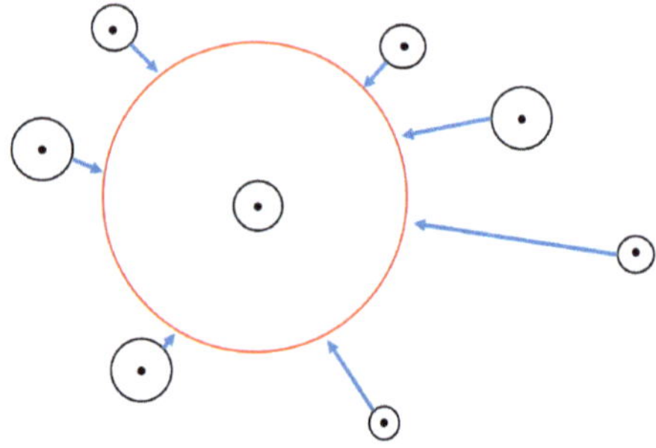

Fig. 4.10 Modeling idea for correlated radii: First, each point is assigned a 'small' random radius. Then, to obtain correlated radii, for each point, the radius is modeled as sum of its own (small) radius and the (small) radii of a fixed number of nearest neighboring points

Let $\{\{S_n^{\prime(z)}\}, z \geq 0\}$ be a configuration of midpoints and let $\{R_n^{(z)}, n \geq 1, z \geq 0\}$ be the radii associated with these midpoints. For some $k \geq 1$, let $\{\widetilde{R}_n^{(z)}, n \geq 1, z \geq 0\}$ be an iid sequence of $\Gamma(q/k, \theta)$-distributed random variables, and let $(z_1, n_1), \ldots, (z_k, n_k)$ for each index (z, n) denote the indices of the k nearest neighbors $S_{n_1}^{\prime(z_1)}, \ldots, S_{n_k}^{\prime(z_k)}$ of $S_n^{\prime(z)}$ (including the point $S_n^{\prime(z)}$ itself). Then, the radius $R_n^{(z)} = \sqrt{3} + \widetilde{R}_{n_1}^{(z_1)} + \ldots + \widetilde{R}_{n_k}^{(z_k)}$ is assigned to the midpoint $S_n^{\prime(z)}$. The reduced radius $R_n^{(z)} - \sqrt{3}$ obtained in this way is $\Gamma(q, \theta)$-distributed. Note that it is reasonable to require a minimum radius of $\sqrt{3}$ since spheres with radii smaller than $\sqrt{3}$ only cover a single voxel or zero voxels when the model is discretized on a voxel grid for later applications.

4.3.3.2 Simulation Model for the Micro-Scale

The simulation model for the micro-scale is used to reinsert all details omitted on the macro-scale. In particular, we differentiate three types of micro-scale components:

1. outer misspecifications (typically small isolated ZnO particles in the polymer phase),
2. interior misspecifications (isolated areas of polymer in the ZnO phase), and
3. boundary misspecifications (typically thin branches of ZnO at the phase boundary).

For each of these components, a suitable model is developed. First, outer misspecifications are modeled by a marked point process (sphere systems), see Fig. 4.11a and 4.11b. Secondly, the boundary misspecifications are modeled by an erosion in dependence of the outer misspecifications (Fig. 4.11c). Finally, interior misspecifications are added by a marked point process (Fig. 4.11d). For details, see [378].

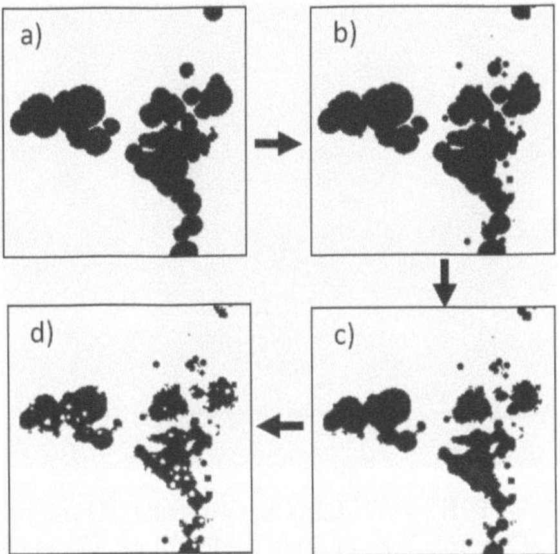

Fig. 4.11 Modeling of micro-scale. a) macro-scale (cutout), b) outer misspecifications added, c) boundary misspecifications corrected, d) interior misspecifications added

4.3.4 Model Fitting

To fit the point process model described in Sect. 4.3.3 to experimental data and determine the parameters of the model we use a minimum contrast method which is widely applied in the literature [73, 179]. Therefore, morphologies are simulated in dependence of a parameter vector $\lambda = (\lambda_1, \ldots, \lambda_n)$ for some $n \geq 1$. Subsequently, image characteristics $F(\lambda)$ of simulated data are compared to their empirical counterparts $\hat{F}$ estimated from experimental image data. The parameter vector λ which minimizes the discrepancy between these image characteristics, defined as the norm $||F(\lambda) - \hat{F}||$, is chosen as a so-called minimum-contrast estimate, i.e. $\widehat{\lambda} = \{\lambda : ||F(\lambda) - \hat{F}|| \text{minimal}\}$.

4.3.5 Model Validation

Fig. 4.12 shows a cutout of the experimental data accompanied by a corresponding simulation and gives a visual impression of the goodness-of-fit that the multi-scale sphere model offers. It is important to note that because we consider a stochastic simulation model it is not intended to match the experimental data, but to match their structural characteristics.

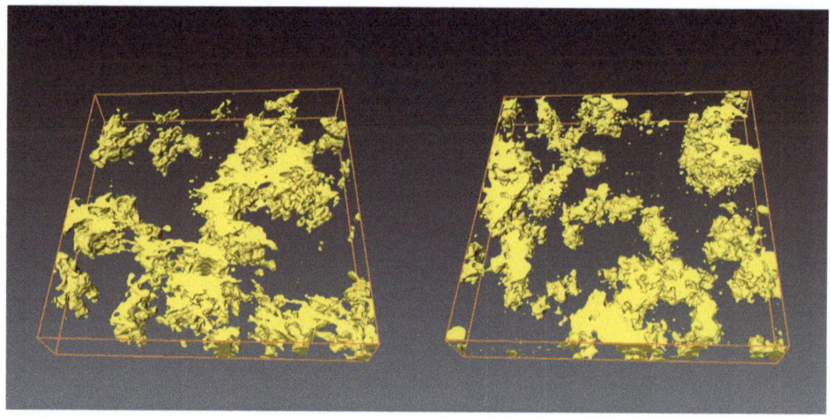

Fig. 4.12 Volume images of P3HT-ZnO. Left: experimental 3D morphology of the 57 nm film (cutout), right: simulated 3D morphology of fitted model for the 57 nm film (cutout). In the images the ZnO volume is yellow and P3HT is transparent against a black background.

Before the model can be applied to improve the understanding of solar cells, it must be assured that it reflects physical properties sufficiently well. Therefore, we compute physically relevant characteristics, like exciton quenching efficiency (relative frequency of excitons being quenched) and connectivity (defined as the fraction of ZnO material connected with the electron collecting top metal electrode), for experimental and simulated data, respectively. In addition, we compute the relative mobility for the P3HT and the ZnO phase, respectively, where relative mobility is the quotient of (electron or hole) mobility in the blend material and mobility in a neat material (consisting of a single phase). Thus, different two-phase morphologies will have different relative mobilities. Note that mobility, in general, is the quotient of the velocity by which charges traverse the system of molecules (in the direction of an electric field) and strength of the electric field. The results displayed in Table 4.1 reveal an excellent correspondence between the experimental and simulated (MSM) morphologies for most of the considered characteristics, where '$\pm$' describes the standard error. Only the values for the mobility in the simulated ZnO phase for the 100 nm and 167 nm films and the connectivity for the 57 nm film differ from the original values computed for experimental image data.

4.3.6 Scenario Analysis

The developed simulation model is fully parameterized. This means that a given solar cell morphology is characterized by the corresponding parameter vector of the

Table 4.1 Characteristics of experimental and simulated (MSM) morphologies

film thickness	connectivity	exciton quenching	relative mobility (polymer phase)	relative mobility (ZnO phase)
57 nm Exper.	0.97	0.43	0.88	0.32
57 nm MSM	0.93 ± 0.01	0.42 ± 0.03	0.89 ± 0.01	0.31 ± 0.01
100 nm Exper.	0.98	0.80	0.76	0.16
100 nm MSM	0.98 ± 0.01	0.81 ± 0.01	0.76 ± 0.02	0.19 ± 0.01
167 nm Exper.	0.97	0.81	0.77	0.15
167 nm MSM	0.96 ± 0.01	0.83 ± 0.01	0.73 ± 0.01	0.11 ± 0.01

model. In extension of the three solar cell morphologies considered so far, the model is fitted to three more solar cell morphologies with film thicknesses of 87 nm ($\omega = 4000$ rpm), 89 nm ($\omega = 2000$ rpm) and 124 nm ($\omega = 1500$ rpm). Thus, altogether the model has been fitted to six different solar cell morphologies fabricated with varying spin coating velocities, and keeping all other experimental parameters fixed. We now fit regression curves f_i for each component λ_i individually, see Fig. 4.13 for an example of the parameters for the macro-scale.

The benefit of the regression curves is that morphologies can be simulated with arbitrary spin coating velocity. Thus, a series of morphologies is simulated for different spin coating velocities (from 500 rpm to 5250 rpm), using the stochastic simulation model (MSM), where the parameters are chosen according to the regression models f_i. Since the layer thickness decreases with increasing spin coating velocity ω, a regression model is fitted to describe layer thicknesses in dependence of ω [379]. The series of virtually simulated morphologies is analyzed in terms of exciton quenching efficiency, see Fig. 4.13. It is found that quenching efficiency is decreasing with increasing ω, whereas the decrease is quite profound for $\omega > 4000$ rpm. In agreement with experimental image data, there is a general trend for the morphology to become coarser (i.e., larger separated domains of both polymer and ZnO) for increasing ω, especially for $\omega > 4000$ rpm, see Fig. 4.14

4.4 Stochastic 3D Modeling of Non-woven GDL

In this section, we present a parametric stochastic 3D model that describes systems of strongly curved and non-overlapping 3D fibers, where, in addition, some superstructures within the fiber system can be included. The stochastic model describing fiber-based materials is applied to non-woven gas-diffusion layers (GDL) in proton exchange membrane fuel cells (PEMFC). More precisely, the parameters of this model are fitted to 3D synchrotron image data of non-woven GDL.

Before discussing the details of this stochastic modeling approach, we first explain the functionality of GDL in PEMFC and the available data.

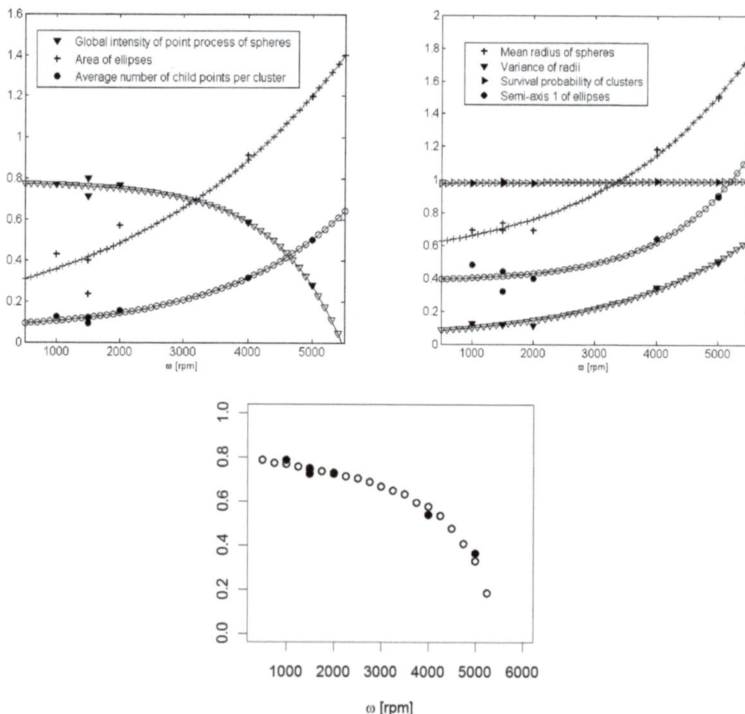

Fig. 4.13 Top: Regression of model parameters of the macro-scale. Bottom: Quenching efficiency in dependence of spin-coating velocity. The open symbols display the regression curves, where the estimated parameters from the experimental data are added by solid symbols

Fig. 4.14 Cutouts of simulated P3HT-ZnO morphologies for $\omega = 1000, 2500, 4000$ and 5000 rpm (from left to right) and window size 355×355 nm^2. Layer thickness varies with ω. In the images the ZnO volume is yellow and P3HT is transparent against a white background.

4.4.1 Data and Functionality

PEMFC are an attractive device for electrical power generation due to their high efficiency and environment-friendly emissions. Moreover, due to their weight, compactness and quick start-up time they are ideal for applications in automobile industry. A key component of PEMFC is their GDL, which has to provide many functions for an efficient operation of the fuel cell: supply of the electrodes with gas, water storage and evacuation, mechanical support of membrane, thermal and electrical conductivity, etc. For further details concerning the functionality of GDL in fuel cells we refer e.g. to [166, 209, 268]. The micro-structure of GDL is closely related to their functionality. For example the porosity of GDL influences several transport processes like the gas flux through the GDL.

Thus, we develop a stochastic 3D micro-structure model for non-woven GDL. The overall goal of this stochastic modeling approach is to provide a methodology which can be used to improve the understanding of the relationship between microstructural characteristics (like porosity, pore size distribution, etc.) and functionality (e.g. mass flow, mechanical stability, etc.).

In our case, we concentrate on the GDL type H2315 produced by the company Freudenberg FFCCT which is a non-woven, carbon fiber-based material where the fibers mainly run in horizontal direction, see Fig. 4.15.

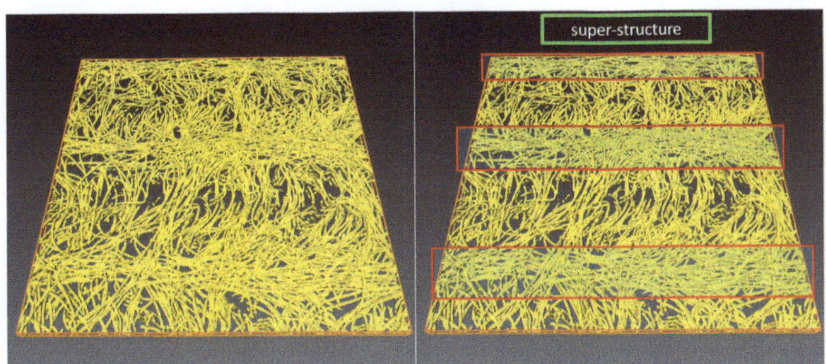

Fig. 4.15 Cut-out of 3D synchrotron data of GDL_{H2315} (left) where its super-structure (right) is highlighted in red

In order to fit the stochastic model such that it describes non-woven GDL, some information about the real micro-structure is needed. Therefore, we consider 3D synchrotron image data of non-woven GDL, see Fig. 4.16 (left). The 3D image data represents a domain of $1250 \times 1250 \times 200 \ \mu m^3$ where the voxel size is equal to $0.833 \ \mu m^3$. In particular, to adequately fit the parameters of the stochastic model to experimental image data of non-woven GDL, we use the following algorithm to extract single fibers from the 3D synchrotron image data.

4.4.2 Data Preprocessing

We briefly describe an algorithm to automatically detect single fibers from 3D tomographic data of fiber-based materials. It is described in detail in [132] and we only describe its basic idea. This algorithm combines tools from image processing and stochastic optimization.

In a first processing step the 3D image is binarized (by global thresholding). Then we focus on the extraction of center lines of single fibers from the binarized image. Due to irregularities like noise or binarization artefacts it is only possible to extract relatively short fragments of the center lines. We thus discuss a stochastic algorithm to accurately connect these parts of the center lines to each other, in order to reconstruct the complete fibers in such a way that the curvature properties of the fibers are represented correctly.

In Fig. 4.16, the result which has been obtained by our extraction algorithm is displayed. The experimental data and the system of fibers extracted from it are in good visual accordance. The detected fibers, represented as polygonal tracks, will be the data basis for fitting the single-fiber model introduced in Sect. 4.4.3.

Fig. 4.16 Cut-out ($830 \times 830 \times 200 \ \mu m^3$) of experimental data gained by synchrotron tomography (left) and extracted fibers (right)

4.4.3 Description of Stochastic Model

The idea for the construction of the stochastic micro-structure model is based on a two-stage approach (c.f. Fig. 4.17), where in a first step, a germ-grain model is used to generate a system of overlapping 3D fibers, see Fig. 4.17 (a). Thereby, the single 3D fibers are modeled by 3D VAR processes describing 3D polygonal tracks, see Section 4.2.2 for the definition of the VAR processes. The germs form a stationary

Poisson point process in 3D and the grains are given by the spherically dilated 3D single-fiber model. In the second step, this system of potentially overlapping fibers is transformed into a system of non-overlapping fibers in the following way: First the fibers are iteratively translated such that they are evenly spread in space (Fig. 4.17, (b,c)). Secondly, as proposed in [4], an iterative avoidance algorithm is applied to the translated fiber system in order to eliminate overlaps between the fibers (Fig. 4.17, (d)).

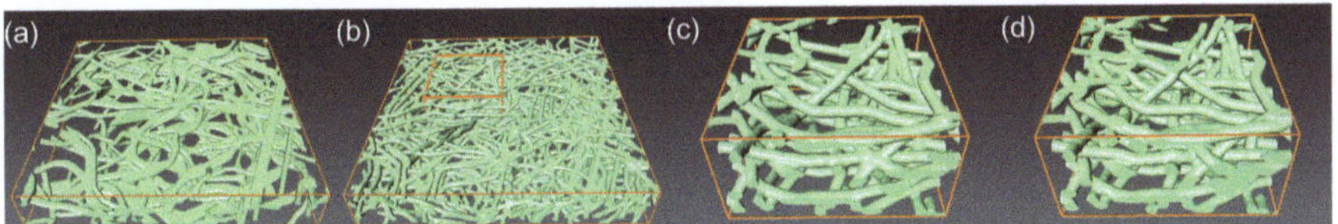

Fig. 4.17 (a) generation of 3D fiber system, (b) iterative translation of fibers such that they are evenly spread in space, (c) cut-out of (b), (d) de-overlapping of fibers

To begin with, we first introduce a stochastic single-fiber model, which describes the typical course of those fibers that have been extracted from 3D synchrotron data using the extraction algorithm described in Sect. 4.4.2. Recall that the extracted fibers are available as polygonal tracks. Thus, our idea for modeling the courses of strongly curved fibers is to consider random polygonal tracks which are based on multivariate time series.

4.4.3.1 Modeling of Single Fibers and Bundles

The following incremental representation of polygonal tracks is useful. Instead of describing a polygonal track $p = \{p_0, p_1, \ldots, p_n\}$ by the endpoints $p_i, p_{i+1} \in \mathbb{R}^3$ of its line segments (p_i, p_{i+1}) we consider an angle-length representation, where we regard the first line segment (p_0, p_1), separately. The further segments of the polygonal track p can then be described by the lengths $\ell_1, \ell_2, \ldots$ of the consecutive line segments and the angles $\alpha_1, \alpha_2, \ldots$ and $\beta_1, \beta_2, \ldots$, where α_i (β_i) denotes the change of direction from the i-th to the $(i+1)$-th segment with respect to the azimuthal (polar) angle. Thus, under the condition that the first line segment is given, a polygonal track is uniquely described by the sequence of vectors $(\alpha_1, \beta_1, \ell_1)^\top, (\alpha_2, \beta_2, \ell_2)^\top, \ldots$, see Fig. 4.18.

Based on the incremental representation of polygonal tracks, we introduce the single-fiber model. The main idea for stochastic modeling of single fibers is to describe polygonal tracks (representing the fibers) by means of a stationary 3D VAR processes (c.f. Definition 4.15) By the usage of multivariate time series we are in a position to include cross-correlations of consecutive line segments into the single-fiber model.

Regarding the incremental representation of a polygonal track, some natural regularity conditions have to be assured, i.e., the changes of directions of consecu-

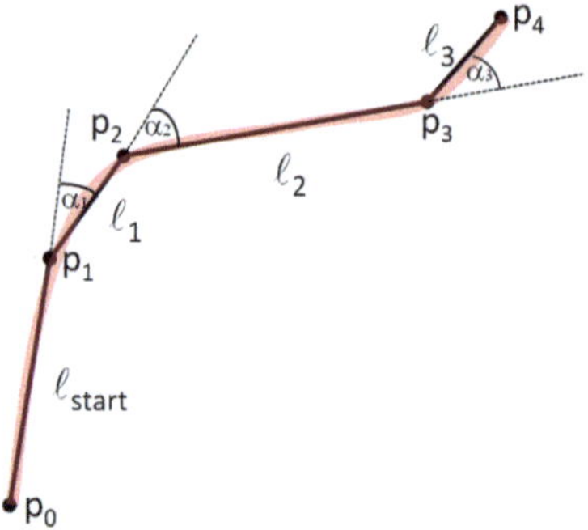

Fig. 4.18 Planar fiber (red) represented by a 2D polygonal track

tive line segments have to be in the interval $[-\pi, \pi)$ and the lengths of line segments have to be non-negative. Thus, we do not directly consider a VAR process as the incremental representation of a random polygonal track, but a modified version $\{F_i, i \geq 1\}$ of it where these conditions are assured, i.e., $F_i = \Psi(\widetilde{F}_i), i \geq 1$, where $\{\widetilde{F}_i, i \geq 1\}$ is a weakly stationary 3D VAR process of order q with parameters $\eta, A_1, \ldots, A_q$ and Σ and the function $\Psi : \mathbb{R}^3 \to [-\pi, \pi)^2 \times \mathbb{R}$ is given by

$$\Psi(r, s, t) = (r - 2k_1\pi, s - 2k_2\pi, \max\{0, t\}) \tag{4.6}$$

if $(2k_1 - 1)\pi \leq r < (2k_1 + 1)\pi$ and $(2k_2 - 1)\pi \leq s < (2k_2 + 1)\pi$ for some $k_1, k_2 \in \mathbb{Z}$.

Thus, the incremental representation of the random polygonal track is given by the modeling components F_0, and $\{F_i, i \geq 1\}$ where $F_0 = (F_{01}, F_{02}, F_{03})$ is the starting line segment ℓ_{start} given in spherical coordinates and $\{F_i, i \geq 1\}$ represents the changes of directions and the lengths of the successive line segments.

By means of F_0, and $\{F_i, i \geq 1\}$, the 3D single-fiber model is provided. It is defined by a finite sequence of dilated line segments $F^S = \{F_0^S, \ldots, F_N^S\}$ in $\mathbb{R}^3$ representing a random 3D fiber of length ℓ and fiber radius r. In other words, the single-fiber model $F^S = \{F_0^S, \ldots, F_N^S\}$ with $N = \min\{k : F_{03} + \sum_{i=1}^k F_{i3} > \ell\}$ is considered, where its dilated line segments F_i^S are given by

$$F_i^S = \begin{cases} (o, L_1) \oplus B(o, r), & \text{if } i = 0, \\ (L_i, L_{i+1}) \oplus B(o, r), & \text{else}, \end{cases} \tag{4.7}$$

with $L_1 = T_{\text{Euclidean}}(F_0)$, $L_{i+1} = L_i + T_{\text{Euclidean}}\left(\sum_{j=0}^i F_{j1}, \sum_{j=0}^i F_{j2}, F_{i3}\right)$ for $i \geq 1$ and $T_{\text{Euclidean}}(\varphi, \theta, s) = (s\sin(\theta)\cos(\varphi), s\sin(\theta)\sin(\varphi), s\cos(\theta))^\top$ being the transformation from spherical to Euclidean coordinates.

In addition, we introduce a model for bundles of fibers where a fiber-bundle F^B consists of a set of parallel single fibers on the basis of the single-fiber model $F^S = \{F_0^S, \ldots, F_N^S\}$ described above. It is defined by

$$F^B = \bigcup_{k=0}^{M} \left(F^S + \delta_k \right) , \tag{4.8}$$

where M indicates the random number of parallel fibers with respect to F^S which is assumed to have a Poisson distribution with some parameter κ. Furthermore, we assume that $\delta_0 = (0,0,0)^\top$ and that the distribution of the random vector $(\delta_1, \ldots, \delta_M)$ is constructed by considering independent and uniformly distributed random vectors $\delta_1, \delta_2, \ldots$ on the boundary of the disc with midpoint o and radius $2r$ under the condition that

$$\min \left\{ \delta_i - \delta_j : i, j \in \{1, \ldots, M\}, \ i \neq j \right\} \geq 2r , \tag{4.9}$$

where the disc is orthogonally orientated with respect to the first line segment F_0^S of the single fiber F^S. Note that the condition considered in (4.9) can be fulfilled with high probability provided that κ is small. It ensures that fibers within bundles do not overlap. If it is not possible to sample δ_j for any $j \in \{1, \ldots, M\}$ such that there occurs no overlapping within the fiber bundle, the radius of the disc is iteratively increased until all M fibers can be placed without overlapping.

4.4.3.2 Stochastic Model for Non-woven GDL

Based on the bundle model introduced in Sect. 4.4.3.1, we now develop a stochastic micro-structure model for the morphology of non-woven GDL. First, we consider a system of overlapping fibers generated by a germ-grain model, where the germs form a stationary Poisson point process in 3D (c.f. Definition 4.4) and the grains are given by the bundle model. Since it is physically impossible that fibers in non-woven GDL overlap mutually, we transform the system of overlapping fibers into a system of spatially regularly distributed, non-overlapping fibers by an iterative translation procedure.

The construction of the 3D micro-structure model for non-woven GDL is motivated by the two-phase superstructure of the experimental GDL data, which is shown in Fig. 4.19 (left). We can clearly see that there is a superlattice of horizontally oriented fibers running parallel to the x-axis in periodic distances. Thus, we subdivide the micro-structure in fiber-channels with some width h where the fibers proceed randomly, and in fiber-bars with some width b where the fibers are mainly running parallel to the x-axis. The basic modeling idea is to first consider a uniformly distributed random variable U on the interval $[0, b + h]$ indicating the location of the starting point of the periodic sequence of fiber-channels and fiber-bars. Then, a stationary Poisson process $\{S_i, i \geq 1\}$ is generated in 3D describing the locations of fibers. Subsequently, each point S_i is marked either with a bundle of fibers drawn from the bundle model introduced in (4.8) if S_i is located in a fiber-channel, or with line segments parallel to the x-axis if S_i is located in a fiber-bar. Note that this is just an approximation of the experimental fiber morphology within the fiber-bars. However, as shown in Sect. 4.4.5, the complete model describes the important morphological and physical characteristics sufficiently well.

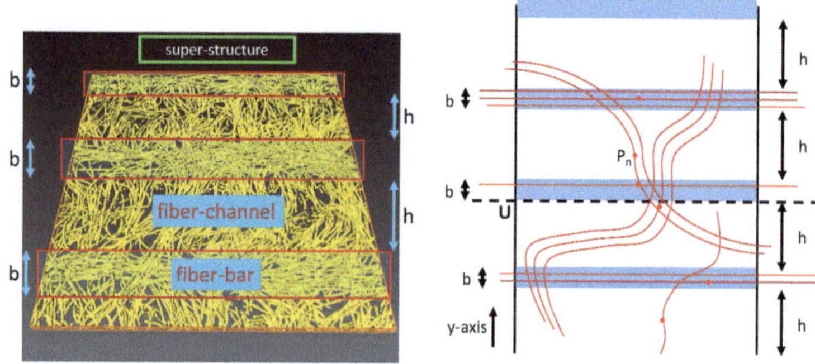

Fig. 4.19 Cut-out of 3D synchrotron data (left), basic idea of bar-channel modeling (right)

The system of overlapping fibers featured with a superlattice is described by a germ-grain model which is given by the set union

$$\Xi = \bigcup_{i=1}^{\infty} \left(X^{(i)} + S_i \right), \tag{4.10}$$

where the germs $S_i = (S_{i1}, S_{i2}, S_{i3})$ form a Poisson point process $\{S_i\}$ in $\mathbb{R}^3$ with some intensity $\lambda > 0$. The grains $X^{(i)}$ are given by

$$X^{(i)} = \begin{cases} F^{B,i}, & \text{if } S_{i2} \in \bigcup_{j\in\mathbb{Z}}[U + j(b+h) - h, U + j(b+h)), \\ F^{C,i}, & \text{else}, \end{cases}$$

where the $F^{B,i}$ are independent copies drawn from the bundle model F^B introduced in (4.8). Furthermore, we define $F^{C,i}$ by

$$F^{C,i} = \bigcup_{j=0}^{M^{(i)}} \left(C + S_i + \delta_j^{(i)} \right),$$

where C is the line segment $C = [-(l/2,0,0)^\top, (l/2,0,0)]$ and $M^{(i)}$ as well as $\delta_k^{(i)}$ are independent copies of M and δ_k considered in (4.8).

Thus, besides the parameters of the bundle model considered in Sect. 4.4.3.1, the model for the system of overlapping fibers has three further parameters λ, b and h which have to be specified.

Subsequently, the transformation of the system Ξ of these potentially overlapping fibers into a system of non-overlapping fibers is discussed. Since the fibers of non-woven GDL are extremely regularly distributed in space, i.e., there exist no large volumes without fibers, we first have to improve the spatial formation of the fibers in Ξ to reduce the amount of vacant volumes in Ξ. This is done by an iterative translation procedure, described in detail in [134].

The resulting fiber system is further transformed to a system of non-overlapping fibers. Therefore, we apply the iterative avoidance algorithm introduced in [4]. The set union $\widetilde{\Xi}$ of the final system of non-overlapping fibers within the sampling window W is denoted by $\widetilde{\Xi} \cap W$.

4.4.4 Model Fitting

The parameters of the VAR process $\{\widetilde{F}_i, i \in \mathbb{Z}\}$ describing the courses of single fibers are estimated by the maximum-likelihood technique, introduced in [133] and further analyzed in [131]. Following the AIC criterion, which is widely applied to estimate the order q of autoregressive processes, see also [131], we obtain $q = 2$ which yields

$$\eta = \begin{pmatrix} 0.003 \\ 0.0008 \\ 23.5 \end{pmatrix}, \ A_1 = \begin{pmatrix} 0.214 & 0.061 & -0.0002 \\ -0.00004 & -0.091 & 0.00002 \\ -0.53 & 1.569 & 0.114 \end{pmatrix},$$

$$A_2 = \begin{pmatrix} 0.106 & 0.039 & 0.0002 \\ -0.001 & -0.11 & -0.00005 \\ -0.321 & 3.846 & 0.025 \end{pmatrix} \text{ and } \Sigma = \begin{pmatrix} 0.08 & 0.00007 & -0.02 \\ 0.00007 & 0.0018 & -0.016 \\ -0.02 & -0.016 & 207 \end{pmatrix}.$$

Furthermore, from the production process of the considered type of non-woven GDL it is known that $r = 4.75 \ \mu m$ and $\ell = 50,000 \ \mu m$. The initial line segment ℓ_0 is chosen equal to $\ell_0 = (0, 1, 0)^\top$. Additionally, it is known from the manufacturer that the width of the fiber-channels is given by $h = 500 \ \mu m$, and the width of the fiber-bars by $b = 70 \ \mu m$. Moreover, it is known that the porosity is 0.765 and, consequently, the volume fraction of the fiber system is 0.235. Based on the extracted fiber system, see Sect. 4.4.2, the expected number of parallel fibers κ is put to $\kappa = 2$ as it is explained in details in [133]. We choose the intensity λ of the 3D Poisson point process representing the locations of fiber-bundles such that the volume fraction of the fiber system $\widetilde{\Xi} \cap W$ coincides with the known volume fraction of 0.235. Therefore, the minimum-contrast method is used in order to estimate the intensity λ of fiber-bundles, i.e.,

$$\lambda = \mathrm{argmin}_\lambda \left| \frac{\nu_3(\widetilde{\Xi} \cap W)}{\nu_3(W)} - 0.235 \right|,$$

where we get that $\lambda = 1.65 \times 10^{-7}$.

4.4.5 Model Validation

In Fig. 4.20 (right) a realization of the fitted GDL model is shown. The synchrotron data, see Fig. 4.20 (left), shows clews of fibers in horizontal direction. They are represented in the stochastic GDL model by horizontal fiber bundles as shown in Fig. 4.20 (right). Although, the fiber-bars modeled by straight fiber bundles are a simplification of the micro-structure of non-woven GDL, the visual agreement between the simulated and experimental images is quite good given the complexity of the fiber system.

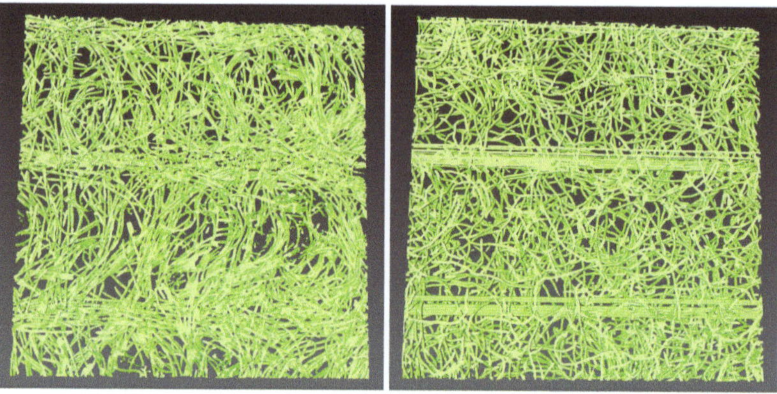

Fig. 4.20 3D synchrotron data (left) and simulated non-woven GDL (right)

We now check more formally if the stochastic 3D model describes the microstructure of the non-woven GDL sufficiently well. Therefore, we consider transport-relevant characteristics computed for the 3D image gained by synchrotron tomography, and for realizations of the non-woven GDL model. The goal is to show that this type of characteristics which have not been used for model fitting match those of the (real) synchrotron image.

As an example of a transport-relevant characteristic, the spherical contact distribution function $H : [0, \infty] \rightarrow [0, 1]$ is computed for experimental and simulated data where $H(r)$ denotes the probability that the minimum distance from a randomly chosen location of the pore phase to the fiber phase is not larger than r. The results of these calculations are given in Fig. 4.21, where we clearly see that there is a quite good accordance between the results obtained for experimental and simulated data. In other words, we see that our model adequately represents this structural characteristic. In [134], further image characteristics are taken into account for model validation, where each characteristic is accurately described by our stochastic model. Thus, the stochastic GDL model proposed in the present section provides a reasonable fit to the experimentally measured 3D image of the GDL., see also [134].

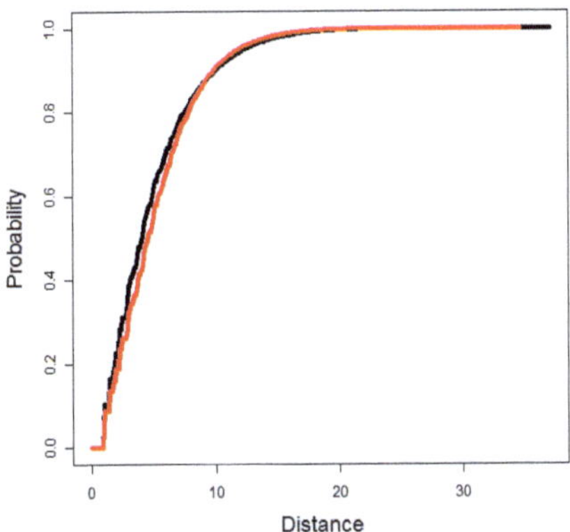

Fig. 4.21 Distribution function of spherical contact distances for experimental data (black) and simulated data (red) drawn from the fitted GDL model

4.5 Stochastic 3D Model for Uncompressed Graphite Electrodes in Li-Ion Batteries

Finally, a stochastic model is presented for efficient simulation of isotropic three-dimensional morphologies consisting of two different phases. We apply this stochastic simulation model to graphite electrodes in Li-ion batteries. In particular, we fit its parameters to 3D image data gained by synchrotron tomography that describes the micro-structure of such graphite electrodes. The model is based on a hybrid approach, where in a first step a random network model is developed using ideas from stochastic geometry. Subsequently, the two-phase morphology model is built by applying simulated annealing to the network model which kind of 'dilates' the network.

4.5.1 Data and Functionality

Li-ion batteries can store electrochemical energy and consist, simplified, of an anode, a cathode and an electrolyte. During discharging, Li atoms release an electron at the anode and the so-formed Li-ions then diffuse from the anode to the cathode via the electrolyte. The electrons also move from the anode to the cathode. During charging, Li-ions move back to the anode and are embedded in the graphite phase. This connection is called *intercalation*. For more details on functionality of Li-ion batteries, we refer to [95].

In particular, the morphology of the anode of Li-ion batteries - as the graphite electrode being studied - is strongly correlated to the electrochemical performance of the battery. For a large capacity as well as a high performance, a high specific surface area of the graphite electrode as well as a large total volume of the graphite phase are beneficial.

We consider as experimental data a 3D tomographic image data displaying the micro-structure of a graphite electrode in Li-ion batteries, see Fig. 4.22 (left). The considered experimental data set has a size of $624 \times 159 \times 376$ voxels where each voxel represents $215 \ nm^3$. Thereby, the considered electrode material consisted of about 95% of graphite and of about 5% of a mixture of conductive carbon and PVDF (polyvinylidendifluoride) as binder material. Note that we model the morphology of the electrode as a two-phase system consisting of solid phase and pore-phase. Thus, we do not distinguish between graphite and binder. Instead, as simplification, we just speak of 'graphite-phase' and 'pore-phase', where the 'graphite-phase' always refers to the material combination consisting of graphite and binder. The overall thickness of the measured sample was about 113 μm, i.e., the anode material had a thickness of about 50 μm on both sides of the copper foil. The material was not loaded with lithium, i.e., raw material. The copper foil was not removed for the measurement in order to avoid any influence caused by sample preparation. For more information, we refer to [380].

4.5.2 Data Preprocessing

In order to fit the stochastic model for two-phase micro-structures introduced in this section to the micro-structure of the tomographic image data described in Sect. 4.5.1, we first extract a 3D network from this data set. Subsequently, we fit the parameters of the stochastic network model that we will introduce to the properties of the extracted network. The network extraction has been performed by a skeletonization tool of Avizo Standard (version *6.3*), see [124, 406], using the default settings. Its idea is to change voxels representing the graphite-phase to pore-phase voxels such that a thin line with thickness one remains. Furthermore, the skeletonization is homotopic, i.e., connectivity-preserving. In a next step, the skeleton is transformed into vector data, i.e., it is approximated by polygonal tracks, see also Fig. 4.22. These polygonal tracks are systems of line segments. The representation by systems of line segments can be interpreted as a spatial network, where the start- and endpoints of the line segments form the set of vertices V and the line segments themselves the set of edges E. Note that for analysis purposes, an edge correction has been performed, where only those line segments are considered whose start- and endpoints are both contained within the image (bounding box).

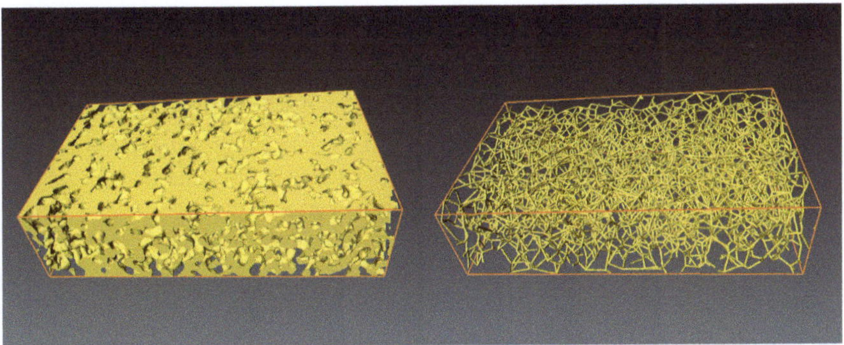

Fig. 4.22 3D image of experimental data (left) and extracted network (right)

4.5.3 Description of Stochastic Model

4.5.3.1 Spatial Random Network Model

The basic idea of the modeling approach is to first consider a class of random 3D networks which describes the essential structural properties of 3D micro-structures consisting of two different phases, and then to 'dilate' the network by simulated annealing. A spatial random network $G = (V, E)$ can be described by a random set of vertices $V = \{S_n, n \geq 1\}$, where S_i is the random location of the ith vertex in $\mathbb{R}^3$, and a random set of edges $E = \{(S_{i_1}, S_{j_1}), (S_{i_2}, S_{j_2}), \ldots\}$ describing the line segments between two adjacent vertices.

The network model is constructed by a two-stage approach, where in the first step, a point process model is used to describe the random positions of the vertices and in a second step, the vertices are connected via edges to form a 3D network.

Stochastic Modeling of Vertices
In this section, a point process model for the vertices of the network is introduced. Note that the random network shall describe the main structural aspects of the morphology, wherefore the network should exhibit some phase separation. Consequently, an appropriate point process model should also separate the two-phase morphology by a corresponding arrangement of points. Thus, the point process model must exhibit a clustering of points which yields a 'rough' phase-separation when connecting neighboring points to form a network. Besides this clustering property, the point process model should also exhibit some repulsion of points for *small* distances. Thus, it is desirable that the edges, put between neighboring points, have a minimum length. Also note that edges whose lengths are smaller than the voxel resolution, when discretizing the model on a voxel grid, cannot be displayed. Thus, some sort of minimum distance between pairs of points (hard-core distance) should be included in the model. To display both properties, repulsion of points for *small* distances and clustering of points for *medium* distances, a modulated Matérn hard-core point process appears suitable, see also Definition 4.8.

We choose the following version of the modulated Matérn hard-core point process. First, a random sphere system $\Xi = \bigcup_{n=1}^{\infty} B(S_n^{(0)}, R_n)$ is considered, where $\left\{ S_n^{(0)}, n \geq 1 \right\}$ is a stationary Poisson process in $\mathbb{R}^3$ with intensity $\lambda_0 > 0$ and $\{R_n, n \geq 1\}$ are independent and identically distributed random radii following some distribution function $F : (0, \infty) \to [0, 1]$. In our case, we assume that $\{R_n, n \geq 1\}$ are uniformly distributed random variables with parameters $0 < r_1 < r_2 < \infty$, i.e., $R_n \sim \mathrm{Unif}(r_1, r_2)$. The union of this random sphere system defines the random area in which points of a stationary Matérn hard-core process $\left\{ S_n^{(1)}, n \geq 1 \right\}$ in $\mathbb{R}^3$ with intensity λ_1 and hard-core radius $r_h > 0$ are released. The modulated Matérn hard-core process $\{S_n, n \geq 1\}$ is then given by

$$\{S_n, n \geq 1\} = \left\{ S_n^{(1)}, n \geq 1 \right\} \cap \Xi.$$

For the intensity λ of $\{S_n, n \geq 1\}$, it holds that

$$\lambda = \lambda_1 \left(1 - \exp(-\lambda_0 \frac{4}{3} \pi (r_2^4 - r_1^4)) \right). \tag{4.11}$$

In this way, a clustering of points for *medium* distances is achieved, while a repulsion of points for *small* distances is assured. The parameters of this point process model are given by $\lambda, \lambda_1, r_1, r_2$ and r_h. The intensity λ_0 of the Poisson process $\left\{ S_n^{(0)}, n \geq 1 \right\}$ is then computed as the solution of (4.11).

Exercise 4.6. Prove formula (4.11).

Stochastic Modeling of Edges

So far, we developed a stochastic model for the random set of vertices $V = \{S_1, S_2, ...\}$. Now we introduce a stochastic model for to put edges between adjacent vertices. The model is similar to an ℓ nearest-neighbor network where for each vertex S_n and $\ell \in \{1, 2, ...\}$, edges are put between S_n and its ℓ nearest neighboring points. However, we modify the ℓ nearest-neighbor network, such that it is possible to control the angles between edges emanating from the same vertex. More precisely, assume that $\{(S_n, S_{n_1}), \ldots, (S_n, S_{n_{k(n)}})\}$ are edges emanating from S_n, where $k(n)$ is the number of these edges. For all pairs of these edges, we consider their angle. Let X_n be the minimum angle, then we define the *minimum angle distribution* as distribution of X_n. The aim is that the edge model has some control on the distribution of these minimum angles. We therefore propose the following model to put edges.

For each vertex $S_n \in V$, we consider its $\ell \geq 1$ nearest neighbors $\{S_{n,(1)}, \ldots, S_{n,(\ell)}\}$, where $S_{n,(i)}$ denotes the ith nearest neighbor of S_n, ordered according to their increasing distance to S_n. We then put edges between S_n and (some of) its ℓ nearest neighbors according to the following rule, where E_n denotes the set of accepted edges, see also Fig. 4.23.

1. Accept the shortest edge $(S_n, S_{n,(1)})$ and put $E_n = \{(S_n, S_{n,(1)})\}$.

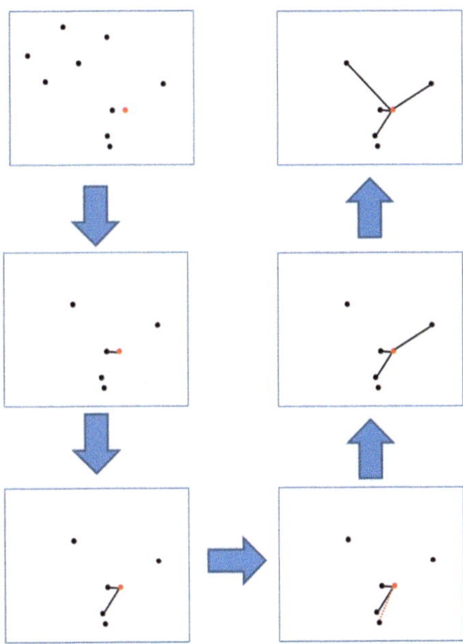

Fig. 4.23 First stage of modeling of edges: connect each point (here: red point) with its nearest ℓ neighbors (here: $\ell = 5$), starting with nearest neighbor, where no edge is put if the angle undercuts a preset threshold (see red, dahsed line)

2. Consider the next nearest neighbor $S_{n,(2)}$. If the angle between $(S_n, S_{n,(2)})$ and every edge in E_n is larger than a certain threshold γ_1, with $\gamma \in (0, \pi)$, then $(S_n, S_{n,(2)})$ is accepted and added to E_n, otherwise rejected.
3. Iteratively, repeat step 2 for $S_{n,(3)}, S_{n,(4)}, \ldots, S_{n,(\ell)}$.

This procedure is accomplished for every vertex S_n, which yields the set $E = \bigcup_{n=1}^{\infty} E_n$ of edges. As we want to implement periodic boundary conditions for the simulated annealing algorithm, this is already done for the network model. Therefore, instead of considering the usual Euclidean distance, a modulo distance is used for computing the distance between two vertices. This means that edges hitting the boundary of the observation window are continued on the opposite site.

In addition to this rule for putting edges as described above, we still perform a certain *post-processing* of edges, see Fig. 4.24. Such a post-processing is reasonable since edges have been placed independently. Thus, an edge $(S_i, S_j) \in E_i$ that has been put from the point of view of the vertex S_i, may interfere with other edges that have been placed from vertex S_j.

To solve this problem we consider the following thinning of edges. Let $(S_i, S_j) \in E$ be an arbitrary (undirected) edge. Then we perform a Bernoulli experiment in order to decide whether (S_i, S_j) is added to a list L of edges that are going to be deleted. Thus, putting $L = \emptyset$ at the beginning, we proceed as follows.

1. The angles between (S_i, S_j) and all edges of the form (S_i, S_f) and $(S_j, S_g) \in E$, where $f \neq i, g \neq j$, are calculated.

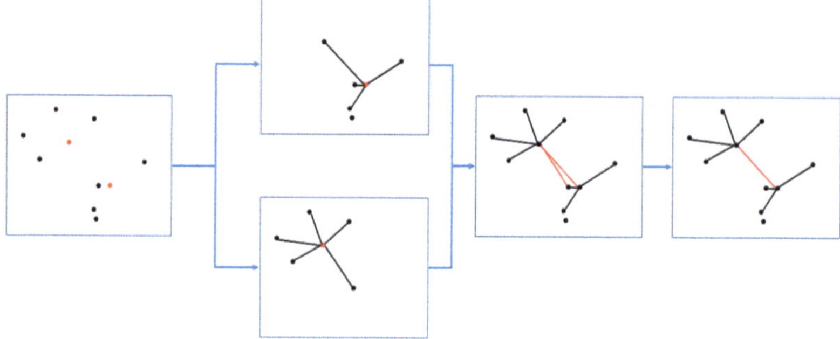

Fig. 4.24 Second stage of modeling of edges: perform postprocessing of edges, where edges that create angles smaller than a certain threshold, are deleted with some probability p

2. If at least one of these angles is less than a certain threshold γ_2, then (S_i, S_j) is added to L with probability of $p \in (0,1)$.
3. Repeat steps 1 and 2 for each edge $(S_i, S_j) \in E$.
4. Take $E^* = E \setminus L$ as the final edge set.

4.5.3.2 Network-based Simulated Annealing

Simulated annealing is a well-established stochastic optimization algorithm with a wide field of applications, such as the traveling-salesman problem, image segmentation, and network partitioning; see [237] for an introduction to this field. It is also a standard method to generate two-phase (or multiple-phase) morphologies on a voxel lattice that are often used as input of physical simulations, see e.g. [409] for an application to organic solar cells. In this section, we briefly describe the basic idea of the simulated annealing algorithm and its specific implementation for the generation of 3D morphologies.

Standard Algorithm
The basic idea of standard simulated annealing is to start with a random distribution of black and white voxels (representing the respective phases of the micro-structure) on a voxel lattice W, e.g. $W = \{1, 2, \ldots, 100\}^3$, with a specified volume fraction of white voxels. Given this initial configuration, a Markov chain Monte Carlo (MCMC) algorithm, see e.g. [237], is used to coarsen the morphology such that a certain value of an image characteristic is met.

In this section, the MCMC algorithm is used to coarsen the morphology such that the specific surface area of the foreground phase matches the specific surface area of some experimental image data. The image characteristic that is used for the coarsening of the blend of black and white voxels is called *cost function*. The MCMC algorithm works as follows, see Chap. 13: two (normally neighboring) voxels are picked at random and exchanged or swapped and the values of the cost function before and after the swap are computed. If the cost function decreases due to the exchange, i.e., the morphology is coarsened, the exchange is accepted, otherwise it is only accepted with a certain acceptance probability. The acceptance probability decreases with time such that swaps that 'refine' the morphology instead of coarsening it (i.e., an increase of the value of the cost function) become less likely. This decrease of the acceptance probability is interpreted as *cooling* of the material and specified by a so-called *cooling schedule*, which is a triple (T, M, c) consisting of an initial temperature T and the number of steps M after which the temperature is decreased by a factor c. The great benefit of simulated annealing – in general – is that it allows changes that temporarily 'worsen' the simulation. But thereby, it is possible to escape from local minima. The coarsening by the MCMC algorithm leads to a structure where voxels are ordered in a special way depending mostly on the chosen cost function but also on the cooling schedule.

In general, an MCMC algorithm is used to generate pseudo random elements $x_0, x_1, \ldots$ following some probability function $\pi : A \to [0, 1]$ for some arbitrary, but finite state space A. This is accomplished by constructing a Markov chain $\{X_n, n \geq 0\}$ with state space A and an (aperiodic and irreducible) transition probability matrix such that π is the limiting distribution of the ergodic Markov chain $\{X_n, n \geq 0\}$, see [50]. After simulating the trajectory of the Markov chain for a larger number of steps, say N, one can assume that X_N is approximately distributed according to π. Given this general setting, simulated annealing, used for the generation of micro-structures, can be specified as follows. Let $W \subset \mathbb{Z}^3$ be a voxel lattice with a finite number $|W|$ of voxels, and $\{0, 1\}^W$ be the set of all binary images on W, where 0 represents a black and 1 a white voxel. A binary image $x \in \{0, 1\}^W$ is then given by $x = (x(v), v \in W)$ with $x(v) \in \{0, 1\}$. Furthermore, let $\beta(x)$ be the specific surface area given the configuration of black and white voxels x and let $\beta_0 > 0$ be some reference value for the specific surface area. For the computation of the (specific) surface area on the voxel grid, an algorithm described in [304] is used. Furthermore, let $p_o \in [0, 1] \cap (\frac{1}{|W|}\mathbb{Z})$ be some reference value for the volume fraction of white voxels. Then the state space A is given by $A = \{(x(v), v \in W) : \sum_{v \in W} x(v) = p_o|W|\}$. The aim of simulated annealing is to generate a configuration of black and white voxels such that the specific surface area matches some references value β_0. Let $A' = \{x \in A : \beta(x) = \beta_0\}$, then the goal is to simulate a configuration $x' \in A'$.

Given an initial configuration $x \in A$, let $\beta(x)$ denote the corresponding specific surface area and let $p_o(x) = \frac{1}{|W|}\sum_{v \in W} x(v)$ be the volume fraction of white voxels of x. We denote $x^{a,b} = (x^{a,b}(v), v \in W)$ where $x^{a,b}(a) = x(b), x^{a,b}(b) = x(a)$, and $x^{a,b}(v) = x(v)$ for $v \neq a, b$. Given this notation, simulated annealing can be described as follows.

1. Generate an initial configuration $x_0 \in A$: Define $x_0 = (x_0(v) = 0, v \in W)$. While $p_o(x) < p_o$, pick a random voxel $v \in W$, and put $x_0(v) = 1$. This generates an initial configuration of black and white voxels where the specific surface area is typically much higher than the reference value β_0.
2a. Put $q = 1$ and repeat steps 2b to 2d until $q = M$.
2b. Pick two neighboring voxels $v, w \in W$ at random such that $x(v) \neq x(w)$

2c. Let $x_n \in A$ be the current state. If $\beta(x_n^{v,w}) - \beta(x_n) \leq 0$, update x_n by $x_{n+1} = x_n^{v,w}$. If $\beta(x_n^{v,w}) - \beta(x_n) > 0$, put $x_{n+1} = x_n^{v,w}$ only with probability $\exp(-(\beta(x^{x,y}) - \beta(x))/T)$, otherwise put $x_{n+1} = x_n$. Put $n = n + 1$.

2d. Put $q = q + 1$ and continue with step 2b

2e. After M steps, put $T = c \cdot T$. Go to step 2a if $\beta(x) > \beta_0$.

Note that especially for larger window sizes run-times are rather large and simulated annealing only provides a limited control of the resulting micro-structure. Therefore, with standard simulated annealing, only small cut-outs of 3D micro-structures can be simulated with reasonable computational effort and improvements of the algorithm are desirable. In the next section, we propose an approach which enables us to simulate 3D micro-structures for window sizes of $100 \times 100 \times 100$ voxels much faster than this is possible with standard simulated annealing.

Combination of Network Model and Simulated Annealing

Standard simulated annealing as described above, can be used to generate 3D morphologies, but run-times are rather large. Moreover, since only two parameters can be adjusted (the values p_o and β_0 of volume fraction and cost function, here: specific surface area, respectively), this algorithm offers only limited control of the resulting morphology of white and black voxels.

We therefore propose another, more efficient approach, which we call *network-based simulated annealing* (NBSA), where first a random 3D network is simulated as explained in Sect. 4.5.3.1, which is then combined with simulated annealing. Thereby the network describes the essential morphological aspects of the micro-structure and serves as a backbone for the simulated annealing algorithm.

Initial Configuration

Instead of throwing uniformly distributed white voxels into the sampling window W, an initial configuration of white voxels is constructed with the previously simulated network. The idea is as follows: The simulated 3D network is discretized on the lattice W, i.e., we put $x(v) = 1$ for those voxels $v \in W$ that belong to the network, and $x(v) = 0$ for those voxels that do not belong to the network. This discretized network indicates voxels around which further white voxels will be located until the volume fraction p_o is reached.

To take the most important advantage of our network-based approach into account, we require that each white voxel of the initial configuration is connected to the network. Therefore, we first choose a voxel $v \in W$ at random. Then, we choose a random direction, either in positive or negative x-, y-, or z-direction. Thus, there are six directions where each of them can be chosen with probability $1/6$. Along the selected direction, we move from $v \in W$ until we either reach a white voxel representing the network or another (white) voxel that has been placed there in an earlier step (and therefore is connected to the network). If v does not hit another white voxel, which may indeed occur, it is rejected and the procedure is repeated with another random voxel $w \in W$. Finally, we put the voxel at the currently reached location to 'white'. This procedure is continued until $\frac{1}{|W|} \sum_{v \in W} x(v) = p_o$.

In this way, we get an initial admissible configuration where every white voxel is connected to the network. In Fig. 4.25, we can see the difference to the initial configuration of the standard simulated annealing algorithm, where voxels are thrown completely at random into the window according to the uniform distribution.

Fig. 4.25 Initial configuration of standard (left) and network-based (right) simulated annealing

Note that the value of the cost function corresponding to the initial configuration described above is typically much closer to the value β_0 of the cost function corresponding to some experimental image data than the value of the cost function corresponding to a purely random initial configuration. This is the main reason why network-based simulated annealing is much faster than the standard version of this algorithm.

Description of Network-Based Algorithm
Besides the different ways to create initial, admissible configurations, there are two further important differences between standard and network-based simulated annealing. First, (white) voxels representing the discretized network may not be changed. Therefore, the discretized network serves as 'rock' and white voxels tend to cluster around it, where the network forms the skeleton of the morphology to be simulated. In this way, i.e., first simulating a random 3D network and then applying simulated annealing, one can nicely control the properties of the resulting morphology.

Consider a window W (here: $W = 100 \times 100 \times 100$), and $x = (x(v), v \in W)$ a binary image on W which displays the network, i.e., $x(v) = 1$ if v belongs to the network and $x(v) = 0$ otherwise. Furthermore, let p_o be the volume fraction (white voxels) of some experimental image data, and β_0 the specific surface area (which we use as cost function). Analogously, let $p_o(x)$ be the volume fraction of image $x \in A$ and $\beta(x)$ the specific surface area. As initial temperature we choose a value (e.g. $T = 0.3$), such that enough changes are accepted. The number M of iterations per step is chosen proportional to the window size (in our case $M = 0.1 \cdot |W|$), and the cooling factor c for the temperature T is put equal to $c = 0.98$ [239].

Finally, as used above for the standard simulated annealing algorithm, we write $x^{a,b} = (x^{a,b}(v), v \in W : x^{a,b}(a) = x(b), x^{a,b}(b) = x(a), x^{a,b}(v) = x(v)$ for $v \neq a,b)$. Then, the network-based simulated annealing algorithm can be described as follows:

1. Generate an admissible configuration $x_0 \in A$ as described above, i.e., a configuration $x_0 = (x_0(v), v \in W)$ with volume fraction p_o such that all white voxels are connected to the network.
2a. Put $q = 1$, $\beta' = \beta(x_n)$, and repeat steps 2b to 2d until $q = M$.
2b. Pick two neighboring voxels $v, w \in W$ at random such that $x(v) \neq x(w)$ and such that neither of the voxels belong to the network.
2c. Let $x_n \in A$ be the current state. If $\beta(x_n^{v,w}) - \beta(x_n) \leq 0$, update x_n by $x_{n+1} = x_n^{v,w}$. If $\beta(x_n^{v,w}) - \beta(x_n) > 0$, put $x_{n+1} = x_n^{v,w}$ only with probability $\exp(-(\beta(x^{x,y}) - \beta(x))/T)$, otherwise put $x_{n+1} = x_n$. Put $n = n + 1$.
2d. Put $q = q + 1$ and continue with step 2b
2e. After M steps, put $T = c \cdot T$ if $(\beta' - \beta(x_n))/\beta' < 5 \cdot 10^{-6}$. Go to step 2a if $\beta(x_n) > \beta_0$.

Note that in contrast to the standard simulated annealing algorithm, we postulate a slightly different condition for the decrease of the temperature T. It is not necessarily changed after M steps but only if the additional condition that $(\beta' - \beta(x_n))/\beta' < 5 \cdot 10^{-6}$ is fulfilled [239]. Recall that periodic boundary conditions are implemented, i.e., swaps over the boundary of the sampling window W are possible. Note that in each iteration step, the surface area has to be calculated to evaluate if a swap of voxels is desired. Here, it is sufficient to only calculate the surface area for a small cut-out, which considerably enhances run-time.

4.5.4 Model Fitting

4.5.4.1 Fitting of Vertex Model

We interpret the vertices of the extracted network as a realization of a stochastic 3D point process, see also Fig. 4.26 for a visualization of the set of vertices.

Hence, the parameters of the vertex model are fitted by choosing appropriate parameters of the modulated hard-core point process $\{S_n\}$ considered in Sect. 4.5.3.1. In total, five parameters have to be estimated: $\lambda, \lambda_1, r_1, r_2$ and r_h. The intensity λ can be easily estimated using the consistent estimator

$$\widehat{\lambda} = \frac{\text{total number of extracted vertices}}{\text{volume of sampling window}},$$

see also [179] for statistical properties of this kind of estimators. Since r_h is the minimum distance between point pairs, we put this model parameter equal to the

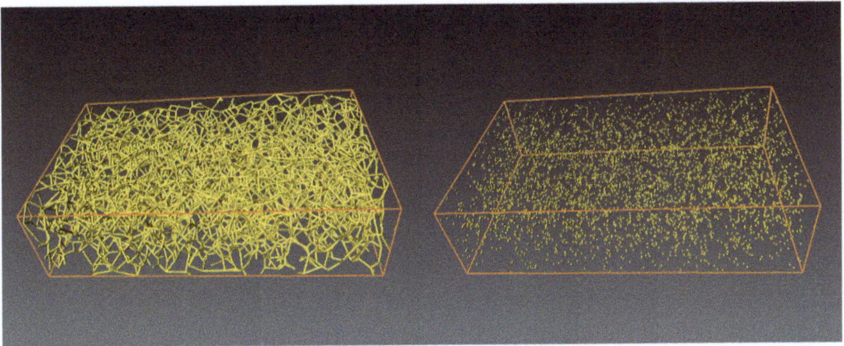

Fig. 4.26 3D extracted network (left) and corresponding vertices (right)

smallest distance between two vertices of the extracted network. The remaining three parameters λ_1, r_1, r_2 are estimated by the minimum-contrast method with respect to the pair-correlation function, i.e., λ_1 and r_1, r_2 are chosen such that the discrepancy $\int_{r'}^{r''} (g(u) - g_{(\lambda_1, r_1, r_2)}(u))^2 \, du$ between the pair-correlation function g computed for the extracted vertices and its model counterpart $g_{(\lambda_1, r_1, r_2)}$ is minimized, where $(r', r'') = (1.2, 16)$ is a suitably chosen interval. Note that the value $g(r)$ of the pair-correlation function is proportional to the relative frequency of point pairs with distance r compared to a stationary Poisson process with the same intensity, see e.g. [21] and Chap. 2. The pair-correlation function g of the point patterns have been computed using a Gaussian kernel density estimator with a bandwidth of 0.04, see Fig. 4.27. The values of the parameters of the fitted point process are given by $\lambda = 1,34 \cdot 10^{-4}, \lambda_1 = 1.94 \cdot 10^{-4}, r_1 = 5, r_2 = 6, r_h = 2$. For small values r, the peak of the estimated pair-correlation function $g(r)$ of the fitted point process is not as high as for the experimental data set. This peak indicates strong clustering of points with medium distances from each other. Nevertheless, the pair-correlation function of the fitted point process is in a good accordance to the experimental one.

4.5.4.2 Fitting of Edge Model

The stochastic edge model introduced in Sect. 4.5.3.1 has four parameters: ℓ, γ_1, γ_2, and p, which have been determined using the minimum-contrast method with respect to the distributions of edge lengths, edge angles, coordination numbers and spherical contact distances. As result we obtain $\ell = 10, \gamma_1 = \pi/2, \gamma_2 = 8/18\pi$ and $p = 0.05$.

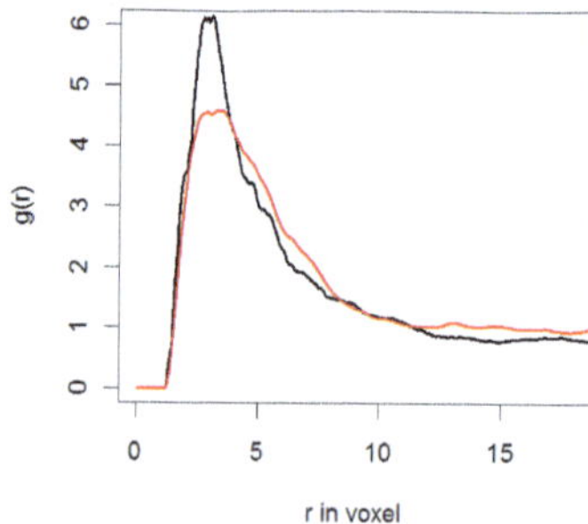

Fig. 4.27 Estimated pair-correlation function for the fitted point process model (red) and extracted vertices (black)

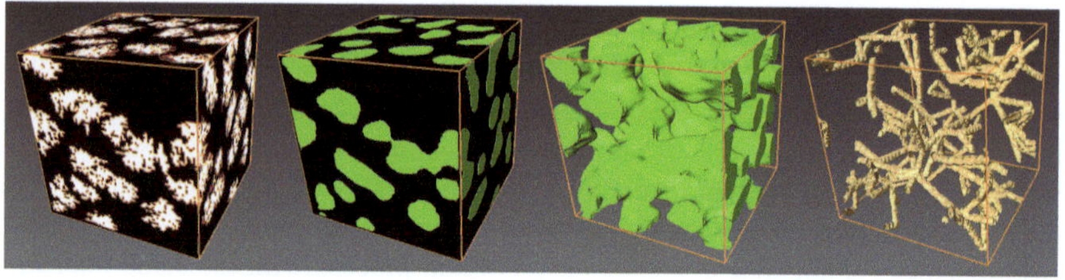

Fig. 4.28 Initial configuration (left) and final result (both center) of network-based simulated annealing and underlying network (right); in the first two images on the left-hand side, only the structures at the boundary of the cube are visualized, whereas in the third and fourth image, the whole structures is visualized (using a transparent pore-phase)

4.5.5 Model Validation

The goal of network-based simulated annealing is to efficiently simulate the microstructure of graphite electrodes as displayed in Fig. 4.29 (left). We therefore combined the simulation of random 3D networks with simulated annealing, where we were matching the volume fraction p_o and the specific surface area β_0 of the experimental image data.

As explained before, the network model can be used to create an improved initial configuration for the simulated annealing algorithm. In Fig. 4.28 an initial configuration of network-based simulated annealing is compared with the corresponding (final) image obtained by this algorithm. We see that in the final simulation result, the clusters of the graphite-phase are at the same locations as in the initial configuration. Thus, the network does indeed serve as backbone of the micro-structure. Furthermore, it is clearly visible in Fig. 4.28 that the network-based algorithm has nicely coarsened the initial configuration. For a visual impression of the goodness-of-fit of network-based simulated annealing, see Fig. 4.29.

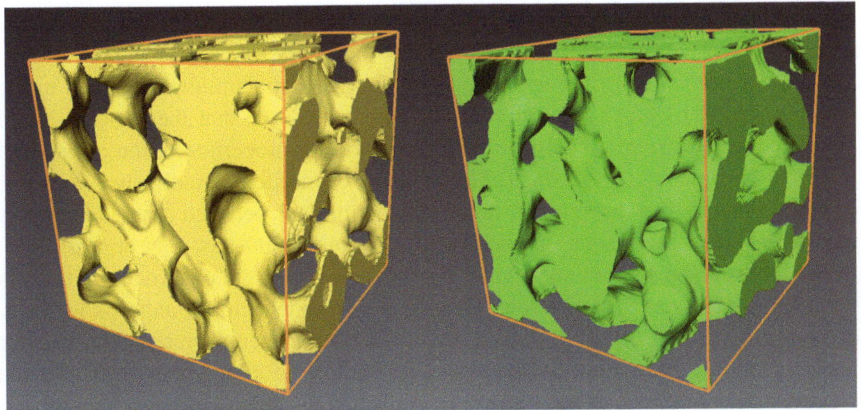

Fig. 4.29 Experimental data (left) and 3D micro-structure obtained by network-based simulated annealing (right)

To formally validate the result of network-based simulated annealing with respect to the 3D morphology of graphite electrodes, we compare several structural characteristics for both experimental and simulated data. To begin with, we compute the distribution functions of spherical contact distances from the pore-phase to the foreground, and vice versa, c.f. Sect. 4.4.5. These characteristics describe the spatial extent of the pore-phase and graphite-phase, respectively. The results displayed in Fig. 4.30 (left and center) show an excellent agreement between experimental and simulated data.

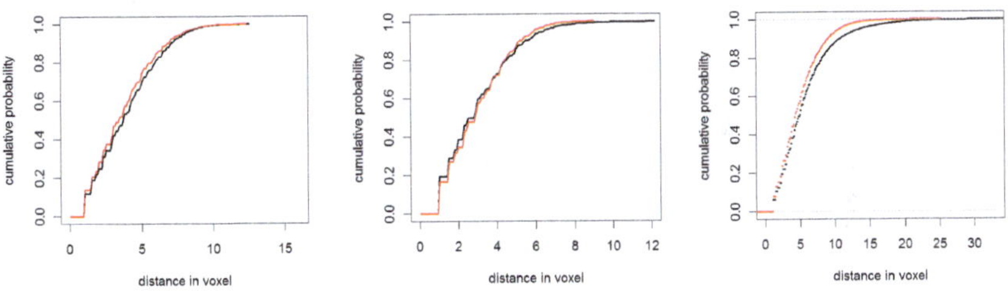

Fig. 4.30 Distribution function of spherical contact distances from pore-phase to foreground (left) and vice versa (center) for simulated (red) and experimental (black) data. Right: spherical contact distances from the edges of the network to the pore-phase

Next, we compare the distribution functions of spherical contact distances from the network to the boundary of the graphite-phase. More precisely, for each (white) voxel from the network, we compute the distance to the nearest pore-phase (i.e., black) voxel. Thus, we describe the spatial extent of the micro-structure, from the point of view of the network. The results in Fig. 4.30 (right) show that the overall agreement is quite good, only large distances are slightly underestimated by the model.

Last but not least, for the application of the stochastic simulation model to the micro-structure of graphite electrodes, it must be assured that the network-based simulated annealing resembles the main connectivity properties of the considered material. Note that the solid phase of the graphite electrode is completely connected. In the 3D image data, however, it can occur that bridges between graphite particles are smaller than the resolution and therefore, isolated clusters may appear. Also, isolated clusters at the boundary may be connected with the electrode via bridges outside of the observation window. Therefore, we made a cluster analysis for both, a cut-out of $130 \times 130 \times 130$ voxels of the 3D experimental image data and for a corresponding simulation. It turns out that 95.54% of experimental 3D graphite electrode is connected, in comparison to 100% for the stochastic model. Note that the underlying spatial random network is connected (with respect to cyclic boundary conditions), as unconnected realizations are rejected. Applying simulated annealing to this connected network will yield a 3D morphology that exhibits values of connectivity close to 100%.

4.5.6 Further Numerical Results

Network-based simulated annealing, as proposed in this chapter, enhances run-time compared to standard simulated annealing. To analyze the difference in computational effort, we consider the surface area of the simulated micro-structure in dependence of the number of iterations used in the algorithm. This is reasonable since the generation of the random network as well as the initial configuration take a negligible amount of time and thus, the coarsening of the morphology is the main factor driving run-time. The results, displayed in Fig. 4.31, show that the network-based simulated annealing reduces the computational effort mainly by the improved initial configuration. Network-based simulated annealing reduces computational effort (given the desired value for the surface area) to 5.89% of the effort required by the standard simulated annealing algorithm.

4.6 Conclusion

This chapter gave an overview how the stochastic 3D modeling of functional materials can be organized. Thereby, three classes of stochastic models are presented describing the micro-morphology of functional materials by means of methods from stochastic geometry. The structure of these materials strongly differ where we consider organic solar cells being a composite of two materials, non-woven gas-diffusion layers in proton exchange membrane fuel cells consisting of a system of

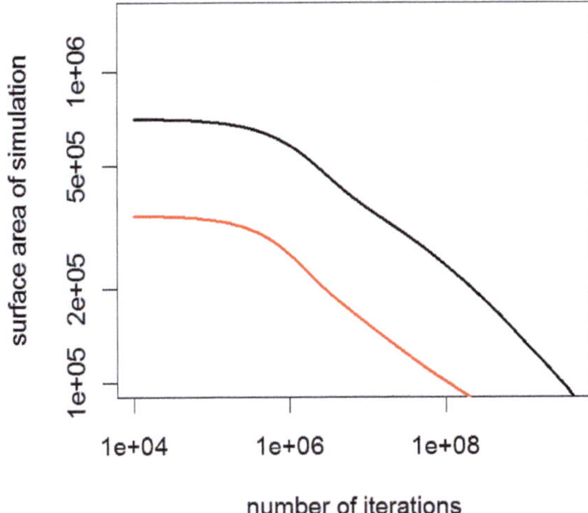

Fig. 4.31 Surface area of simulated micro-structure vs. the number of steps in the algorithm for standard simulated annealing (black) and network-based simulated annealing (red). On both axes, a log-scale is applied

curved carbon fibers and graphite electrodes in Li-ion batteries which consists of a porous two-phase micro-structure.

Since the power of computers increases day by day, simulation-based analysis will be a key aspect in research in the next years. Among other things, functional materials optimization will play a major role in future. We have provided an outlook how spatial stochastic models can be used for material optimization with respect to its functionality.

Acknowledgements This work was partially supported by the German Research Foundation under the priority programme 'Elementary Processes of Organic Photovoltaics' (SPP 1355) and German Federal Ministry of Education and Research in the framework of the programme 'Mathematics for Innovations in Industry and Services'.

Chapter 5
Boolean Random Functions

Dominique Jeulin

Abstract The notion of Boolean random functions is considered which is a generalization of Boolean random closed sets. Their construction is based on the combination of a sequence of primary random functions by the operation $\vee$ (supremum) or $\wedge$ (infimum), and their main properties (among which the supremum or infimum infinite divisibility) are given in the case of scalar random functions built on Poisson point processes. Examples of applications to the modeling of rough surfaces are given.

5.1 Introduction

This chapter reviews a family of random functions (RF), the so-called Boolean random function (BRF) which is an extension of the binary Boolean model and of wide use for applications.

This family owns the interesting property of supremum (or infimum, according to the chosen type of construction) infinite divisibility. Note that these models are particularly interesting for applications in physics, such as in fracture statistics [193, 197, 198]. The basic idea of the BRF was born about the modelling of rough surfaces in 1979, by a generalization of the Boolean model of G. Matheron, where the first presentations and applications are given in [204, 361]. In [190] an anisotropic version of the BRF is developed. Out of the field of materials science, other examples of applications are given e.g. for biomedical images [315, 316, 344], for scanning electron microscope images [364], and for solving problems of exploitation of oceanographic reserves [67]. The first theoretical studies of the BRF are given in [190, 204, 361, 362]. In [363, 364], a general BRF model has been introduced, connected to a non stationary Poisson point process in $\mathbb{R}^{d+1}$. In [315, 317],

Dominique Jeulin
Centre de Morphologie Mathématique, MINES Paris Tech, 77305 Fontainebleau, France, e-mail: dominique.jeulin@mines-paristech.fr

© Springer International Publishing Switzerland 2015
V. Schmidt (ed.), *Stochastic Geometry, Spatial Statistics and Random Fields*,
Lecture Notes in Mathematics 2120, DOI 10.1007/978-3-319-10064-7_5

some characteristic properties of the BRF have been proved, which are useful for the identification of a model from images. Finally, a generalization of the BRF at two levels was proposed [192, 193]: the introduction of Boolean varieties RF (including the Poisson point process as a particular case), and of the multivariate case.

In what follows, a reminder on random closed sets and on semi-continuous RF is given. Then we review the properties of the Boolean RF model. Finally we end this chapter by some practical applications of this model and of some extended models to the simulation of the topography of real rough surfaces.

5.2 Some Preliminaries

The heterogeneity of materials can be handled through a probabilistic approach, which enables us to generate models and simulations of the microstructures. Two-phase media can be modelled by realizations of random closed sets. More general microstructures involve the use of random functions.

5.2.1 Random Closed Sets

When considering two-phase materials (for instance a set of particles $A \subset \mathbb{R}^d$ embedded in a matrix A^c), we use the model of a random closed set (RACS) A, see e.g. [73, 196, 265, 267, 277, 361], fully characterized from a probabilistic point of view by its Choquet capacity $T(K)$ defined on the family of compact sets $K \subset \mathbb{R}^d$, see (5.1) below, where $\mathbf{P}$ denotes a probability measure:

$$T(K) = \mathbf{P}(K \cap A \neq \emptyset) = 1 - Q(K) = 1 - \mathbf{P}(K \subset A^c). \tag{5.1}$$

In the Euclidean space $\mathbb{R}^d$, the Choquet capacity is related to the dilation operation $A \oplus \check{K}$, where $A \oplus \check{K} = \{x - y : x \in A, y \in K\}$. We have

$$T(K_x) = \mathbf{P}(K_x \cap A \neq \emptyset) = \mathbf{P}\left(x \in A \oplus \check{K}\right)$$

where $K_x = K + x$. In practice, if $d = 2$ or $d = 3$, $T(K)$ can be estimated by area fraction measurements on 2D images, or from volume fraction estimation on 3D images (from true microstructures, or from simulations), after a morphological dilation of the set A by the set K [196, 265, 267, 361], or calculated for a given theoretical model. Equation (5.1) can be used for the identification of a model (estimation of its parameters, and test of its validity). Particular cases of morphological properties expressed in terms of $T(K)$ are the volume fraction V_v, the covariance (a useful tool to detect the presence of scales or anisotropies), the distribution of distances of a point in A^c to the boundary of A. The access to 3D images of microstructures by means of X-ray microtomography [98] makes it possible to use 3D compact sets K (like balls $B(r)$ with various radii r) to characterize the random set.

5.2.2 *Upper Semi-Continuous Random Functions*

We consider semi-continuous (upper, lower) random functions $\{Z(x), x \in \mathbb{R}^d\}$, for which the changes of supports by $\vee$ (supremum) or by $\wedge$ (infimum) provide random variables [266]:

$$Z_\vee(K) = \vee_{x \in K}\{Z(x)\},$$

$$Z_\wedge(K) = \wedge_{x \in K}\{Z(x)\}.$$

Note that the domain of a random function can be an arbitrary Borel set $E \subset \mathbb{R}^d$. Random functions (RF) and random sets are related by means of their subgraph and their overgraph.

Definition 5.1. The subgraph Γ^φ of the function $\varphi : E \to \mathbb{R}$ is made of the pairs $\{x, z\}$, $x \in E \subset \mathbb{R}^d$, $z \in \overline{\mathbb{R}}$, with $z \leq \varphi(x)$. The overgraph Γ_φ is made of the pairs $\{x, z\}$, $x \in E$, $z \in \overline{\mathbb{R}}$, with $z \geq \varphi(x)$.

We have the following results connecting semi-continuous functions and closed sets [74].

Proposition 5.1. *The function φ is lower semi-continuous (lsc) if and only if its overgraph Γ_φ is a closed set in $E \times \overline{\mathbb{R}}$; φ is upper semi-continuous (usc) if and only if its subgraph Γ^φ is a closed set in $E \times \overline{\mathbb{R}}$.*

Theorem 5.1. *An upper semi-continuous random function $Z(x)$ defined in $\mathbb{R}^d$ is characterized by its Choquet capacity $T(g)$ defined over lower semi-continuous functions $g : \mathbb{R}^d \to \mathbb{R}$ with a compact support K, where*

$$T(g) = \mathbf{P}(x \in D_Z(g)) = 1 - Q(g), \tag{5.2}$$

$$D_Z(g)^c = \{x \in \mathbb{R}^d : Z(x+y) < g(y) \text{ for all } y \in K\}.$$

Particular cases of the Choquet capacity are obtained from (5.2), depending on the choice of the test function g.

1. Let $g(x_i) = z_i$ for x_i ($i = 1, 2, ..., d$), and $g(x) = +\infty$ else, then we have

$$T(g) = 1 - \mathbf{P}(Z(x_1) < z_1, ..., Z(x_d) < z_d),$$

where $1 - T(g)$ gives the spatial law. In what follows, by $A_Z(z)$ we denote the random closed set obtained by thresholding the RF $Z(x)$ at level z, i.e.

$$A_Z(z) = \{x \in \mathbb{R}^d, Z(x) \geq z\}.$$

2. Let $g(x) = z$ if $x \in K$, and $g(x) = +\infty$ if $x \notin K$, and

$$D_Z(g)^c = \{x \in \mathbb{R}^d : Z(x+y) < z \text{ for all } y \in K\} = \left[A_{Z_\vee(K)}(z)\right]^c.$$

Then, we have

$$Z_\vee(K)(x) < z \text{ iff } K_x \subset (A_Z(z))^c \text{ iff } x \in A_Z(z)^c \ominus \check{K}$$

and

$$Z_\vee(K)(x) \geq z \text{ iff } x \in (A_Z(z)^c \ominus \check{K})^c = A_Z(z) \oplus \check{K}$$

and therefore

$$A_{Z_\vee(K)}(z) = D_Z(g) = A_Z(z) \oplus \check{K}.$$

For this type of test function g, we have that

$$T(g) = \mathbf{P}(x \in D_Z(g)) = 1 - Q(g) = 1 - \mathbf{P}(Z_\vee(K) < z).$$

The Choquet capacity gives the probability distribution of the RF $Z(x)$ after a change of support by $\vee$ over the compact set K.

Let $Z_1(x)$ and $Z_2(x)$ be two upper semi-continuous RF and $Z(x) = Z_1(x) \vee Z_2(x)$. Then, we have

$$\begin{aligned}
(D_{Z_1 \vee Z_2}(g))^c = D_Z(g)^c &= \{x \in \mathbb{R}^d : Z_1(x+y) \vee Z_2(x+y) < g(y) \text{ for all } y \in K\} \\
&= \{x \in \mathbb{R}^d : Z_1(x+y) < g(y) \text{ and } Z_2(x+y) < g(y) \text{ for all } y \in K\} \\
&= D_{Z_1}(g)^c \cap D_{Z_2}(g)^c.
\end{aligned}$$

Therefore

$$D_{Z_1 \vee Z_2}(g) = D_{Z_1}(g) \cup D_{Z_2}(g) \tag{5.3}$$

and

$$T_{Z_1 \vee Z_2}(g) = \mathbf{P}(x \in D_Z(g)) = 1 - Q(g) = \mathbf{P}(x \in D_{Z_1}(g) \cup D_{Z_2}(g)), \tag{5.4}$$

$$Q(g) = \mathbf{P}(x \in D_{Z_1}(g)^c \cap D_{Z_2}(g)^c). \tag{5.5}$$

Furthermore we have that

$$\begin{aligned}
A_Z^c(z) = \{x \in \mathbb{R}^d : Z(x) < z\} &= \{x \in \mathbb{R}^d : Z_1(x) \vee Z_2(x) < z\} \\
&= \{x \in \mathbb{R}^d : Z_1(x) < z, \, Z_2(x) < z\} = A_{Z_1}^c(z) \cap A_{Z_2}^c(z),
\end{aligned}$$

and

$$A_{Z_1 \vee Z_2}(z) = A_{Z_1}(z) \cup A_{Z_2}(z). \tag{5.6}$$

If the RF $Z_1(x)$ and $Z_2(x)$ are independent, then the random sets $D_{Z_1}(g)$ and $D_{Z_2}(g)$ are also independent, and (5.5) writes as

$$\begin{aligned}
Q(g) = \mathbf{P}(x \in D_Z(g)^c) &= \mathbf{P}(x \in D_{Z_1}(g)^c \cap D_{Z_2}(g)^c) \\
&= \mathbf{P}(x \in D_{Z_1}(g)^c)\,\mathbf{P}(x \in D_{Z_2}(g)^c) = Q_1(g)Q_2(g).
\end{aligned}$$

More generally, if $Z(x) = \vee_{i=1}^{i=d} Y_i(x)$ and if the $Y_i(x)$ are independent copies of the same RF $Y(x)$ with $\mathbf{P}(x \in D_Y(g)^c) = Q_Y(g)$, we get that

$$Q_Z(g) = \mathbf{P}\left(x \in D_Z(g)^c\right) = \mathbf{P}\left(x \in \cap_{i=1}^{i=d} D_{Yi}(g)^c\right) = Q_Y(g)^d. \qquad (5.7)$$

5.2.3 Principle of Random Sets and of Random Function Modeling

The main steps to go when designing a random structure model are as follows:

1. Choice of basic assumptions,
2. Computation or estimation of the Choquet's capacity functionnal $T(K)$.

The functionals $T(K)$ and $T(g)$ are obtained as functions depending on

1. the assumptions,
2. the parameters of the model,
3. the compact set K or the function g.

For a given model, the functional T is obtained by theoretical calculations or statistically, either on simulations, or on real structures. This gives access to a possible estimation of the parameters from the "experimental" T, and to tests of the validity of assumption for model identification.

5.3 Boolean Random Functions

In what follows, we review the main properties of the BRF built on Poisson point processes.

5.3.1 Construction of the BRF

We are concerned in this section by BRF with support in the Euclidean space $\mathbb{R}^d$. By $\mu_d(dx)$ and $\theta(dt)$ we denote the Lebesgue measure in $\mathbb{R}^d$ and a σ-finite measure on $\mathbb{R}$ (such that $\int_B \theta(dt)$ remains finite for every bounded Borel set B in $\mathbb{R}$). We consider:

1. a Poisson point process $\mathcal{P}$, with the intensity measure $\mu_d(dx) \otimes \theta(dt)$ in $\mathbb{R}^d \times \mathbb{R}$;
2. a family of independent lower semi-continuous primary RF $\{Z'_t(x), x \in \mathbb{R}^d\}$, with a subgraph $\Gamma^{Z'_t} = A'(t)$ having almost surely compact sections $A_{Z'_t}(z)$.

Definition 5.2. The Boolean random function (BRF) with the primary function $Z'_t(x)$ and with the intensity $\mu_d(dx) \otimes \theta(dt)$ is the RF $\{Z(x), x \in \mathbb{R}^d\}$ obtained by

$$Z(x) = \vee_{(t_k, x_k) \in \mathcal{P}} \{Z'_{t_k}(x - x_k)\}. \qquad (5.8)$$

We can notice the following points.

1. This definition, given in [193], is more general than the one proposed by J. Serra in [363, 364]; it covers the previous definitions:

   1.1. the Boolean islands, for which the measure $\theta(dt)$ is the Dirac distribution concentrated in a point t of $\mathbb{R}$: $\theta(dt) = \theta\delta_0(t)$ [190, 204, 361];

   1.2. the "generalized" BRF, where we have $Z'_t(x) = Y'_t(x) + t$, where $Y'_t(x)$ is a family of primary RF. The addition of t comes from a definition of the BRF as a non stationary Boolean RACS in $\mathbb{R}^{d+1}$ with Poisson germs in $\mathbb{R}^{d+1}$ and with primary random sets $A'(t)$ defined at the origin (o,o) of the coordinates of $\mathbb{R}^{d+1}$. To introduce BRF on more general lattices, where the addition is not necessarily defined, this construction process cannot be used.

2. The parameter t, which can be assimilated to z in the definition [363, 364], as for the examples given in Sect. 5.3.8, can also be interpreted as a time, leading to the notion of sequential RF. In these conditions, for the time interval $(t, t + dt)$ is defined an infinitesimal BRF.

3. It is possible to parametrize the primary functions by $t \in \mathbb{R}^k$, with a σ−finite measure $\theta(dt)$ on $\mathbb{R}^{d-k}$.This enables us to introduce a primary function depending on several indexes. Instead of $\mathbb{R}^k$, an abstract space E and a measure θ defined on E can be chosen. Similarly, the Lebesgue measure on $\mathbb{R}^d$ can be replaced by a σ-finite measure $\theta(dx)$ on $\mathbb{R}^d$, dropping the stationarity in $\mathbb{R}^d$. This process can be used to build multiscale RF, as illustrated in Sect. 5.4.

4. From (5.8), we get that the "floor" value of $Z(x)$ is $-\infty$. This value can be bounded (z_0) by use of primary functions such that $A_{Z'_t}(z_0) = \mathbb{R}^d$, or by taking $Y(x) = z_0 \vee Z(x)$.

5. From lower semi-continuous primary functions $Z'(t)$ (with overgraph $\Gamma_{Z'_t}$), it is possible to build a $\wedge$ BRF [193], by replacing in (5.8) the operation $\vee$ by $\wedge$, and starting from a $+\infty$ ceiling value of $Z(x)$. It is equivalent to build a $\vee$ BRF Y from the primary RF $Y'_t(x) = -Z'_t(x)$ and to consider as a $\wedge$ BRF $Z(x) = -Y(x)$. For this reason, we limit this presentation mainly to the $\vee$ BRF given by (5.8).

6. From the point of view of subgraphs (closed in $\mathbb{R}^{d+1}$ for lower semi-continuous functions), the relation (5.8) involves:

$$\Gamma^Z = \cup_{(t_k,x_k)\in\mathcal{P}}A'(t_k)_{x_k}. \tag{5.9}$$

By definition, Γ^Z is a Boolean RACS in $\mathbb{R}^d$ with primary grain $A'(t)$.

For illustration, some examples of simulations of BRF are shown in Fig. 5.1.

5.3.2 BRF and Boolean Model of Random Sets

Using (5.6) and (5.8), we have for the BRF $Z(x)$ that

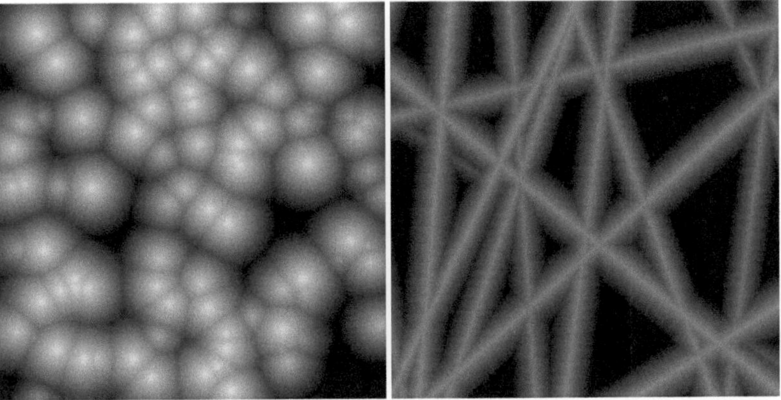

Fig. 5.1 Examples of realizations of a Boolean random function with cone primary functions (left), and built from Poisson lines (right).

$$A_Z(z) = \cup_{(t_k,x_k)\in\mathcal{P}} A_{Z'_{t_k}}(z)_{x_k}.$$

As a consequence, we obtain the following result.

Proposition 5.2. *Every RACS $A_Z(z)$ obtained by thresholding a BRF $\{Z(x), x \in \mathbb{R}^d\}$ at level z is a Boolean random set with primary grain $A_{Z'_t}(z)$.*

This property will be useful for the identification of BRF models, since available tools for the Boolean model can be used for this purpose.

5.3.3 Choquet Capacity of the BRF

As mentioned in Theorem 5.1, we can characterize a BRF by means of the functional $T(g)$ defined on lower semi-continuous functions g with a compact support K, i.e.,

$$T(g) = \mathbf{P}(x \in D_Z(g)); D_Z(g)^c = \{x \in \mathbb{R}^d : Z(x+y) < g(y) \text{ for all } y \in K\}.$$

Since $D_{Z_1 \vee Z_2}(g) = D_{Z_1}(g) \cup D_{Z_2}(g)$, we get for a BRF $Z(x)$ that

$$D_Z(g) = \cup_{(t_k,x_k)\in\mathcal{P}} D_{Z'_{t_k}}(g)_{x_k} \tag{5.10}$$

and $D_Z(g)$ is a Boolean RACS with the primary grain $D_{Z'_t}(g)$. Since $D_Z(g)$ corresponds to the event $A^c(Z) = \{Z(x+y) \geq g(y) \text{ for some } y \in \mathbb{R}^d\}$, the following results are true.

Theorem 5.2. *Consider a BRF $\{Z(x), x \in \mathbb{R}^d\}$ and a lower semi-continuous function g translated in x. The number of primary functions Z'_t for which the event $A^c(Z'_t)$ is satisfied, follows a Poisson distribution with parameter $\int_{\mathbb{R}} \overline{\mu}_d(D_{Z'_t}(g))\, \theta(dt)$.*

Theorem 5.3. *The Choquet capacity of the BRF $\{Z(x), x \in \mathbb{R}^d\}$ is given by*

$$1 - T(g) = Q(g) = \exp\left(-\int_{\mathbb{R}} \overline{\mu}_d(D_{Z'_t}(g))\, \theta(dt)\right). \tag{5.11}$$

For the Boolean islands model with $\theta(dt) = \theta\,\delta_0(t)$ and $Z'_0 = Z'$, it holds that

$$1 - T(g) = Q(g) = \exp\left(-\theta\overline{\mu}_d(D_{Z'}(g))\right). \tag{5.12}$$

As particular functions g, let us examine the following cases:

1. If $g(x_i) = z_i$ for points x_i $(i = 1, 2, ..., n)$, and $g(x) = +\infty$ else, we obtain the spatial law of the BRF with

$$1 - T(g) = \mathbf{P}\left(Z(x_1) < z_1, ..., Z(x_d) < z_d\right)$$
$$= \exp\left(-\int_{\mathbb{R}} \overline{\mu}_d(A_{Z'_t}(z_1)_{x_1} \cup ... \cup A_{Z'_t}(z_d)_{x_d})\, \theta(dt)\right). \tag{5.13}$$

For a single point x, the cumulative distribution function $F(z)$ is obtained, where

$$F(z) = \mathbf{P}(Z(x) < z) = \exp\left(-\int_{\mathbb{R}} \overline{\mu}_d(A_{Z'_t}(z))\, \theta(dt)\right). \tag{5.14}$$

For two points x and $x + h$, formula (5.13) gives the bivariate distribution $F(h, z_1, z_2)$ as a function of the cross geometrical covariogram $K(h, z_1, z_2, t)$ between the two sets $A_{Z'_t}(z_1)$ and $A_{Z'_t}(z_2)$, where

$$F(h, z_1, z_2) = \mathbf{P}(Z(x) < z_1, Z(x+h) < z_2)$$
$$= \exp\left(-\int_{\mathbb{R}} \overline{\mu}_d(A_{Z'_t}(z_1) \cup A_{Z'_t}(z_2)_{-h})\, \theta(dt)\right)$$
$$= F(z_1)F(z_2)\exp\left(\int_{\mathbb{R}} \overline{\mu}_d(A_{Z'_t}(z_1) \cap A_{Z'_t}(z_2)_{-h})\, \theta(dt)\right)$$
$$= F(z_1)F(z_2)\exp\left(\int_{\mathbb{R}} K(h, z_1, z_2, t)\, \theta(dt)\right).$$

From (5.15), it is clear that for the BRF we always have $F(h, z_1, z_2) \geq F(z_1)F(z_2)$, so that no negative correlation can occur.

2. If $g(x) = z$ for $x \in K$ and $g(x) = +\infty$ else, K being a compact set, formula (5.11) enables us to calculate the distribution of $Z(x)$ after a change of support by the operator $\vee$ taken over the compact set K $(Z_\vee(x) = \vee_{x \in K}\{Z(x)\})$; we have in that case $D_{Z'_t}(g) = A_{Z'_t}(z) \oplus \check{K}$ and

$$\mathbf{P}\left(Z_\vee(K) < z\right) = \exp\left(-\int_{\mathbb{R}} \overline{\mu}_d(A_{Z_t'}(z) \oplus \check{K})\, \theta(dt)\right). \qquad (5.15)$$

From (5.8) and (5.10), the following result is obtained.

Proposition 5.3. *The RF* $\{Z_\vee(x), x \in \mathbb{R}^d\}$ *is a BRF with the primary function* $\{Z_\vee'(x), x \in \mathbb{R}^d\}$.

All previous results can be specialized to the Boolean islands version of the model, when $\theta(dt) = \theta \delta_0(t)$ and $Z_0' = Z'$. Starting form the Choquet capacity given in (5.12), for the spatial law we have that

$$
\begin{aligned}
1 - T(g) &= \mathbf{P}\left(Z(x_1) < z_1, ..., Z(x_d) < z_d\right) \\
&= \exp\left(-\theta \overline{\mu}_d(A_{Z'}(z_1)_{x_1} \cup ... \cup A_{Z'}(z_d)_{x_d})\, \theta(dt)\right).
\end{aligned}
$$

The bivariate distribution is given by

$$F(h, z_1, z_2) = F(z_1)F(z_2)\exp\left(\theta K(h, z_1, z_2)\right)$$

and the change of support by the operator $\vee$ follows

$$\mathbf{P}\left(Z_\vee(K) < z\right) = \exp\left(-\theta \overline{\mu}_d(A_{Z'}(z) \oplus \check{K})\right).$$

5.3.4 Supremum Stability and Infinite Divisibility

Let $Z_1(x)$ and $Z_2(x)$ be independent BRF with the primary functions Z_{1t}' and Z_{2t}', and the intensities $\theta_1(t)$ and $\theta_2(t)$. From (5.9) we get that

$$\Gamma^Z = \Gamma^{Z_1} \cup \Gamma^{Z_2} = \cup_{(t_k, x_k) \in \mathcal{P}_1} A_1'(t_k)_{x_k} \,\cup_{(t_k, x_k) \in \mathcal{P}_2} A_2'(t_k)_{x_k},$$

and therefore Γ^Z is a Boolean model in $\mathbb{R}^{d+1}$; as a consequence, $Z(x)$ is a BRF with intensity $\theta(t) = \theta_1(t) + \theta_2(t)$ and with a mixture of primary functions.

Proposition 5.4. *Every supremum of a family of independent BRF* $Z_i(x)$ *is a BRF with intensity* $\theta(t) = \sum_i \theta_i(t)$. *The BRF is stable with respect to the supremum.*

The supremum stability property in Proposition 5.4 of the BRF is shared with more recent RF models, namely so-called max-stable processes.

As a consequence of the infinite divisibility of the Boolean model with respect to $\cup$, we get the following result.

Theorem 5.4. *Every BRF* $Z(x)$ *is infinite divisible for* $\vee$, *i.e., for all* $n \geq 1$ *it holds that* $Z(x) \equiv \vee_{k=1}^{k=d} Z_k(x)$ *where the* Z_k *are independent and identically distributed BRF.*

This results immediately from the expression for the Choquet capacity of the BRF given in (5.11) and from the Choquet capacity of $\vee_{k=1}^{k=d} Z_k(x)$ derived from (5.7): for any integer d, we have

$$1 - T(g) = Q(g) = \exp\left(-\int_{\mathbb{R}} \overline{\mu}_d(D_{Z'_t}(g)) \, \frac{\theta(dt)}{d} \right)^d .$$

5.3.5 Characteristics of the Primary Functions

Some characteristics of the pair (intensity, primary function) can be determined from information on the BRF $Z(x)$. These characteristics are directly deduced from the Choquet capacity given in (5.11) and from the derived properties in (5.13-5.15).

5.3.5.1 Transformation by Anamorphosis

Let φ be an anamorphosis transformation (namely a monotonous non-decreasing transformation mapping the real line $\mathbb{R}$ into $\mathbb{R}$).

Proposition 5.5. *Every anamorphosis transformation $Y = \varphi(Z)$ of a BRF Z with primary function Z', is a BRF with the same intensity $\theta(t)$ and with the primary function $Y' = \varphi(Z')$.*

Proof. We have

$$
\begin{aligned}
A_{\varphi(Z)}(z) &= \{x \in \mathbb{R}^d : \varphi(Z(x)) \geq z\} \\
&= \{x \in \mathbb{R}^d : Z(x) \geq \varphi^{-1}(z)\} \\
&= A_Z(\varphi^{-1}(z)) \\
&= \cup_{(t_k,x_k)\in\mathcal{P}} A_{Z'_{t_k}}(\varphi^{-1}(z))_{x_k} \\
&= \cup_{(t_k,x_k)\in\mathcal{P}} A_{Y'_{t_k}}(z)_{x_k} \\
&= A_Y(z).
\end{aligned}
$$

This result enables us to restrict our study to strictly positive BRF, since it is always possible to transform any function Z into a positive function $Y = \varphi(Z)$ (consider for instance the anamophosis obtained by an exponential transformation).

5.3.5.2 Moments of $Z'_\vee(K)$ and Mathematical Expectation of the Anamorphosed of $Z'_\vee(K)$

We now consider positive BRF.

Proposition 5.6. *It holds that*

$$M(i,K) = -\int_{\mathbb{R}} z^{i-1} \log\left(\mathbf{P}\left(Z_\vee(K) < z\right)\right) dz \tag{5.16}$$

$$= \frac{1}{i} \int_{\mathbb{R}^+} \mathbf{E}\left(\int_{\mathbb{R}^d} (Z'_{t\vee}(K)(x))^i \, dx\right) \theta(dt). \tag{5.17}$$

Let $\Phi(z)$ a strictly positive function with $\Phi(z) = \int_0^z \varphi(u) \, du$. Then,

$$-\int_{\mathbb{R}} \varphi(z) \log\left(\mathbf{P}\left(Z_\vee(K) < z\right)\right) dz$$
$$= \int_{\mathbb{R}^+} \mathbf{E}\left(\int_{\mathbb{R}^d} \Phi(Z'_{t\vee}(K)(x)) \, dx\right) \theta(dt). \tag{5.18}$$

Proof. By $\mathbf{1}_{Z'_t \geq z}(x)$, we denote the indicator function of the set $A_{Z'_t}(z)$ at point x (i.e. $\mathbf{1}_{Z'_t \geq z}(x) = 1$ if $Z'_t(x) \geq z$, and $\mathbf{1}_{Z'_t \geq z}(x) = 0$ else). For a given realization of the primary function we have that

$$\mu_d(A_{Z'_t}(z)) = \int_{\mathbb{R}^d} \mathbf{1}_{Z'_t \geq z}(x) \, dx$$

and

$$\int_{\mathbb{R}^+} z^{i-1} \mathbf{1}_{Z'_t \geq z}(x) \, dz = \int_0^{Z'_t(x)} z^{i-1} dz = \frac{(Z'_t(x))^i}{i}$$

By integration over $\mathbb{R}^d$ we obtain

$$\int_{\mathbb{R}^d} \frac{(Z'_t(x))^i}{i} \, dx = \int_{\mathbb{R}^+} z^{i-1} \mu_d(A_{Z'_t}(z)) \, dz$$

and, by taking the mathematical expectation,

$$\mathbf{E}\left(\int_{\mathbb{R}^d} \frac{(Z'_t(x))^i}{i} \, dx\right) = \int_{\mathbb{R}^+} z^{i-1} \overline{\mu}_d(A_{Z'_t}(z)) \, dz.$$

The moment $M(i)$ is deduced by integration of the last expression with respect to the measure $\theta(dt)$, and similarly for the moment $M(i,K)$ after replacing Z and Z' by $Z_\vee(K)$ and by $Z'_\vee(K)$. Similarly, we have

$$\int_{\mathbb{R}^+} \varphi(z) \mathbf{1}_{Z'_t \geq z}(x) \, dz = \int_0^{Z'_t(x)} \varphi(z) \, dz = \Phi(Z'_t(x))$$

and, by integration over $\mathbb{R}^d$,

$$\int_{\mathbb{R}^+} \varphi(z) \mu_d(A_{Z'_t}(z)) \, dz = \int_{\mathbb{R}^d} \Phi(Z'_t(x)) \, dx.$$

After taking the mathematical expectation and after integration over $\theta(dt)$ the expression in (5.18) is immediate for Z and for $Z_\vee(K)$. $\qquad\square$

5.3.5.3 Geometrical Covariogram of the Primary Function

Starting from the bivariate distribution given in (5.15), for $z = z_1 = z_2$ we obtain for a positive RF $\{Z(x), x \in \mathbb{R}^d\}$, that

$$\int_0^\infty \log\left(\mathbf{P}\left(Z(x) < z, Z(x+h) < z\right) / (F(z))^2\right)$$
$$= \int_\mathbb{R}\int_0^\infty K(h,z,z,t)\theta(dt)\,dz = \int_\mathbb{R} K(h,t)\theta(dt) \tag{5.19}$$

with the notation $K(h,t) = \mu_{d+1}(A'_+(t) \cap A'_+(t)_{-h})$ for the geometrical covariogram in $\mathbb{R}^{d+1}$ of the positive part of the subgraph of Z'_t, $A'_+(t)$. Formula (5.19) may be useful for the identification of primary functions from $K(h,t)$, often simpler for calculations than the bivariate distribution deduced from the cross geometrical covariogram $K(h,z_1,z_2,t)$.

5.3.6 Some Stereological Aspects of BRF

As for the Boolean model, a BRF defined in $\mathbb{R}^d$, generates by section in $\mathbb{R}^k$ ($k < d$), a BRF with induced intensity and primary functions. This is a property connected to the Poisson point process. For some families of primary functions (for instance when the positive part of the subgraph is made in $\mathbb{R}^{d+1}$ of spheres, similar cylinders, or similar parallelepipeds,...), it is possible to estimate the properties of the primary functions (up to the intensity), from the sole bivariate distribution known on profiles, through the function $\int_\mathbb{R} K(h,t)\theta(dt)$. As far as these primary functions are well suited to real data, it can be relatively easy to implement them in applications.

5.3.7 BRF and Counting

In this section, we consider digital images with support in $\mathbb{R}^2$, modelled by Boolean island BRF.

As in [363, 364], we assume that the integral $V = \int_0^\infty \overline{\mu}_2(A_{Z'}(z))\,dz$ is known from a preliminary study. When considering a topographical surface in $\mathbb{R}^3$, V is the volume covered by the primary function Z'. We wish to estimate θ for images considered as realizations of BRF with intensity θ. From the distribution function $F(z)$, we get that

$$-\int_0^\infty \log(F(z))\,dz = \theta V. \tag{5.20}$$

This counting algorithm is very convenient, since it does not require any segmentation or any choice of a threshold. Well suited to Boolean textures, it is weakly sensitive to noise, but it is sensitive to illumination conditions (through V), which should remain strictly constant between a standard experiment (to estimate V) and an image acquisition for counting.

5.3.8 Identification of a BRF Model

To identify a BRF from data, both the family of primary function $Z'_t(x)$ and the measure $\theta(dt)$ must be known. However, the Choquet capacity given in (5.11), experimentally estimated from realizations of BRF, depends on the product of two factors: the intensity and a measure on the primary function. It is therefore not possible to know these two terms separately from their product, so that we have to face an indetermination. To raise it, we rely on the following results proved by M. Schmitt and F. Preteux [316, 317, 345]. We note (Z'_t, θ) the BRF defined by a choice of the primary function Z'_t and of the intensity $\theta(dt)$.

Proposition 5.7 (Characterization of BRF). *Consider a BRF (Z'_t, θ). If $\theta(\mathbb{R}) = \theta < \infty$, then the BRF admits a unique representation as a Boolean island $(Z', \theta\delta)$, where Z' is centered on the projection on the plane $z = 0$ of the center m of the sphere in $\mathbb{R}^{d+1}$ circumscribing the maxima of the primary function. If $\theta(\mathbb{R}) = +\infty$, then the BRF can be uniquely represented by (Z'_t, θ), where the Z'_t are centered in m and where $Z(x) = \vee_{(t_k, x_k) \in \mathcal{P}} \{Z'_{t_k}(x - x_k) + t_k\}$.*

From experimental data, we always access to a bounded range of variation for $Z(x)$. We can therefore mostly consider Boolean islands. It will be the same situation for simulations. However, at the level of a theoretical model, it is often interesting to consider the case ii) with $\theta(\mathbb{R}) = +\infty$. For instance we can use the following BRF:

1. The *Weibull model* is obtained by implantation of primary functions $Z'_t(x)$ with a point support ($Z'_t(x) = t\delta(x)$, $\delta(x)$ being the Dirac distribution in $\mathbb{R}^d$) and $\theta(dt) = \theta m(z_0 - t)^{m-1}$ for $t \le z_0 \le 0$. With this definition, the BRF differs from $-\infty$ on points of a Poisson process. It cannot be characterized by its spatial law, which is equal to zero. Use must be made of the Choquet capacity given in (5.11) for functions g having a support with non zero measure in $\mathbb{R}^d$. For instance, the distribution function of $Z_\vee(K)$ can be derived from (5.15):

$$\mathbf{P}(Z_\vee(K) < z) = \exp\left(-\int_{-\infty}^{z_0} \theta m(z_0 - t)^{m-1} \overline{\mu}_d(A_{Z'_t}(z) \oplus \check{K})\, dt\right)$$

with $A_{Z'_t}(z) \oplus \check{K} = \check{K}$ for $t \ge z$, else $A_{Z'_t}(z) \oplus \check{K} = \emptyset$. We thus get that

$$\mathbf{P}(Z_\vee(K) < z) = \exp\left(-\theta(z_0 - z)^m \mu_d(K)\right) \tag{5.21}$$

In fracture statistics, the variable of interest is $Z > 0$ (the fracture stress), and use is made of the BRF $Y(x) = -Z(x)$, which can be directly obtained with the intensity $\theta(dt) = \theta m(t - z_0)^{m-1} dt$ ($t \ge z_0$), by means of the operator $\wedge$ instead of $\vee$, and starting from the value $+\infty$ outside of the Poisson point process in $\mathbb{R}^d$. For $z \geq z_0$

$$\mathbf{P}(Z_\wedge(K) \geq z) = \exp\left(-\theta(z-z_0)^m \mu_d(K)\right) \tag{5.22}$$

2. The *Pareto model* is obtained with the same construction as the Weibull model, with the intensity $\theta(dt) = \dfrac{-\theta dt}{t}$ for $t \leq z_0 \leq 0$, and $\theta(dt) = 0$ else. We have

$$\mathbf{P}(Z_\vee(K) < z) = \exp\left(\int_{-\infty}^{z_0} \theta\overline{\mu}_d(A_{Z_t'}(z) \oplus \check{K})\frac{dt}{t}\right) \tag{5.23}$$

$$= \exp\left(\theta\mu_d(K)\int_{-z}^{-z_0}\frac{dt}{t}\right) = \left(\frac{z_0}{z}\right)^{\theta\mu_d(K)}. \tag{5.24}$$

Using the operator $\wedge$ instead of $\vee$, and starting from the value $+\infty$ outside of the Poisson point process in $\mathbb{R}^d$. For $z \geq z_0$

$$\mathbf{P}(Z_\wedge(K) \geq z) = \left(\frac{z_0}{z}\right)^{\theta\mu_d(K)}. \tag{5.25}$$

Exercise 5.1 (BRF with cylindrical primary random functions). Consider a primary RF defined in two steps: start with a compact random set A_0'. To every realization of A_0' a random variable Z' with distribution function $G(z) = \mathbf{P}(Z' < z)$ is associated. A Boolean islands RF $Z(x)$ with intensity θ is built from this cylinder primary function. 1. Show that the univariate and bivariate distribution functions of $Z(x)$ are given by

$$F(z) = \mathbf{P}(Z(x) < z) = \exp\left(-\theta\overline{\mu}_d(A_0')(1-G(z))\right)$$

and

$$F(h,z_1,z_2) = \mathbf{P}(Z(x) < z_1, Z(x+h) < z_2)$$
$$= F(z_1)F(z_2)F(z_1 \wedge z_2)^{-r(h)},$$

where $r(h) = \dfrac{\overline{\mu}_d(A_0' \cap A_{0-h}')}{\overline{\mu}_d(A_0')}$. 2. Show that the distribution function of $Z_\vee(K)$ for this model is given by

$$\mathbf{P}(Z_\vee(K) < z) = \exp\left(-\theta\overline{\mu}_d(A_0' \oplus \check{K})(1-G(z))\right)$$
$$= F(z)^{\frac{\overline{\mu}_d(A_0' \oplus \check{K})}{\overline{\mu}_d(A_0')}}.$$

5.3.9 Tests of BRF

Procedures proposed for testing the BRF model are derived from tests proposed for the Boolean model. In a first step, it is possible to work on sets obtained by applying thresholds on $Z(x)$ at different levels z_i, which are Boolean models with intensity $\int_{\mathbb{R}} (1 - G_t(z_i))\theta(dt)$, where $G_t(z)$ is the distribution function of the maximum of $Z'_t(x)$. Other tests can be directly applied to the function $Z(x)$. They involve the following criteria: convexity of the sections $A_{Z'_t}(z)$, change of support on convex sets [193].

5.3.9.1 Convexity of $A_{Z'_t}(z)$

This is the most often used test used in applications until now. It is based on an additional assumption, the convexity of the sections of the primary function, $A_{Z'_t}(z)$. This is not satisfied in the general case. The test makes use of the Steiner formula to the distribution function $\mathbf{P}(Z_\vee(K) < z)$ given in (5.15) when K is a compact convex set. Under these conditions, $\log(\mathbf{P}(Z_\vee(\lambda K) < z))$ and similarly $\int_{\mathbb{R}} \log(\mathbf{P}(Z_\vee(\lambda K) < z))\, dz$ are polynomials of degree k in λ for $K \subset \mathbb{R}^k$.

It is easy to implement these tests, since they only require the estimation of the distribution functions after change of support by the operator $\vee$ on convex sets with increasing sizes λK. The first test, based on a threshold z, is the same as for the Boolean random set model. The second test may be the source of numerical difficulties, since we may obtain $\mathbf{P}(Z_\vee(\lambda K) < z) \simeq 0$ for small values of z. In that case, we have to set a lower value z_0 for the numerical integration of the integral.

Examples of applications of these tests are given in [67, 190, 204, 315, 316, 364]. In [190], the test was satisfactory for change of support on segments with increasing lengths; primary functions of different shapes were used for the simulation of the rough surface of steel plates: cylinders, paraboloids, cones.

5.3.9.2 Change of Support on Convex Sets

Again we consider $Z_\vee(\lambda K)$ with K convex, and λ is chosen in such a way that $\mu_d(\lambda K) \gg \overline{\mu}_d(A_{Z'_t}(z))$. We do not need to make any assumption about the convexity of $A_{Z'_t}(z)$ for the proposed asymtotic tests [205].

Let z be such that $\vee_t\{G_t(z)\} < 1$ and consider two convex sets $K_1 \subset \mathbb{R}^{d_1}$, $K_2 \subset \mathbb{R}^{d_2}$ with $d_1 \leq d, d_2 \leq d$. Then we have that

$$H(\lambda_1, \lambda_2) = \frac{\log(\mathbf{P}(Z_\vee(\lambda_1 K_1) < z))}{\log(\mathbf{P}(Z_\vee(\lambda_2 K_2) < z))} = \frac{\int_{\mathbb{R}} \overline{\mu}_d(A_{Z'_t}(z) \oplus \lambda_1 \check{K}_1)\theta(dt)}{\int_{\mathbb{R}} \overline{\mu}_d(A_{Z'_t}(z) \oplus \lambda_2 \check{K}_2)\theta(dt)}. \tag{5.26}$$

For $\lambda_1 \to +\infty$ and $\lambda_2 \to +\infty$, from (5.26) we get that

$$H(\lambda_1, \lambda_2) = \frac{\lambda_1^{d_1} \mu_{d_1}(K_1)}{\lambda_2^{d_2} \mu_{d_2}(K_2)} \tag{5.27}$$

For instance in $\mathbb{R}^3$, the following formulae are true.

1. If K_1 is the cube with edge length 1 and if K_2 is the square with edge length 1, then

$$H(\lambda_1, \lambda_2) = \frac{\lambda_1^3}{\lambda_2^2} \quad \text{and} \quad H(\lambda, \lambda) = \lambda.$$

2. If K_1 is the cube with edge length 1 and if K_2 is a segment with length 1, then

$$H(\lambda_1, \lambda_2) = \frac{\lambda_1^3}{\lambda_2} \quad \text{and} \quad H(\lambda, \lambda) = \lambda^2.$$

3. If K_1 is the square with edge length 1 and if K_2 is a segment with length 1, then

$$H(\lambda_1, \lambda_2) = \frac{\lambda_1^2}{\lambda_2} \quad \text{and} \quad H(\lambda, \lambda) = \lambda.$$

In practice, it is also possible to set λ_1 and λ_2 constant and to vary z. The two curves $\log(\mathbf{P}(Z_\vee(\lambda_1 K_1) < z))$ and $\log(\mathbf{P}(Z_\vee(\lambda_2 K_2) < z))$ must be proportional, with a slope equal to $H(\lambda_1, \lambda_2)$.

These tests, not based on the assumption of convexity of the sections $A_{Z_t'}(z)$, can be implemented after a first change of support over a non convex set K (for instance $\{x, x+h\}$, or K made of any number of points), provided that λ stays larger than the range of $Y(x)$, deduced from $Z(x)$ by this first transformation.

5.3.10 BRF and Random Tessellations

Boolean random functions can be used for the generation of models of random tessellations [202]. A large class of random tessellation models combines a point process and a distance to the points. For instance, attaching to every Poisson point x_k a primary random function $Z_k'(x)$ defined according to the Euclidean distance, the standard Voronoi model can be deduced from a $\wedge$ BRF with primary function made of an increasing paraboloid of revolution, i.e.

$$Z(x) = \wedge_k Z_k'(x - x_k). \tag{5.28}$$

Sections of primary functions at level z are balls defined by the corresponding metric. Define

$$B_k'(z) = \{x \in \mathbb{R}^d : Z_k'(x) < z\}.$$

From (5.28) we have

$$B(z) = \{x \in \mathbb{R}^d : Z(x) < z\} = \cup_{x_k} B'_k(z)_{x_k}. \tag{5.29}$$

By (5.29), $B(z)$ is a Boolean random set with convex primary grains $B'_k(z)$. Consider a compact set K and the infimum $Z_\wedge(K) = \wedge_{y \in K}\{Z(y)\}$. Then, for the stationary case, we have that

$$\mathbf{P}(Z_\wedge(K) \geq z) = \exp\left(-\theta\mathbf{E}(\mu_d(B'_k(z) \oplus \check{K}))\right). \tag{5.30}$$

More general random tessellations can be generated in the same way, starting from a BRF with any primary random function $Z'(x)$. We consider that the realization k of $Z'(x)$ owns simply connected compact sections $B'_k(z)$, such that $B'_k(z_1) \subset B'_k(z_2)$ for $z_2 > z_1$. We consider primary random functions reaching their minimum $Z'(o)$ for $x = 0$. We associate to $Z'_k(x)$ the floor set A'_k defined by

$$A'_k = \{x \in \mathbb{R}^d : Z'_k(x) = Z'_k(o)\}. \tag{5.31}$$

If we have $A' = \{o\}$, we can define the class C_k of the random tessellation, generated by the germ x_k and the primary random function $Z'(x)$ by

$$C_k = \{x \in \mathbb{R}^d, Z'_k(x - x_k) < Z'_l(x - x_l), x_k \in \mathcal{P}, x_l \in \mathcal{P}, l \neq k\}. \tag{5.32}$$

For the simulation of random tessellations, we just need to simulate realizations of the Boolean random function with primary functions Z'_k corresponding to the model. The boundaries of the tessellation are provided by the crest lines of the random functions, obtained by the watershed of the random function using as markers apparent markers. By construction of the Boolean random functions, the location of crest lines, and therefore the boundaries of the classes of the resulting tessellation are invariant by a non decreasing transformation Φ (anamorphosis) of the values of $Z'_k(x)$ (for instance using $Z'^P_k(x)$ instead of $Z'_k(x)$), that is compatible with the order relationship, namely such that $z_1 < z_2$ implies $\Phi(z_1) < \Phi(z_2)$.

An alternative extraction of classes is given by their labels C_k. Starting from the simulation, and from the germs x_k, we generate in each point x a set of labels $L(x)$, i.e.

$$L(x) = \{k, Z(x) = Z'_k(x - x_k)\}. \tag{5.33}$$

Points x with the single label k generate the interior of cell C_k. Points with two labels k and l are on the boundaries between cells C_k and C_l. In $\mathbb{R}^3$, points with three labels are on the edges of the tessellation, and points with four labels are its vertices. More details about the properties of such random tessellations are given in [202].

5.4 Multiscale Boolean Random Functions

In many practical situations, there is a non-homogeneous dispersion of objects in a matrix, and possibly arrangement of aggregates at different scales [199, 200, 201].

A convenient way to account for theses situations is to replace the Poisson point process by a Cox process, generating a multi scale Cox BRF.

In a first step, we can replace in the construction of the BRF the intensity measure $\mu_d(dx) \otimes \theta(dt)$ in $\mathbb{R}^d \times \mathbb{R}$ by the intensity $\theta(dx, dt)$, dropping the stationarity of the Poisson point process. In a second step, we use for $\theta(dx, dt)$ a realization of a positive random function, generating a Cox point process. The Choquet capacity is then given by

$$T(g) = 1 - \mathbf{E}_\theta \left(\exp(-\mathbf{E}_{(}Z'_t)\{\int_{\mathbb{R}} \theta(D_{Z'_t}(g), dt)\}) \right) = 1 - \varphi_g(1)$$

with $\varphi_g(\lambda)$ the Laplace transform of the positive random variable

$$\mathbf{E}_{Z'_t} \left(\int_{\mathbb{R}} \theta(D_{Z'_t}(g), dt) \right)$$

and $\mathbf{E}_{Z'_t}$ being the expectation with respect to the random function Z'_t.

A typical example is given by a constant intensity θ inside a first random set A_1 (such as a stationary Boolean model of spheres with a large radius $\mathbb{R}$). We keep the points of a Poisson point process contained in A_1, as germs for centers of primary RF. We have $\theta(dx) = \theta\mathbf{1}_{A_1}(x)dx$, where $\mathbf{1}_{A_1}(x)$ is the indicator function of the set A_1. Then

$$T(g) = 1 - \varphi_g(\theta)$$

where $\varphi_g(\lambda)$ is the Laplace transform of the positive random variable

$$\mathbf{E}_{Z'_t} \left(\int_{\mathbb{R}} \overline{\mu}_d(D_{Z'_t}(g) \cap A_1) \, \theta(dt) \right).$$

For a deterministic primary function grain Z'_t, we have to use the change of support of the random set A_1 over the compact set $D_{Z'_t}(g))$, which is easily estimated from simulations. In [200, 201] use is made of the Beta distribution for the Cox-Boolean model.

For the test function $g(x) = z$ if $x \in K$, and $g(x) = +\infty$ if $x \notin K$, we obtain the distribution of the supremum $Z_\vee(K)$ of $Z(x)$ over the compact set K, i.e.

$$1 - T(K, z) = \varphi_K(z, 1)$$

where $\varphi_K(z, \lambda)$ is the Laplace transform of the positive random variable

$$\mathbf{E}_{Z'_t} \left(\int_{\mathbb{R}} \overline{\mu}_d(A_{Z'_t}(z) \oplus \check{K} \cap A_1) \, \theta(dt) \right).$$

An alternative way to generate multiscale BRF is to use a hierarchical model built from a random tessellation.

Exercise 5.2 (Hierarchical BRF model). Consider a RF Z which is built in two steps. A random stationary tessellation π of the space $\mathbb{R}^d$ delimits classes C_i. In

every class C_i a stationary Boolean island Z_i is considered with primary RF $Z'(x)$ and with the random intensity Y_i. For two classes C_i and C_j the random vectors (Z_i, Y_i) and (Z_j, Y_j) are independent. Consider the RF Z as a function of the statistical properties of π and of the Laplace transform Φ of the positive random variable Y.
1. Show that the distribution function $F(z) = \mathbf{P}(Z(x) < z)$ is given by

$$F(z) = \Phi(\overline{\mu}_d(A_{Z'}(z))).$$

Hint. Note that $F(z) = \mathbf{P}(x \in A_Z(z))$. The restriction of $A_Z(z)$ to every class C_i is a Boolean random set with primary grain $A_{Z'}(z)$ and with intensity Y_i. Given $Y_i = y$, it holds that

$$F(z) = \exp(-y\overline{\mu}_d(A_{Z'}(z)).$$

2. Show that the bivariate distribution function $F(h, z_1, z_2)$ is given by

$$F(h, z_1, z_2) = r(h)\Phi(\overline{\mu}_d(A_{Z'}(z_1) \cup A_{Z'}(z_2)_h))$$
$$+ (1 - r(h))\Phi(\overline{\mu}_d(A_{Z'}(z_1)))\Phi(\overline{\mu}_d(A_{Z'}(z_2))).$$

Hint. For $x \in C_i$ and $x + h \in C_i$, consider the bivariate distribution of the BRF with the intensity Y_i. If $x \in C_i$ and $x + h \in C_j$ (with $i \neq j$), the random variables $Z(x)$ and $Z(x + h)$ are independent, with univariate distribution functions $F(z_1) = \exp(-y_1\overline{\mu}_d(A_{Z'}(z_1)))$ and $F(z_2) = \exp(-y_2\overline{\mu}_d(A_{Z'}(z_2)))$. Denote the probability of $x \in C_i$ and $x + h \in C_i$. After deconditioning over π and over Y, the assertion follows by

$$r(h) = \frac{\overline{\mu}_d(C \cap C_{-h})}{\overline{\mu}_d(C)}.$$

5.5 Application of BRF to Modeling of Rough Surfaces

The original idea of the BRF model is coming from the simulation of rough surfaces. We illustrate this model by some industrial applications to rough surfaces providing simulations of textures with support in R^2. The first example is a simulation of the shot peening process [203]. The second one simulates the transfer of roughness on steel sheets during the rolling process [208]. Finally, the third example reproduces the EDT (electro discharge texture) roughness of rolls used in steel industry [206, 207], where it is shown that deviations from the BRF have to be implemented for realistic simulations of this type of texture. Further details and other models of random surfaces are also available in [195].

5.5.1 Simulation of the Evolution of Surfaces and of Stresses during Shot Peening

We consider the changes of surface morphologies during deformation process (like shot peening, or indentation). These aspects can be explored by means of models involving simulations on a computer. Shot peening is a process involving random impacts of high velocity particles on metallic surfaces. By shot peening, it is possible to generate residual stresses on the surface, to increase the fatigue life of parts. In a simplified study of this process [203], it was possible to propose a model simulating the mechanical effect of the random impacts on a surface of an elastic-plastic material. This type of simulations brings information on a possible heterogeneity of residual stresses at a small scale.

Since the impacts by the particles are generated at random, the covered area forms a Boolean random set model with discs as primary grains, see Fig. 5.2. This has been tested with respect to the covariance of the image, for various processing times and metallic surfaces with different hardnesses [203].

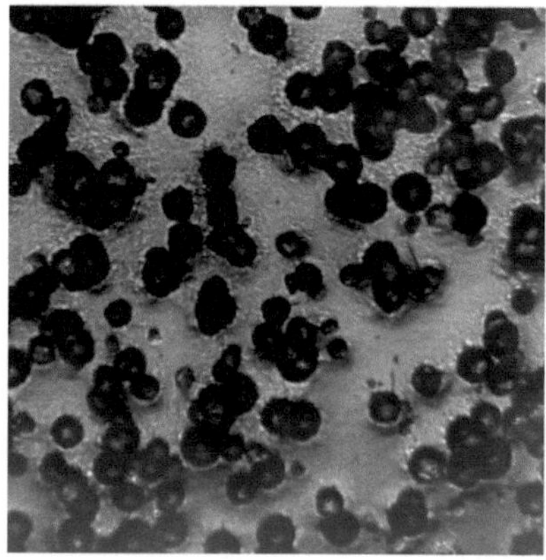

Fig. 5.2 Microscopic image of the surface obtained by shot peening. The dark zone was covered by random impacts.

Moreover, in the simulation we map the maximum Hertz elastic pressure generated by the penetration of spherical particles and the plastic deformation of the surface. This leads to a Boolean random function with spheroid primary functions, see Fig. 5.3, assuming a quasi-static approach due to low velocities involved in the process [211]. The radius a of the zone in contact at the maximal depth of the ball and the maximum contact pressure p_0 are obtained from the Hertz theory of the elastic contact between a ball and a plane [96]. For a ball of radius R_b, E being the effective stiffness of the combination ball-surface, and ρ_b the mass per unit volume of the ball, we have that

$$a = 1/(\frac{5}{2}\pi\rho_b\frac{V^2}{E})^{1/5}R_b \tag{5.34}$$

and

$$p_0 = (\frac{5}{2}\pi\rho_b V^2 E^4)^{1/5}. \tag{5.35}$$

From the Hertz theory, the distribution of the pressure contact within the contact zone is ellipsoidal, according to

$$p(r) = p_0(1 - (r/a)^2)^{1/2}. \tag{5.36}$$

The sizes and shapes of the spheroid primary functions are deduced from (5.34) – (5.36).

Fig. 5.3 Simulation by a BRF of shot peening; results after 3000 impacts: maximal pressure

For every pixel of the impact, the plastic deformation is obtained from the available local inelastic energy, accounting for the local hardening in the stress-strain relation. This results in a model which is a non linear transformation of a dilution random function, where the supremum operator used in the construction of the BRF is replaced by an addition.

From the obtained maps of local mechanical properties, as in Fig. 5.3, statistical information such as the evolution of the distribution functions of the maximal pressures and of the plastic deformations, are available. The evolution with the number of impacts of the corresponding empirical distribution function of maximum pressures is given in Fig. 5.4. This is in agreement with the prediction of the theory, see (5.13), which can also predict higher order statistics . A validation of this model could be obtained from experimental maps of micro-hardness measurements made on specimens.

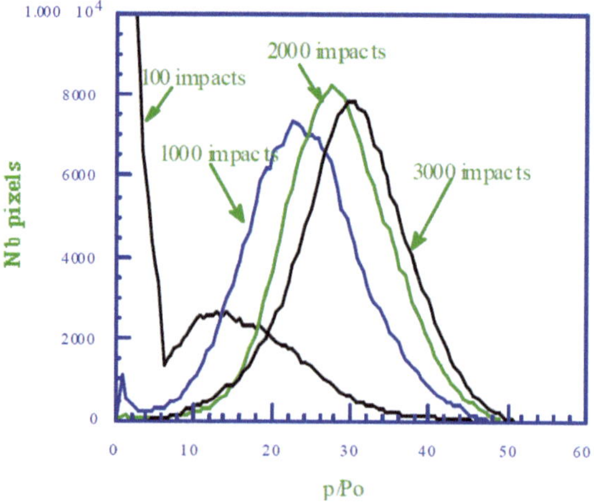

Fig. 5.4 Empirical probability density functions of the local maximum contact pressure estimated from simulations similar to Fig. 5.3.

5.5.2 Simulation of the Roughness Transfer on Steel Sheets

The evolution of the topography of surfaces was simulated during the rolling process [208]. The roughness of steel sheets used in the automotive industry is a consideration for their quality. Therefore it is important to control the roughness transfer during a cold rolling operation like the skin-pass. This involves a complex indentation process during the penetration of the roll cylinder. Micro mechanical models of indentation are limited to single indenters with a simple geometry. A morphological model for the simulation of the roughness transfer from the cylinder to the steel sheet was developed according to the following steps. For every increment of the cylinder penetration, new indenters (peaks of the cylinder topography) come in contact, while previous indenters generate rims with an overall conservation of the volume, assuming a purely plastic deformation of the steel. At the distance r from the boundary of an indentation, the vertical displacement in the rim, $B(r)$, is given by:

$$B(r) = kr\exp(-\beta r^2) \tag{5.37}$$

The parameter β of this primary function depends on the volume of the indentation. The parameter k, depending on every indentation, insures the conservation of volume. The displacements produced by neighboring indentations are cumulated according to (5.37). The law of evolution of β was determined from punching experiments, and maps obtained at the same location for increasing pressures. The model is implemented in an incremental way, for a progressive indentation of the sheet by the roll, generating a modified BRF to account for the conservation of the volume. An excellent agreement was found between simulations and punching experiments, as illustrated in [208].

5.5.3 Modeling of Electro Discharge Textures (EDT)

The above mentionned skin-pass operation has to be carefully monitored in order to ensure a roughness corresponding to the further use of steel sheets. The EDT (electro discharge texturing) process engraves a controlled roughness by electrical discharges between an electrode and the roll. They create craters during the heating of the surface which are made of spherical cavities surrounded by a rim (Fig. 5.5).

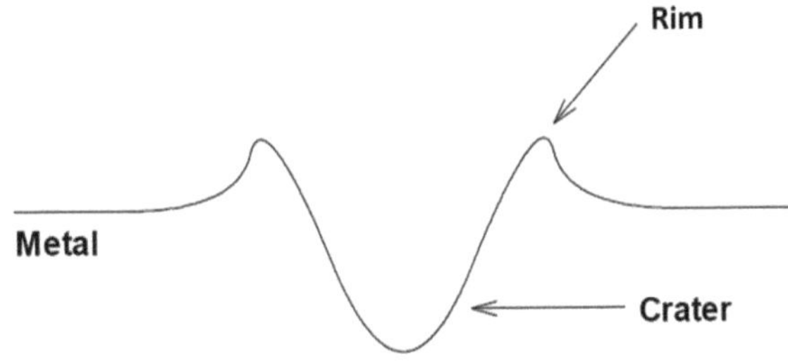

Fig. 5.5 Schematic profile of an EDT crater.

Mappings of the surface are obtained with a 3D stylus instrument, using a 5 μm diameter sensor that records height variations. The mappings are presented as $512 \times 512 \times 16$ bits digital images (Fig. 5.6). On the edge of the roll, the density of impacts is so low that we could map non overlapping craters that were stored in a library. Their average diameter is about 100 μm and their depth can reach 5 μm. The mappings of craters are used as primary functions in random function models simulating the EDT process in a simplified way [195, 206, 207].

The texture of the roll is the result of a combination of elementary patterns made by craters and rims. Their implantation on the surface is made at random by discharges. In a first step, a Poisson point process is a good candidate to model these locations of impacts.

The first model to be tested is the Boolean random function (BRF). An example of a realization obtained by the supremum of craters taken with a uniform probability in the library is shown in Fig. 5.7 (left). Its negative is the corresponding $\wedge$ BRF (Fig. 5.7, right). A high intensity of Poisson points was used to cover the field of the simulation, since the EDT process is stopped after engraving the full surface of the roll. It is clear from this simulation that the Boolean texture is not correct for this process: the rims are enhanced in the supremum version, while the craters are

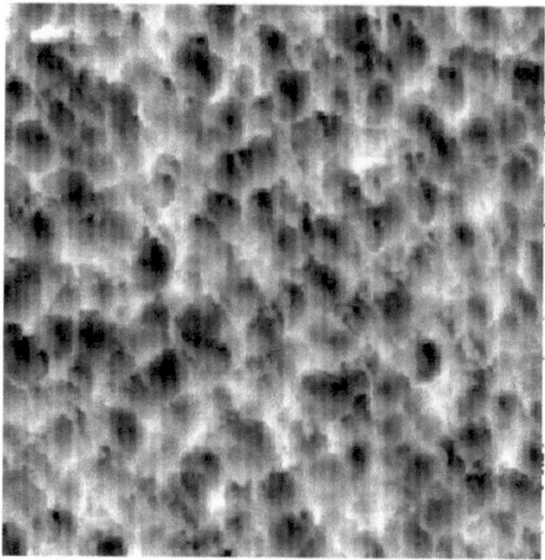

Fig. 5.6 Topographical map of an EDT texture.

enhanced in the infimum version. Moreover, the rims can cover cavities and the cavities can cover rims, which is not observed in reality. As seen later, these overlaps can be prevented by modification of the point process with a repulsion distance.

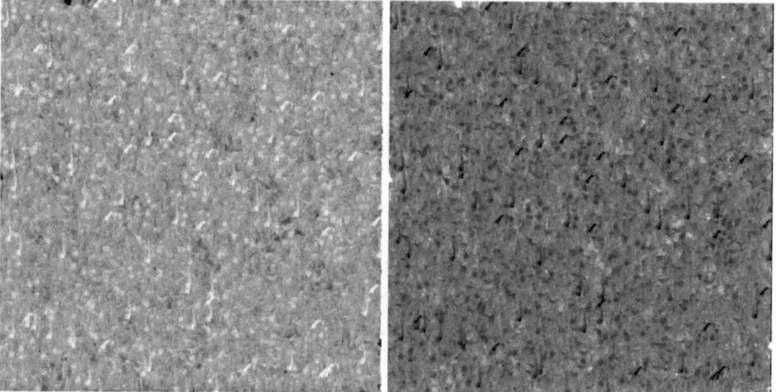

Fig. 5.7 Realization of a $\vee$ BRF (left) and its negative ($\wedge$ BRF on the right), using experimental topographic images of craters.

The second version of the model is based on the dead leaves random function (DLRF) [191]. For this model, the primary functions are introduced sequentially, and cover the previous ones. In Fig. 5.8 (left) the texture in a realization of a DLRF is shown which is obtained for a long time sequence ($t \rightarrow +\infty$). From a visual comparison, it is clear that it looks more similar to the real texture than the texture in Fig. 5.7. However discontinuities appear on the boundaries of the primary functions, which are not in agreement with the physical process.

A third version of the model is an alternating sequential RF ([189, 193, 194]), combining deposition and abrasion of matter. To take a better account of the rims surrounding the craters, we can split the primary function in two parts: craters and rims. They can be sequentially combined in two different ways: craters Z'^1_t are implanted on Poisson points x_k by the infimum, and rims Z'^2_t on the same points x_k by the supremum:

$$Z_{t+dt}(x) = [Z_t(x) \bigwedge Z'^1_{t_k}(x - x_k)] \bigvee Z'^2_{t_k}(x - x_k) \tag{5.38}$$

Iterations of this process simulate the formation of cavities by the electrical arcs and of rims in a sequential way. In fact, replacing the operator $\bigvee$ by an addition gives better results in the present case by simulating the deposition of molten metal. The final model is given by

$$Z_{t+dt}(x) = [Z_t(x) \bigwedge Z'^1_{t_k}(x - x_k)] + Z'^2_{t_k}(x - x_k) \tag{5.39}$$

A realization of this model is shown in Fig. 5.8 (right). Visual inspection shows that the obtained structure is closer to the actual texture of the roll than the results of previous models.

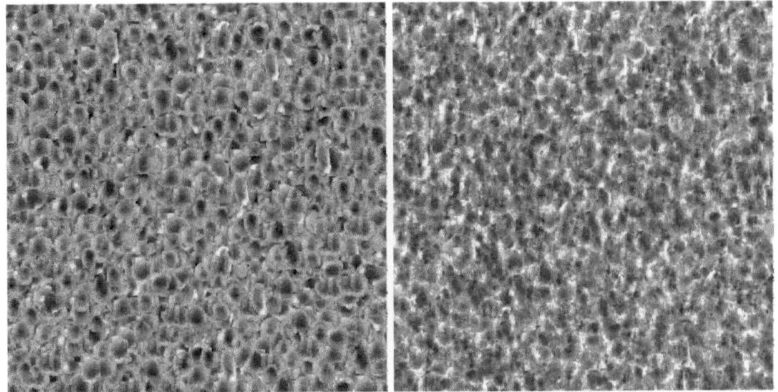

Fig. 5.8 Realization of a dead leaves RF (left) and of an alternate sequential RF (right)

The parameters of the model are as follows:

1. the intensity of the Poisson point process
2. the frequency distribution of the primary functions

To validate the model more formally, we can use e.g. the following criteria: comparison between the actual and the simulated texture, from the statistical height distribution, and the variogram. Additional probabilistic properties can be used: bivariate distribution, trivariate distribution, and more generally the Choquet capacity.

In particular, to make a step further in the comparison between simulated and actual textures, we use the variogram $\gamma(h)$ as a structural tool:

$$2\gamma(h) = \mathbf{E}\left((Z(x+h) - Z(x))^2\right) \tag{5.40}$$

In the stationary case and for a finite variance of the RF $Z(x)$, the variogram is deduced from the covariance $C(h) = \mathbf{E}\left(Z(x)Z(x+h)\right)$:

$$\gamma(h) = C(0) - C(h)$$

In Fig. 5.9 (left) the experimental variogram of the surface presents values above its sill, meaning the presence of negative correlations between $Z(x)$ and $Z(x+h)$ for some specific distance. As already said before, this effect cannot be observed for a standard BRF based on the Poisson point process. In the present case, this is the result of a repulsion effect between elementary patterns that prevents a perfect overlap to occur. This "hole" effect can be reproduced by a repulsion distance (hard-core) in the simulations. The location of the maximum of the variogram is an indication on the average size of primary grains which is useful for the choice of the distribution function of the primary functions in the simulations. Together with the experimental distribution function of the heights of the surface, it was used to monitor the size distribution of the grains in the simulation. The final model makes use of a size distribution favouring largest craters and of a grain depending repulsion distance, equal to the diameter of the grains. The variogram of the final model (Fig. 5.9, right) is close to the variogram of the surface. A simulated texture is shown on Fig. 5.10, which is very similar to the real surface (Fig. 5.6).

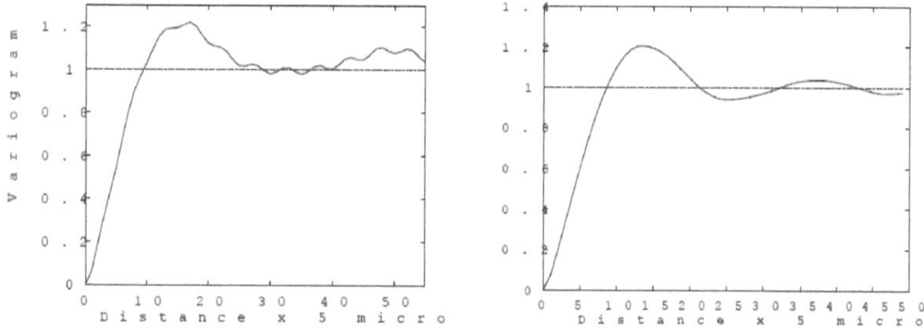

Fig. 5.9 Variogram of the EDT surface (left) and of a realization of the random model (right)

To conclude this section on rough surfaces, note that the BRF model and its modifications provide a versatile collection of random function models to generate random textures where some random motif (the primary function) appears at different locations. This generalization of grain models already known for random set is very flexible to handle real data.

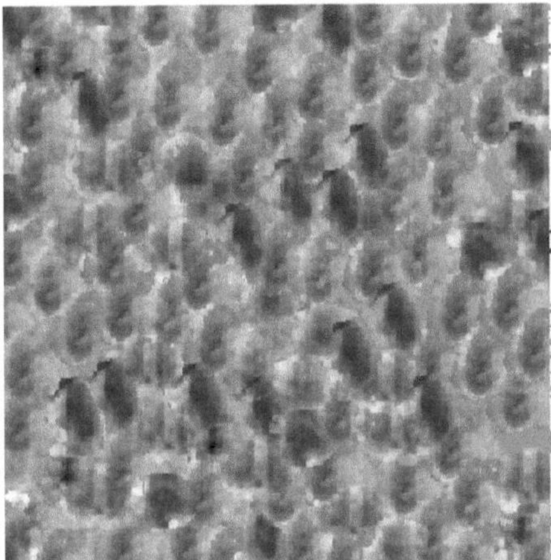

Fig. 5.10 Realization of the model of EDT texture.

Chapter 6
Random Marked Sets and Dimension Reduction

Viktor Beneš, Jakub Staněk, Blažena Kratochvílová and Ondřej Šedivý

Abstract In this chapter, random marked closed sets are investigated. Special models with integer Hausdorff dimension are presented based on tessellations and numerical solutions of stochastic differential equations. Statistical analysis is developed which invovles the random-field model test and estimation of first and second-order characteristics. Real data analyses from neuroscience (track modeling marked by spiking intensity) and materials research (grain microstructure with disorientations of faces) are presented. Dimension reduction of point processes with Gaussian random fields as covariates was recently studied in the literature. In the present chapter this research is generalized in three different ways. Marked fibre and surface processes with covariates are subject to dimension reduction, where we restrict to the sliced inverse regression method. Slicing is suggested based on geometrical marks. In a refined model for dimension reduction the second-order central subspace is analyzed. Numerical results on estimation and testing the central subspace are presented based on simulations.

Viktor Beneš
Charles University in Prague, Faculty of Mathematics and Physics, 18675 Praha 8, Czech Republic
e-mail: benesv@karlin.mff.cuni.cz

Jakub Staněk
Charles University in Prague, Faculty of Mathematics and Physics, 18675 Praha 8, Czech Republic
e-mail: stanekj@karlin.mff.cuni.cz

Blažena Kratochvílová
Palacky University, Faculty of Science, 77146 Olomouc, Czech Republic e-mail: blaza.kratochvilova@gmail.com

Ondřej Šedivý
Charles University in Prague, Faculty of Mathematics and Physics, 18675 Praha 8, Czech Republic
e-mail: sedivy@karlin.mff.cuni.cz

© Springer International Publishing Switzerland 2015
V. Schmidt (ed.), *Stochastic Geometry, Spatial Statistics and Random Fields,*
Lecture Notes in Mathematics 2120, DOI 10.1007/978-3-319-10064-7_6

6.1 Preliminaries

Let $(\mathbb{R}^d, \mathcal{B}^d)$ be the d-dimensional Euclidean space with Borel σ-algebra, and $\mathcal{B}_0$ the subsystem of bounded Borel sets in $\mathbb{R}^d$. For $A \in \mathcal{B}_0$ its Lebesgue measure is denoted by $|A|$. By $B_r(x) = \{y \in \mathbb{R}^d,\ ||x - y|| \leq r\}$, we denote the d-dimensional ball with centre at $x \in \mathbb{R}^d$ and radius $r > 0$, where $||.||$ is the Euclidean norm, and $\omega_d = |B_1(o)|$ denotes the volume of the unit ball.

6.1.1 Random Measures and Random Marked Closed Sets

Let $(\Omega, \mathcal{F}, \mathbf{P})$ be a probability space. Let $\mathcal{M}^d$ be the set of all locally finite measures on $(\mathbb{R}^d, \mathcal{B}^d)$ equipped with the smallest σ-algebra $\mathfrak{M}^d$ which makes the mapping $\mu \mapsto \mu(A)$, $\mu \in \mathcal{M}^d$, measurable for all $A \in \mathcal{B}^d$. Furthermore, let $\Psi : \Omega \times \mathcal{B}^d \to (\mathcal{M}^d, \mathfrak{M}^d)$ be a random measure on $\mathbb{R}^d$ and $\Lambda(\cdot) = \mathbf{E}\Psi(\cdot)$ is the intensity measure of Ψ. The *Campbell measure* $\mathcal{C}$ on $\mathbb{R}^d \times \mathcal{M}^d$ is defined as

$$\mathcal{C}(A \times \mathcal{U}) = \mathbf{E}\left(\Psi(A)\mathbf{1}_{\mathcal{U}}(\Psi)\right),$$

where $A \in \mathcal{B}_0, \mathcal{U} \in \mathfrak{M}^d$. For $z \in \mathbb{R}^d$, let t_z denote the shift operator on $\mathcal{M}^d$ defined by

$$t_z\mu(B) = \mu(B - z), \quad B \in \mathcal{B}^d.$$

The random measure Ψ is called *stationary* if $t_z\Psi$ has the same distribution as Ψ for any $z \in \mathbb{R}^d$.

Let $\mathcal{H}^k$ be the Hausdorff measure of order k in $\mathbb{R}^d$. In [428] the concept of random $\mathcal{H}^k$-sets in $\mathbb{R}^d$ was introduced as random closed sets which are $\mathcal{H}^k$-rectifiable. A random $\mathcal{H}^k$-set Y such that $\Psi_Y(.) = \mathcal{H}^k(Y \cap .)$ is a locally finite random measure in $\mathbb{R}^d$ will be called a point, fibre, or surface process for $k = 0, 1, d - 1$, respectively.

The l-th moment measure $\mu^{(l)}$ of a random measure Ψ is given by

$$\mu^{(l)}(A_1 \times \cdots \times A_l) = \mathbf{E}\left(\Psi(A_1)\ldots\Psi(A_l)\right)$$

for $A_1, \ldots, A_l \in \mathcal{B}_0$. The l-th order intensity function λ_l is defined by

$$\mu^{(l)}(ds_1 \times \cdots \times ds_l) = \lambda_l(s_1, \ldots, s_l)ds_1 \ldots ds_l \tag{6.1}$$

provided that it exists. In the case $k = 0$ (point processes) there is the l-th factorial moment measure $\mu^{(l)!}$ (see [73]) on the left hand side of (6.1) instead of $\mu^{(l)}$. The notation $\lambda = \lambda_1$ will be used. In doubly stochastic models the intensity measures (functions) are random, in this case they are called driving intensity measures (functions), respectively. E.g. conditionally given a realization λ of a driving intensity $\tilde{\lambda}$ of a Cox point process, we have a Poisson point process with intensity function λ, see e.g. [404].

Following [25] define

$$\Phi_{usc} = \{(X,f) : X \subset \mathbb{R}^d \text{ is closed}, \ f : X \to \mathbb{R} \text{ is upper semi} - \text{continuous}\},$$

$$\mathcal{U}_{cl} = \{A \subset \mathbb{R}^d \times \mathbb{R} \text{ closed, for all } x \in \mathbb{R}^d \text{ for all } t \in \mathbb{R} : (x,t) \in A \implies$$

$\{x\} \times [-\infty,t] \subset A\}$. Then

$$\tau : \Phi_{usc} \to \mathcal{U}_{cl}, \quad (X,f) \mapsto \{(x,t) \in X \times \mathbb{R} : t \le f(x)\}$$

is a bijection. Let $(Y,\Gamma) : \Omega \to \Phi_{usc}$ be a mapping such that

$$\{\omega \in \Omega : \tau(Y,\Gamma) \cap B \ne \emptyset\} \in \mathcal{F}$$

for each compact $B \in \mathbb{R}^d \times \mathbb{R}$. Then (Y,Γ) is called a *random marked closed set* (RMCS). It can be viewed as a random function Γ defined on a random domain Y.

A RMCS is called *stationary* if

$$\mathbf{P}(\tau(Y,\Gamma) + (x,0) \in .) = \mathbf{P}(\tau(Y,\Gamma) \in .)$$

for all $x \in \mathbb{R}^d$. It is *isotropic* if

$$\mathbf{P}(\theta \tau(Y,\Gamma) \in .) = \mathbf{P}(\tau(Y,\Gamma) \in .)$$

for all $\theta \in SO_{d+1}$ (group of rotations on $\mathbb{R}^{d+1}$) with $\theta(\mathbb{R}^d \times \{0\}) = \mathbb{R}^d \times \{0\}$.

A concept closely related to RMCSs is that of a weighted random measure (see [73]). Let Ψ be a random measure in $\mathbb{R}^d$, $\mathcal{C}$ its Campbell measure, W be a locally compact space and w a measurable mapping (weight function)

$$w : \text{supp } \mathcal{C} \to W$$

(consider the product σ-algebra on the support supp $\mathcal{C} \subset \mathbb{R}^d \times \mathcal{M}^d$). Then we call the tuple (Ψ,w) a *weighted random measure* (WRM) in $\mathbb{R}^d$ with weight space W. A weighted random measure induces a random measure $\tilde{\Psi}$ on $\mathbb{R}^d \times W$ which is given by

$$\tilde{\Psi}(B \times D) = \Psi\{x \in B : w(x,\Psi) \in D\}, \quad B \in \mathcal{B}^d, \ D \in \mathcal{B}(W).$$

We say that WRM (Ψ,w) is *stationary* if Ψ is stationary and $w(x,\mu) = w(x+z,t_z\mu)$ for any $(x,\mu) \in \text{supp } \mathcal{C}$ and $z \in \mathbb{R}^d$. The relation between RMCS and WRM for a $\mathcal{H}^k$-set $y \subset \mathbb{R}^d$ is expressed by the following diagram

$$
\begin{array}{ccc}
\Phi_{usc} = \{(y,f)\} & & \\
\uparrow & \searrow & \\
(\Omega,\mathcal{F}) & \to & \xi = \{(\psi_y,w)\}
\end{array}
$$

which commutes for $\psi_y(.) = \mathcal{H}^k(y \cap .)$ and $w(x,\psi_y) = f(x)$ on y. Thus for a random marked $\mathcal{H}^k$-set (Y,Γ), $k = 0,1,d-1$ we have the notions of a marked point, fibre or surface process, respectively.

A random field is a stochastic process on $\mathbb{R}^d$. For more information about random fields the reader is referred to Chapter 11. An important concept is the random-field model which is related to a special class of RMCSs. Let Γ' be an upper semicontinuous random field on $\mathbb{R}^d$ and RMCS (Y,Γ) be such that $\Gamma = \Gamma'$ on Y (often we denote Γ and Γ' by the same letter). If Y and Γ' are stochastically independent then RMCS (Y,Γ) is called a *random-field model*. If a given RMCS is a random-field model then separate investigation of Y and Γ is possible which leads to a substantial reduction of effort in statistical inference.

6.1.2 Second-order Characteristics

The reduced second moment measure $\mathcal{K}$ of a stationary random measure Ψ can be expressed as

$$\mathcal{K}(B) = \frac{1}{\lambda^2 |A|} \mathbf{E}\left(\int_A \Psi(B-x)\Psi(dx) \right), \tag{6.2}$$

where the value of the expression in (6.2) does not depend on $A \in \mathcal{B}_0$, $|A| > 0$. The K-function is defined as $K(r) = \mathcal{K}(B_r(0))$. The concept of second-order intensity reweighted stationarity introduced for point processes (cf. [179]) can be generalized to random measures. Let Ψ be a random measure in $\mathbb{R}^d$ with intensity function $\lambda > 0$. Put

$$M(D,B) = \mathbf{E}\left(\int_D \frac{\Psi(dy)}{\lambda(y)} \int_B \frac{\Psi(dy)}{\lambda(y)} \right), \quad B,D \in \mathcal{B}^d.$$

We say that Ψ is *second-order intensity reweighted stationary* (SOIRS) if

$$M(D,B) = M(D+x, B+x)$$

for all $x \in \mathbb{R}$, $B,D \in \mathcal{B}^d$. Then define an inhomogeneous reduced second moment measure

$$\mathcal{K}_{\mathrm{inhom}}(B) = \frac{1}{|A|} \mathbf{E}\left(\int_A \int \frac{\mathbf{1}_B(x-y)}{\lambda(x)\lambda(y)} \Psi(dx)\Psi(dy) \right).$$

A class of second-order characteristics of a RMCS (Y,Γ), for $x,y \in \mathbb{R}^d$, is introduced as follows ([25]). For $\varepsilon > 0$ define a random field $\mathcal{K}_{\mathrm{inhom}}$ by the random field $\{Z_\varepsilon(x), x \in \mathbb{R}^d\}$ by

$$Z_\varepsilon(x) = \begin{cases} \max_{y \in Y \cap B_\varepsilon(x)} \Gamma(y), & \text{if } x \in Y_{\oplus\varepsilon}, \\ \\ 0 & \text{else.} \end{cases}$$

For a right-continuous function $f : \mathbb{R}^2 \mapsto \mathbb{R}$ define

$$\kappa_f(x,y) = \lim_{\varepsilon \downarrow 0} \mathbf{E}\left(f(Z_\varepsilon(x), Z_\varepsilon(y)) | x,y \in Y_{\oplus\varepsilon} \right) \tag{6.3}$$

whenever $\mathbf{P}(x,y \in Y_{\oplus \varepsilon}) > 0$ for all $\varepsilon > 0$. Otherwise κ_f is undefined. Here

$$Y_{\oplus \varepsilon} = Y \oplus B_\varepsilon(0)$$

for Minkowski addition $\oplus$. In the case when Γ is a restriction of a random field Γ' on $\mathbb{R}^d$ then we do not use Z_ε and put $f(\Gamma'(x),\Gamma'(y))$ instead of $f(Z_\varepsilon(x),Z_\varepsilon(y))$ in (6.3).

Common choices of f are

$$e(u,v) = u, \quad c(u,v) = uv, \quad v(u,v) = u^2.$$

Then define the conditional mean mark and the mark covariance function by

$$E(x) = \kappa_e(x,y), \qquad C_Y(x,y) = \kappa_c(x,y) - \kappa_e(x,y)\kappa_e(y,x).$$

Furthermore the mark correlation function is given by

$$\mathrm{cor}(x,y) = \frac{\kappa_c(x,y) - \kappa_e(x,y)\kappa_e(y,x)}{(\kappa_v(x,y) - \kappa_e(x,y)^2)^{1/2}(\kappa_v(y,x) - \kappa_e(y,x)^2)^{1/2}}$$

and the mark variogram by $\gamma(x,y) = \frac{1}{2}(\kappa_v(x,y) + \kappa_v(y,x)) - \kappa_c(x,y)$. Another characteristics is Stoyan's k_{mm}-function given by

$$k_{mm}(x,y) = \bar{m}^{-2}\kappa_c(x,y), \qquad \bar{m} = \mathbf{E}\left(\Gamma(x)|x \in Y\right).$$

6.2 Statistical Methods for RMCS

In this section basic statistical inference for a RMCS (Y,Γ) is developed under various assumptions. We assume that a single realization of (Y,Γ) (which we denote by the same letters) is observed in a bounded window $W \subset \mathbb{R}^d$.

6.2.1 Random-field Model Test

Consider statistical testing of the null-hypothesis

$$H_0 : (Y,\Gamma) \text{ is a random field model} \tag{6.4}$$

against the alternative hypothesis

$$H_A : (Y,\Gamma) \text{ is not a random field model}.$$

In [343] such tests are developed for stationary marked point processes. The test based on the mark-weighted K-function K_Γ is generalized here to RMCS (Y,Γ)

with random $\mathcal{H}^k$-set Y, $0 \le k < d$, nonnegative mark Γ and the corresponding random measure Ψ_Y being second-order intensity reweighted stationary with intensity function λ. The algorithm consists of the following steps:

1. choose n test points $x_i \in Y \cap W$, evaluate $\Gamma(x_i)$, $i = 1, \ldots, n$, $\bar{\Gamma}(W) = \frac{1}{n} \sum_{i=1}^{n} \Gamma(x_i)$,
2. evaluate an estimate $\hat{\lambda}$ of the intensity λ,
3. evaluate

$$\hat{K}_{\Gamma}(r) = \frac{1}{n|W|} \sum_{x_i} \frac{\Gamma(x_i)}{\bar{\Gamma}(W)} \int_Y \frac{\mathbf{1}(||x_i - y|| < r)}{\lambda(x_i)\lambda(y)} \mathcal{H}^k(dy), \tag{6.5}$$

4. random reallocation: in step 3 make q random permutations of $\{\Gamma(x_i), i = 1, \ldots n\}$, evaluate $\hat{K}_{\Gamma}(r)$ for each permutation, denote $\hat{K}_{\max}$, $\hat{K}_{\min}$ the pointwise maximum and minimum from all permuted cases, respectively.

5. transform to L-function: $\hat{L}. = \left(\frac{\hat{K}.}{\omega_d}\right)^{\frac{1}{d}}$, draw envelopes

$$\hat{L}_{\max}(r) - \hat{L}_{\Gamma}(r), \ \hat{L}_{\min}(r) - \hat{L}_{\Gamma}(r). \tag{6.6}$$

If the horizontal axis lies between the envelopes, the null hypothesis cannot be rejected on an approximate significance level $\frac{2}{q+1}$. This holds only for fixed r, whereas a global test is suggested in [149].

In order to determine $\hat{K}_{\Gamma}(r)$ consider a partition of $Y \cap W$ into disjoint subsets B_j, points $z_j \in B_j$, $j = 1, 2, \ldots, e$ (typically $e > n$) and let $\mathcal{H}^k(B_j) = \triangle_{z_j}^k$. Then the estimator $\hat{K}_{\Gamma}(r)$ of $K_{\Gamma}(r)$ is approximately given by

$$\hat{K}_{\Gamma}(r) \approx \frac{1}{n|W|} \sum_{x_i} \frac{\Gamma(x_i)}{\bar{\Gamma}(W)} \sum_j \triangle_{z_j}^k \frac{\mathbf{1}(||x_i - z_j|| < r)}{\hat{\lambda}(x_i)\hat{\lambda}(z_j)}$$

and it suffices to estimate λ at the points x_i, z_j, e.g. using a kernel estimator.

6.2.2 Estimation of Characteristics

For a stationary and isotropic RMCS (Y, Γ) the second-order characteristics depend on the scalar $r = ||x - y||$ only, for any pair $(x, y) \in \mathbb{R}^{2d}$. To estimate them let $\varepsilon > 0$, consider the extension $(Y_{\oplus \varepsilon}, \Gamma)$ and let $U \subset W$ be a finite test set (e.g. a fixed grid of points). Let $r > 0$ be such that

$$N_U^{\varepsilon}(r) = \{(x, y) \in (Y_{\oplus \varepsilon} \cap U)^2 : ||x - y|| = r\} \tag{6.7}$$

is nonempty. Then we suggest an estimator of $\kappa_f(x, y) = \kappa_f(r)$ of the form

$$\hat{\kappa}_f(r) = \frac{1}{|N_U^{\varepsilon}(r)|} \sum_{(x,y) \in N_U^{\varepsilon}(r)} f(\Gamma(x), \Gamma(y)). \tag{6.8}$$

The following result holds.

Lemma 6.1. *For a random-field model (Y, Γ), $r, \varepsilon > 0$ such that $\mathbf{P}(N_T^\varepsilon(r) \neq \emptyset) > 0$ it holds that*

$$\mathbf{E}\hat{\kappa}_f(r) = \mathbf{E}\left(f(\Gamma(x), \Gamma(y)) | x, y \in Y_{\oplus \varepsilon}\right), \quad ||x - y|| = r.$$

Proof. It holds that

$$\mathbf{E}\hat{\kappa}_f(r) = \mathbf{E}\frac{1}{|N_U^\varepsilon(r)|} \sum_{(x,y) \in N_U^\varepsilon(r)} f(\Gamma(x), \Gamma(y))$$

$$= \mathbf{E}[\mathbf{E}\left(\frac{1}{|N_U^\varepsilon(r)|} \sum_{(x_j, y_j) \in N_U^\varepsilon(r)} f(\Gamma(x_j), \Gamma(y_j)) | N_U^\varepsilon = \{(x_1, y_1) \ldots, (x_k, y_k)\}\right)]$$

$$= \mathbf{E}\left(\frac{1}{k} \sum_{j=1}^{k} f(\Gamma(x_j), \Gamma(y_j))\right)$$

$$= \int f(m_1, m_2) Q_{\varepsilon;x,y}(d(m_1, m_2))$$

$$= \mathbf{E}\left(f(\Gamma(x), \Gamma(y)) | x, y \in Y_{\oplus \varepsilon}\right),$$

where $Q_{\varepsilon;x,y}$ is the two-point mark distribution of the weighted random measure $(\Psi_\varepsilon, \Gamma)$, and $\Psi_\varepsilon = |. \cap Y_{\oplus \varepsilon}|$, is the random volume measure associated with $Y_{\oplus \varepsilon}$, see [25]. $\qquad \square$

6.3 Modeling of RMCS; Simulation Results

In order to build RMCS models with dependence between the random set and its mark we start with a simple example. We assume that $d = 2$ throughout this section.

Example 6.1. Consider the ball $\mathcal{A} = B_1(0)$ in $\mathbb{R}^2$, let Y be a union of two radius segments in $\mathcal{A}$ with independent random orientations φ, ψ uniformly distributed on $[0, 2\pi)$. Then the smaller angle between segments $\sphericalangle(\varphi, \psi) = V$ is uniformly distributed on $[0, \pi)$. Let Γ be given by a random variable $\mathcal{V}$ uniformly distributed on $[0, c]$, $c > 0$, and the dependence of Γ on Y be completely described by a bivariate distribution on $[0, c] \times [0, \pi)$ with marginals $\mathcal{V}$ and V. In the following, we will consider three cases: independence between $\mathcal{V}$ and V, and the functional dependences

$$\mathcal{V} = c - \frac{c}{\pi}V, \qquad \mathcal{V} = \frac{c}{\pi}V \tag{6.9}$$

which are called lower and upper Frechet-Hoeffding bounds (LFHB, UFHB, respectively), cf. [129]. Thus three models of RMCS (Y, Γ) are obtained.

Exercise 6.1. Consider the previous example, and evaulate the characteristics $E(x) = \kappa_e(x, y)$, $C_Y(x, y)$ and $k_{mm}(x, y)$ for all three models of the RMCS (Y, Γ). Show that the following formulae hold

	$E(x)$	$C_Y(x,y)$	$k_{mm}(x,y)$
UFHB	$\frac{ca}{\pi}$	0	$4\left(\frac{a}{\pi}\right)^2$
LFHB	$c\left(1-\frac{a}{\pi}\right)$	0	$4\left(1-\frac{a}{\pi}\right)^2$
independence	$\frac{c}{2}$	$\frac{c^2}{12}$	$\frac{4}{3}$

where $x = (t\cos\alpha, t\sin\alpha)$, $y = (s\cos\beta, s\sin\beta)$, $0 < s, t \le 1$, and a denotes the angle $\sphericalangle(\alpha,\beta)$ in radians.

More complex models of RMCS with dependencies between the random $\mathcal{H}^k$-set in $\mathbb{R}^d$ and the mark are hardly tractable analytically. For $k = d$ the level sets of Gaussian random fields are studied in [25]. Simulation based results for $k < d$ are presented in the following, cf. [374].

6.3.1 Tessellation Models

For some integer $p \ge 2$, consider a stationary Gaussian random field

$$X(s) = (X_1(s),\ldots,X_p(s)), \ s \in \mathbb{R}^2 \tag{6.10}$$

with independent components of zero mean. The covariance function of each component is equal to

$$R_i(s,t) = \alpha \exp(-\sigma \|s-t\|), \ \alpha, \sigma > 0, \ i = 1,\ldots,p.$$

Let $\lambda(s) = a\exp(X_1(s))$, $a > 0$, be the driving intensity function of a Cox point process ([170]) $\Phi \in \mathbb{R}^2$ and Y be the union of edges of the 2D Voronoi tessellation generated by Φ. In Fig. 6.1a.,b., the simulated realizations of RMCSs (Y, X_i), $i = 1, 2$, are shown, and in Fig. 6.1c.,d., there are RMCSs $(Y_{\oplus\varepsilon}, X_i)$. Since Y depends on X_1 we observe small edges in the areas of high values of X_1 and vice versa in Fig.6.1a). This is not the case in Fig. 6.1b. where Y does not depend on X_2.

Furthermore, we consider a nonnegative mark $\Gamma_i = \exp X_i$, $i = 1, 2$. Numerical results of the estimation of characteristics are shown in Fig. 6.2. We observe a higher conditional expectation and a more rapid decrease of the covariance function in the dependent case. Further in Fig. 6.3 results of the random-field test are presented. The results are as expected, in the dependent case a) $i = 1$ the random-field hypothesis is rejected which is not the case in the independent case b) $i = 2$.

Exercise 6.2. Modify the random-field model test according to [149] to achieve a global envelope test. Simulate the tessellation model with Gaussian random field described above. In step 4 of the algorithm described in Sect. 6.2.1 put

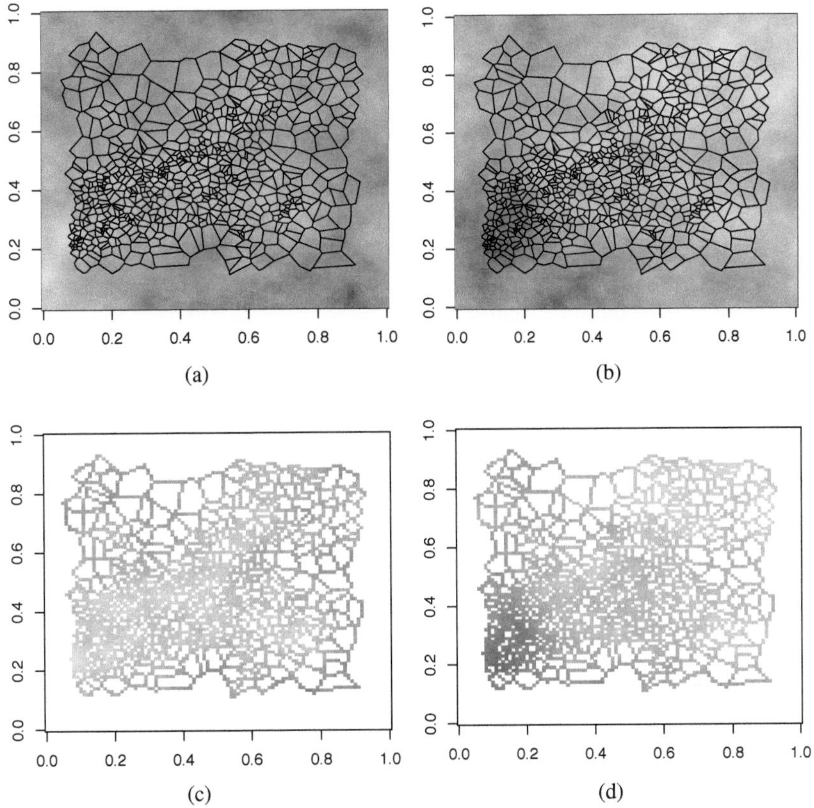

Fig. 6.1 Tessellation models: Simulated realizations of RMCS's (Y, X_i), a) $i = 1$, b) $i = 2$, c) RMCS $(Y_{\oplus\varepsilon}, X_1)$, d) RMCS $(Y_{\oplus\varepsilon}, X_2)$, $\varepsilon > 0$. Random field values are increasing from black to white grey level.

$$\hat{K}_{\max}(r) = \max\{\hat{K}_1(r), \ldots, \hat{K}_q(r)\}, \quad \hat{K}_{\min}(r) = \min\{\hat{K}_1(r), \ldots, \hat{K}_q(r)\}$$

for each $r > 0$. Let s denote the number of those j-th permutations such that there exists r with $\hat{K}_{\max}(r) = \hat{K}_j(r)$ or $\hat{K}_{\min}(r) = \hat{K}_j(r)$. When the horizontal axis lies between the envelopes built in step 5, the null hypothesis cannot be rejected on an approximate significance level $\frac{s}{q}$. How large q should be in order to achieve significance level 0.1?

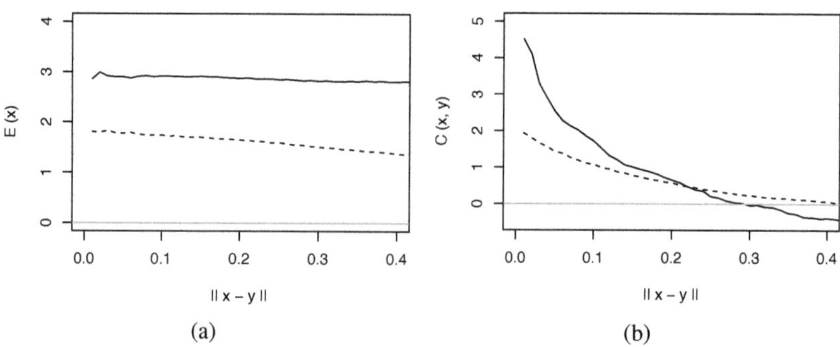

(a)   (b)

Fig. 6.2 Tessellation models: Graphs of estimated RMCS characteristics for (Y,Γ_1) (solid line) and (Y,Γ_2) (dotted line). a): Conditional mean mark $E(x)$, b): Mark covariance function $C_Y(x,y)$.

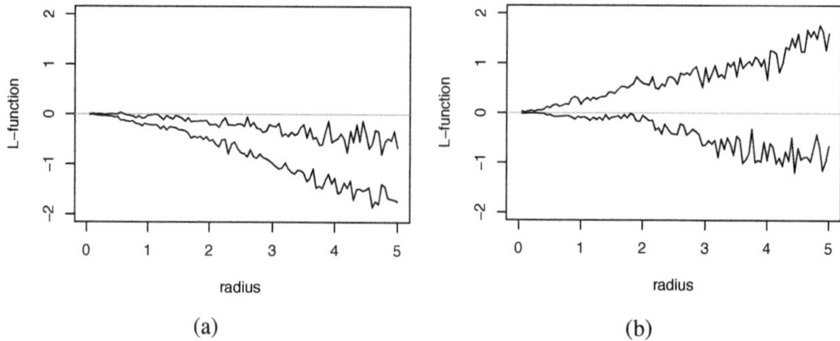

(a)   (b)

Fig. 6.3 Tessellation models: Results of the random-field test for (Y,Γ_i), $i = 1,2$. a) ($i = 1$): Horizontal axis lies outside the envelopes, H_0 in (6.4) is rejected. b) ($i = 2$): Horizontal axis between normalized L-function envelopes, therefore we do not reject H_0.

6.3.2 Fibre Process Based on Diffusion

As a model of a $\mathcal{H}^1$-set in $\mathbb{R}^2$, we will consider numerical solutions of a stochastic differential equation (SDE) for $Y_t = (Y_t^{(1)}, Y_t^{(2)}) \in \mathbb{R}^2$, $t \geq 0$. Let

$$dY_t^{(1)} = -\frac{a}{2} Y_t^{(1)} dt + b(Y_t^{(1)}, Y_t^{(2)}) h(Y_t^{(1)}, Y_t^{(2)}) dW_t^{(1)}, \tag{6.11}$$

$$dY_t^{(2)} = -\frac{a}{2} Y_t^{(2)} dt + b(Y_t^{(1)}, Y_t^{(2)}) h(Y_t^{(1)}, Y_t^{(2)}) dW_t^{(2)},$$

where

$$b(y) = b_1 \exp\left(b_2 g(y)\right), \; b_1, b_2 \in \mathbb{R}, \; g : \mathbb{R}^2 \to \mathbb{R}, \tag{6.12}$$

$$h(y) = (1 - ||y||^k), \; y \in \mathbb{R}^2$$

and $W_t = (W_t^{(1)}, W_t^{(2)})$ is the two-dimensional Brownian motion. In this subsection let $C = B_1(0) \subset \mathbb{R}^2$. The conditions under which the solution of (6.11) remains in C are known.

Theorem 6.1. *Let the function* $g : \mathbb{R}^2 \to \mathbb{R}$ *be bounded on C. Then for an arbitrary initial condition* $y_0 \in C$, *there exists a solution* $\{Y_t, \, t \geq 0\}$ *of* (6.11) *such that* $Y_0 = y_0$ *and* $\{Y_t, \, t \geq 0\} \subset C$ *almost surely. Moreover, if* $g : \mathbb{R}^2 \to \mathbb{R}$ *is Lipschitz continuous on* $B_{1+\varepsilon}(0)$ *for some* $\varepsilon > 0$ *then* $\{Y_t, \, t \geq 0\} \subset C$ *almost surely for an arbitrary solution of* (6.11) *with initial condition* $Y_0 = y_0 \in C$.

Proof. See [405]. $\qquad\square$

Exercise 6.3. Let $L_t = (Y_t^{(1)})^2 + (Y_t^{(2)})^2 = ||Y_t||^2$. Use Itô's formula to compute the stochastic differential dL_t. What is the effect of the relation between the obtained differential dL_t and the statement of Theorem 6.1?

Consider two probability spaces $(\Omega_i, \mathcal{F}_i, \mathbf{P}_i)$, $i = 1, 2$, the two-dimensional Brownian motion $W_t = (W_t^{(1)}, W_t^{(2)})$ on $(\Omega_1, \mathcal{F}_1, \mathbf{P}_1)$ and X_i, $i = 1, 2$, random fields in $\mathbb{R}^2$ on $(\Omega_2, \mathcal{F}_2, \mathbf{P}_2)$. Denote by $(\Omega, \mathcal{F}, \mathbf{P}) = (\Omega_1 \times \Omega_2, \mathcal{F}_1 \otimes \mathcal{F}_2, \mathbf{P}_1 \otimes \mathbf{P}_2)$ the product probability space, and for all $\omega_2 \in \Omega_2$, let $Y_t = (Y_t^{(1)}, Y_t^{(2)})$ be a solution of the stochastic differential equation given in (6.11). Further, denote by

$$Y_t^\alpha = (Y_t^{\alpha,(1)}, Y_t^{\alpha,(2)}) = (Y_t^{(1)}, Y_t^{(2)}) \begin{pmatrix} \cos\alpha & \sin\alpha \\ -\sin\alpha & \cos\alpha \end{pmatrix}$$

the α-rotation of the process Y, and analogously, X_i^α is α-rotation of the field X_i; $\alpha \in \mathbb{R}$.

Now we show that the RMCS (Y, X_i) is isotropic.

Theorem 6.2. *Let X be an isotropic random field. Then for arbitrary* $\alpha \in \mathbb{R}$, (Y, X) *and* (Y^α, X^α) *have the same distribution.*

Proof. Let $\omega_2 \in \Omega_2$. Then by the Itô formula we get that

$$dY_t^{\alpha,(1)} = \cos(\alpha)dY_t^{(1)} - \sin(\alpha)dY_t^{(2)}$$

$$= -\frac{a}{2}(\cos(\alpha)Y_t^{(1)} - \sin(\alpha)Y_t^{(2)})dt + b(Y_t)h(Y_t)(\cos(\alpha)dW_t^{(1)}$$

$$- \sin(\alpha)dW_t^{(2)})$$

$$= -\frac{a}{2}Y_t^{\alpha,(1)}dt + b(Y_t)h(Y_t^\alpha)dW_t^{\alpha,(1)} = -\frac{a}{2}Y_t^{\alpha,(1)}dt + b^\alpha(Y_t^\alpha)h(Y_t^\alpha)dW_t^{\alpha,(1)}$$

$$dY_t^{\alpha,(2)} = -\frac{a}{2}Y_t^{\alpha,(2)}dt + b^\alpha(Y_t^\alpha)h(Y_t^\alpha)dW_t^{\alpha,(2)},$$

where $b^\alpha(y) = b_1 \exp\left(b_2 X^\alpha(y, \omega_2)\right)$. Since W^α is also a Brownian motion, $(Y|X)$ and $(Y^\alpha|X^\alpha)$ have the same distribution. Finally, since X and X^α have the same distribution, the proof is complete. $\qquad\square$

In the following we assume that $g(s)$, $s \in C$, is a linear combination $\mathcal{L}_X$ of realizations of the components of a vector Gaussian random field $\{X(s), s \in C\}$, see (6.10). In order to simulate a realization of the random set $Y \in C$ we first simulate $g(s)$ and then simulate a trajectory $Y = \{Y_t, 0 \leq t \leq T\}$ of the SDE solution by means of the Euler method (conditionally on $g = \mathcal{L}_X$). From Theorem 6.1 we get that the theoretical solution of SDE remains in the circular region. However, the numerical solution may cross the boundary, therefore a condition is added that in this case the trajectory is projected on the boundary. Simulated RMCS's (Y, X_i), $i = 1, 2$, where $g = X_1$, are shown in Fig. 6.4. In the dependent case the high values of X_1 imply higher speeds of the motion and longer segments in the numerical solution.

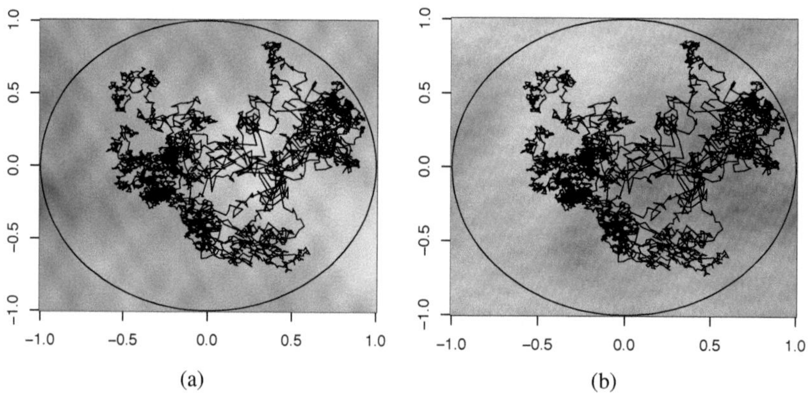

(a)   (b)

Fig. 6.4 Gaussian random fields a) X_1 and b) X_2, and simulated fibre process Y (a numerical solution of the SDE given in (6.11) given $g = X_1$). In a) the high values of X_1 (tending to white) imply higher speed of the motion and longer segments in the numerical solution.

Let $x, y \in \mathbb{R}^2$ such that $\|x\| = \|y\| = r > 0$. From the isotropy of (Y, X_i) proved in Theorem 2, the second-order characteristics based on (6.3) depend only on the angle α between vectors x and y. Thus, we will write $k_f(r, \alpha), E(r, \alpha)$ and $C_Y(r, \alpha)$. To estimate the second-order characteristics from a realization of RMCS in C using a finite test set U, $\varepsilon > 0$, let N_U in (6.7) have form

$$N_U^\varepsilon(r, \alpha) = \{(x, y) \in (Y_{\oplus \varepsilon} \cap U)^2 : \|x\| = r, \, x^\alpha = y \text{ or } x^{-\alpha} = y\}$$

and the estimator of $\hat{\kappa}_f(r, \alpha)$ be given as in (6.8).

We can study the temporal evolution of the characteristics $\kappa_f(r, \alpha)$ in this model. To do this, we define $Y(t, .)$ as a union of $n = 100$ different trajectories $\{Y_s, t \leq s \leq t + \delta t\}$, where $\delta t = 0.5$. The estimated characteristics are shown in Fig.6.5 at times $t = 0$ and $t = 10$ and with $r = 0.3$ and $r = 0.7$.

All dotted lines in Fig.6.5 describe characteristics for a random-field model, therefore they correspond to properties of the Gaussian random field (zero mean).

The decrease of the mark covariance function in Fig.6.5a is more rapid than in Fig.6.5c, which is explained by the fact that the radius r is larger in (a) than in (c) and the horinzotal axis is in radians. Concerning the dependent case (full lines) for the conditional mean mark in Figs. 6.5b and 6.5d the larger times (heavy full line) emphasize smaller values of the mark than smaller times do. This is because the former yields more time to reach the circle with given radius r even at small values of the random field than the latter. This difference is more considerable from larger r, therefore the conditional mean mark in (b) is larger than in (d).

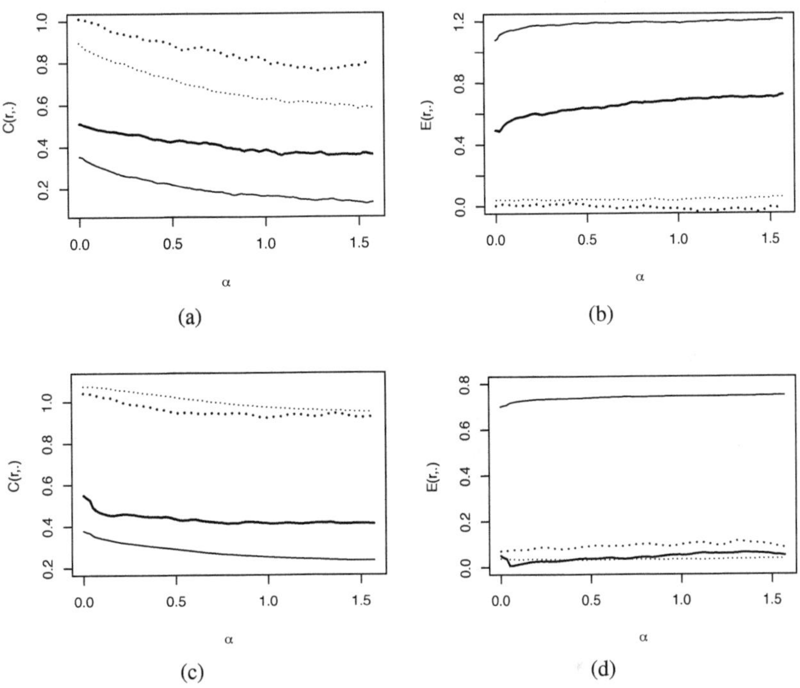

Fig. 6.5 Spatio-temporal model. Plots of estimated RMCS characteristics for (Y, X_1) (solid line) and (Y, X_2) (dotted line). Thick lines describe characteristics at time $t = 10$, thin lines describe characteristics at time $t = 0$. The characteristics are a) the mark covariance function $C_Y(r, \alpha)$ with fixed $r = 0.7$, b) the conditional mean mark $E(r, \alpha)$ with fixed $r = 0.7$, c) the mark covariance function $C_Y(r, \alpha)$ with fixed $r = 0.3$ and d) the conditional mean mark $E(r, \alpha)$ with fixed $r = 0.3$.

6.4 Real Data Analyses

Throughout this section we again assume that $d = 2$. Two applications of RMCS in neurophysiology and materials science are presented, where the models are similar to those simulated above. First a track of a rat is described by a planar fibre process marked by the spiking intensity in a neurophysiological experiment, see [52] for more about track modelling. Secondly the random tessellation of the grain microstructure in granular materials is marked by the disorientation in a crystallographic sense.

6.4.1 Random-field Model Test in a Neurophysiological Experiment

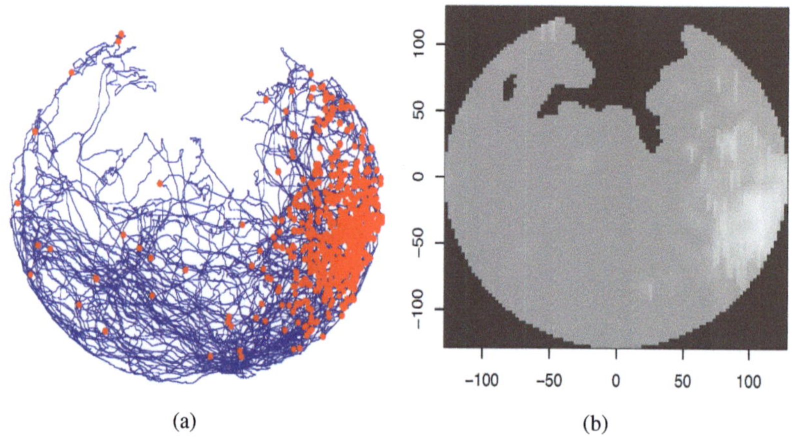

Fig. 6.6 A neurophysiological experiment. a) The lines describe the track Y of a rat moving in a circular area $\mathcal{A}$ and points (spikes) are locations where the neuron fired. b) The mark Γ is the estimate given in (6.13) of spiking intensity decreasing in grey level (white is the largest value). Its domain $\mathcal{A}' \subset \mathcal{A}$ lies outside the black area.

A real data specimen from [128] which involves a planar fibre system is analyzed. A spatio-temporal point pattern of action potentials (called spikes) of a cell place of hippocampus of a rat looking for food in a circular area $\mathcal{A} \subset \mathbb{R}^2$ is shown in Fig. 6.6a. The experiment lasted $T = 614$ seconds, altogether 1096 spikes were recorded. The location of the rat was recorded each $\frac{1}{60}$ sec, so that we observe a track, which is modelled by a random $\mathcal{H}^1$-set Y. Even if there is a temporal element in the experiment, we solve a purely spatial problem within our context.

Fig. 6.7 Test of the random-field model in the neurophysiological experiment. Here Y is the track and Γ the spiking intensity. After $n = 39$ permutations the horizontal axis lies inside the envelopes given in (6.6), therefore we do not reject the hypothesis of the random-field model.

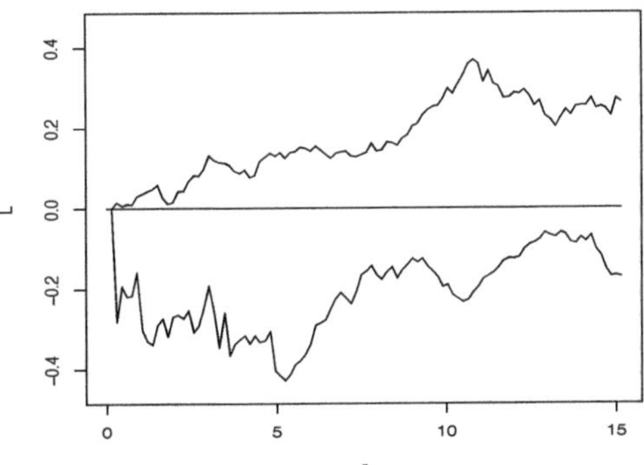

The aim is to find out whether there is a significant dependence between the spiking activity of the neuron and the track of the rat. A RMCS (Y,Γ) is considered, where

$$\Gamma(s) = \frac{\sum_{s_i \in \mathcal{A}} k(\|s - s_i\|)}{\hat{\lambda}(s)}, \quad s \in \mathcal{A}' \tag{6.13}$$

is the planar spiking intensity kernel estimate, s_i is the location of the $i-$th spike, and $\mathcal{A}' \subset \mathcal{A}$ denotes the domain, see Fig. 6.6b. The kernel function is given by

$$k(r) = \begin{cases} (b^2 - r^2)\frac{2}{\pi b^4}, & \text{if } r \le b \\ 0, & \text{if } r > b. \end{cases}$$

A kernel estimator of the (track) intensity λ of the random measure Ψ_Y is given by

$$\hat{\lambda}(s) = \sum_{t_i \in H} k(\|s - Y(t_i)\|)\triangle_t,$$

$s \in \mathcal{A}'$, H is a set of equidistant times, $\triangle_t = t_{i+1} - t_i$, and $Y(t)$ is the location of the rat at time t.

Fig. 6.6b shows the spiking intensity estimate Γ, where the bandwidth was chosen $b = 10$. The result of the random-field test from Sect. 6.2.1 is visualized in Fig. 6.7. The hypothesis of independence of the spiking activity and the random track cannot be rejected at significance level 0.05 for each fixed r.

6.4.2 Second-order RMCS Analysis of Granular Materials

In microstructural research of crystallic materials, it is an interesting task to find the link between the granular microstructure and macroscopical properties of the material. Second-order analysis is a useful tool for characterizing the spatial distribution of grain boundaries. In what follows we deal with data from two-dimensional electron backscatter diffraction (EBSD) which is a progressive method allowing to directly measure the crystallic orientations at the surface of the material ([325]).

During an automatic scanning procedure three Euler angles characterizing the crystal lattice orientation are measured at each grid point. The difference in orientations of two neighbouring points can be expressed in angle/axis notation - one orientation can be transformed to another by rotation around an axis $[uvw]$ by an angle θ. Because of non-uniqueness of this transformation we only consider the solution with minimum angle θ and we call it disorientation of two lattice points. Grain boundaries are then identified as interfaces where the disorientation angle between two neighbouring points exceeds a given limit.

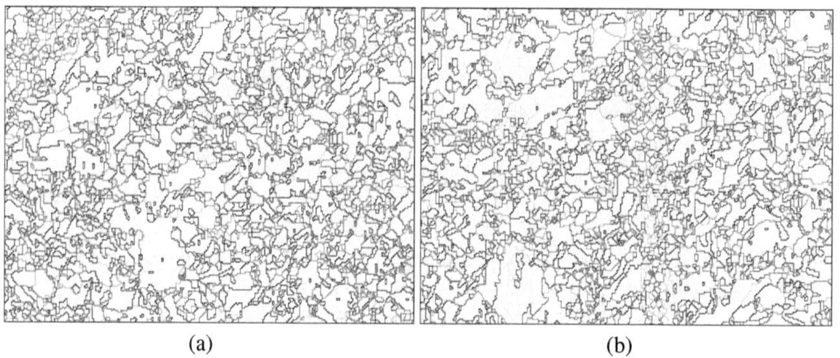

(a)   (b)

Fig. 6.8 EBSD images of grain structure with darkness of the boundaries corresponding to the disorientation angle: $\theta_{\min} \approx$ white, $\theta_{\max} \approx$ black. Samples of copper processed by eight ECAP passes and annealed for 10h in a) 373K, b) 423K.

It is well known that physical properties of ultrafine-grained materials are related to the distribution of different grain boundary types. The grain boundary network in a planar section can be considered as a marked fibre process with marks related to disorientations of the boundaries. The simplest way of marking is that one which is based on the disorientation angle $\theta \in (\theta_{\min}, \theta_{\max})$ where the lower bound is chosen conventionally as $\theta_{\min} = 5°$ and the upper bound is given by crystal lattice symmetries; in our case of material with cubic crystal symmetry it is $\theta_{\max} \approx 62.8°$. In Fig. 6.8 the grain boundaries in two samples are visualized with respect to their disorientation angle - the larger is the angle, the darker is the corresponding line. Consider the RMCS (Y, Γ) where Y is the fibre process given by intersection of the grain boundaries with a 2D section (surface of the sample) and Γ is the restriction of the disorientation angle on Y. Fig. 6.9 compares the mark correlation function and mark variogram of pure copper processed by eight passes of equal-channel angular pressing (ECAP is a production method based on pressing, cf. [180]) and two different times of subsequent annealing.

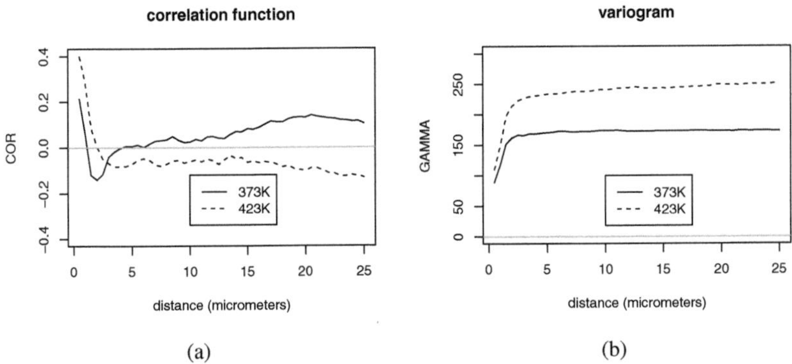

Fig. 6.9 Mark correlation function (a) and mark variogram (b) of disorientation angles of grain boundaries in Cu after 8 ECAP passes with two different temperatures of subsequent annealing.

The behaviour of the functions near zero is influenced by the fact that the disorientation angle is nearly constant along an edge. However, for larger distances we can observe differencies related to the origin of the samples. While in the first sample (373K) the correlation function remains near the zero level, in the second sample (423K) the marks are positively correlated even for larger distances and they embody smaller variability expressed by the variogram.

6.5 Dimension Reduction

The dimension reduction problem in statistics concerns the situation where we have a response Y (random variable) which depends on a $p-$dimensional random column vector of covariates X and the aim is to reduce the number of covariates in order to use only the most significant ones. For vector data, [251] suggested using the sliced inverse regression (SIR) method. The idea to regress X on Y inversely instead of the direct regression of Y on X stems from the fact that in such a way we obtain p univariate regressions instead of a multivariate one. Further methods of dimension reduction are described in e.g. [82, 249, 250].

The dimension reduction problem for point processes was formulated by [156]. In our setting, Y is a random point, fibre or surface process in $\mathbb{R}^d$ and X is a p-dimensional stationary Gaussian random field in $\mathbb{R}^d$, $d = 2, 3$. Without loss of generality, we assume that X is standardized, i.e. $\mathbf{E}X(s) = o$ and $\mathbf{cov}X(s) = I_p$ for each $s \in \mathbb{R}^d$, where I_p is the unit matrix of order p. Let $A^\top$ be the transpose of a real-valued matrix A and $\mathcal{S}(A)$ the linear subspace spanned by the columns of A.

Definition 6.1. Let Y be conditionally independent of X given $B^\top X$ for a $p \times c$ matrix B, $c \leq p$. Then $\mathcal{S}(B)$ is called a *sufficient dimension reduction subspace*. Let

$\mathcal{S}_{Y|X}$ be the intersection of all such subspaces. Assume that $\mathcal{S}_{Y|X}$ is also a sufficient dimension reduction subspace, which is then called the *central subspace*.

The above definition says that $B^\top X = \{B^\top X(s), s \in \mathbb{R}^d\}$ contains all information of X about Y. Let c be the dimension of the central subspace.

A refined analysis of dimension reduction is based on the following definition. Assume that the l−th order intensity functions, cf. (6.1), exist for all $l \geq 1$ (depending on X, and being random functions).

Definition 6.2. Let $l \in \mathbb{N}$ and assume that the relation

$$\lambda_l(s_1,\ldots,s_l) = f_l(B^\top X(s_1),\ldots,B^\top X(s_l)), \tag{6.14}$$

holds for some measurable function f_l and a $p \times c$ matrix B, $c \leq p$. Then $\mathcal{S}_l(B)$ is called the l−*th order sufficient intensity dimension reduction subspace* and the intersection of all such subspaces $\mathcal{S}_l = \cap \mathcal{S}_l(B)$ is called the l−*th order central intensity subspace* (if it is also an l−th order sufficient intensity dimension reduction subspace).

The central subspace defined above can be expressed as

$$\mathcal{S}_{Y|X} = \cup_{l \geq 1} \mathcal{S}_l,$$

cf. [157]. The aim is to investigate the structure of the k−th order sufficient intensity dimension reduction subspaces.

We concentrate mostly on the sliced inverse regression (SIR) method, which can be described in the following steps: (i) slicing the random set Y according to a suitable mark Γ, (ii) finding the slice means of the random field X, (iii) applying the principal components analysis of slice means.

The first c directions from (iii) generate the central subspace of the corresponding order. Notice that Y plays a role in step (i) only.

6.6 Theoretical Results

We start with a lemma needed for the refined analysis of the dimension reduction.

Lemma 6.2. *Let $C \subset \mathbb{R}^d$ be a compact convex set. a) Assume that (6.14) holds for $l = 1$. Then*

$$\int_C \mathbf{E}\,(X(s)\lambda(s))\,ds = \mathbf{E} \int_{s \in Y \cap C} X(s)\mathcal{H}^k(ds).$$

b) Assume that (6.14) holds for $l = 2$. Then

$$\int_C \int_C \mathbf{E}\left(X(s)X(t)^\top \lambda_2(s,t)\right) ds dt = \mathbf{E} \int_{s,t \in Y \cap C} X(s)X(t)^\top \mathcal{H}^k(ds)\mathcal{H}^k(dt).$$

Proof. Assertions a) and b) follow from the first and second order Campbell theorem, respectively, see e.g. [73]. For a) we have

$$\mathbf{E}\int_{s\in Y\cap C}X(s)\mathcal{H}^k(ds) = \int\int_{s\in Y\cap C}x(s)\mathcal{H}^k(ds)P(d\psi)$$
$$= \int\int\int_{s\in Y\cap C}x(s)\mathcal{H}^k(ds)P_\Lambda(d\psi)Q(d\Lambda)$$
$$= \int\int_C\int x(s)\Lambda(d(x,s))Q(d\Lambda)$$
$$= \int_C\int\int x(s)\lambda(s)M_s(dx)Q(d\Lambda)ds$$
$$= \int_C\mathbf{E}(X(s)\lambda(s))\,ds,$$

where P is the distribution of the marked process $(Y,X(s))$ with random intensity measure

$$\Lambda(d(x,s)) = \lambda(s)M_s(dx)ds = f(B^\top x(s))M_s(dx)ds$$

and the distribution M_s of the mark $X(s)$ at a point s. Note that Q is the distribution of Λ and P_Λ is the conditional distribution of the marked process given Λ. $\qquad\square$

For $k = 0$ and a point process Y we have

$$\mathbf{E}\sum_{s\in Y\cap C}X(s),\quad \mathbf{E}\sum_{s,t\in Y\cap C}^{\neq}X(s)X(t)^\top,$$

on the right-hand sides of the formulas in Lemma 6.2 in a), b), respectively.

6.6.1 Investigation of $\mathcal{S}_1$

In this section we assume that for each $s \in \mathbb{R}^d$ it holds that

$$\lambda(s) = f(B^\top X(s)) \tag{6.15}$$

for a matrix B of size $p \times c$, $c \leq p$. Let the range of the real-valued mark Γ of Y be divided into m disjoint intervals $J_1,\ldots,J_m$ called slices. This induces a partition of Y into m disjoint subsets $(Y^1,\ldots,Y^m)$ which are assumed to be random $\mathcal{H}^k$-sets again. Alternatively, we may consider the Y^i as processes with first-order intensities $\lambda^{(i)}$, where

$$\int_A \lambda^{(i)}(s)ds = \mathbf{E}\Psi_{Y^i}(A),\ A \in \mathcal{B}^d,\ i = 1,\ldots,m, \tag{6.16}$$

respectively, corresponding to each slice, so that $\lambda = \sum_{i=1}^m \lambda^{(i)}$.

Lemma 6.3. *Consider the mark* $\Gamma(s) = g(B^\top X(s))$ *for a measurable function g. Then*

$$\lambda^{(j)}(s) = f^j(B^\top X(s))$$

for some nonnegative measurable functions f^j, $j = 1, \ldots, m$.

Proof. For $|ds| \downarrow 0$ we have

$$
\begin{aligned}
\lambda^{(j)}(s)|ds| &= \mathbf{E}\left(Y^j(ds)|X(s)\right) \\
&= \mathbf{E}\left(\mathcal{H}^k(\{y \in Y \cap ds; \ \Gamma(y) \in J_j\})|X(s)\right) \\
&= \mathbf{E}\left(\mathcal{H}^k(\{y \in Y \cap ds; \ \Gamma(y) \in J_j\})|B^\top X(s)\right) \\
&= f^j(B^\top X(s))|ds|
\end{aligned}
$$

$\square$

Consider a convex compact window $C \subset \mathbb{R}^d$ and the statistic

$$
\hat{V}_1 = \frac{1}{\Psi_Y(C)} \sum_{j=1}^{m} \frac{1}{\Psi_{Yj}(C)} \int_{Y^j \cap C} X(s)\mathcal{H}^k(ds) \left[\int_{Y^j \cap C} X(s)\mathcal{H}^k(ds)\right]^\top. \tag{6.17}
$$

Assume that Y and $\{X(s), \ s \in Y\}$ are ergodic. Then, from Lemma 6.3, for each $j = 1, \ldots, m$ we get that

$$
\frac{1}{\Psi_{Yj}(C)} \int_{Y^j \cap C} X(s)\mathcal{H}^k(ds) \to \frac{\int_{\mathbb{R}^d} \mathbf{E}\lambda^{(j)}(s)X(s)ds}{\int_{\mathbb{R}^d} \mathbf{E}\left(\lambda^{(j)}(s)\right)ds} \tag{6.18}
$$

in probability when $C \uparrow \mathbb{R}^d$. This limit is defined as ratio of the limits

$$
\lim_{C \uparrow \mathbb{R}^d} \frac{1}{|C|} \int_C \mathbf{E}\left(\lambda^{(j)}(s)X(s)\right)ds, \quad \lim_{C \uparrow \mathbb{R}^d} \frac{1}{|C|} \int_C \mathbf{E}\lambda^{(j)}(s)ds.
$$

Their finiteness can be verified easily e.g. in our case when X is a stationary random field. Then the limit in (6.18) is equal to

$$
\frac{\mathbf{E}\left(\lambda^{(j)}(.)X(.)\right)}{\mathbf{E}\lambda^{(j)}(.)}. \tag{6.19}
$$

Let the theoretical counterpart of the expression given in (6.17) be

$$
V_1 = \frac{1}{\int_{\mathbb{R}^d} \mathbf{E}\lambda(s)ds} \sum_{j=1}^{m} \frac{\int_{\mathbb{R}^d} \mathbf{E}\left(\lambda^{(j)}(s)X(s)\right)ds \int_{\mathbb{R}^d} \mathbf{E}\left(\lambda^{(j)}(s)X(s)\right)^\top ds}{\int_{\mathbb{R}^d} \mathbf{E}\lambda^{(j)}(s)ds}. \tag{6.20}
$$

Theorem 6.3. *Under the above assumptions it holds that $\mathcal{S}(V_1) \subset \mathcal{S}_1$.*

Proof. Let B be a matrix with $\mathcal{S}(B) = \mathcal{S}_1$ and $j \in \{1, \ldots, m\}$. From Lemma 6.3 we have that $\lambda^{(j)}(s) = f^j(B^\top X(s))$ for some measurable function f^j. It is enough to show that

$$
\mathcal{S}(\int \mathbf{E}\left(f^j(B^\top X(s))X(s)\right)ds \int \mathbf{E}\left(f^j(B^\top X(s))X(s)\right)^\top ds) \subset \mathcal{S}_1.
$$

Using the argument given in [157], p.385, (A.1), for the projection matrix $P_B = B(B^\top B)^{-1}B^\top$ we obtain

$$\int \mathbf{E}\left(f^j(B^\top X(s))X(s)\right)ds \int \mathbf{E}\left(f^j(B^\top X(s))X(s)\right)^\top ds$$

$$= P_B \int \mathbf{E}\left(f^j(B^\top X(s))X(s)\right)ds \int \mathbf{E}\left(f^j(B^\top X(s))X(s)\right)^\top ds P_B.$$

Since this holds for each $j = 1, \ldots, m$, we have $\mathcal{S}(V_1) \subset \mathcal{S}(B) = \mathcal{S}_1$. $\qquad\square$

From the proof of Theorem 6.3, it can be seen that for each j-th slice, $j = 1, \ldots, m$, the vector given in (6.19) belongs to $\mathcal{S}_1$. Principal component analysis is applied to find c vectors among them. Define the slice means as conditional expectations

$$m_j = \mathbf{E}(X(s) \mid s \in Y^j) \tag{6.21}$$

and let $p_j = P(x \in Y^j \mid x \in Y)$, $j = 1, \ldots, m$. The weighted covariance matrix

$$V = \sum_{j=1}^{m} p_j m_j m_j^\top \tag{6.22}$$

of size $p \times p$ has eigenvalues $\xi_1 \geq \cdots \geq \xi_p$. Then the eigenvectors η_l of V corresponding to the c largest eigenvalues form the columns of the matrix B. When dealing with data observed in a compact window C, the matrix V can be estimated by

$$\hat{V}_1 = \sum_{j=1}^{m} \frac{\Psi_{Y^j}(C)}{\Psi_Y(C)} \frac{\int_{Y^j \cap C} X(s)\mathcal{H}^k(ds)}{\Psi_{Y^j}(C)} \frac{\left(\int_{Y^j \cap C} X(s)\mathcal{H}^k(ds)\right)^\top}{\Psi_{Y^j}(C)},$$

c.f. (6.17) and (6.22). For more information on the estimation of V, see Sect. 6.7.

6.6.2 Investigation of $\mathcal{S}_2$

Let $l = 2$ in Definition 6.2 and assume that

$$\lambda_2(s,t) = f_2(B^\top X(s), B^\top X(t)) \tag{6.23}$$

for a matrix B of size $p \times c$, $c \leq p$. The aim is to estimate the subspace $\mathcal{S}_2 = \mathcal{S}(B)$. Assume that Y and $\{X(s)X(t)^\top, s,t \in Y\}$ are ergodic, $C \subset \mathbb{R}^d$ is a convex compact window. Then from Lemma 6.3 we get that

$$\hat{M}_2 = \frac{\int_{s,t \in Y \cap C} X(s)X(t)^\top \mathcal{H}^k(ds)\mathcal{H}^k(dt)}{\Psi_Y(C)^2} \to \frac{\int\int \mathbf{E}\left(\lambda_2(s,t)X(s)X(t)^\top\right)dsdt}{\int\int \mathbf{E}\lambda_2(s,t)dsdt} = M_2 \tag{6.24}$$

in probability when $C \uparrow \mathbb{R}^d$, where the limit in (6.24) is defined as ratio of the limits

$$\lim_{C\uparrow\mathbb{R}^d}\frac{1}{|C|^2}\int_C\int_C\mathbf{E}\left(\lambda_2(s,t)X(s)X(t)^\top\right)dsdt,\quad \lim_{C\uparrow\mathbb{R}^d}\frac{1}{|C|^2}\int_C\int_C\mathbf{E}\lambda_2(s,t)dsdt.$$

$$(6.25)$$

Their finiteness can be verified e.g. in the case when X is a stationary or a stationary and isotropic random field. Then the integrands depend only of the difference of the variables and on the modul of the difference of the variables, respectively.

For a point process ($k=0$) we consider the estimator

$$\hat{M}_2 = \frac{\sum_{s,t\in Y\cap C}^{\neq}X(s)X(t)^\top}{\Psi_Y(C)(\Psi_Y(C)-1)}.$$

Theorem 6.4. *Let* $Q_B = I_p - P_B$. *Then it holds that*

$$M_2 = M_2^P + M_2^Q = \frac{P_B\int\int\mathbf{E}\left(f_2(B^\top X(s),B^\top X(t))X(s)X(t)^\top\right)dsdt\,P_B}{\int\int\mathbf{E}\left(f_2(B^\top X(s),B^\top X(t))\right)dsdt} \tag{6.26}$$
$$+\frac{\int\int\mathbf{E}\left(f_2(B^\top X(s),B^\top X(t))\right)\mathbf{E}\left(Q_B X(s)X(t)^\top Q_B\right)dsdt}{\int\int\mathbf{E}\left(f_2(B^\top X(s),B^\top X(t))\right)dsdt}.$$

Proof. Write $X(s) = P_B X(s) + Q_B X(s)$. Then

$$\mathbf{E}\left(\lambda_2(s,t)X(s)X(t)^\top\right) = \mathbf{E}\Big(\mathbf{E}[f_2(B^\top X(s),B^\top X(t))\left(P_B X(s)X(t)^\top P_B\right.$$
$$+P_B X(s)X(t)^\top Q_B + Q_B X(s)X(t)^\top P_B$$
$$+Q_B X(s)X(t)^\top Q_B\Big)\,|P_B X(s),P_B X(t)]\Big)$$
$$= P_B\mathbf{E}\left(f_2(B^\top X(s),B^\top X(t))X(s)X(t)^\top\right)P_B$$
$$+\mathbf{E}\left(f_2(B^\top X(s),B^\top X(t))P_B X(s)\mathbf{E}\left(X(t)^\top Q_B|P_B X(s),P_B X(t)\right)\right)$$
$$+\mathbf{E}\left(f_2(B^\top X(s),B^\top X(t))\mathbf{E}[Q_B X(s)|P_B X(s),P_B X(t)]X(t)^\top P_B\right)$$
$$+\mathbf{E}\left(f_2(B^\top X(s),B^\top X(t))\mathbf{E}\left(Q_B X(s)X(t)^\top Q_B|P_B X(s),P_B X(t)\right)\right).$$

The inner expectation in the second and third term is equal to zero by the assumptions, and

$$\mathbf{E}\left(Q_B X(s)X(t)^\top Q_B|P_B X(s),P_B X(t)\right) = \mathbf{E}\left(Q_B X(s)X(t)^\top Q_B\right).$$

Thus the assertion follows. $\qquad\square$

If the second term M_2^Q on the right-hand side of (6.26) were zero, then

$$\mathcal{S}(M_2 M_2^\top) \subset \mathcal{S}(M_2) \subset \mathcal{S}_2. \tag{6.27}$$

We are interested in situations where M_2^Q is negligible w.r.t. M_2^P, i.e.

$$||M_2^Q|| << ||M_2^P|| \qquad (6.28)$$

for the Euclidean matrix norm. This means that approximately $\mathcal{S}_2$ can be estimated by the SIR method applied to $M_2 M_2^\top$. Typically

$$\mathbf{E}\left(Q_B X(s) X(t)^\top Q_B \right)$$

is negligible for $||s-t|| \to \infty$, while in some models (e.g. repulsive point processes) $\lambda_2(s,t)$ is close to zero for $||s-t||$ small. Moreover, when there is a positive correlation between $P_B X(s) X(t)^\top P_B$ and $\lambda_2(s,t)$, then intuitively (6.28) may hold.

We make this reasoning precise in the following two examples, where we consider the Gaussian random field $X = (X_1, X_2)$ in $\mathbb{R}^2$ with independent components of correlation functions

$$\zeta(s,t) = \exp(-||s-t||^2).$$

Furthermore, let $B = (1,0)^\top$, so that $\lambda_2(s,t) = f_2(X_1(s), X_1(t))$. Note that the 2×2 matrix functions

$$P_B \mathbf{E}\left(\lambda_2(s,t) X(s) X(t)^\top \right) P_B \qquad (6.29)$$

and

$$\mathbf{E}(\lambda_2(s,t)) \mathbf{E}\left(Q_B X(s) X(t)^\top Q_B \right) \qquad (6.30)$$

are the integrands in the numerators of (6.26). Nonzero elements of these matrices are in the upper left and lower right corner, respectively. These elements are evaluated as functions of the variable $z = ||s-t||$. The expectations are evaluated w.r.t. the bivariate Gaussian probability density

$$g(x,y) = \frac{1}{2\pi\sqrt{1-\zeta(s,t)}} \exp\left(-\frac{1}{2(1-\zeta(s,t))}(x^2 + y^2 - 2xy\sqrt{\zeta(s,t)}) \right), \ x,y \in \mathbb{R}.$$

Example 6.2. Consider a stationary Poisson point process Φ with intensity ρ. Let Y be a simple inhibition point process such that each pair of points $s,t \in \Phi$ satisfying

$$\max(Z(s), Z(t)) \geq ||s-t|| \qquad (6.31)$$

is removed, where $Z(s) = g(X_1(s))$ is a nonnegative function of X_1. The process Y has second-order intensity given by (cf. [102])

$$\lambda_2(s,t) = \begin{cases} \rho^2 \exp(-\rho U(Z(s), Z(t))), & \text{if } \max(Z(s), Z(t)) \leq ||s-t||, \\ 0, & \text{else,} \end{cases}$$

where $U(Z(s), Z(t))$ is the area of the union of balls centered in s and t with radii $Z(s)$ and $Z(t)$, respectively. In this case the assumption (6.23) holds with $B = (1,0)^\top$. Put

$$Z(s) = a + b\mathbf{1}_{[X_1(s)<0]}, \ a,b > 0. \qquad (6.32)$$

It holds that $\lambda_2(s,t) = 0$ if either $[X_1(s) < 0 \vee X_1(t) < 0] \wedge ||s-t|| < a+b$ or $[X_1(s) \geq 0 \wedge X_1(t) \geq 0] \wedge ||s-t|| < a$. In the opposite case there are the following variants:

$$\begin{aligned}
X_1(s) \geq 0,\ X_1(t) \geq 0,\ ||s-t|| \geq a, &\qquad \lambda_2(s,t) = \rho^2 e^{-\rho U_2(s,t,a,a)}, \\
X_1(s) \geq 0,\ X_1(t) < 0,\ ||s-t|| \geq a+b, &\qquad \lambda_2(s,t) = \rho^2 e^{-\rho U_2(s,t,a,a+b)}, \\
X_1(s) < 0,\ X_1(t) \geq 0,\ ||s-t|| \geq a+b, &\qquad \lambda_2(s,t) = \rho^2 e^{-\rho U_2(s,t,a+b,a)}, \\
X_1(s) < 0,\ X_1(t) < 0,\ ||s-t|| \geq a+b, &\qquad \lambda_2(s,t) = \rho^2 e^{-\rho U_2(s,t,a+b,a+b)}.
\end{aligned}$$

See also Fig. 6.10a for illustration.

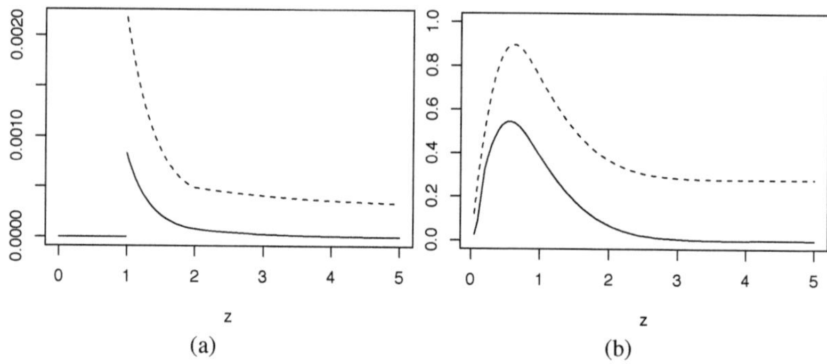

Fig. 6.10 Comparison of nonzero elements of the matrix functions of variable $z = ||s-t||$ given in (6.29) (dashed line) and (6.30) (solid line); (a) Example 6.2 (simple inhibition) for $a = b = \rho = 1$. (b) Example 6.3 (determinantal process) for $\alpha = 1$. Values of the dashed graph are much larger than those of the solid one in both cases.

Example 6.3. The stationary determinantal point process Y has second-order intensity equal to the determinant

$$\lambda_2(s,t) = \begin{vmatrix} C_0(0) & C_0(s-t) \\ C_0(t-s) & C_0(0) \end{vmatrix}$$

where C_0 is a covariance function. For parameters $\alpha, \rho > 0$ we use a covariance function with finite range α, where

$$C_0(x) = \frac{2\rho}{\pi} \left(\arccos \frac{||x||}{\alpha} - \frac{||x||}{\alpha} \sqrt{1 - \left(\frac{||x||}{\alpha}\right)^2} \right) \mathbf{1}_{[||x|| < \alpha]}.$$

The parameter ρ is randomized and it depends on the first component of X, where

$$\rho = \frac{4}{\pi^2 \alpha^2} \left(\arctan(X_1(s)X_1(t)) + \frac{\pi}{2} \right).$$

Under this scaling, given X the determinantal process always exists, cf. [246]. See also Fig. 6.10b for illustration.

Exercise 6.4. Construct a RMCS model where the first and the second order sufficient intensity dimension reduction subspaces will be different (which is not the case in the previous example). *Hint*: Consider again the determinantal point process ([246]) with suitable randomization of each parameter α, ρ by components of Gaussian random fields.

Theorem 6.4 also enables us to understand the availability of a slicing procedure in the analysis of $\mathcal{S}_2$. In this case, the Cartesian product $Y \times Y$ should be marked. Let

$$
\begin{aligned}
\mathcal{Y} &= Y \times Y, & \Psi_{\mathcal{Y}}(C) &= \Psi_Y(C)^2, & &\text{if } k > 0, \\
\mathcal{Y} &= \{(s,t); s \in Y, t \in Y, s \neq t\}, & \Psi_{\mathcal{Y}}(C) &= \Psi_Y(C)(\Psi_Y(C)-1), & &\text{if } k = 0.
\end{aligned}
\tag{6.33}
$$

Consider a mark $\Gamma : \mathcal{Y} \to \mathbb{R}$ which is a measurable symmetric function, i.e. $\Gamma(s,t) = \Gamma(t,s)$ for each $(s,t) \in \mathcal{Y}$. Let the range of Γ be divided into m disjoint intervals called slices. This induces a (random) partition of $\mathcal{Y}$ into m disjoint subsets $(\mathcal{Y}^1, \ldots, \mathcal{Y}^m)$. Let

$$
\Psi_{\mathcal{Y}^j}(C) = \int \int_{\mathcal{Y}^j \cap C^2} \mathcal{H}^k(ds)\mathcal{H}^k(dt).
$$

Define the conditional expectation matrices (slice means)

$$
o_j = \mathbf{E}(X(s)X(t)^\top | (s,t) \in \mathcal{Y}^j)
\tag{6.34}
$$

and let $q_j = \mathbf{P}((s,t) \in \mathcal{Y}^j | (s,t) \in \mathcal{Y})$, $j = 1, \ldots, m$. The matrix $U_2 = \sum_{j=1}^m q_j o_j o_j^\top$ is then subject to the principal component analysis. An empirical version of the matrix U_2 calculated from data is given by

$$
\begin{aligned}
\hat{U}_2 &= \frac{1}{\Psi_{\mathcal{Y}}(C)} \sum_{j=1}^m \Psi_{\mathcal{Y}^j}(C)\hat{M}_2^j[\hat{M}_2^j]^\top \\
&= \frac{1}{\Psi_{\mathcal{Y}}(C)} \sum_{j=1}^m \frac{1}{\Psi_{\mathcal{Y}^j}(C)} \int \int_{\mathcal{Y}^j \cap C^2} X(s)X(t)^\top \mathcal{H}^k(ds)\mathcal{H}^k(dt) \\
&\quad \times \left[\int \int_{\mathcal{Y}^j \cap C^2} X(s)X(t)^\top \mathcal{H}^k(ds)\mathcal{H}^k(dt) \right]^\top.
\end{aligned}
\tag{6.35}
$$

In the special case $k = 0$ we can express $\hat{M}_2^j$ as

$$
\hat{M}_2^j = \frac{1}{\Psi_{\mathcal{Y}^j}(W)} \sum_{\mathcal{Y}^j \cap C^2} X(s)X(t)^\top.
$$

6.7 Statistical Methods

When we deal with simulated or real data, the theoretical quantities from Sect. 6.6 are estimated by their empirical analogues. Furthermore, the dimension c of the central subspace is not known and hypotheses on this quantity are tested.

6.7.1 Estimation

Generally, for a stationary Gaussian p-dimensional random field $\tilde{X}$ observed in a bounded window $C \subset \mathbb{R}^d$ with Lebesgue measure $|C|$, put

$$\bar{X} = \frac{1}{|C|} \int_C \tilde{X}(s)ds, \quad \Sigma = \frac{1}{|C|} \int_C [\tilde{X}(s) - \bar{X}][\tilde{X}(s) - \bar{X}]^\top ds.$$

Then the standardized p-dimensional random field X is given by

$$X(s) = \Sigma^{-1/2}[\tilde{X}(s) - \bar{X}].$$

Based on the observation of the random field X on a set of grid points G, the empirical analogues of $\bar{X}$ and Σ are given by

$$\widehat{\bar{X}} = \frac{1}{\operatorname{card} G} \sum_{s \in G} \tilde{X}(s)ds, \quad \widehat{\Sigma} = \frac{1}{\operatorname{card} G} \sum_{s \in G} [\tilde{X}(s) - \widehat{\bar{X}}][\tilde{X}(s) - \widehat{\bar{X}}]^\top ds$$

and the empirical standardized random field $\widehat{X}$ at an arbitrary point $s \in C$ is given by

$$\hat{X}(s) = \widehat{\Sigma}^{-1/2}[\tilde{X}(s_G) - \widehat{\bar{X}}]$$

where $s_G \in G$ is the nearest grid point to s.

The characteristics of $\mathcal{H}^k$-sets can be estimated by choosing a finite set

$$T = \{t_i\}_{i=1}^n \subset Y \cap C \tag{6.36}$$

of random test points. Generally, samples T_p of complementary dimension $d - k$ are used to get test points as intersections $T = T_p \cap Y \cap C$, cf. [73]. Let

$$T_j = T \cap Y^j, \ j = 1,\ldots,m, \quad n_j = \operatorname{card} T_j$$

and (cf. (6.33))

$$\mathcal{T} = \{(s,t) \in T \times T, \ s \neq t\}, \quad \mathcal{T}_j = \mathcal{T} \cap \mathcal{Y}^j,$$

$l_j = \operatorname{card} \mathcal{T}_j$, $\sum_{j=1}^m l_j = n(n-1)$. We have the estimators

$$\hat{p}_j = \frac{n_j}{n}, \quad \hat{q}_j = \frac{l_j}{n(n-1)}, \tag{6.37}$$

$$\hat{m}_j = \frac{1}{n_j} \sum_{t \in T_j} X(t), \quad \hat{o}_j = \frac{1}{l_j} \sum_{(s,t) \in T_j} X(s) X(t)^\top. \tag{6.38}$$

Using these estimates, we put

$$\hat{V} = \sum_{j=1}^{m} \hat{p}_j \hat{m}_j \hat{m}_j^\top, \quad \hat{U} = \sum_{j=1}^{m} \hat{q}_j \hat{o}_j \hat{o}_j^\top. \tag{6.39}$$

The eigenvectors $\hat{\eta}_l$ of $\hat{V}$ and $\hat{U}$ corresponding to the c largest eigenvalues are evaluated and transformed to

$$\hat{\beta}_l = \hat{\Sigma}^{-1/2} \hat{\eta}_l, \; l = 1, \ldots, c. \tag{6.40}$$

The vectors $\hat{\beta}_l$ form the columns of an estimator $\hat{B}$ of the matrix B for the dimension reduction problem of $(\tilde{X}, Y)$ under the assumptions (6.15) and (6.23), respectively.

Exercise 6.5. Sliced average variance estimation (SAVE) [82] is another inverse regression method based on the matrix $M_{\mathrm{SAVE}} = (\Sigma - I_p)^2$, where

$$\Sigma = \mathbf{cov}\,(X(s) \mid s \in Y) = \frac{\int \mathbf{E}\left(\lambda(s) X(s) X^\top(s)\right) ds}{\int \mathbf{E}\lambda(s) ds} - B_{SIR} B_{SIR}^T$$

and $B_{\mathrm{SIR}} = \frac{\int \mathbf{E}(\lambda(s) X(s)) ds}{\int \mathbf{E}\lambda(s) ds}$. Write down an estimator $\hat{M}_{\mathrm{SAVE}}$ based on slices. Prove that when X is a p-dimensional stationary Gaussian random field, then under standard ergodicity assumptions it holds that $\mathcal{S}(M_{\mathrm{SAVE}}) \subset \mathcal{S}_1$, cf. [157].

Exercise 6.6. Recently, in [249], a directional regression method (DR) has been developed which combines SIR and SAVE. In the spatial setting, cf. [157], it is based on the use of the $p \times p$ matrix

$$M_{\mathrm{DR}} = 2\mathbf{E}\left(\mathbf{E}^2(X(s)X(s)^\top - I_p|s \in Y)\right) + 2\mathbf{E}^2\left(\mathbf{E}(X(s)|s \in Y)\mathbf{E}(X(s)^\top|s \in Y)\right) \tag{6.41}$$

$$+ 2\mathbf{E}\left(\mathbf{E}(X(s)^\top|s \in Y)\mathbf{E}(X(s)|s \in Y)\right)\mathbf{E}\left(\mathbf{E}(X(s)|s \in Y)\mathbf{E}(X(s)^\top|s \in Y)\right),$$

where I_p is the unit matrix of order p. Write down an estimator $\hat{M}_{\mathrm{DR}}$ based on slices. Show that from the c_0 largest eigenvectors $\hat{\eta}_l$ of $\hat{M}_{\mathrm{DR}}$ the estimator of the matrix B as in (6.40) is obtained.

Note that in [157] the following estimation error

$$\triangle(B, \hat{B}) = ||B(B^\top B)^{-1}B^\top - \hat{B}(\hat{B}^\top \hat{B})^{-1}\hat{B}^\top||_{\max} \tag{6.42}$$

has been suggested to compare the estimated and true matrix of the central subspace, where $||A||_{\max}$ denotes the maximum of the absolute singular value of a matrix A.

Having n data sets and getting the estimators $\hat{\beta}_l^i$, $i = 1,..,n$, from (6.40) we want to obtain an estimator of β_l, $l = 1,\ldots,c$, which uses this information. In fact, the directions of the vectors are crucial. Consider an arbitrary unit vector $w \in \mathbb{R}^d$ and let $\hat{\beta}^1,\ldots,\hat{\beta}^n$ be iid unit random vectors. Let $\langle .,. \rangle$ be the inner product and

$$\hat{\beta} = \underset{v:||v||=1,\langle w,v\rangle \geq 0}{\arg\max} \sum_{i=1}^{n} |\langle \hat{\beta}^i, v \rangle|. \tag{6.43}$$

This estimator is unbiased in the following sense.

Proposition 6.1. *Let the distribution of $\hat{\beta}^i$ be symmetric with respect to the axis given by $\tilde{\beta}$, i.e. $\hat{\beta}^i$ has the same distribution as $-\hat{\beta}^i + 2\tilde{\beta}\langle \tilde{\beta}, \hat{\beta}^i \rangle$. Then there exists $a \in [0,1]$ such that*

$$\mathbf{E}\hat{\beta} = a\tilde{\beta}.$$

Proof. Without loss of generality, consider $\tilde{\beta} = (1,0,..,0)$. Furthermore, introduce $^sv = (v_1, -v_2,\ldots,-v_d)$, $^s\hat{\beta} = (\hat{\beta}_1, -\hat{\beta}_2,\ldots,-\hat{\beta}_d)$ and $^s\hat{\beta}^i = (\hat{\beta}_1^i, -\hat{\beta}_2^i,\ldots,-\hat{\beta}_d^i)$ for all $i = 1,\ldots,n$. Then $\hat{\beta}^i$ and $^s\hat{\beta}^i$ have the same distribution. Obviously, $|^sv| = |v| = 1$, $\langle ^sv, \tilde{\beta} \rangle = \langle v, \tilde{\beta} \rangle$ and $\langle ^sv, ^s\hat{\beta}^i \rangle = \langle v, \hat{\beta}^i \rangle$. Therefore $\hat{\beta}$ and $^s\hat{\beta}$ have the same distribution, and hence $\mathbf{E}\hat{\beta} = a\tilde{\beta}$ for some $a \in [0,1]$. $\qquad\square$

6.7.2 Statistical Testing

Generally, the dimension c of the central subspace is not known. A starting point would be the test of the null hypothesis

$$H_0 : c = 0 \quad \text{vs.} \quad H_A : c > 0 \tag{6.44}$$

where by $c = 0$ we mean the independence of X and Y. Consider the coefficient (cf. [251])

$$R^2 = R^2(\hat{\beta}_1) = \max_{\beta \in \mathcal{S}_{Y|X}} \frac{(\hat{\beta}_1^\top \beta)^2}{\hat{\beta}_1^\top \hat{\beta}_1 \beta^\top \beta}, \tag{6.45}$$

where $\hat{\beta}_1$ is given by (6.40). From the independence of X and Y it follows that $R^2 = 0$. Therefore, if we reject the null hypothesis of orthogonality

$$H_0 : R^2 = 0 \quad \text{vs.} \quad H_A : R^2 > 0, \tag{6.46}$$

then also the null hypothesis in (6.44) has to be rejected. The test of orthogonality for a known distribution of X can proceed in the following steps:
1. calculate R^2 from observed data Y and X,
2. calculate R^2 from observed Y and each of n independently simulated realizations of X, thus we have $R^2{}_j$, $j = 1,\ldots,n$,
3. the p-value of the test is $\frac{\text{card}\{R^2{}_j \geq R^2\}+1}{n+1}$.

In practice, we cannot simulate independent realizations of X since its distribution is unknown. On a rectangular or circular region $W \subset \mathbb{R}^2$ we can use this testing algorithm with the assumption that under H_0 the joint distribution of Y and X is invariant w.r.t. translation and rotation of Y, respectively. Then instead of independent realizations of X we use n systematic translations and rotations, respectively, of X w.r.t. fixed Y. For translations, the window is wrapped on a cylinder.

For complete estimation of the dimension c, the statistics

$$\hat{\Lambda}_c = n \sum_{i=c+1}^{p} \xi_i \qquad (6.47)$$

might be of use, where $\xi_1 \geq \xi_2 \geq .. \geq \xi_p$ are the eigenvalues of the weighted covariance matrix $\hat{V}$ or $\hat{U}$ given by (6.39). The number of slices m must be chosen larger then $c+1$. To estimate c we start with $c_0 = 0$. If the null hypothesis in (6.44) is rejected, we increase $c_0 = c_0 + 1$ and repeat the same procedure sequentially until

$$H_0 : c = c_0 \quad \text{vs.} \quad H_A : c > c_0$$

is not rejected or $c_0 = p$. Under the validity of H_0, it has been proved in [81] that $\hat{\Lambda}_c$ asymptotically has a chi-square distribution with $(p - c_0)(m - c_0 - 1)$ degrees of freedom when $X(t_i)$, $i = 1, \ldots, n$ are iid. This is not the case for Gaussian random fields, so the test is an approximate one here and can be tried only when the t_i are rather sparse. An analogous reasoning is necessary when thinking of other sampling properties of SIR (consistency, etc.) as summarized in [250] for iid observations of random vectors X.

6.8 Simulation Studies

Three simulation studies are presented to demonstrate the sliced inversed regression method for dimension reduction in stochastic geometry models with covariates. Numerical results are presented and interpreted.

6.8.1 Description of the Simulation

Let $X = (X_1, X_2, X_3)$ be a threedimensional Gaussian random field in $\mathbb{R}^d$ with independent components which have zero mean and covariance function given by

$$\zeta(s,t) = \exp(-\gamma\|s - t\|^{\alpha}), \; s,t \in \mathbb{R}^d, \; 1 \leq \alpha \leq 2, \; \gamma > 0. \qquad (6.48)$$

The method GaussRF from the RandomFields library in software R was used for generating realizations $X = \{X(s), \; s \in J_d = [0,1]^d\}$ for either $d = 2$ or $d = 3$. A random set Y is simulated so that it depends on a linear combination $\mathcal{L}_X$ of components of the Gaussian random field X. Obviously, for $\mathcal{L}_X = \sum_{i=1}^{3} e_i X_i$, $e_i \in \mathbb{R}$ we have that

$$\lambda(s) = f(B^\top X(s)), \ B = (e_1, e_2, e_3)^\top$$

and the dimension of the central subspace is $c = 1$. Slicing of Y is based on its geometrical properties.

In the simulation study I given the Gaussian random field X we consider an nonstationary Poisson point process Y_p with intensity

$$\lambda(s) = a \exp(\mathcal{L}_X(s)), \ s \in J_d, \ a > 0, \ d = 2 \text{ or } 3. \tag{6.49}$$

Note that Y_p is in fact the log-Gaussian Cox process ([287]), the slicing is based on the nearest neighbour distance as the mark Γ_p. Furthermore a Poisson-Voronoi tessellation is simulated in J_d with germs corresponding to events of Y_p, cf. Fig. 6.1a for $d = 2$. In $\mathbb{R}^2$ the system of edges forms a random fibre process Y. A piecewise constant mark Γ at a point of Y is the length of the corresponding edge ($\mathcal{H}^1$-a.s. unique). In $\mathbb{R}^3$, the system of faces forms a random surface process Y. A piecewise constant mark Γ at a point of Y is the area of the corresponding face ($\mathcal{H}^2$-a.s. unique). For the estimation, test points T are centroids of edges and faces randomly chosen with probability proportional to the length of the edge and the surface area of the face, respectively. Besides the basic dependent case as in Fig. 6.1a we consider $n - 1$ independent cases where the same realization of Y is sampled independent of each component of X, cf. Fig. 6.1b. The whole procedure is repeated in order to get a number q of simulated sets of data.

In the simulation study II given the Gaussian random field X we evaluate a fibre process Y based on diffusion from Sect. 6.3.2. The linear combination $g(s) = \mathcal{L}_X(s)$ enters in (6.12). In the numerical solution of (6.11) using the Euler method with fixed temporal step the curve Y (a fibre process) is formed by segments whose length is proportional to the speed of the motion, cf. Fig. 6.4a. In any point of Y the length of the corresponding segment is the mark Γ ($\mathcal{H}_1$-a.s. unique). In each of q simulations of X, Y a number of n systematic rotations of X are taken in angular steps $\frac{2\pi j}{n}$, $j = 0, 1, \ldots, n - 1$. The number of test points along Y is equal to s (taken equidistantly in time).

The simulation study III corresponds to the theoretical Example 6.2 in Sect. 6.6.2. Given the Gaussian random field X a simple inhibition point process Y with second-order intensity λ_2 as in (6.32) was simulated on J_2. We consider two choices of Z, let (Z_1) be given by $Z_1(s) = a(\arctan(X_1(s)) + \frac{\pi}{2})$, $a > 0$, and (Z_2) as in (6.32). Slicing is performed in the same way as explained in (6.33), where the criterion for slicing (the mark) is the theoretical second-order intensity $\lambda_2(s,t)$. Its value is calculated for each pair of points and the range is divided into several slices with approximately equal cardinality. Finally the matrix $\hat{U}_2$ described in (6.35) is a subject for the principal components method.

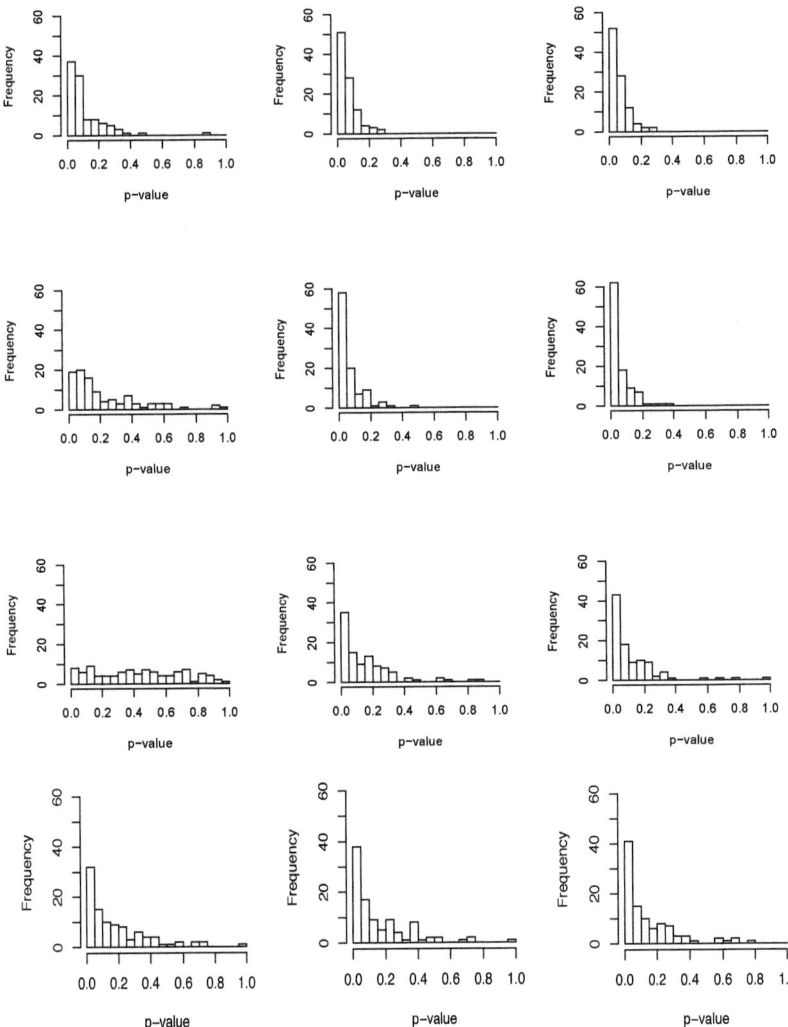

Fig. 6.11 The SIR method, histograms of p-values for the orthogonality test for the number of slices $m = 1$ (first column), $m = 2$ (second column), $m = 4$ (third column), based on $q = 100$, $n = 20$, $p = 3$, $B = (1, 0, 0)$. From simulation study I we present a marked point process of tessellation generators in $\mathbb{R}^2$ (first row), a fibre process of tessellation edges in $\mathbb{R}^2$ (second row), a surface process of tessellation faces in $\mathbb{R}^3$ (third row) and from simulation study II a fibre process based on diffusion (fourth row). The mean number of generators is 1000 in $\mathbb{R}^2$ and 10000 in $\mathbb{R}^3$.

6.8.2 Numerical Results

In Fig. 6.11 the histograms of p−values for the test of orthogonality considered in (6.46) are presented for both simulation studies I and II. Each row represents one type of a random marked set, namely a point, fibre (twice) and surface process. We observe how the power of the test increases with the number of slices (from left to right). This is more apparent in the upper two rows.

The question of an optimal number of slices arises. It was investigated by means of the estimation errors criterion given in (6.42). The following table summarizes the results for five models of random marked sets. By (B1), (B2) we denote the choice $B = (1,0,0)^\top$, $B = (1,1,0)^\top$, respectively. Two different values of parameter b_2 in (6.12) are considered in simulation study II. Both alternatives (Z1) with $a = 0.02$ and (Z2) with $a = b = 0.04$, $\rho = 100$ are considered for the Simulation III. In each case $q = 100$ simulations were realized and the sample means of $\triangle(B,\widehat{B})$ with different numbers of slices were computed. The results vary, while in simulation study I optimal values are below 10, in simulation studies II and III the opposite is true, as can be seen in Tab. 6.1.

Table 6.1 Sliced inverse regression. The error of estimation of central subspace, dependence on the number of slices.

slices	points in $\mathbb{R}^2$		edges in $\mathbb{R}^2$		faces in $\mathbb{R}^3$		$b_2 = 0.3$	$b_2 = 0.9$	Z_1	Z_2
							Simulation II		Simulation III	
	B_1	B_2	B_1	B_2	B_1	B_2	B_1	B_1	B_1	B_1
1	0.296	0.243	0.418	0.329	0.640	0.538	0.543	0.390	0.582	0.463
2	0.236	0.216	0.214	0.220	0.341	0.346	0.504	0.352	0.343	0.407
4	0.231	0.212	0.211	0.207	0.309	0.315	0.426	0.336	0.195	0.405
8	0.227	0.212	0.212	0.208	0.309	0.307	0.408	0.331	0.167	0.385
16	0.230	0.214	0.220	0.210	0.318	0.309	0.406	0.331	0.159	0.378
32	0.244	0.220	0.231	0.215	0.340	0.318	0.435	0.345	0.160	0.361

The estimators $\hat{\beta}_1$ from (6.43) based on simulation study II are presented graphically in Fig. 6.8.2. The centre of the circle represents the true vector B, each vector $\hat{\beta}_1^i$ is represented by a point x, its distance from the centre corresponds to the angle between $\hat{\beta}_1^i$ and B. We observe that the spread of $x's$ is smaller when the dependence between X_1 and Y is bigger, cf. Figs 6.29a and b. The method works also for the general vector B considered in Fig. 6.29c. The triangle represents $\hat{\beta}_1$, which in each case is closely located to the true vector.

In order to demonstrate the estimation of the dimension c of the central subspace using the statistic $\hat{\Lambda}_c$ given in (6.47), in simulation study II we made $q = 50$ simulations of X with $\alpha = 1$ and Y as above (without rotations). The number of slices was chosen $m = 4$, since $p = 3$, the maximal value of c_0 we can use for testing $H_0 : c = c_0$ vs. $H_A : c > c_0$ is $c_0 = 2$, whereas the true value is $c = 1$. The numbers of accepted hypotheses for different cases are given in Tab. 6.2. It turns out that for a larger number s of test points as well as for a smaller coefficient γ in the covariance

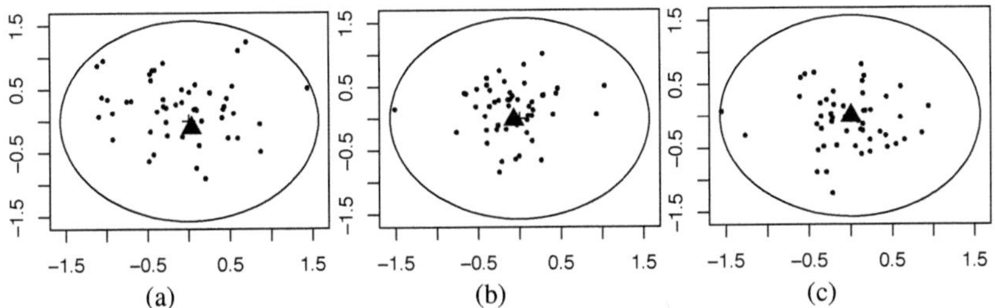

Fig. 6.12 Simulation II: Figures (a) and (b) show estimators of $B = (1,0,0)$, where we have weaker ($b_2 = 0.3$ in (a)) and stronger ($b_2 = 0.9$ in (b)) dependence between X_1 and Y. Figure (c) presents $b_2 = 0.9$ and estimators of $B = (\frac{1}{2}, -1, \frac{1}{2})$. Points correspond to 50 simulations (estimators given by (6.40)), the triangle is the final estimator given in (6.43).

Table 6.2 Testing the dimension of the central subspace.

	$\gamma = 1.666$		$\gamma = 10$	
	$s = 50$	$s = 100$	$s = 50$	$s = 100$
$c = 1$	39	31	41	37
$c = 2$	11	19	9	13

function given in (6.48), the test points involve more dependence and the conclusion is false. In the opposite case their dependence decreases and the approximative test yields better results.

Acknowledgements The work was supported by the Czech Science Foundation, project P201/10/0472 Stochastic geometry - inhomogeneity, marking, dynamics and stereology.

Chapter 7
Space-Time Models in Stochastic Geometry

Viktor Beneš, Michaela Prokešová, Kateřina Staňková Helisová and Markéta Zikmundová

Abstract Space-time models in stochastic geometry are used in many applications. Mostly these are models of space-time point processes. A second frequent situation are growth models of random sets. The present chapter aims to present more general models. It has two parts according to whether the time is considered to be discrete or continuous. In the discrete-time case we focus on state-space models and the use of Monte Carlo methods for the inference of model parameters. Two applications to real situations are presented: a) evaluation of a neurophysiological experiment, b) models of interacting discs. In the continuous-time case we discuss space-time Lévy-driven Cox processes with different second-order structures. Besides the well-known separable models, models with separable kernels are considered. Moreover fully nonseparable models based on ambit processes are introduced. Inference for the models based on second-order statistics is developed.

Viktor Beneš
Charles University in Prague, Faculty of Mathematics and Physics, 18675 Praha 8, Czech Republic, e-mail: benesv@karlin.mff.cuni.cz

Michaela Prokešová
Charles University in Prague, Faculty of Mathematics and Physics, 18675 Praha 8, Czech Republic, e-mail: prokesov@karlin.mff.cuni.cz

Kateřina Staňková Helisová
Czech Technical University in Prague, Faculty of Electrical Engineering, 166 27 Praha 6, Czech Republic e-mail: helisova@math.feld.cvut.cz

Markéta Zikmundová
Charles University in Prague, Faculty of Mathematics and Physics, 18675 Praha 8, Czech Republic, e-mail: zikmundm@karlin.mff.cuni.cz

V. Schmidt (ed.), *Stochastic Geometry, Spatial Statistics and Random Fields*, Lecture Notes in Mathematics 2120, DOI 10.1007/978-3-319-10064-7_7

7.1 Discrete-Time Modelling

In this section we describe the state-space model in a Bayesian setting and introduce the problem of the estimation of posterior distributions. Sequential Monte Carlo methods [104] are described, namely the particle filter and the particle Markov chain Monte Carlo (MCMC) [5]. In the following sections these methods are used for the estimation of parameters of space-time models in stochastic geometry.

7.1.1 State-space Models and Sequential Monte Carlo

Let $\mathbb{R}^d$ be the d-dimensional Euclidean space, $\mathbb{N}$ the set of integers, $\mathbb{N}_0 = \mathbb{N} \cup \{0\}$. For any Borel set $A \subset \mathbb{R}^d$ let $|A| = \nu_d(A)$ denote its d-dimensional Lebesgue measure. Furthermore, let $(\Omega, \mathcal{F}, \mathbf{P})$ be an arbitrary probability space. The notation $X_{0:t} = \{X_0, \ldots, X_t\}$ will be used for a sequence of $t + 1$ random vectors with values in $\mathbb{R}^d$, analogously for non-random $x_{0:t}$.

Consider the following state-space model. Let

$$X = \{X_t, t \in \mathbb{N}_0\} \tag{7.1}$$

be a Markov process with state space $\mathbb{R}^d$, initial distribution with density $p(x_0)$, transition probability density $p(x_t \mid x_{t-1})$. The index t is interpreted as time. Assume that instead of X we observe the random variables $\{Y_t, t \in \mathbb{N}_0\}$ which are conditionally independent given $\{X_t, t \in \mathbb{N}_0\}$. The aim is to draw samples from the posterior distribution $p(x_{0:t} \mid y_{0:t})$, and to evaluate expectations of functions f_t on $\mathbb{R}^{d(t+1)}$ of the form

$$\mathbf{E} f_t = \int f_t(x_{0:t}) p(x_{0:t} \mid y_{0:t}) dx_{0:t}.$$

Recall that by the Bayes theorem we have that

$$p(x_{0:t} \mid y_{0:t}) = \frac{p(y_{0:t} \mid x_{0:t}) p(x_{0:t})}{\int p(y_{0:t} \mid x_{0:t}) p(x_{0:t}) dx_{0:t}}.$$

Using this formula we obtain that the so-called filtering density $p(x_t \mid y_{0:t})$ satisfies the recursion equations

$$p(x_t \mid y_{0:t-1}) = \int p(x_t \mid x_{t-1}) p(x_{t-1} \mid y_{0:t-1}) dx_{t-1} \tag{7.2}$$

and

$$p(x_t \mid y_{0:t}) = \frac{p(y_t \mid x_t) p(x_t \mid y_{0:t-1})}{\int p(y_t \mid x_t) p(x_t \mid y_{0:t-1}) dx_t},$$

since we have $p(x_{0:t} \mid y_t, y_{0:t-1}) \propto p(y_t \mid x_{0:t}) p(x_{0:t} \mid y_{0:t-1})$ from the assumed conditional independence.

Analytical evaluation of the equation system (7.2) for large t is impossible. Therefore Monte Carlo methods were developed. Among them the importance sampling is a basic tool. Let $q(x_{0:t} \mid y_{0:t})$ be a proposal distribution such that for its support $\mathrm{supp}\{p(x_{0:t} \mid y_{0:t})\} \subset \mathrm{supp}\{q(x_{0:t} \mid y_{0:t})\}$ holds. Then

$$\mathbf{E} f_t = \frac{\int f_t(x_{0:t})w(x_{0:t})q(x_{0:t} \mid y_{0:t})dx_{0:t}}{\int w(x_{0:t})q(x_{0:t} \mid y_{0:t})dx_{0:t}},$$

where

$$w(x_{0:t}) = \frac{p(x_{0:t} \mid y_{0:t})}{q(x_{0:t} \mid y_{0:t})}$$

is the importance weight. Simulating M independent identically distributed (iid) particles $\{X_{0:t}^i,\ i = 1,\ldots,M\}$ according to $q(x_{0:t} \mid y_{0:t})$, we obtain the Monte Carlo estimate

$$\hat{\mathbf{E}}_M f_t = \sum_{i=1}^{M} f_t(X_{0:t}^i)\tilde{w}_t^i, \tag{7.3}$$

with normalized importance weights

$$\tilde{w}_t^i = \frac{w(X_{0:t}^i)}{\sum_{j=1}^{M} w(X_{0:t}^j)}, \qquad i = 1,\ldots,M.$$

Sequential Monte Carlo (SMC) methods enable us to draw samples from the posterior recursively and thus to evaluate long time-series data. Among these methods we will use the particle filter (PF) which is described below as Algorithm 7.1. The convention is that whenever the index k is used, we mean that it stands for all $k = 1,\ldots,M$, where M is the number of particles. Furthermore, $\tilde{w}_t = (\tilde{w}_t^1,\ldots,\tilde{w}_t^M)$ are normalized importance weights at time t, $\mathcal{F}(.\mid \tilde{w}_t)$ is a discrete probability distribution with atoms proportional to the weights $\tilde{w}_t$ and A_{t-1}^k represents the index of the parent at time $t-1$ of particle $X_{0:t}^k$ for $t \geq 2$. Typically a finite time sequence $t = 1,2,\ldots,T$ is considered.

Algorithm 7.1 (PF)

 1. *Sample $x_0 \sim p(x)$.*
 2. *At time $t = 1$:*

    a. sample $X_1^k \sim q(.\mid y_1)$,
    b. compute and normalize weights

$$w(X_1^k) = \frac{p(X_1^k \mid x_0)p(y_1 \mid X_1^k)}{q(X_1^k \mid y_1)}, \quad \tilde{w}_1^k = \frac{w(X_1^k)}{\sum_{m=1}^{M} w(X_1^m)}.$$

 3. *At times $t = 2,\ldots,T$:*

    a. sample $A_{t-1}^k \sim \mathcal{F}(.\mid \tilde{w}_{t-1})$,

b. *sample* $X_t^k \sim q(.\,|\,y_t, X_{t-1}^{A_{t-1}^k})$ *and put* $X_{0:t}^k = (X_{0:t-1}^{A_{t-1}^k}, X_t^k)$,

c. *compute and normalize weights*

$$w(X_{0:t}^k) = \frac{p(X_t^k\,|\,X_{t-1}^{A_{t-1}^k})p(y_t\,|\,X_t^k)}{q(X_t^k\,|\,y_t, X_{t-1}^{A_{t-1}^k})}, \quad \tilde{w}_t^k = \frac{w(X_{0:t}^k)}{\sum_{m=1}^{M} w(X_{0:t}^m)}. \tag{7.4}$$

In the case when the proposal density q in the denominator of (7.4) is chosen equal to the transition density, we have in particular that $w_t(X_{0:t}^k) = p(y_t\,|\,X_t^k)$. When the algorithm is completed, (7.3) applies.

7.1.2 The PMMH Method

The state-space models can be used in stochastic geometry to describe the time evolution of random sets. The corresponding space-time models usually depend on further auxiliary parameters, where we denote their vector by θ. A class of PMCMC (particle Markov chain Monte Carlo) algorithms can be used for the estimation of both $x_{0:T}$ and θ. Here we describe the particle marginal Metropolis-Hastings (PMMH) algorithm, where the parameters are estimated iteratively using MCMC and the proposal distribution of $x_{0:t}$ is generated by means of the particle filter.

The state-space model is considered with transition density $p_\theta(x_t\,|\,x_{t-1})$ depending on an unknown auxiliary parameter θ. The PMMH algorithm can sample from the joint posterior density $p(\theta, x_{0:t}\,|\,y_{0:t})$ as follows, see [5].

Algorithm 7.2 (PMMH)

1. Initialization: $i = 0$,

    a. put $\theta(0)$ arbitrarily,

    b. run the sequential Monte Carlo (SMC) Algorithm 7.1 targeting $p_{\theta(0)}(x_{0:t}\,|\,y_{0:t})$, use the estimator $\hat{p}_{\theta(0)}(.\,|\,y_{0:t})$ to sample $X_{0:t}$ and let $\hat{p}_{\theta(0)}(y_{0:t})$ denote the marginal likelihood estimate.

2. For iteration $i \geq 1$:

    a. sample $\theta^* \sim q(.\,|\,\theta(i-1))$,

    b. run the sequential Monte Carlo (SMC) Algorithm 1.1 targeting $p_{\theta^*}(x_{0:t}\,|\,y_{0:t})$,

    c. sample $X_{0:t}^* \sim \hat{p}_{\theta^*}(.\,|\,y_{0:t})$ and let $\hat{p}_{\theta^*}(y_{0:t})$ denote marginal likelihood estimate,

    d. with probability

$$1 \wedge \frac{\hat{p}_{\theta^*}(y_{0:t})p(\theta^*)}{\hat{p}_{\theta(i-1)}(y_{0:t})p(\theta(i-1))} \frac{q(\theta(i-1)\,|\,\theta^*)}{q(\theta^*\,|\,\theta(i-1))}$$

put $\theta(i) = \theta^*$, $X_{0:t}(i) = X_{0:t}^*$ *and* $\hat{p}_{\theta(i)}(y_{0:t}) = \hat{p}_{\theta^*}(y_{0:t})$, *otherwise put*
$\theta(i) = \theta(i-1)$, $X_{0:t}(i) = X_{0:t}(i-1)$ *and* $\hat{p}_{\theta(i)}(y_{0:t}) = \hat{p}_{\theta(i-1)}(y_{0:t})$.

Here $p(\theta)$ is a prior density, and $q(\theta \mid \theta^*)$ a proposal density for θ. The estimate of the marginal likelihood $p_\theta(y_{0:t})$ is given by

$$\hat{p}_\theta(y_{0:t}) = \hat{p}_\theta(y_0) \prod_{t=1}^{T} \hat{p}_\theta(y_t \mid y_{0:t-1}), \quad \hat{p}_\theta(y_j \mid y_{j-1}) = \frac{1}{M} \sum_{k=1}^{M} w(X_{0:j}^k).$$

After a large enough number of iterations the algorithm is stopped.

7.2 Application: Firing Activity of Nerve Cells

Space-time models in stochastic geometry are used in many applications. Mostly these are models of space-time point processes, cf. the review papers [101, 348]. A second frequent situation are growth models of random sets [212]. In the following real data application, the state-space model represents the evolution of parameters of the conditional intensity of a temporal point process [112]. Since the events of the point process are monitored in a bounded planar set, the space-time nature of the experiment follows.

7.2.1 Space-Time Model and Particle Filter

Experimental data of occurence times of action potentials (spikes) of a hippocampal neuron together with the track of a rat in a circular arena $S \subset \mathbb{R}^2$ are investigated, see Fig. 7.1. Spatial receptive fields of neurons in the CA1 region of the rat hippocampus evolve as the animal executes a spatial learning task and the aim is to detect this evolution.

The conditional intensity λ^* of a temporal point process N_t for $t \in [0,T]$ is defined by (see [93])

$$\lambda^*(t \mid N_{0:t}) = \lim_{h \to 0} \frac{1}{h} \mathbf{P}(N_{t+h} - N_t = 1 \mid N_{0:t}),$$

where N_t is the random number of spikes in $(0,t]$ and

$$N_{0:t} = \{u_1, \ldots, u_j;\ 0 < u_1 < \cdots < u_j \leq t\}$$

are the spike times.

In discrete time let $K \in \mathbb{N}$, $K = \frac{T}{\triangle}$ for a step size $\triangle > 0$. Denote

$$N_{1:k} = \{\triangle N_1, \ldots, \triangle N_k\}, \quad J_k = ((k-1)\triangle, k\triangle], \qquad k = 1, \ldots K,$$

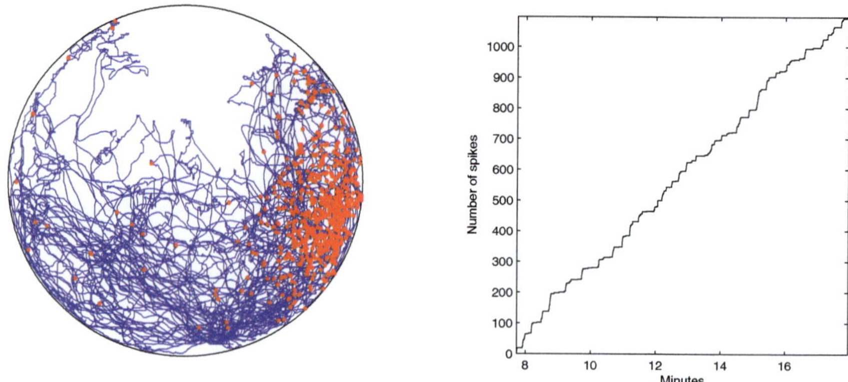

Fig. 7.1 Neurophysiological experiment, see [128]. Left: rat's track (line) and spikes (points) in an arena S (cf.Fig. 6.6), right: temporal evolution of the total number of spikes N_t counted from some time t_0

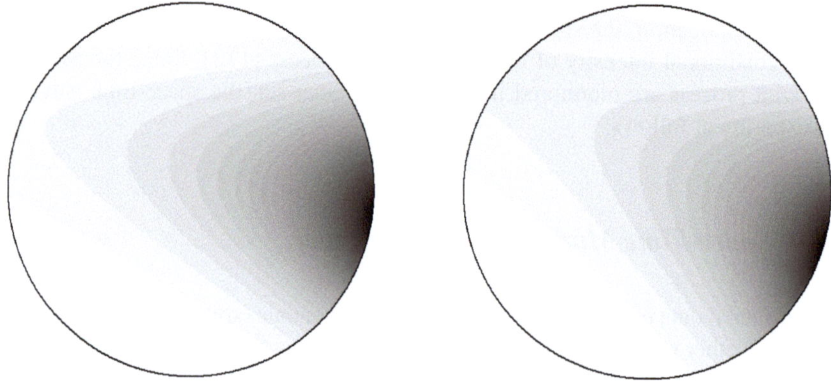

Fig. 7.2 The time interval of data from Fig. 7.1 is split in two halfs and the conditional intensity function (increasing with grey level) is computed in each half from the average in time of the coefficients ψ_k^l

where $\triangle N_k = 0$ or 1 being the indicator of a spike lying in J_k. For an observation of $N_{0:t}$ we choose $\triangle$ small enough such that there is at most one event (spike) in each subinterval J_k.

Denote the position of the rat in S at time t by v_t. In discrete time we shortly write v_k for time $k\triangle$. A parametric model for the conditional intensity suggested in [112] is given by

$$\lambda_k^* = \lambda^*(k\triangle \mid \psi_k, N_{1:k-1}) = \exp\left(\sum_{l=1}^{6} \psi_k^l Z_l(v_k)\right), \tag{7.5}$$

where the state vector $\psi_k = (\psi_k^l)$ is to be estimated and in polar coordinates $v = (r, \varphi)$ we have

$$Z(v) = (Z_l(v))_{l=1}^{6} = (1, r\cos\varphi, r\sin\varphi, 2r^2 - 1, r^2\cos 2\varphi, r^2\sin 2\varphi).$$

The transition density is modeled by a Gaussian random walk

$$\psi_k = \psi_{k-1} + \eta_k, \quad \eta_k \sim N(0, \sigma^2 I),$$

with I being the unit matrix of size 6, and ψ_0, σ^2 are chosen fixed. The likelihood is approximated by

$$p(\triangle N_k \mid \psi_k, N_{1:k-1}) \approx \exp(\triangle N_k \log(\lambda_k^*) - \lambda_k^*\triangle).$$

Then the particle filter described in Algorithm 7.1 can be used for the numerical evaluation of the posterior. Given the estimated coefficients ψ, a planar plot of the conditional intensity given in (7.5) can be obtained, see Fig. 7.2. It is interpreted as a measure of firing activity in space and time, a slight change in time is apparent.

7.2.2 Model Checking

The justification of the model can be based on residual analysis (cf. [94]). For $C \subset S$ and a predictable random function h in continuous time consider the scaled innovation

$$V_C = \int_0^T \mathbf{1}_C(v_t)h(t)[N(dt) - \lambda_t^* dt].$$

Then it holds that

$$\mathbf{var}\, V_C = \mathbf{E}\left(\int_0^T \mathbf{1}_C(v_t)h(t)^2\lambda_t^* dt\right).$$

In particular for the Pearson innovation for $D \subset [0,T]$ we put $h(t) = \mathbf{1}_D(t)[\lambda_t^*]^{-\frac{1}{2}}$, getting

$$V_C = \int_D \mathbf{1}_C(v_t)([\lambda_t^*]^{-\frac{1}{2}}N(dt) - [\lambda_t^*]^{\frac{1}{2}}dt)$$

with $\mathbf{var}\, V_C = |\{t \in D; v_t \in C\}|$. The estimator of the Pearson residual at time $k\triangle$ in $C \subset S$ is given by

$$\hat{R}_P(k,C) = \sum_{\substack{j \leq k \\ \triangle N_j = 1, v_j \in C}} [\hat{\lambda}_j^*]^{-\frac{1}{2}} - \triangle\sum_{j=1}^{k} \mathbf{1}_C(v_j)[\hat{\lambda}_j^*]^{\frac{1}{2}}. \tag{7.6}$$

Note that Pearson residuals can be plotted as a diagnostic tool at times $k\triangle, k = 1,\ldots,K$. When $\hat{R}_P(k,C)$ lies within the bounds $2\sigma_k$ where

$$\sigma_k = |\{t \in \mathbb{R}, 0 < t \le k\triangle, y_t \in C\}|^{\frac{1}{2}}, \tag{7.7}$$

we say that the model fits well. Numerical results for the filtering and residual analysis of the present neurophysiological data are shown in Fig. 7.3.

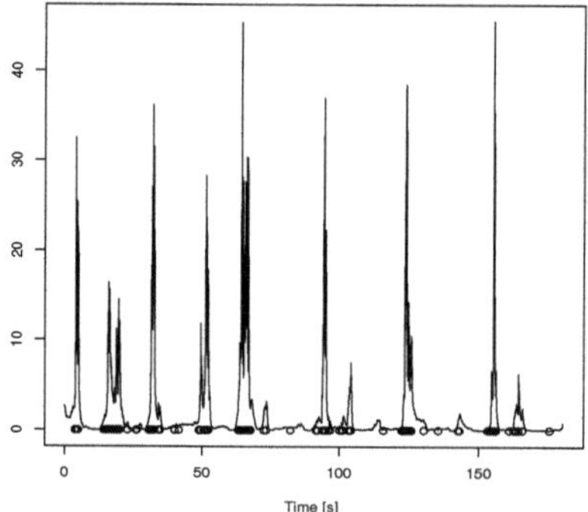
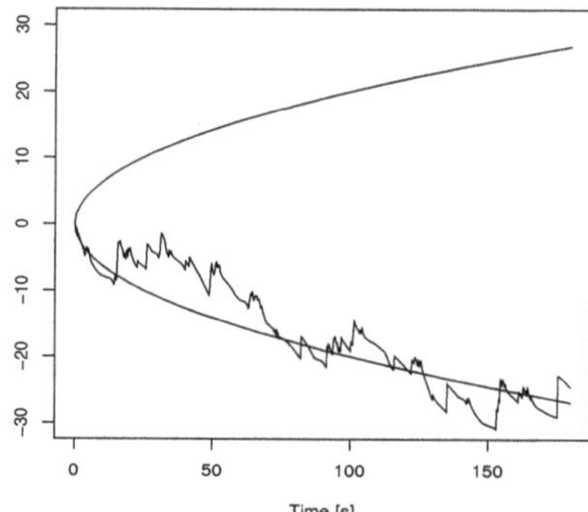

Fig. 7.3 Left: Conditional intensity model (7.5) evaluated in time by means of the particle filter, right: Residual analysis based on the graph of Pearson residuals (7.6) with bounds (7.7) for $C = S$

7.3 Systems of Interacting Discs

In this section, a space-time germ-grain model where the state-space model presents the evolution of parameters of a probability density is investigated. While in the literature mostly growth models of random sets are presented, the model of interacting discs enables us to describe the temporal evolution of various geometrical properties. Among those derived from Minkowski functionals of integral geometry, we consider the volume, surface area (perimeter) and connectivity in the following.

7.3.1 Background in Space

Let $S \subset \mathbb{R}^2$ be a bounded region, $\tilde{Y}$ a germ-grain model with grain centres $z \in S$ and circular grains $B_r(z)$ with random radii $r > 0$. Denote by $y = \{B_{r_1}(z_1), \dots, B_{r_m}(z_m)\}$ a configuration of m discs and U_y the corresponding union of these discs. In [283] a probability density

$$p(y \mid x) = c_x^{-1} \exp(\langle x, G(U_y) \rangle), \tag{7.8}$$

is considered with respect to a given reference Poisson point process Ψ of discs with centres in S and intensity measure $\lambda(z)dzQ(dr)$, where Q is a probability measure on $\mathbb{R}_+$, and λ a non-negative function. Here $x = (x^{(1)}, \dots, x^{(d)})$ is a vector of unknown real parameters, c_x a normalizing constant, $G(U_y) \in \mathbb{R}^d$ is a vector of geometrical characteristics of U_y and $\langle .,. \rangle$ denotes the inner product in $\mathbb{R}^d$. For example, for $d = 3$ and $d = 4$ let

$$G(U_y) = (A(U_y), L(U_y), \chi(U_y)), \tag{7.9}$$

$$G(U_y) = (A(U_y), L(U_y), N_{cc}(U_y), N_h(U_y)), \tag{7.10}$$

respectively, where A is the total area, L the total perimeter, χ the Euler–Poincaré characteristic, N_{cc} the number of connected components and N_h the number of holes (recall that it holds $\chi = N_{cc} - N_h$). If $d = 2$, then under a technical regularity condition, the expression given in (7.8) with the canonical sufficient statistics given in (7.9) is a regular exponential family with parameter space

$$\{x = (x^{(1)}, x^{(2)}) \in \mathbb{R}^2 : \int \exp(\pi x^{(1)} r^2 + 2\pi x^{(2)} r) Q(dr) < \infty\}. \tag{7.11}$$

Exercise 7.1. Show that under condition (7.11), the density given in (7.8) is well defined.

An MCMC method for the simulation of $\tilde{Y}$ and a maximum likelihood method (ML) for the estimation of parameter x have been developed in [276]. The ML method using MCMC simulations is based on finding $\hat{x} = \text{argmax}_{x \in \mathbb{R}^d} \, p(y \mid x)$, where the data y are represented by the vector $G(U_y)$. However, since c_x has no explicit expression, $p(y \mid x)/p(y \mid x^0)$ for fixed $x^0 \in \mathbb{R}^d$ is maximized instead, because in that case, we can use importance sampling for the approximation of the ratio of normalizing constants. The log-likelihood ratio is then given by

$$l_{x^0}(x) = \log \frac{p(y \mid x)}{p(y \mid x^0)} = (x - x^0) \cdot G(U_y) - \log \frac{c_x}{c_{x^0}}$$

$$\approx (x - x^0) \cdot G(U_y) - \log \frac{1}{n} \sum_{i=1}^{n} \exp\{(x - x^0) \cdot G(U_{z_i})\}, \tag{7.12}$$

where z_i, $i = 1, \ldots, n$ for some n, are realizations drawn from $p(. \mid x^0)$ by MCMC simulations. The function given in (7.12) has a simple analytical form, so the maximum likelihood estimate is obtained as

$$\hat{x} = \text{argmax} \, l_{x^0}(x). \tag{7.13}$$

Exercise 7.2. By G_j and x_j denote the j-th component of the vector G and x, respectively. Show that 1. if $G_j(U_y) \leq G_j(U_{z_i})$ for all $i = 1, \ldots, R$ and $G_j(U_y) < G_j(U_{z_i})$ for at least one i, then $l_{x^0}(x)$ is decreasing function in x_j, 2. if $G_j(U_y) \geq G_j(U_{z_i})$ for all $i = 1, \ldots, R$ and $G_j(U_y) > G_j(U_{z_i})$ for at least one i, then $l_{x^0}(x)$ is increasing function in the item x_j.

7.3.2 Space-Time Model

A generalization of the germ-grain model $\tilde{Y}$ considered in Section 7.3.1 to a space-time random set in discrete time is given by

$$Y = \{\tilde{Y}_{0:T}\}. \tag{7.14}$$

Here $\tilde{Y}_t$, $0 \le t \le T$, are germ-grain models with density given in (7.8) where the d-dimensional parameter x is developing in time so that it is a realization of a Markov process $X_{0:t}$ in $\mathbb{R}^d$, cf. (7.1). More precisely, let $X_t \in \mathbb{R}^d$ develop as

$$X_t = X_{t-1} + \eta_t, \quad t = 1, \ldots, T, \tag{7.15}$$

where X_0 (deterministic or random) is given and η_t are iid Gaussian random variables with distribution $N(a, \sigma^2 I_d)$. Here we consider $\theta = (x_0, a, \sigma) \in \mathbb{R}^{2d+1}$ as the unknown auxiliary parameter.

For $d = 3$, a simulated realization of the space-time model Y with canonical statistics given in (7.9) is shown in Fig. 7.4. We start the simulation of the time evolution of the process Y so that we choose a fixed x_0, and according to (7.15), we simulate parameter vectors $x_t, t = 1, 2 \ldots, T$. Furthermore, using the birth-death Metropolis-Hastings algorithm [286], we simulate a realization y_0 of the process $\tilde{Y}_0$ which is given by the density (7.8) where we put $x = x_0$. In this part of the simulation, if $y_0^{(j)}$ is the state at iteration j, we generate a proposal which is either a "birth" $y_0^{(j)} \cup \{b\}$ of a new disc b with center z and radius r, or a "death" $y_0^{(j)} \setminus \{b_i\}$ of an existing disc $b_i \in y_0^{(j)}$. In case of a birth-proposal, z and r are independent, z has a density proportional to the intensity function $\rho(z)$ and r follows the distribution Q of the reference process. In case of a death-proposal, b_i is a uniformly selected disc from $y_0^{(j)}$, and each of these two proposals may happen with equal probability $\alpha = 1/2$. Acceptance depends on the Hastings ratio $H_x(y_0^{(j)}, b)$ or $H_x(y_0^{(j)} \setminus \{b_i\}, b_i)$, respectively. Then, for $t = 1, 2 \ldots, T$, we simulate realizations y_t of the processes Y_t which are given by the density (7.8) with $x = x_t$. Here the Y_t are conditionally independent given X_t, $0 \le t \le T$.

In the next step the canonical sufficient statistics are extracted from simulations using a computer program from [283]. The aim is to suggest an estimator of all parameters $x_{0:t}$ and θ. Besides individual ML estimators at each time, in [429] the state-space model was considered with the parameter $x_{0:t}$ and inference was based on the particle filter with identity function f_t in (7.3). The auxiliary parameter θ was estimated with the help of individual ML estimators and regression techniques. Below we present numerical results from [430] where also the Algorithm 1.2 with the posterior mean estimator was investigated.

There is a possibility to involve temporal dependence in the random set within its simulation algorithm as follows. Again the birth-death Metropolis-Hastings algorithm described above is used, but with the difference that when adding a disc it also depends on the previously simulated configuration y_{t-1}. This dependence is

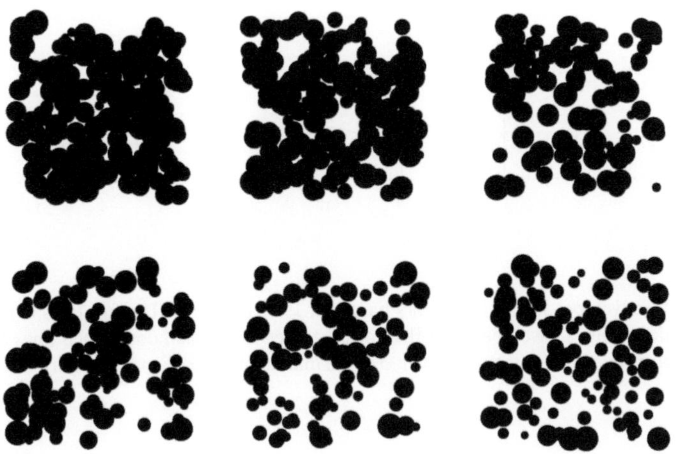

Fig. 7.4 Simulated evolution of the germ-grain model in times $t = 0, 5, 10, 15, 20, 25$, $x_0 = (1, -0.5, -1)$, $a = (-0.07, 0.035, 0.07)$, $\sigma^2 = 0.001$, where S is a square of size 10×10. Observe that the first negative component of a forces the total volume to be decreasing in time, while the other positive components imply increasing total perimeter and Euler number

ensured so that the proposal distribution Prop_t at time t is a mixture

$$\mathrm{Prop}_t = (1 - \beta) \cdot \mathrm{Prop}^{(RP)} + \beta \cdot \mathrm{Prop}_{t-1}^{(emp)}, \quad \beta \in (0, 1),$$

where $\mathrm{Prop}^{(RP)}$ is the distribution of the reference process and $\mathrm{Prop}_{t-1}^{(emp)}$ is the empirical distribution obtained from the configuration y_{t-1}.

This method evokes the question how to determine the probability α of adding a disc for $t = 1, \ldots, T$. With probability β, we choose a disc from a finite set of discs and it may happen that the disc has already been involved in the configuration $y_t^{(j-1)}$, so adding such a disc in the j-th iteration does not change the configuration.

Lemma 7.1. *The choice of α in the j-th iteration so that the probabilities of proposing to delete a disc and that of proposing to add a disc are the same is given by*

$$\alpha^{(j)} = \cfrac{1}{2 - \left(\beta \dfrac{n_{used\ discs}^{(j)}}{n_{y_{t-1}}} \right)}, \tag{7.16}$$

where $n_{used\ discs}^{(j)}$ is the number of discs from the configuration y_{t-1} which are already obtained in the configuration $y_t^{(j)}$ in the j-th iteration and $n_{y_{t-1}}$ is the total number of discs in the configuration y_{t-1}.

Proof. See [429]. $\qquad\qquad\qquad\qquad\qquad\qquad\qquad\qquad\qquad\qquad\qquad\qquad\qquad\square$

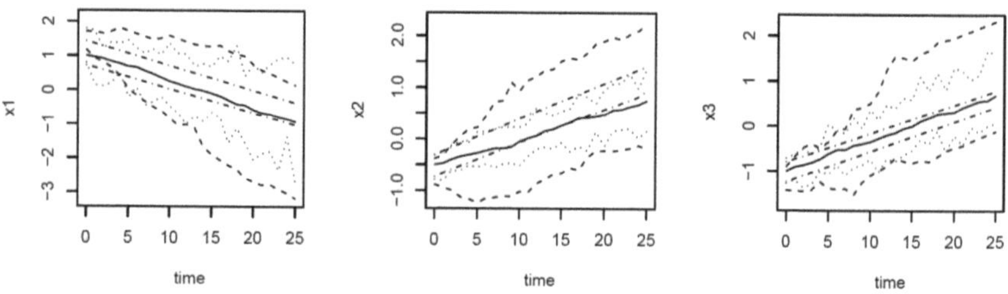

Fig. 7.5 The envelopes for all estimators based on 19 realizations. The full lines denote the true evolution, dotted lines envelopes for MLE, dashed lines for PF and dot-dashed lines for PMMH. The left plot corresponds to the parameter $x^{(1)}$, the middle one to $x^{(2)}$ and the right plot corresponds to $x^{(3)}$.

7.3.3 Model Checking

The model fit can be based on two summary statistics, the contact distribution function and the covariance. Given a compact convex set $B \subset \mathbb{R}^2$ and a germ-grain model $\tilde{Y}$ define

$$D = \inf\{r \geq 0 : \tilde{Y} \cap rB \neq \emptyset\},$$

Assuming that $\mathbf{P}(D > 0) > 0$, the contact distribution function $H_B : [0,\infty] \to [0,1]$ with structuring element B is given by

$$H_B(r) = \mathbf{P}(D \leq r \mid D > 0), \quad r \geq 0.$$

A non-parametric estimator of H_B for stationary $\tilde{Y}$ including edge-effect correction is given by

$$\hat{H}_B(r) = \frac{\sum_{u \in L} \mathbf{1}[u \notin \tilde{Y}, \, u+rB \subset S, \, (u+rB) \cap \tilde{Y} \neq \emptyset]}{\sum_{u \in L} \mathbf{1}[u \notin \tilde{Y}, \, u+rB \subset S]},$$

where L is a regular lattice of test points in $\mathbb{R}^2$.

Recall that the following covariance of a planar random set $\tilde{Z}$ is defined as

$$C_{sp}(u,v) = \mathbf{P}(u \in \tilde{Z}, \, v \in \tilde{Z}), \quad u,v \in \mathbb{R}^2.$$

Additionally, one can consider a covariance function of two temporal arguments of a space-time random set $Z = \{\tilde{Z}_{0:T}\}$:

$$C_{ti}(s,t) = \mathbf{P}(u \in \tilde{Z}_s, \, u \in \tilde{Z}_t) \quad s,t \in \{0,\dots,T\}. \tag{7.17}$$

Assuming planar stationarity of each $\tilde{Z}_t$, the function given in (7.17) does not depend on the choice of $u \in \mathbb{R}^2$. An unbiased estimator of the covariance $C_{ti}(s,t)$ considered in (7.17) using the lattice L as above is given by

$$\hat{C}_{ti}(s,t) = \frac{\sum_{u \in L} \mathbf{1}[u \in Z_s, u \in Z_t]}{|L|}. \tag{7.18}$$

The checking of model fit using the contact distribution function is done for several different times separately. On the contrary the covariance function given in (7.17) is evaluated with respect to all pairs of times. In the case of conditional independence the covariance is useless since then it is simply equal to the product of area fractions at times s and t.

Finally the mean integrated square error (its theoretical counterpart is the squared L_2-distance)

$$\text{MISE} = \frac{1}{l} \sum_{j=1}^{l} \sum_{i=0}^{T} (\hat{x}_{i,j}^{(m)} - x_{i,true}^{(m)})^2, \quad m = 1,\dots,d, \tag{7.19}$$

is a criterion for quality of any estimator $\hat{x}_{0:t}$ of $x_{0:t}$, where $x_{i,true}$ is the true value at time i and l is the number of simulated realizations of Y (7.14).

Table 7.1 Square root of MISE computed for all three methods (MLE, PF and PMMH) used in the simulation study. The estimates are based on $l = 19$ sets of characteristics

	$x^{(1)}$	$x^{(2)}$	$x^{(3)}$
MLE	2.646	1.086	1.551
PF	4.010	2.530	2.504
PMMH	1.081	0.936	0.593

7.3.4 Simulation Study

In a simulation study the suggested estimation methods MLE, PF and PMMH are compared for systems of interacting discs. Let S be a square window of size 10×10, $\rho = 1$ and let the distribution Q be uniform on $[0.2, 0.7]$. Here 19 realizations of the process $Y = \{\tilde{Y}_{0:25}\}$ were simulated (with conditional independence) using the model given in (7.9) with $x_0 = (1, -0.5, -1)$, $a = (-0.07, 0.035, 0.07)$ and $\sigma^2 = 0.001$. A typical simulation run is shown in Fig. 7.4.

In Table 7.1 the values of the MISE (7.19) for the methods based on PMMH, PF and MLE are given. Here it is obvious that the best results are obtained by PMMH which gives the smallest MISE for all parameters $x^{(1)}$, $x^{(2)}$ and $x^{(3)}$.

It can be seen in Fig. 7.5 that except for a few cases ($x^{(2)}$ at later times), all the envelopes (given by the pointwise maximum and minimum of estimates over 19 realizations) cover the true evolution of parameters and the envelopes given by PMMH are the narrowest. This means that the PMMH method gives the best results in the sense of estimation variability.

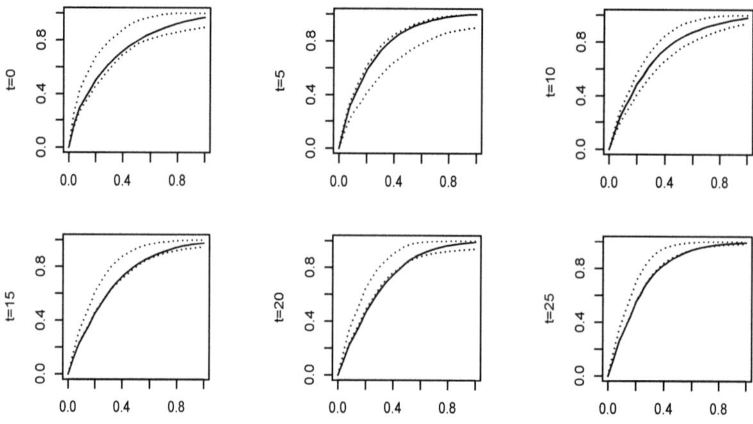

Fig. 7.6 Simulation with PMMH estimates. The full lines denote the estimate of the spherical contact distribution function obtained from the realization $\mathcal{R}$. The envelopes for the spherical contact distribution function at times $t = 0, 5, 10, 15, 20, 25$ are based on additional 19 realizations $\mathcal{R}_{new}$.

In order to check the model we used the PMMH estimation of $x_{0:25}$ for a single realization $\mathcal{R}$ from Fig. 7.4 to simulate a set $\mathcal{R}_{new}$ of 19 new realizations of $\{\tilde{Y}_{0:25}\}$. For $\mathcal{R}$ the estimator $\hat{H}_B(r)$ of the spherical contact distribution function was computed at times $t = 0, 5, 10, 15, 20$ and 25. Edge-effects were avoided by the using subwindow $[0.7, 9.3] \times [0.7, 9.3]$. The envelopes of $\hat{H}_B(r)$, obtained from $\mathcal{R}_{new}$, given by taken the pointwise maximum and minimum are presented in Fig. 7.6.

7.4 Continuous-Time Modelling

In this section we discuss a flexible class of space-time point process models with continuous time: Lévy-driven Cox point processes. The driving field (i.e. the random intensity function) of these processes can be expressed in terms of an integral of a weight function with respect to a Lévy basis [170]. Depending on our choice of the ingredients of the model we can obtain a wide range of stationary and nonstationary, separable, partially separable and nonseparable models.

7.4.1 Space-Time Point Processes

First we recall the basic notions for space-time point processes. A more detailed information on space-time point processes can be found in [101]. We consider a space-time point process Y without multiple points as a random locally finite subset of $\mathbb{R}^2 \times \mathbb{R}$, where a point $(u,t) \in Y$ corresponds to an event $u \in \mathbb{R}^2$ in space occurring at a time $t \in \mathbb{R}$. We assume that the process Y has well defined first- and second-order intensity functions λ and $\lambda^{(2)}$, i.e., for any non-negative Borel functions h_1 on $\mathbb{R}^2 \times \mathbb{R}$ and h_2 on $(\mathbb{R}^2 \times \mathbb{R}) \times (\mathbb{R}^2 \times \mathbb{R})$ it holds that

$$\int h_1(u,t)\lambda(u,t)d(u,t) = \mathbf{E} \sum_{(u,t)\in Y} h_1(u,t) \, ,$$

$$\int \int h_2((u,t),(v,s))\lambda^{(2)}((u,t),(v,s))d(u,t)d(v,s) = \mathbf{E} \sum_{(u,t),(v,s)\in Y}^{\neq} h_2((u,t),(v,s)).$$

We call a space-time process Y stationary if its distribution is invariant with respect to shifts in $\mathbb{R}^2 \times \mathbb{R}$, i.e., by stationarity we really mean space-time stationarity. We will also consider nonstationary processes fulfilling the second-order intensity reweighted stationarity assumption (SOIRS) introduced in [22], i.e., processes for which the (non-stationary) pair correlation function

$$g((u,t),(v,s)) = \frac{\lambda^{(2)}((u,t),(v,s))}{\lambda(u,t)\,\lambda(v,s)}, \tag{7.20}$$

depends only on the difference $(v-u, s-t)$. This is the most common type of nonstationarity used for point processes [318].

The simplest way to obtain a SOIRS point process is to use location dependent thinning. In particular let Y be a stationary point process on a domain $D \subset \mathbb{R}^d$ and $f : D \to [0,1]$ an non-stationarity function. Define the thinned point process $\tilde{Y} = \{y \in Y : Z_y < f(y)\}$ where the Z_y are i.i.d. random variables uniformly distributed on $[0,1]$. Then $\tilde{Y}$ is SOIRS with intensity function λ proportional to f and the same pair correlation function g as the stationary process Y.

Exercise 7.3. Show that $\tilde{Y}$ is a SOIRS point process with intensity function λ proportional to f and the same pair correlation function g as the stationary process Y.

Besides the g-function we will also need the K-function in the sequel. The space-time K-function [282] is defined by

$$K(r,t) = \int \mathbf{1}(\|u\| \le r, |s| \le t)g(u,s)duds, \tag{7.21}$$

where $\|u\|$ denotes the length of the vector u. Thus for the stationary space-time process Y, the K-function determines the number of further points of the process in the "cylinder" neighbourhood $B_r(u) \times [s-t, s+t]$ of a point of Y at the space-time location (u,s) normalized by the intensity. For nonstationary SOIRS processes the influence of the nonstationary first order intensity function is compensated by the normalization of the g-function considered in (7.20). If Y is a Poisson process (i.e. a Cox process with a nonrandom driving field, see Section 7.4.2) then $g = 1$ and $K(r,t) = 2\pi r^2 t$.

7.4.2 Space-Time Lévy-Driven Cox Point Processes

Cox point processes are natural models for clustered point patterns, see also [404].
A Cox point process on a set $A \subseteq \mathbb{R}^d$ is a doubly-stochastic process which conditionally on the realization of the random driving field $\{\Lambda(x), x \in A\}$ is a Poisson
process with intensity function Λ. Lévy-driven Cox point processes have a driving
field of the form

$$\Lambda(x) = \int_A k(x,y) L(dy),$$

where k is a kernel function and L is a nonnegative Lévy basis, i.e. an independently
scattered infinitely divisible random measure. For ease of exposition we restrict ourselves to the case of a Lévy jump basis since in this case Λ has an equivalent shot-
noise representation [280]. A more detailed introduction into Lévy-driven Cox point
processes can be found in [170].

Thus a space-time Lévy-driven Cox point process (or a shot-noise Cox point
process) Y is a Cox process whose driving field Λ given by

$$\Lambda(u,t) = \sum_{(r,v,s) \in \Pi_U} r k((u,t),(v,s)), \qquad (u,t) \in \mathbb{R}^2 \times \mathbb{R}, \qquad (7.22)$$

where Π_U is a Poisson measure on $\mathbb{R}^+ \times \mathbb{R}^2 \times \mathbb{R}$ with intensity measure U and k
is a smoothing kernel, i.e., a non-negative function integrable in both coordinates.
Under some basic integrability assumptions the driving field Λ considered in (7.22)
is an almost surely locally integrable field and Y is a well-defined Cox process (see
[170, 280] for details).

The shot-noise Cox process Y is a stationary space-time process if the kernel k is
just a function of the difference of the two arguments, i.e., $k((u,t),(v,s)) = k((v -
u, s - t))$, and the measure U has the form $U(d(r,v,s)) = c_\mu V(dr) d(v,s)$ where $c_\mu >
0$ and V may be an arbitrary measure on $\mathbb{R}^+$ satisfying the integrability assumption
$\int_{\mathbb{R}^+} \min(1,r) V(dr) < \infty$.

Exercise 7.4. Show that under the above stated assumptions, Y is a stationary space-
time process.

Since the class of stationary shot-noise Cox processes is still a very broad class
of models, let us discuss three particular examples suitable for parametric modelling
(for further examples see e.g. [170, Sect. 4]).

Example 7.1 (Poisson cluster process). If $V(dr) = \delta_1(dr)$ is the Dirac measure
concentrated at 1, then Y is a Poisson cluster process with cluster centers forming a
stationary Poisson process on $\mathbb{R}^2 \times \mathbb{R}$ with intensity c_μ (given by Π_U). Conditionally
on the positions of the cluster centers the particular clusters are independent with
Poisson distributed numbers of points (with mean value $\int k(v,s) d(v,s)$), and the
points within the clusters are distributed independently according to the normalized
density k around the cluster center. Thus in this case we get a class of Neyman-Scott
processes (see e.g. [179, Sect. 6.3.2]). An example of a realization of such a point

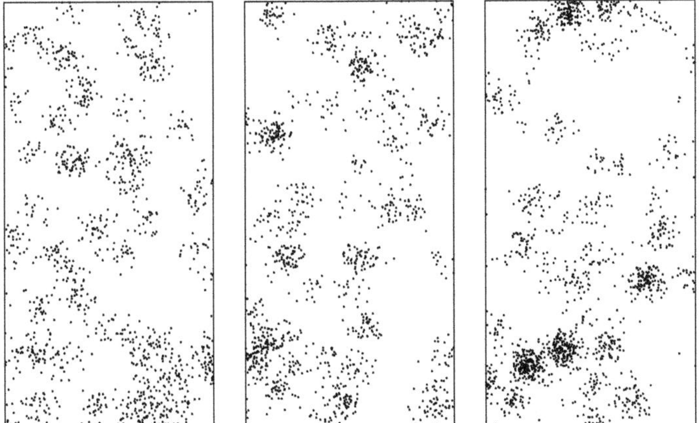

Fig. 7.7 Examples of realizations of different planar shot-noise Cox processes.

process (in the plane) can be seen in the left panel of Fig. 7.7. It is the well-known Thomas process [393].

In fact all shot-noise Cox processes can be viewed as generalized cluster processes. The measure V determines the distribution of the weight (i.e. the mean number of points) of the clusters. By choosing an appropriate measure V we can obtain much more variable clustered point patterns than in Example 7.1.

Example 7.2 (Gamma shot-noise Cox process). Let V correspond to the gamma Lévy basis $V(dr) = r^{-1}\exp(-\theta r)$ where $\theta > 0$ is some parameter. Note that V is not integrable in the neighbourhood of 0. Thus the corresponding shot-noise Cox process Y is not a cluster process in the classical sense [179, Sect. 6.3] since the number of "clusters" in any compact set is infinite. However because the weights of the majority of the clusters are very small, Y is still a well-defined Cox process. A realization of such a point process (in the plane) is shown in the middle panel of Fig. 7.7. It has the same kernel k like in Example 7.1, but the higher variability in the distribution of clusters is obvious.

Example 7.3 (Inverse-Gaussian shot-noise Cox process). Let V correspond to the inverse-Gaussian Lévy basis $V(dr) = \frac{1}{\sqrt{\pi}} r^{-\frac{3}{2}} \exp(-\theta r)$ where $\theta > 0$ is some parameter. Then we obtain even more variable point patterns as exemplified in the right panel of Fig. 7.7.

The moments of shot-noise Cox processes are easily available [170, Sect. 4], in particular for the intensity function we have

$$\lambda(u,t) = c_\mu \int_{\mathbb{R}^+} rV(dr) \int_{\mathbb{R}^2 \times \mathbb{R}} k((u,t),(v,s))\,dv\,ds, \qquad (7.23)$$

and for the pair correlation function

$$g((u,t),(v,s)) = 1 + \frac{c_\mu \int_{\mathbb{R}^+} r^2 V(dr) \int_{\mathbb{R}^2 \times \mathbb{R}} \int_{\mathbb{R}^2 \times \mathbb{R}} k((u,t),(w,\tau)) k((v,s)(w,\tau)) dw d\tau}{\lambda(u,t)\lambda(v,s)}.$$

$$(7.24)$$

For the parametric examples considered above both integrals with respect to V from (7.23) and (7.24) are simple functions of the model parameters. To simplify the notation we will write $V_1 = \int_{\mathbb{R}^+} r V(dr)$ and $V_2 = \int_{\mathbb{R}^+} r^2 V(dr)$ in the sequel.

Exercise 7.5. Derive the formulas (7.23) and (7.24).

When we apply the location dependent thinning with non-stationarity function $f(u,t)$ to a stationary space-time shot-noise Cox process specified by c_μ, V and k, a new SOIRS shot-noise Cox process is obtained with the same c_μ and V and a new kernel function given by $\tilde{k}((u,t),(v,s)) = f(u,t)k(v-u,s-t)$. In the sequel we will consider such SOIRS shot-noise Cox processes and we will prefer the parametrisation by the stationary kernel $k(v-u,s-t)$ and the non-stationarity function f (instead of using the non-stationarity kernel $\tilde{k}$).

7.5 Separability and Space-Time Point Processes

When it comes to statistical analysis of space-time point processes, separability is a popular assumption. The obvious reason is the simplification of the inference: if the process (or at least its moment measures) is separable the inference about the quite complicated space-time model can be based on the properties of the lower dimensional and easier to handle spatial and temporal marginal processes.

7.5.1 Separability

The strongest separability property may be defined by the requirement that the distribution of the space-time point process is equal to the product of the distributions of the marginal processes. However this separability is equal to space-time independence in the point process and does not provide very interesting models. Weaker notions of separability are characterized by the product form of the factorial moment measures or, equivalently, by the product form of the intensity functions (of different orders) when they are well-defined. This separability does not generally imply independence of the spatial and temporal marginal processes however it can still simplify inference. An excellent discussion of the meaning and testing of this kind of separability can be found in [282].

Inspired by an example from [282] we define a nonstationary space-time shot-noise Cox process with separable first-order intensity function but nonseparable second-order intensity function. Thus the process has nontrivial space-time dependencies but still the statistical inference may be based on the marginal spatial and temporal processes, see Sect. 7.7.

Thus let $W \times [0, T]$ be a space-time observation window, where $W \subset \mathbb{R}^2$ is a compact set with area $|W| > 0$ and $[0, T]$ is a bounded time interval of length $T > 0$. Let Y be a space-time SOIRS shot-noise Cox process observed on $W \times [0, T]$ specified by the constant $c_\mu > 0$, the measure V on $\mathbb{R}^+$, the non-stationarity function $f : W \times [0, T] \to [0, 1]$ and the stationary kernel $k(u, t)$ on $\mathbb{R}^2 \times \mathbb{R}$ such that k integrates to 1 (i.e., k is a probability density on $\mathbb{R}^2 \times \mathbb{R}$).

We assume that $f(u, t) = f_1(u) f_2(t)$ where $f_1 : W \to [0, 1]$ is the spatial non-stationarity function and $f_2 : [0, T] \to [0, 1]$ is the temporal non-stationarity function. Moreover we assume $\max_W f_1 = 1 = \max_{[0,T]} f_2$, which implies that $\max_{W \times [0,T]} f = 1$ and prevents overparametrisation of the model. Then for the (first order) intensity function it follows from (7.23) that

$$\lambda(u, t) = f_1(u) f_2(t) c_\mu V_1, \quad (u, t) \in W \times [0, T], \tag{7.25}$$

i.e., the intensity function is separable since it is a product of a function depending on t and a function depending on u.

Furthermore, we assume a product structure of the kernel k, i.e., $k(u, v) = k_1(u) k_2(v)$ where k_1 and k_2 are density functions on $\mathbb{R}^2$ and $\mathbb{R}$, respectively. From (7.24) it follows that

$$g((u, t)(v, s)) = 1 + \frac{V_2}{c_\mu (V_1)^2} \int_{\mathbb{R}^2} k_1(u - w) k_1(v - w) dw \int_{\mathbb{R}} k_2(t - \tau) k_2(s - \tau) d\tau, \tag{7.26}$$

for $(u, t), (v, s) \in W \times [0, T]$. Obviously, neither g nor $\lambda^{(2)}$ have a space-time product structure and the process Y has nontrivial spatio-temporal interactions.

7.5.2 Spatial and Temporal Projection Processes

Let us now investigate projections of the process Y onto the spatial and temporal coordinates:

$$Y_{\text{space}} = \{u : (u, t) \in Y \cap (W \times [0, T])\}, \quad Y_{\text{time}} = \{t : (u, t) \in Y \cap (W \times [0, T])\}. \tag{7.27}$$

Since we assume Y to be observed on a compact window $W \times [0, T]$ the projection processes are sufficient for the inference. We prefer to use Y_{space} and Y_{time} instead of the full marginal processes on $\mathbb{R}^2$ and $\mathbb{R}$, respectively, since the marginal processes may not have well-defined first- and second-order properties. On the other hand from the existence of $\lambda^{(2)}$ it follows that for any pair of distinct points $(u, t) \neq (v, s)$ from Y we have $u \neq v$ and $s \neq t$ with probability 1. Thus the processes considered in (7.27) are well-defined almost surely simple point processes.

Moreover the intensity functions of the projection processes are obtained easily from the intensity functions of Y by integration. Namely under our model assumptions it holds that

$$\lambda_{\text{time}}(t) = c_\mu V_1 f_2(t) \int_W f_1(w)dw, \quad \lambda_{\text{space}}(u) = c_\mu V_1 f_1(u) \int_0^T f_2(\tau)d\tau. \quad (7.28)$$

For the pair correlation functions we get that

$$g_{\text{time}}(t,s) = 1 + C_{\text{time}} \frac{V_2}{c_\mu (V_1)^2} \int_{\mathbb{R}} k_2(t-\tau)k_2(s-\tau)d\tau, \quad (7.29)$$

$$g_{\text{space}}(u,v) = 1 + C_{\text{space}} \frac{V_2}{c_\mu (V_1)^2} \int_{\mathbb{R}^2} k_1(u-w)k_1(v-w)dw, \quad (7.30)$$

where the constants $C_{\text{time}}, C_{\text{space}}$ are defined by

$$C_{\text{time}} = \frac{1}{(\int_W f_1(w)dw)^2} \int_W \int_W \int_{\mathbb{R}^2} f_1(u)f_1(v)k_1(u-w)k_1(v-w)dwdudv, \quad (7.31)$$

$$C_{\text{space}} = \frac{1}{(\int_0^T f_2(\tau)d\tau)^2} \int_0^T \int_0^T \int_{\mathbb{R}} f_2(s)f_2(t)k_2(s-\tau)k_2(t-\tau)d\tau dsdt. \quad (7.32)$$

Thus, eventhough neither $\lambda^{(2)}$ nor g of Y are separable, our model assumptions imply an important simplification for g_{space} and g_{time}. The pair correlation function of the temporal projection process depends on the "spatial" part of the model (i.e. f_1 and k_1) only through the constant C_{time} and, analogously, the pair correlation function of the spatial projection process depends on f_2 and k_2 only through C_{space}.

Formulas for the corresponding K-functions of the projection processes

$$K_{\text{space}}(r) = \int_{\|u\| \le r} g_{\text{space}}(u)du, \quad K_{\text{time}}(s) = \int_{-s}^s g_{\text{time}}(t)dt,$$

are obtained by plugging-in formulas (7.29) and (7.30).

Exercise 7.6. Check the validity of formulas (7.28) – (7.32).

7.6 Ambit Sets and Nonseparable Kernels

Besides the separable kernel $k(u,t) = k_1(u)k_2(t)$ introduced in Sect. 7.5.2, which is popular mainly in epidemiological applications, a different class of space-time Lévy-driven Cox processes may be defined by using so-called ambit sets. The idea of ambit sets was used before e.g. in Lévy modelling of turbulence fields or in growth models [212]. The dependency on the past at time $t \in \mathbb{R}$ and position $u \in \mathbb{R}^2$ may be modelled using an ambit set $A_t(u)$ satisfying

$$(u,t) \in A_t(u) \quad \text{and} \quad A_t(x) \subseteq \mathbb{R}^2 \times (-\infty, t].$$

See Fig. 7.8 for an illustration of the idea. The space-time process $\Lambda(u,t)$ (the ambit process) is then for each point (u,t) defined as an integral of the kernel $\tilde{k}((u,t),(v,s))$ over an attached ambit set $A_t(u)$ with respect to the Lévy basis, i.e.

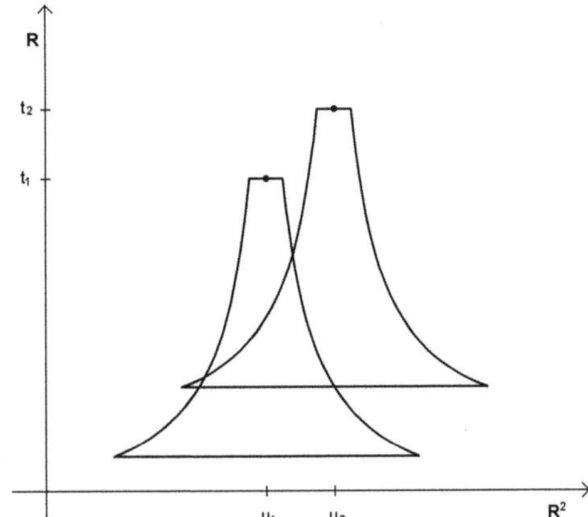

Fig. 7.8 Illustration of the idea of an ambit set. Two ambit sets $A_{t_1}(u_1), A_{t_2}(u_2)$ are shown.

$$\Lambda(u,t) = \int_{A_t(u)} \tilde{k}((u,t),(v,s))L(d(v,s)).$$

Thus the ambit set $A_t(u)$ determines the part of L that influences the behaviour of the driving field Λ at (u,t). In the special case when $A_t(u) = \{(v,s) : (v-u,s-t) \in A_0(0)\}$ for all (u,t) the family of ambit sets is stationary.

Now, if we denote

$$k((u,t),(v,s)) = \mathbf{1}((v,s) \in A_t(u))\tilde{k}((u,t),(v,s)), \tag{7.33}$$

we have the same shot-noise Cox representation like in (7.22). When the ambit sets are stationary and $\tilde{k}$ is a function of the difference $(v-u,s-t)$, only so is k and we obtain a stationary shot-noise Cox processes. The main difference between this model and the one considered in Sect. 7.5.2 is, unless that $A_0(0)$ is a cylinder, the kernel k cannot be a product of a spatial and a temporal kernel. On the other side, the variety of dependence structures which can be modelled by using kernel functions on ambit sets is much wider than what we can obtain by using separable kernels from Sect. 7.5.1.

Example 7.4. Let the ambit sets be stationary and let $\tilde{k}((u,t)(v,s)) = 1/|A_0(0)|$. Then from (7.24) we get that

$$g((u,t),(v,s)) = 1 + \frac{V_2}{c_\mu(V_1)^2}\frac{|A_t(u) \cap A_s(v)|}{|A_0(0)|^2},$$

and in particular, $g((u,t),(v,s)) = 1$ when $A_t(u) \cap A_s(v) = \emptyset$.

The general formula for the pair correlation function of the SOIRS shot-noise Cox process with kernel (7.33) is

$$g((u,t),(v,s)) = 1 + \frac{V_2}{c_\mu(V_1)^2} \int_{A_0(0)\cap A_{s-t}(v-u)} \tilde{k}(w,\tau)\tilde{k}(w-(v-u),\tau-(s-t))d(w,\tau),$$

$$(7.34)$$

and it is not possible to derive formulas similar to (7.29) and (7.30) which could be used for estimation based on the projection processes. A formula for the K-function is obtained by integration, using equation (7.21).

7.7 Estimation Procedures

For Lévy driven Cox processes (like for any other spatial or space-time Cox processes) the likelihood function does not have a tractable form, but the moment characteristics have a much simpler form (cf. [286]). The preferred method for SOIRS nonstationary Poisson cluster processes and also for Lévy-driven Cox processes is thus a two-step estimation procedure with minimum contrast estimation based on the g- or K-function in the second step [282, 319, 407].

In this section we discuss such an estimation procedure based on the space-time K-function. The quality of the obtained estimates of course strongly depends on the quality of the estimated empirical K-function. Because of the high dimensionality of the space-time K-function this could be a problem for general SOIRS shot-noise Cox processes like those in Sect. 7.6, where the nonparametric estimate of the K-function is highly variable unless we use a very large amount of data. For the model considered in Sect. 7.5.1 a better method was developed in [319] which uses the particular structure of the model and thus enables to base the estimation on the lower dimensional characteristics of the projection processes.

Let Y be a space-time SOIRS shot-noise Cox process observed on a window $W \times [0,T]$. We assume that the measure V is parametrized by the parameter $\theta \in \Theta \subset \mathbb{R}$ and the stationary kernel function k is parametrized by the (vector) parameter $\omega \subset \mathbb{R}^p$ and k integrates to 1. Furthermore, we assume separability of the nonstationarity function f, i.e., $f(u,t) = f_1(u)f_2(t)$ with $\max_W f_1 = 1 = \max_{[0,T]} f_2$.

The two-step estimation procedure goes as follows: In the first step the nonstationary intensity function λ is estimated (either parametrically or nonparametrically) and, in the second step, conditionally on the knowledge of λ the space-time K-function is estimated from the data and used for the minimum contrast estimation.

7.7.1 Estimation of the Space-Time Intensity Function

For estimation of the space-time intensity function we may take advantage of the separability (7.25) of the first-order intensity function. Namely from (7.25) and (7.28) we get the equation

$$\lambda_{\text{space}}(u)\lambda_{\text{time}}(t) = \lambda(u,t)\cdot\int_W\int_0^T c_\mu V_1 f_1(v) f_2(s)\,ds\,dv = \lambda(u,t)\cdot\mathbf{E}|Y\cap(W\times[0,T])| \tag{7.35}$$

for any $(u,t)\in W\times[0,T]$. A natural (unbiased) estimator of the mean number of points in $Y\cap(W\times[0,T])$ is the actual observed number of points in $Y\cap(W\times[0,T])$. Therefore we define the estimate of the space-time intensity function $\widehat{\lambda}$ by

$$\widehat{\lambda}(u,t) = \frac{\widehat{\lambda}_{\text{space}}(u)\widehat{\lambda}_{\text{time}}(t)}{|Y\cap(W\times[0,T])|}, \qquad (u,t)\in W\times[0,T], \tag{7.36}$$

where $\widehat{\lambda}_{\text{space}}$ and $\widehat{\lambda}_{\text{time}}$ are estimators of the intensity functions of the projection processes. Thus the dimensionality of the problem is reduced and, moreover, $\widehat{\lambda}$ is a ratio-unbiased estimator of λ if $\widehat{\lambda}_{\text{space}}$ and $\widehat{\lambda}_{\text{time}}$ are unbiased.

The intensities $\widehat{\lambda}_{\text{space}}$ and $\widehat{\lambda}_{\text{time}}$ may be estimated either nonparametrically or a model for the thinning functions f_1 and f_2 may be specified. The nonparametric estimation is usually implemented by the kernel estimate

$$\widehat{\lambda}_{\text{space}}(u) = \sum_{y\in Y_{\text{space}}} h_b(u-y)/w_{b,W}(y), \qquad u\in W, \tag{7.37}$$

where $h_b(u) = h(\frac{u}{b})b^{-2}$ is a kernel with bandwidth $b>0$, i.e., h is a given probability density function. Then edge correction factors $w_{b,W}$ are defined by

$$w_{b,W}(y) = \int_W h_b(u-y)\,du,$$

so that $\int_W \widehat{\lambda}_{\text{space}}(u)\,du = |Y_{\text{space}}|$ which implies the approximate unbiasedness of the estimator $\widehat{\lambda}_{\text{space}}$. An analogous kernel estimator can be used for $\widehat{\lambda}_{\text{time}}$. For further information about kernel estimation for point processes see e.g. [286, Sect. 4.3].

A disadvantage of kernel estimation is its dependence on the choice of the bandwidth b — for different values of b we get estimators of different quality. The accuracy of $\widehat{\lambda}, \widehat{\lambda}_{\text{space}}$ and $\widehat{\lambda}_{\text{time}}$ is not important just for its own sake, but even more because the inverted values of the intensity functions are used in the estimators of the non-stationary pair-correlation function or the K-function in the second step. High variability of the estimator of λ may make the estimators of the second-order characteristics too unstable to provide good estimators of the model parameters in the minimum contrast estimation of the second estimation step. Therefore it is recommendable to use the more stable parametric estimator of λ when possible (see [22] for detailed discussion of this issue).

The most popular log-linear form of intensity function may be parametrized by

$$\lambda_{\text{space}}(u) = \lambda_{\text{space}}(c_{\text{space}}, \beta_{\text{space}}; u) = c_{\text{space}}\exp(\beta_{\text{space}}\cdot Z(u)), \tag{7.38}$$

$$\lambda_{\text{time}}(t) = \lambda_{\text{time}}(c_{\text{time}}, \beta_{\text{time}}; t) = c_{\text{time}}\exp(\beta_{\text{time}}\cdot Z(t)), \tag{7.39}$$

where $Z(u), u \in W$, and $Z(t), t \in [0, T]$, are the spatial and temporal covariates, $\beta_{\text{space}}, \beta_{\text{time}}$ are the vector parameters and $c_{\text{space}}, c_{\text{time}} > 0$. When ignoring the interactions in the first estimation step the non-stationarity parameters $\beta_{\text{space}}, \beta_{\text{time}}$ may be estimated by means of the Poisson likelihood score estimating function like in [407]. For Y_{space} the Poisson loglikelihood score function is given by

$$U(c_{\text{space}}, \beta_{\text{space}}) = \sum_{u \in Y_{\text{space}}} \frac{\lambda'_{\text{space}}(c_{\text{space}}, \beta_{\text{space}}; u)}{\lambda_{\text{space}}(c_{\text{space}}, \beta_{\text{space}}; u)} - \int_W \lambda'_{\text{space}}(c_{\text{space}}, \beta_{\text{space}}; v) dv,$$

(7.40)

where ρ'_{space} denotes the vector of first derivatives with respect to $(c_{\text{space}}, \beta_{\text{space}})$. The estimators are obtained as solutions of the equation $U(\beta_{\text{space}}, c_{\text{space}}) = 0$. Analogous formulas are used to estimate c_{time} and β_{time}. For further details, see e.g. Chap. 3.3.4 in [21].

7.7.2 Estimation by Means of the Space-Time K-Function

Having obtained an estimator of the intensity function $\widehat{\lambda}$ we are ready to estimate the K-function. An approximately unbiased non-parametric estimator of the K function is given by

$$\widehat{K}(r, t) = \frac{1}{|W||T|} \sum^{\neq}_{(u, \tau), (v, s) \in Y \cap W \times [0, T]} \frac{\mathbf{1}(\|u - v\| \leq r, |\tau - s| \leq t)}{w_1(u, v) w_2(\tau, s) \widehat{\lambda}(u, \tau) \widehat{\lambda}(v, s)}, \quad (7.41)$$

where $\sum^{\neq}$ means the sum over pairs of distinct points and w_1, w_2 are edge correction factors. For the planar case w_1 can be either the translation $\frac{|W| \cap |W + (u - v)|}{|W|}$ or an isotropic edge correction factor (see [179, Chap. 4.3.3]). The temporal edge correction factor $w_2(\tau, s)$ was defined in [103] as

$$w_2(\tau, s) = |\{\tau - |\tau - s|, \tau + |\tau - s|\} \cap [0, T]|/2.$$

Exercise 7.7. Show that even if we knew the intensity function λ exactly, then

$$\widehat{K}^*(r, t) = \frac{1}{|W||T|} \sum^{\neq}_{(u, \tau), (v, s) \in Y \cap W \times [0, T]} \frac{\mathbf{1}(\|u - v\| \leq r, |\tau - s| \leq t)}{w_1(u, v) w_2(\tau, s) \lambda(u, \tau) \lambda(v, s)},$$

would not be an unbiased estimator of K. How large is the "approximation error" with which $\widehat{K}^*(r, t)$ would be unbiased?

Let us denote $\alpha = \frac{V_2}{c_\mu (V_1)^2}$. Assume moreover that the K-function is isotropic in space, i.e., $K(u, t) = K(\|u\|, t)$ (this is a quite frequent assumption in the literature). Like in the purely spatial case (cf. [407]) the parameters ω and α may be estimated as arguments of minima of the contrast

$$\int_0^{t_{\max}} \int_0^{r_{\max}} \left((K(r,s))^{1/2} - (\widehat{K}(r,s))^{1/2} \right)^2 dr ds. \tag{7.42}$$

Here $t_{\max}$, $r_{\max}$ are user-specified constants which determine the scale at which the behaviour of K is of interest. They should be chosen close to the (assumed or observed) interaction range of the point process, since using K for larger values only increases variability of the estimators (of course the values of $t_{\max}$ and $r_{\max}$ will be generally different). The exponent $1/2$ is used for variance stabilization.

Since the parameters of the Lévy basis c_μ, θ are nonidentifiable from a single value $\widehat{\alpha}$ we need to use the intensity (7.23) of the point process to get a second equation. Namely the equation for the total number of points

$$|Y \cap W \times [0,T]| = c_\mu V_1 \int_W \widehat{f}_1(u) du \int_{[0,T]} \widehat{f}_2(t) dt. \tag{7.43}$$

Here the estimates of the non-stationary functions f_1 and f_2 are obtained naturally from the estimators of the intensity functions in the first step. Namely,

$$\widehat{f}_1(u) = \widehat{\lambda}_{\text{space}}(u) \Big/ \left(\max_{v \in W} \widehat{\lambda}_{\text{space}}(v) \right), \quad \widehat{f}_2(t) = \widehat{\lambda}_{\text{time}}(t) \Big/ \left(\max_{s \in [0,T]} \widehat{\lambda}_{\text{time}}(s) \right).$$

The advantage of the presented method is that it is simple and applicable to any model described in Sect. 7.5 and 7.6. The disadvantage is that a very large amount of data is needed to make the estimator of the space-time K-function stable enough for the method to be practically useable. Therefore for the special model Sect. 7.5.1 the method discussed in Sect. 7.7.3 is preferable.

7.7.3 Estimation by Means of Projection Processes

For the model considered in Sect. 7.5.1, a more practical estimation procedure was suggested in [319]. The first step of the estimation procedure is the same as above but in the second step the projection processes are used for the estimation.

We assume the same parametrization as above and, moreover, let $\omega = (\omega_1, \omega_2)$ where ω_1 parametrizes the spatial kernel k_1 and ω_2 parametrizes the temporal kernel k_2. For the minimum contrast estimation the functions K_{time} and g_{space} are used. They are estimated from the data as follows:

$$\widehat{K}_{\text{time}}(t) = \frac{1}{|T|} \sum_{\tau,s \in Y_{\text{time}}}^{\neq} \frac{\mathbf{1}(|\tau - s| \le t)}{w_2(\tau,s)\widehat{\lambda}_{\text{time}}(\tau)\widehat{\lambda}_{\text{time}}(s)}, \tag{7.44}$$

and

$$\widehat{g}_{\text{space}}(u) = \sum_{z,y \in Y_{\text{space}}}^{\neq} \frac{h_b(u - z + y)}{w_1(z,y)\widehat{\lambda}_{\text{space}}(z)\widehat{\lambda}_{\text{space}}(y)}, \tag{7.45}$$

where h_b is a suitable kernel function with bandwidth $b > 0$ and w_1, w_2 are the same edge correction factors as above.

The disadvantage of using $\widehat{g}$ is the necessity of choosing the bandwith $b > 0$. No bandwiths are necessary for the estimation of K. This is the reason why minimum contrast estimation with the K-function is more popular (see e.g. [282, 407]). However it was shown in [109, 155] that for the planar Neyman-Scott cluster processes (both stationary and non-stationary) minimum contrast estimation with the pair correlation function provides better estimators of the cluster parameters than minimum contrast estimation with the K-function. Therefore the spatial clustering parameters are estimated by minimizing the contrast

$$\int_0^{r_{\max}} \left(\left(g_{\text{space}}(u) \right)^{1/2} - \left(\widehat{g}_{\text{space}}(u) \right)^{1/2} \right)^2 du. \tag{7.46}$$

Here $r_{\max}$ is a user-specific parameter and it is again preferable to choose it close to the (expected) range of interaction of the underlying point process. Note that the pair correlation function g_{space} has the same form as the pair-correlation function of a Neyman-Scott process, see (7.30), and we minimize (7.46) with respect to ω_1 and the constant $\alpha_1 = C_{\text{space}} \frac{V_2}{c_\mu(V_1)^2}$.

For estimation of the temporal clustering parameters the minimum contrast estimation with the K-function proved more stable than with the g-function, see [319]. Therefore ω_2 and the constant $\alpha_2 = C_{\text{time}} \frac{V_2}{c_\mu(V_1)^2}$ is estimated by minimizing the contrast

$$\int_0^{t_{\max}} \left(\left(K_{\text{time}}(s) \right)^{1/2} - \left(\widehat{K}_{\text{time}}(s) \right)^{1/2} \right)^2 ds. \tag{7.47}$$

The same holds for $t_{\max}$ like for $r_{\max}$ considered above.

With the knowledge of the estimators $\widehat{\omega}_1$ and $\widehat{\omega}_2$ we can plug them into formulas (7.31), (7.32) and obtain an estimator of $\alpha = \frac{V_2}{c_\mu(V_1)^2}$. Finally, estimators for θ and c_μ are obtained from $\widehat{\alpha}$ and equation (7.43) for the total number of observed points.

Note that we can actually obtain two different estimators of α – one by using $\widehat{\alpha}_1$ and $\widehat{C}_{\text{space}}$ and another by using $\widehat{\alpha}_2$ and $\widehat{C}_{\text{time}}$. From the simulation results stated in [319] it follows that $\widehat{\alpha}_1$ is more stable than $\widehat{\alpha}_2$ and consequently it leads to substantially better estimators of c_μ and θ.

Example 7.5 (Estimation of non-stationary space-time gamma shot-noise Cox processes). A SOIRS gamma shot-noise Cox process was simulated on the unit cube $W \times [0, T] = [0, 1]^3$. The non-stationarity functions were log-linear, i.e.,

$$f_1(u) \propto \exp(0.5u_1 - u_2), \qquad f_2(t) \propto \exp(0.7t),$$

the spatial kernel k_2 was the density function of a zero-mean bivariate radially symetric normal distribution $N(o, \sigma^2 I)$ with scale parameter $\sigma = \omega_2$ and the temporal kernel was rectangular, i.e., $k_1(t) = \frac{1}{t_0} \mathbf{1}(t \in [0, t_0])$ with scale parameter $t_0 = \omega_1$. For the gamma basis we have $V_1 = 1/\theta$ and $V_2 = 1/\theta^2$. Thus $\alpha = 1/c_\mu$. Different values of the Lévy-basis parameters $\theta \in \{1/20, 1/30, 1/40\}$, $c_\mu \in \{50, 75, 100\}$, and of the kernel parameters $\sigma \in \{0.01, 0.02\}$, $t_0 \in \{0.015, 0.03\}$

were used. The average number of observed points ranged from 350 to 1400. The estimation procedure described above with parametric estimation of the intensity function was used.

For the three models with the maximal, minimal and moderate number of observed points, Table 7.2 shows the results, the relative biases and relative mean squared errors (MSE) of the estimated parameters (by relative we mean normalized by the true value of the parameter). The full version of the simulation study can be found in [319].

Table 7.2 Relative biases and MSE of the estimator of the non-stationary space-time gamma shot-noise Cox process

true values				rel bias				rel MSE			
c_μ	θ	t_0	σ	$\widehat{c_\mu}$	$\widehat{\theta}$	$\widehat{t_0}$	$\widehat{\sigma}$	$\widehat{c_\mu}$	$\widehat{\theta}$	$\widehat{t_0}$	$\widehat{\sigma}$
50	1/20	0.015	0.01	0.169	0.104	-0.028	0.099	0.188	0.247	0.022	0.088
			0.02	0.181	0.192	-0.033	-.0.047	0.116	0.325	0.021	0.010
		0.03	0.01	0.179	0.110	-0.047	0.094	0.103	0.285	0.036	0.079
			0.02	0.155	0.215	-0.062	-0.042	0.099	0.365	0.042	0.011
75	1/30	0.015	0.01	0.109	0.107	-0.027	0.087	0.045	0.189	0.015	0.073
			0.02	0.122	0.167	-0.028	-0.043	0.070	0.209	0.014	0.011
		0.03	0.01	0.097	0.116	-0.048	0.088	0.045	0.193	0.028	0.070
			0.02	0.136	0.177	-0.054	-0.043	0.074	0.215	0.032	0.008
100	1/40	0.015	0.01	0.081	0.087	-0.031	0.017	0.032	0.144	0.013	0.032
			0.02	0.089	0.117	-0.024	-0.031	0.057	0.164	0.012	0.010
		0.03	0.01	0.070	0.105	-0.046	0.036	0.029	0.128	0.029	0.040
			0.02	0.124	0.126	-0.054	-0.041	0.060	0.154	0.027	0.010

Exercise 7.8. For the situation of Example 7.5, determine estimators for the non-stationarity parameters $\beta_{\text{space},1}, \beta_{\text{space},2}, \beta_{\text{time}}$? Derive formulas for $K_{\text{space}}, K_{\text{time}}$, as functions of the parameters $\omega_1, \omega_2, \alpha$. *Hint.* Compare the solution with [319].

We see from Table 7.2 that it is easier to estimate the kernel parameters t_0 and σ (they are estimated with higher precision) than the Lévy-basis parameters c_μ and θ. The quality of the estimates improves with a higher number of observed points (i.e. intensity). It is interesting to note that the quality of the estimators of c_μ, θ is better for tighter spatial clusters (i.e. smaller σ). But this is not so clear-cut for the tightness of the temporal clusters (i.e. the value of t_0). Related to this issue is the question what is lost if we are using only the projection processes for inference, not the whole space-time process? The problem is the overlapping of distinct clusters when we only use projections. Thus pairs of points which in the space-time setting could be far apart from each other may get very close in one or the other projection process. This is where we loose efficiency when using only projection processes. And the problem could be more serious for X_{time} since we reduce two dimensions by the projection. Nevertheless, despite this problem, according to the simulation results the estimators perform very well.

Acknowledgements The work was supported by the Czech Science Foundation, project P201/10/0472 Stochastic geometry - inhomogeneity, marking, dynamics and stereology.

Chapter 8
Rotational Integral Geometry and Local Stereology - with a View to Image Analysis

Eva B. Vedel Jensen and Allan Rasmusson

Abstract This chapter contains an introduction to rotational integral geometry that is the key tool in local stereological procedures for estimating quantitative properties of spatial structures. In rotational integral geometry, focus is on integrals of geometric functionals with respect to rotation invariant measures. Rotational integrals of intrinsic volumes are studied. The opposite problem of expressing intrinsic volumes as rotational integrals is also considered. It is shown how to express intrinsic volumes as integrals with respect to geometric functionals defined on lower dimensional linear subspaces. Rotational integral geometry of Minkowski tensors is shortly discussed as well as a principal rotational formula. These tools are then applied in local stereology leading to unbiased stereological estimators of mean intrinsic volumes for isotropic random sets. At the end of the chapter, emphasis is put on how these procedures can be implemented when automatic image analysis is available. Computational procedures play an increasingly important role in the stereological analysis of spatial structures and a new sub-discipline, computational stereology, is emerging.

8.1 Rotational Integral Geometry

Although the chapter is self-contained, it can also be read as a continuation of [230]. In particular, the notation used in [230] has been adopted in the majority of cases.

Eva B. Vedel Jensen
Department of Mathematics, Aarhus University, 8000 Aarhus C, Denmark, e-mail: `eva@imf.au.dk`

Allan Rasmusson
Department of Clinical Medicine and Department of Computer Science, Aarhus University, 8000 Aarhus C, Denmark, e-mail: `alras@cs.au.dk`

© Springer International Publishing Switzerland 2015
V. Schmidt (ed.), *Stochastic Geometry, Spatial Statistics and Random Fields*,
Lecture Notes in Mathematics 2120, DOI 10.1007/978-3-319-10064-7_8

Let $\mathcal{K}^d_{\text{conv}}$ denote the set of convex bodies (compact and convex sets) in $\mathbb{R}^d$. In this chapter, we will consider geometric identities of the following general form

$$\int \alpha(K \cap L)\, dL = \beta(K), \tag{8.1}$$

where α, β are geometrical functionals to be defined more precisely below, $K \in \mathcal{K}^d_{\text{conv}}$ is the spatial object of interest, L is the probe (line, plane, sampling window, ...) and dL is the element of a rotation invariant measure on the set of probes L. We will mainly focus on geometric identities for k-dimensional planes L in $\mathbb{R}^d$ passing through the origin o (L is a k-dimensional linear subspace in $\mathbb{R}^d$, called a k-subspace in the following). The choice of origin is an important question in applications; in biomedicine, K is typically a cell and o is the nucleus or a nucleolus of the cell.

8.1.1 Rotational Integrals of Intrinsic Volumes

In this section, rotational integrals of intrinsic volumes will be studied. So α is an intrinsic volume, determined on $K \cap L$, cf. (8.1), and the aim is to find the corresponding β. As we shall see, β involves weighted curvature measures.

Recall that for $K \in \mathcal{K}^d_{\text{conv}}$, we can define $d+1$ intrinsic volumes $V_k(K)$, $k = 0, \ldots, d$. We have

$$V_d(K) = \text{ volume (Lebesgue measure) of } K$$
$$V_{d-1}(K) = 2^{-1} \times \text{ surface area of } K$$
$$V_0(K) = \text{ the Euler-Poincaré characteristic of } K$$

The interpretation of $V_{d-1}(K)$ holds if K has non-empty interior. For non-empty $K \in \mathcal{K}^d_{\text{conv}}$, V_0 is identically equal to 1. The intrinsic volumes can be extended to larger set classes for which V_0 contains interesting topological information.

The intrinsic volumes are examples of real-valued valuations on $\mathbb{R}^d$. They are motion invariant and continuous with respect to the Hausdorff metric. Recall that a real-valued valuation on $\mathbb{R}^d$ is a mapping $f : \mathcal{K}^d_{\text{conv}} \to \mathbb{R}$ satisfying

$$f(K \cup M) + f(K \cap M) = f(K) + f(M),$$

whenever $K, M, K \cup M \in \mathcal{K}^d_{\text{conv}}$. Hadwiger's famous characterization theorem states that any motion invariant, continuous valuation is a linear combination of intrinsic volumes. For more details, see [230, 347] and references therein.

For $k = 0, \ldots, d-1$, $V_k(K)$ can be expressed as integral with respect to principal curvatures. Assume for simplicity of presentation that K is a compact d-dimensional C^2 manifold with boundary. Then, for $k = 0, \ldots, d-1$,

$$V_k(K) = \frac{1}{\omega_{d-k}} \int_{\partial K} \sum_{|I|=d-1-k} \prod_{i \in I} \kappa_i(x)\, \mathcal{H}^{d-1}(dx), \tag{8.2}$$

where $\omega_k = 2\pi^{k/2}/\Gamma(k/2)$ is the surface area of the unit sphere in $\mathbb{R}^k$, ∂K is the boundary of K, the sum runs over all subsets $\{1,\ldots,d-1\}$ with $d-1-k$ elements, $\kappa_i(x)$, $i=1,\ldots,d-1$, are the principal curvatures at $x \in \partial K$ and $\mathcal{H}^{d-1}$ is $(d-1)$-dimensional Hausdorff measure. For $k=d-1$, (8.2) reduces to $V_{d-1}(K) = \frac{1}{2}\mathcal{H}^{d-1}(\partial K)$.

The classical Crofton formula relates intrinsic volumes defined on k-dimensional affine subspaces to intrinsic volumes of the original set

$$\int_{\mathcal{E}_k^d} V_j(K \cap E)\, dE = c_{j,d}^{k,d-k+j} V_{d-k+j}(K), \quad 0 \le j \le k \le d, \tag{8.3}$$

cf. [230, Theorem 2.4]. Here, $\mathcal{E}_k^d$ is the set of k-dimensional affine subspaces in $\mathbb{R}^d$, called k-flats in the following. Any $E \in \mathcal{E}_k^d$ is of the form $E = x+L$, where L is the parallel k-subspace and $x \in L^\perp$. Furthermore, $dE = v_{d-k}(dx)\, dL$, where dL is the element of the rotation invariant probability measure on the set $\mathcal{L}_k^d$ of k-subspaces in $\mathbb{R}^d$ and v_{d-k} is the Lebesgue measure on $L^\perp$. The explicit form of the known constant is

$$c_{j,d}^{k,d-k+j} = \frac{k!\,\tau_k\,(d-k+j)!\,\tau_{d-k+j}}{j!\,\tau_j\,d!\,\tau_d},$$

where $\tau_d = \pi^{d/2}/\Gamma(1+\frac{d}{2})$ is the volume of the unit ball in $\mathbb{R}^d$, cf. [230, formula (2.3)]. Note that for $j=k$, the Crofton formula relates Lebesgue measure on sections $K \cap E$ to Lebesgue measure of the original set K. Likewise, for $j=k-1$, sectional surface area is related to the surface area of K.

Note that in order to ease the reading of this chapter as a continuation of [230], we use above and throughout this chapter the normalized version of the rotation invariant measure on $\mathcal{L}_k^d$ which is a probability measure.

Exercise 8.1. Show that for $d=2$ and $k=1$, (8.3) reduces to

$$\int_{\mathcal{E}_1^2} \mathbf{1}(K \cap E \neq \emptyset)\, dE = \frac{1}{\pi}\mathrm{length}(\partial K),$$

for $j=0$, and

$$\int_{\mathcal{E}_1^2} \mathrm{length}(K \cap E)\, dE = \mathrm{area}(K),$$

for $j=1$.

In rotational integral geometry, the interest is in rotational averages of intrinsic volumes, i.e., integrals of the following form

$$\int_{\mathcal{L}_k^d} V_j(K \cap L)\, dL, \quad 0 \le j \le k \le d \tag{8.4}$$

are considered. These integrals are valuations on $\mathbb{R}^d$. They are rotation invariant, but typically not translation invariant.

Let us first consider the case $j = k$. This is the simplest case where Lebesgue measure is determined on the section. For $k = 1, \ldots, d$, we have

$$\int_{\mathcal{L}_k^d} V_k(K \cap L)\, dL = \frac{\Gamma(d/2)}{\pi^{(d-k)/2}\Gamma(k/2)} \int_K |x|^{-(d-k)} v_d(dx). \qquad (8.5)$$

The proof of this result is based on the Blaschke-Petkantschin formula. This formula exists in many versions. Generally, the Blaschke-Petkantschin formula concerns a decomposition of a product of Hausdorff measures, see [185, Theorem 5.6]. Here, we only need the decomposition of a single copy of Lebesgue measure. In this case, the Blaschke-Petkantschin formula takes the following form

$$\int_{\mathbb{R}^d} f(x)\, v_d(dx) = \frac{\pi^{(d-k)/2}\Gamma(k/2)}{\Gamma(d/2)} \int_{\mathcal{L}_k^d} \int_L |x|^{d-k}\, v_k(dx)\, dL,$$

for any non-negative measurable function f on $\mathbb{R}^d$, see [185, Proposition 4.5]. For $k = 1$ (line sections), the Blaschke-Petkantschin formula is simply polar decomposition in $\mathbb{R}^d$, see also [230, Sect. 2.1.2]. Note that for $j = k = 0$, (8.4) reduces to

$$\int_{\mathcal{L}_0^d} V_0(K \cap L)\, dL = \mathbf{1}_K(o).$$

Example 8.1. For $d = 3$ and $k = 2$, we get, cf. (8.5),

$$\int_{\mathcal{L}_2^3} \text{area}(K \cap L)\, dL = \beta(K),$$

where

$$\beta(K) = \frac{1}{2} \int_K |x|^{-1}\, v_3(dx).$$

The situation is much more complicated, when $j < k$. Assume for simplicity of the presentation that K is a compact $d-$dimensional C^2 manifold with boundary. Then, under mild regularity conditions,

$$\int_{\mathcal{L}_k^d} V_j(K \cap L)\, dL = \int_{\partial K} |x|^{-(d-k)} \sum_{|I|=k-1-j} w_{I,k,j}(x) \prod_{i \in I} \kappa_i(x)\, \mathcal{H}^{d-1}(dx), \qquad (8.6)$$

$0 \le j < k \le d$. The sum runs over all subsets of $\{1, \ldots, d-1\}$ with $k-1-j$ elements and the $w_{I,k,j}$s are weight functions involving hypergeometric functions. This result has been published in [187]. Here, the result was established for the more general set class consisting of sets with positive reach. The proof involves extensive geometric measure theory.

Very recently, the explicit form of the weight functions $w_{I,k,j}$ has been published ([12]). If K is a ball centred at the origin o, then the $w_{I,k,j}$s are constant and $|x|$ is also constant when $x \in \partial K$. Generally, the $w_{I,k,j}$s depend on the angle between x and the outer unit normal $u(x)$ at $x \in \partial K$, *and* the angle between x and the subspace spanned by the principal directions with indices outside I. In [12], it is shown for $j < k$ that $\int_{\mathcal{L}_k^d} V_j(K \cap L) \, dL$ can also be expressed as an integral with respect to flag measures.

The special case $j = k - 1$ gives rise to some simplifications of (8.6), see e.g. [187, Section 4.1]. When $j = k - 1$, $I = \emptyset$, the sum on the right-hand side of (8.6) has only one element and the curvature product disappears. The following result holds for the rotational average of the sectional surface area

$$\int_{\mathcal{L}_k^d} V_{k-1}(K \cap L) \, dL$$

$$= \frac{1}{2} \frac{\Gamma(d/2)}{\pi^{(d-k)/2} \Gamma(k/2)} \int_{\partial K} |x|^{-(d-k)} F_{-\frac{1}{2}, \frac{d-k}{2}; \frac{d-1}{2}}(\sin^2 \beta(x)) \, \mathcal{H}^{d-1}(dx), \quad (8.7)$$

where F is a hypergeometric function and $\beta(x)$ is the angle between $x \in \partial K$ and the unique outer unit normal $u(x)$ to the boundary at $x \in \partial K$ (unique because of the smoothness condition).

The class of hypergeometric functions is parametrized by three parameters and has well-known series expansions as well as integral representations. In particular, we have for $0 < \beta < \gamma$ the following integral representation

$$F_{\alpha, \beta; \gamma}(z) = \frac{1}{B(\beta, \gamma - \beta)} \int_0^1 (1 - zy)^{-\alpha} y^{\beta-1} (1-y)^{\gamma-\beta-1} dy. \quad (8.8)$$

Example 8.2. For $d = 3$ and $k = 2$, we find, using (8.8),

$$F_{-\frac{1}{2}, \frac{d-k}{2}; \frac{d-1}{2}}(\sin^2 \beta(x)) = \frac{2}{\pi} \int_0^{\pi/2} (1 - \sin^2 \beta(x) \sin^2 \varphi)^{1/2} d\varphi$$

$$= \frac{2}{\pi} E(|\sin \beta(x)|, \pi/2),$$

where E is the elliptic integral of the second kind. We find, cf. (8.7),

$$\int_{\mathcal{L}_2^3} \text{length}(\partial K \cap L) \, dL = \beta(K),$$

where

$$\beta(K) = \frac{1}{\pi} \int_{\partial K} |x|^{-1} E(|\sin \beta(x)|, \pi/2) \, \mathcal{H}^2(dx).$$

8.1.2 *Intrinsic Volumes as Rotational Integrals*

In this section, we want to study the 'opposite/inverse' problem of determining the measurement in the section with rotational integral equal to a given intrinsic volume.

So now β is an intrinsic volume and the aim is to find α such that (8.1) is satisfied. This problem has been studied in detail in [11, 154].

More specifically, we want to find a functional $\alpha_{k,j}$, satisfying the following rotational integral equation

$$\int_{\mathcal{L}_k^d} \alpha_{k,j}(K \cap L)\, \mathrm{d}L = V_{d-k+j}(K), \tag{8.9}$$

$k = 1,\ldots,d,\ j = 1,\ldots,k$. From an applied point of view, this question is more interesting than the one studied in the previous section, because $\alpha_{k,j}$ is then the measurement to be performed in the section. This measurement has a rotational average equal to the intrinsic volume considered and can be used to estimate the intrinsic volume in question. Further details will be given in Sect. 8.2.

Let us first consider a simple example in $\mathbb{R}^2$ with $d = 2$ and $j = k = 1$. The aim is then to find a functional $\alpha_{1,1}$ such that

$$\int_{\mathcal{L}_1^2} \alpha_{1,1}(K \cap L)\, \mathrm{d}L = \mathrm{area}\,(K). \tag{8.10}$$

It is easy to find a solution to this problem. Consider an infinitesimal neighbourhood of $x \in X$ of area $v_2(\mathrm{d}x)$. Transforming to polar coordinates in $\mathbb{R}^2$, $x = (r\cos\theta, r\sin\theta)$, gives us the following decomposition of area measure in the plane

$$v_2(\mathrm{d}x) = |r|\,\mathrm{d}r\,\mathrm{d}\theta, \tag{8.11}$$

$r \in \mathbb{R}$, $\theta \in [0,\pi)$. Identifying θ with the line L passing through the origin, having an angle θ with a fixed axis, we have $\mathrm{d}L = \mathrm{d}\theta/\pi$, and (8.11) can equivalently be expressed as

$$v_2(\mathrm{d}x) = \pi\,|x|\,v_1(\mathrm{d}x)\,\mathrm{d}L.$$

It follows that

$$\alpha_{1,1}(K \cap L) = \pi \int_{K\cap L} |x|\,v_1(\mathrm{d}x)$$

is a solution to (8.10).

A solution to the general problem of finding a functional $\alpha_{k,j}$ satisfying (8.9) can be derived by combining the classical Crofton formula with another version of the Blaschke-Petkantschin formula, see [230, Theorem 2.7],

$$\int_{\mathcal{E}_r^d} f(E)\,\mathrm{d}E = \frac{\omega_{d-r}}{\omega_{k-r}} \int_{\mathcal{L}_k^d} \int_{\mathcal{E}_r^L} f(E)\,d(o,E)^{d-k}\,\mathrm{d}E\,\mathrm{d}L, \tag{8.12}$$

where $1 \le r < k \le d-1$, f is a non-negative measurable function on $\mathcal{E}_r^d$ and $\mathcal{E}_r^L$ is the set of r-flats contained in $L \in \mathcal{L}_k^d$.

The general solution to (8.9) is given in the proposition below.

Proposition 8.1 ([11]; [154]). *Let $M \in \mathcal{K}_{\mathrm{conv}}^L$ be a compact and convex subset of $L \in \mathcal{L}_k^d$. Then, for $k = 1,\ldots,d,\ j = 1,\ldots,k$, the functional*

$$\alpha_{k,j}(M) = \frac{\omega_{d-k+1}}{\omega_1}(c_{j-1,d}^{k-1,d-k+j})^{-1} \int_{\mathcal{E}_{k-1}^L} d(o,E)^{d-k} V_{j-1}(M \cap E)\, \mathrm{d}E$$

is a solution to (8.9).

Proof. Using the Blaschke-Petkantschin formula (8.12), we find

$$\int_{\mathcal{L}_k^d} \alpha_{k,j}(K \cap L)\, \mathrm{d}L$$

$$= \frac{\omega_{d-k+1}}{\omega_1}(c_{j-1,d}^{k-1,d-k+j})^{-1} \int_{\mathcal{L}_k^d} \int_{\mathcal{E}_{k-1}^L} d(o,E)^{d-k} V_{j-1}(K \cap L \cap E)\, \mathrm{d}E\, \mathrm{d}L$$

$$= \frac{\omega_{d-k+1}}{\omega_1}(c_{j-1,d}^{k-1,d-k+j})^{-1} \int_{\mathcal{L}_k^d} \int_{\mathcal{E}_{k-1}^L} d(o,E)^{d-k} V_{j-1}(K \cap E)\, \mathrm{d}E\, \mathrm{d}L$$

$$= (c_{j-1,d}^{k-1,d-k+j})^{-1} \int_{\mathcal{E}_{k-1}^d} V_{j-1}(K \cap E)\, \mathrm{d}E$$

$$= V_{d-k+j}(K).$$

At the last equality sign, we have used the Crofton formula (8.3). $\square$

Exercise 8.2. Show that for $d=3$ and $j=k=2$, (8.9) reduces to

$$\int_{\mathcal{L}_2^3} \alpha_{2,2}(K \cap L)\, \mathrm{d}L = v_3(K),$$

where, according to Proposition 8.1,

$$\alpha_{2,2}(K \cap L) = \pi \int_{\mathcal{E}_1^L} d(o,E)\, \mathrm{length}\,(K \cap E)\, \mathrm{d}E.$$

Show also that for $d=3$, $j=1$ and $k=2$, (8.9) takes the form

$$\int_{\mathcal{L}_2^3} \alpha_{2,1}(K \cap L)\, \mathrm{d}L = \frac{1}{2}\, \mathrm{surface\ area}\,(K),$$

where, according to Proposition 8.1,

$$\alpha_{2,1}(K \cap L) = 2\pi \int_{\mathcal{E}_1^L} d(o,E)\, \mathbf{1}\{K \cap E \neq \emptyset\}\, \mathrm{d}E.$$

It was shown in [11] that for $j=k$ and $j=k-1$ the functional $\alpha_{k,j}$ can be considerably simplified and given in more explicit form. The result is presented in the corollary below.

Corollary 8.1. *Let the situation be as in Proposition 8.1. Suppose that $M \in \mathcal{K}_{\mathrm{conv}}^L$ is a compact k-dimensional C^2 manifold with boundary. Then,*

$$\alpha_{k,k}(M) = \frac{\pi^{(d-k)/2}\Gamma(k/2)}{\Gamma(d/2)} \int_M |z|^{d-k} v_k(\mathrm{d}z)$$

and

$$\alpha_{k,k-1}(M) = \frac{1}{2} \frac{\pi^{(d-k)/2}\Gamma(k/2)}{\Gamma(d/2)} \int_{\partial M} |z|^{d-k} F_{-\frac{1}{2},-\frac{d-k}{2};\frac{k-1}{2}}(\sin^2(\beta(z))) \mathcal{H}^{k-1}(dz),$$

where $\beta(z)$ is the angle between $z \in \partial M$ and the unique outer unit normal $u(z)$ to the boundary of M at $z \in \partial M$.

Proof. Using that $E = L + x$, where $x \in L^\perp$, we find

$$\frac{\Gamma((d-k+1)/2)}{\pi^{(d-k+1)/2}}\alpha_{k,k}(M)$$

$$= \int_{\mathcal{E}^L_{k-1}} d(o,E)^{d-k} V_{k-1}(M \cap E)\, dE$$

$$= \int_{\mathcal{L}^k_{k-1}} \int_{L^\perp} |x|^{d-k} V_{k-1}(M \cap (L+x))\, \nu_1(dx)\, dL$$

$$= \int_{\mathcal{L}^k_{k-1}} \int_{L^\perp} \int_{M \cap (L+x)} |x|^{d-k}\, \nu_{k-1}(dy)\, \nu_1(dx)\, dL$$

$$= \int_{\mathcal{L}^k_{k-1}} \int_M |p(z|L^\perp)|^{d-k}\, \nu_k(dz)\, dL$$

$$= \int_M |z|^{d-k} \left(\int_{\mathcal{L}^k_{k-1}} \frac{|p(z|L^\perp)|^{d-k}}{|z|^{d-k}}\, dL \right) \nu_k(dz)$$

$$= \int_M |z|^{d-k} \left(\frac{1}{B(\frac{1}{2},\frac{k-1}{2})} \int_0^1 y^{\frac{d-k-1}{2}} (1-y)^{\frac{k-3}{2}}\, dy \right) \nu_k(dz).$$

At the last equality sign, we have used [185, Proposition 3.9]. The result concerning $\alpha_{k,k}$ now follows immediately. The result concerning $\alpha_{k,k-1}$ is more difficult to show. The details can be found in [11]. Let us here just give a proof sketch. In [11], it is shown that

$$\frac{\Gamma((d-k+1)/2)}{\pi^{(d-k+1)/2}} \cdot c^{k-1,d-1}_{k-2,d} \cdot \alpha_{k,k-1}(M)$$

$$= \frac{1}{2} \int_{\partial M} \int_{\mathcal{L}^k_{k-1}} |p(u(z)|L)|\,|p(z|L^\perp)|^{d-k}\, dL\, \mathcal{H}^{k-1}(dz).$$

The result now follows if we use the following result proved in [11]. For x and y unit vectors in $L \in \mathcal{L}^d_k$ and non-negative integers n, m, we have

$$\int_{\mathcal{L}^k_{k-1}} |p(x|L)|^m |p(y|L^\perp)|^n dL$$

$$= \frac{\omega_{k-1}}{\omega_k} B\left(\frac{n+1}{2}, \frac{m+k-1}{2}\right) F_{-\frac{m}{2},-\frac{n}{2};\frac{k-1}{2}}(\sin^2 \angle(x,y)). \qquad \square$$

In this section we have found a functional $\alpha_{k,j}$ satisfying the rotational integral equation (8.9). A natural question to ask is whether $\alpha_{k,j}$ is unique. If a solution is sought among rotation invariant functionals only, this is indeed the case for $j = k = 1$, cf. [231]. It is an open question whether uniqueness holds for general j and k.

8.1.3 Rotational Integral Geometry of Minkowski Tensors

In this section, we will extend the results obtained so far to tensor valuations. These results are very recent ([13]). We will define so-called integrated Minkowski tensors for which a genuine rotational Crofton formula holds. As we shall see, using integrated Minkowski tensors, the two problems of finding (1) rotational averages of intrinsic volumes and (2) expressing intrinsic volumes as rotational integrals can be given a common formulation.

For non-negative integers r and s, $k = 0,\ldots,d-1$, the Minkowski tensors are

$$\Phi_{k,r,s}(K) = \frac{\omega_{d-k}}{r!\,s!\,\omega_{d-k+s}} \int_{\mathbb{R}^d \times S^{d-1}} x^r u^s \Lambda_k(K, \mathrm{d}(x,u)) \ \text{(surface tensor)}$$

$$\Phi_{d,r,0}(K) = \frac{1}{r!} \int_K x^r V_d(\mathrm{d}x) \ \text{(volume tensor)}$$

Here, x^r is the symmetric tensor of rank r determined by x, while $x^r u^s$ is the symmetric tensor product of x^r and u^s. Furthermore, $\Lambda_k(K,\cdot)$ is the kth support measure or generalized curvature measure of K, $k = 0,\ldots,d-1$. The support measure Λ_k is concentrated on the normal bundle $\mathrm{Nor}\,K$ of K which consists of all pairs (x,u) where $x \in \partial K$ and u is an outer unit normal vector of K at x. The rank of $\Phi_{k,r,s}(K)$ is $r+s$. If K is smooth such that there is a unique outer unit normal $u(x)$ for each $x \in \partial K$, then the surface tensors can be expressed as follows

$$\Phi_{k,r,s}(K) = \frac{1}{r!s!\omega_{d-k+s}} \int_{\partial K} x^r u(x)^s \sum_{|I|=d-1-k} \prod_{i \in I} \kappa_i(x)\,\mathcal{H}^{d-1}(\mathrm{d}x),$$

$k = 0,\ldots,d-1$, r,s non-negative integers.

For $r = s = 0$, we have $\Phi_{k,0,0}(K) = V_k(K)$, the kth intrinsic volume, $k = 0,\ldots,d$. Otherwise, $\Phi_{k,r,s}(K)$ carries information about the position, shape and orientation of K, cf. e.g. [31, 32, 353, 354]. The normalized rank 1 tensor $\Phi_{d,1,0}(K)/V_d(K)$ is equal to the usual centre of gravity of K while $\Phi_{d-1,1,0}(K)/V_{d-1}(K)$ is a boundary centre of gravity. Minkowski tensors of rank two and higher provide additional information about the shape and the orientation of K. For further details, see [188] and references therein.

For the development of rotational integral geometry of Minkowski tensors, we will now introduce the integrated Minkowski tensors. These tensors are weighted integrals of Minkowski tensors defined on lower-dimensional k-flats.

Definition 8.1. For $0 \leq j < k < d$, $t > k - d$ and non-negative integers r and s, the integrated Minkowski tensors are

$$\Phi_{j,r,s}^{k,t}(K) = \int_{\mathcal{E}_k^d} \Phi_{j,r,s}^{(E)}(K \cap E)\, d(o,E)^t \, dE,$$

and

$$\Phi_{k,r,0}^{k,t}(K) = \int_{\mathcal{E}_k^d} \Phi_{k,r,0}^{(E)}(K \cap E)\, d(o,E)^t \, dE,$$

where the integrands $\Phi_{j,r,s}^{(E)}(K \cap E)$ and $\Phi_{k,r,0}^{(E)}(K \cap E)$ are calculated relative to E.

The condition $t > k - d$ ensures that $\Phi_{j,r,s}^{k,t}(K)$ is well-defined. The integrated Minkowski tensors defined in [13] are identical to those given in Definition 8.1, up to multiplication by the constant

$$c_{d,k} = \omega_d \cdots \omega_{d-k+1} / [\omega_k \cdots \omega_1].$$

There are a number of interesting special cases of integrated Minkowski tensors. Using Definition 8.1 for $r = s = t = 0$ we have

$$\Phi_{j,0,0}^{k,0}(K) = c_{j,d}^{k,d-k+j} V_{d-k+j}(K), \quad 0 \leq j \leq k < d \quad \text{(classical Crofton formula)}$$

More generally, using [176, Theorem 2.4 and 2.5], we find

$$\Phi_{j,r,s}^{k,0}(K) = c_{d,k,j,s}\, \Phi_{d-k+j,r,s}(K), \quad 0 \leq j < k < d, s = 0,1, \tag{8.13}$$

$$\Phi_{k,r,0}^{k,0}(K) = \Phi_{d,r,0}(K), \quad 0 < k < d. \tag{8.14}$$

Here,

$$c_{d,k,j,s} = c_{j,d}^{k,d-k+j} \frac{\omega_{j+2}}{\omega_{j+s+2}} \frac{\omega_{d-k+j+s+2}}{\omega_{d-k+j+2}}.$$

In [176], it is also shown for arbitrary non-negative integers s that $\Phi_{j,r,s}^{k,0}$ is a linear combination of Minkowski tensors.

The integrated Minkowski tensors obey a genuine rotational Crofton formula.

Proposition 8.2 (Rotational Crofton formula). *For $0 \leq j < k < p \leq d$, $t > k - d$ and non-negative integers r and s, it holds that*

$$\Phi_{j,r,s}^{k,t}(K) = \frac{\omega_{d-k}}{\omega_{p-k}} \int_{\mathcal{L}_p^d} \Phi_{j,r,s}^{k,d-p+t}(K \cap L)\, dL. \tag{8.15}$$

For $j = k$, (8.15) holds for $s = 0$.

Proof. We use (8.12) with $r = k$ and $k = p$ and find

$$\Phi_{j,r,s}^{k,t}(K) = \int_{\mathcal{E}_k^d} \Phi_{j,r,s}^{(E)}(K \cap E) d(o,E)^t \, dE$$

$$= \frac{\omega_{d-k}}{\omega_{p-k}} \int_{\mathcal{L}_p^d} \int_{\mathcal{E}_k^L} \Phi_{j,r,s}^{(E)}(K \cap E) d(o,E)^{d-p+t} \, dE \, dL$$

$$= \frac{\omega_{d-k}}{\omega_{p-k}} \int_{\mathcal{L}_p^d} \Phi_{j,r,s}^{k,d-p+t}(K \cap L) \, dL.$$

The second statement is proved in exactly the same manner. $\qquad\square$

By choosing the parameters in the rotational Crofton formula appropriately, either the left-hand side or the right-hand side of the formula becomes a classical Minkowski tensor.

Corollary 8.2 (Rotational averages of Minkowski tensors). *For $s \in \{0,1\}$ and $t = p - d$, it holds that*

$$\int_{\mathcal{L}_p^d} \Phi_{m,r,s}^{(L)}(K \cap L) \, dL = c_{p,p-q,m-q,s}^{-1} \frac{\omega_q}{\omega_{d-(p-q)}} \Phi_{m-q,r,s}^{p-q,p-d}(K), \tag{8.16}$$

for $0 < q \le m < p \le d$. If $m = p$, then $s = 0$, and

$$\int_{\mathcal{L}_p^d} \Phi_{p,r,0}^{(L)}(K \cap L) \, dL = \frac{\omega_q}{\omega_{d-(p-q)}} \Phi_{p-q,r,0}^{p-q,p-d}(K), \tag{8.17}$$

for $0 < q < p \le d$.

Proof. Combining (8.13) and (8.15), we find

$$\int_{\mathcal{L}_p^d} \Phi_{m,r,s}^{(L)}(K \cap L) \, dL = c_{p,p-q,m-q,s}^{-1} \int_{\mathcal{L}_p^d} \Phi_{m-q,r,s}^{p-q,0}(K \cap L) \, dL$$

$$= c_{p,p-q,m-q,s}^{-1} \frac{\omega_q}{\omega_{d-(p-q)}} \Phi_{m-q,r,s}^{p-q,p-d}(K).$$

The second statement is proved in exactly the same manner. $\qquad\square$

Note that for $r = s = 0$, the left-hand sides of (8.16) and (8.17) are rotational averages of intrinsic volumes, see Sect. 8.1.1 and [12, 187].

As we shall see, it is more interesting for applications in local stereology to try to find the functional defined on the subspace L_p whose rotational average equals a given classical Minkowski tensor. This problem can again be solved for $s \in \{0,1\}$ by combining the rotational Crofton formula with equations (8.13) and (8.14).

Corollary 8.3 (Minkowski tensors as rotational averages). *For $s \in \{0,1\}$ and $t = 0$, it holds that*

$$\Phi_{d+m-p,r,s}(K) = c_{d,p-q,m-q,s}^{-1} \frac{\omega_{d-(p-q)}}{\omega_q} \int_{\mathcal{L}_p^d} \Phi_{m-q,r,s}^{p-q,d-p}(K \cap L) \, dL, \tag{8.18}$$

for $0 < q \leq m < p \leq d$. If $m = p$, then $s = 0$, and

$$\Phi_{d,r,0}(K) = \frac{\omega_{d-(p-q)}}{\omega_q} \int_{\mathcal{L}_p^d} \Phi_{p-q,r,0}^{p-q,d-p}(K \cap L)\, dL, \tag{8.19}$$

for $0 < q < p \leq d$.

Proof. Combining (8.13) and (8.15), we find

$$\int_{\mathcal{L}_p^d} \Phi_{m-q,r,s}^{p-q,d-p}(K \cap L)\, dL = \frac{\omega_q}{\omega_{d-(p-q)}}\, \Phi_{m-q,r,s}^{p-q,0}(K)$$

$$= \frac{\omega_q}{\omega_{d-(p-q)}}\, c_{d,p-q,m-q,s} \Phi_{d+m-p,r,s}(K).$$

The second statement is proved in exactly the same manner. $\qquad\square$

Exercise 8.3. Show that for $r = s = 0$ and $q = 1$, the result in Corollary 8.3 reduces to (8.9) with $\alpha_{k,j}$ given in Proposition 8.1.

It is clearly of interest to study what kind of geometric information the integrated Minkowski tensors carry about the original set K. In the proposition below, we give such geometric interpretation for $\Phi_{k,r,0}^{k,t}$ and $\Phi_{d-2,r,0}^{d-1,t}$. For a proof, the reader is referred to [13].

Proposition 8.3. *For $0 < k < d$, $t > k - d$ and any non-negative integer r, it holds that*

$$\Phi_{k,r,0}^{k,t}(K) = \frac{1}{r!}\frac{\Gamma(\frac{t+d-k}{2})\Gamma(\frac{d}{2})}{\Gamma(\frac{t+d}{2})\Gamma(\frac{d-k}{2})} \int_K x^r |x|^t\, v_d(dx). \tag{8.20}$$

Furthermore, if K is a compact d-dimensional C^2 manifold with boundary, then for $t > 0$ and a non-negative integer r

$$\Phi_{d-2,r,0}^{d-1,t}(K) = \frac{\omega_{d-1}}{2r!\omega_d} B\left(\tfrac{t+1}{2}, \tfrac{d}{2}\right)$$

$$\times \int_{\partial K} x^r |x|^t F_{-\frac{1}{2},-\frac{t}{2};\frac{d-1}{2}}(\sin^2 \beta(x))\, \mathcal{H}^{d-1}(dx). \tag{8.21}$$

Exercise 8.4. Show that the functional $\alpha_{k,j}$ specified in Proposition 8.1 is a special case of an integrated Minkowski tensor

$$\alpha_{k,j}(M) = \frac{\omega_{d-k+1}}{\omega_1}(c_{j-1,d}^{k-1,d-k+j})^{-1}\Phi_{j-1,0,0}^{k-1,d-k}(M),$$

$M \in \mathcal{K}_{\mathrm{conv}}^L$, $L \in \mathcal{L}_k^d$. Furthermore, show that if in Proposition 8.3 we insert these parameter values, we get the result in Corollary 8.1.

8.1.4 A Principal Rotational Formula

To the best of our knowledge, a principal rotational formula is still not available in the literature. Focusing on intrinsic volumes, such a formula involves integrals of the form

$$\int_{SO_d} V_k(K \cap RM)\,dR, \tag{8.22}$$

$k = 0,\ldots,d$, where SO_d is the special orthogonal group in $\mathbb{R}^d$, K and M are convex and compact subsets of $\mathbb{R}^d$, and dR is the element of the unique rotation invariant probability measure on SO_d. From an applied point of view such a formula is interesting. Here, K is the unknown spatial structure of interest while M is a known 'sampling window' constructed by the observer. The aim is to get information about K from observation of the intersection of K with a randomly rotated version of M. For $k = d$, (8.22) is equal to

$$\frac{1}{\omega_d} \int_0^\infty r^{-(d-1)} \mathcal{H}^{d-1}(K \cap r\mathbb{S}^{d-1}) \mathcal{H}^{d-1}(M \cap r\mathbb{S}^{d-1})\,dr.$$

To see this, we use that

$$\int_{SO_d} V_d(K \cap RM)\,dR = \int_{SO_d} \int_{\mathbb{R}^d} \mathbf{1}_{K \cap RM}(x)\, v_d(dx)\,dR$$

$$= \int_{\mathbb{R}^d} \mathbf{1}_K(x) \left[\int_{SO_d} \mathbf{1}_{RM}(x)\,dR \right] v_d(dx).$$

Since

$$\int_{SO_d} \mathbf{1}_{RM}(x)\,dR = \int_{SO_d} \mathbf{1}_M(R^{-1}x)\,dR$$

$$= \int_{SO_d} \mathbf{1}_M(Rx)\,dR$$

$$= \mathcal{H}^{d-1}(M \cap |x|\mathbb{S}^{d-1})/\mathcal{H}^{d-1}(|x|\mathbb{S}^{d-1})$$

$$= |x|^{-(d-1)} \omega_d^{-1} \mathcal{H}^{d-1}(M \cap |x|\mathbb{S}^{d-1}),$$

we obtain

$$\int_{SO_d} V_d(K \cap RM)\,dR$$

$$= \frac{1}{\omega_d} \int_K |x|^{-(d-1)} \mathcal{H}^{d-1}(M \cap |x|\mathbb{S}^{d-1})\, v_d(dx)$$

$$= \frac{1}{\omega_d} \int_0^\infty r^{-(d-1)} \mathcal{H}^{d-1}(K \cap r\mathbb{S}^{d-1}) \mathcal{H}^{d-1}(M \cap r\mathbb{S}^{d-1})\,dr.$$

At the last equality sign, we have used spherical coordinates and

$$\int_{\mathbb{S}^{d-1}} \mathbf{1}_K(ru)\, \mathcal{H}^{d-1}(\mathrm{d}u) = r^{-(d-1)}\mathcal{H}^{d-1}(K \cap r\mathbb{S}^{d-1}).$$

A result of a similar form involving two terms can be obtained for $k = d-1$. The case of general k is still open.

Exercise 8.5. Suppose that

$$\mathcal{H}^{d-1}(\partial(K \cap RM)) = \mathcal{H}^{d-1}(\partial K \cap RM) + \mathcal{H}^{d-1}(K \cap R\partial M),$$

for almost all R. Show under this assumption that

$$\int_{SO_d} V_{d-1}(K \cap RM)\, \mathrm{d}R = \frac{1}{2\omega_d}\Big[\int_{\partial K} |x|^{-(d-1)} \mathcal{H}^{d-1}(M \cap |x|\mathbb{S}^{d-1})\, \mathcal{H}^{d-1}(\mathrm{d}x)$$
$$+ \int_{\partial M} |x|^{-(d-1)} \mathcal{H}^{d-1}(K \cap |x|\mathbb{S}^{d-1})\, \mathcal{H}^{d-1}(\mathrm{d}x) \Big].$$

8.2 Local Stereology

Local stereology is the branch of stereology, dealing with inference about $K \in \mathcal{K}^d_{\mathrm{conv}}$ from sections $K \cap L$, $L \in \mathcal{L}^d_k$, $0 < k < d$. Usually, the set class considered is not restricted to compact and convex subsets of $\mathbb{R}^d$, but we will here focus on such sets for the sake of simplicity of presentation. A model example of application of local stereology is the case when K is a biological cell, studied via sections of the cell with planes passing through a reference point, usually taken to be the cell nucleus or a nucleolus. In the following, we identify the reference point with the origin o.

The monograph [185] is an introduction to local stereology, where the focus is on Hausdorff measures rather than on intrinsic volumes. In [185], the local stereological procedures are mainly presented from a design-based point of view, where K is regarded as fixed, while the k-subspace L is isotropic random. See also the recent publication [188] where this point of view is taken in relation to estimation of Minkowski tensors.

In this chapter, we will take the dual model-based point of view. We let Z be an isotropic random convex body in $\mathbb{R}^d$ and let $\bar{V}_j(Z) = \mathbf{E}V_j(Z)$, $j = 0,\dots,d$, denote its mean intrinsic volumes. One of our aims is to use the rotational integral geometric identities, developed in the previous section, to derive unbiased local stereological estimators of the mean intrinsic volumes.

The results in Sect. 8.1.1 can be used to relate mean intrinsic volumes $\bar{V}_j(Z \cap L)$ on a k-subspace L to properties of the original random set Z. For this purpose, let

$$\bar{\Phi}_d(Z,A) = \mathbf{E}V_d(Z \cap A), \quad A \in \mathcal{B}(\mathbb{R}^d),$$
$$\bar{\Lambda}_j(Z, A \times B) = \mathbf{E}\Lambda_j(Z, A \times B), \quad A \in \mathcal{B}(\mathbb{R}^d), B \in \mathcal{B}(\mathbb{S}^{d-1}),$$

$j = 0,\dots,d-1$. Here, $\bar{\Lambda}_j(Z,\cdot)$ is the mean jth support measure associated with Z. Note that $\bar{\Phi}_d(Z,\cdot)$ has the following simple expression

$$\bar{\Phi}_d(Z,A) = \int_A p_Z(x)\,v_d(\mathrm{d}x), \quad A \in \mathcal{B}(\mathbb{R}^d),$$

where $p_Z(x) = \mathbf{P}(x \in Z)$, $x \in \mathbb{R}^d$. Using (8.5), we find for $L \in \mathcal{L}_k^d$,

$$\bar{V}_k(Z \cap L) = \int_{\mathcal{L}_k^d} \mathbf{E} V_k(Z \cap M)\,\mathrm{d}M$$

$$= \mathbf{E} \int_{\mathcal{L}_k^d} V_k(Z \cap M)\,\mathrm{d}M$$

$$= \mathbf{E}\,\frac{\Gamma(d/2)}{\pi^{(d-k)/2}\Gamma(k/2)} \int_Z |x|^{-(d-k)}\,v_d(\mathrm{d}x)$$

$$= \frac{\Gamma(d/2)}{\pi^{(d-k)/2}\Gamma(k/2)} \int_{\mathbb{R}^d} |x|^{-(d-k)}\,\bar{\Phi}_d(Z,\mathrm{d}x).$$

At the first equality sign, we have used that V_k and the distribution of Z is invariant under rotations. Likewise, we find for $L \in \mathcal{L}_k^d$, using (8.7),

$$\bar{V}_{k-1}(Z \cap L)$$
$$= \frac{\Gamma(d/2)}{\pi^{(d-k)/2}\Gamma(k/2)} \int_{\mathbb{R}^d \times \mathbb{S}^{d-1}} |x|^{-(d-k)} F_{-\frac{1}{2},\frac{d-k}{2};\frac{d-1}{2}}(\sin^2 \angle(x,u))\,\bar{\Lambda}_{d-1}(Z,\mathrm{d}(x,u)).$$

The result (8.6) can be used to get a general expression for $\bar{V}_j(Z \cap L)$, $j = 0,\dots,k-1$, as an integral with respect to $\bar{\Lambda}_{d-1}(Z,\cdot)$.

Example 8.3. For $d = 3$ and $k = 2$, we get

$$\overline{\mathrm{area}}(Z \cap L) = \frac{1}{2} \int_{\mathbb{R}^3} |x|^{-1}\,\bar{\Phi}_3(Z,\mathrm{d}x)$$

and

$$\overline{\mathrm{length}}(\partial Z \cap L) = \frac{2}{\pi} \int_{\mathbb{R}^3 \times \mathbb{S}^2} |x|^{-1} E(|\sin \angle(x,u)|, \pi/2)\,\bar{\Lambda}_2(Z,\mathrm{d}(x,u)),$$

see also Examples 8.1 and 8.2.

In order to derive unbiased local stereological estimators of mean intrinsic volumes of the original set Z, we need the rotational integral geometric identities derived in Sect. 8.1.2. Using Proposition 8.1, we find for $L \in \mathcal{L}_k^d$

$$\bar{V}_{d-k+j}(Z) = \frac{\omega_{d-k+1}}{\omega_1}(c_{j-1,d}^{k-1,d-k+j})^{-1} \int_{\mathcal{E}_{k-1}^L} d(o,E)^{d-k} \bar{V}_{j-1}(Z \cap E)\,\mathrm{d}E, \quad (8.23)$$

$k = 1,\dots,d$, $j = 1,\dots,k$. In the particular case $j = k$, we get, using Corollary 8.1, the more explicit expression

$$\bar{V}_d(Z) = \frac{\pi^{(d-k)/2}\Gamma(k/2)}{\Gamma(d/2)} \int_L |x|^{d-k}\,\bar{\Phi}_k(Z \cap L,\mathrm{d}x). \quad (8.24)$$

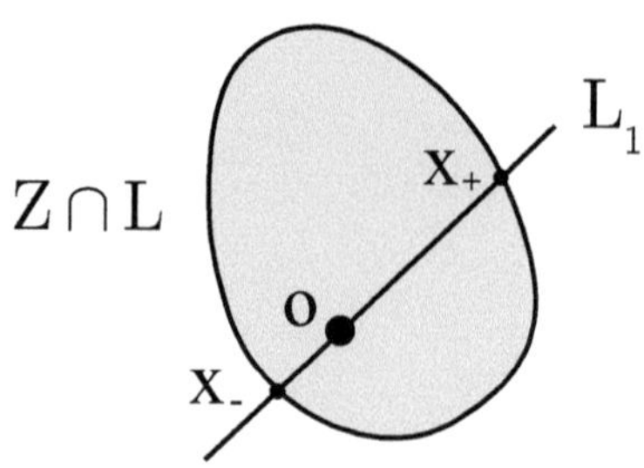

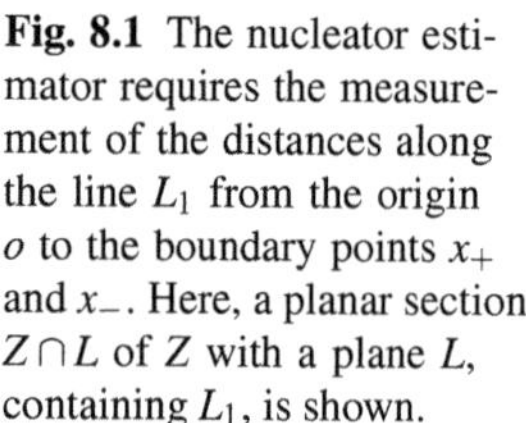

Fig. 8.1 The nucleator estimator requires the measurement of the distances along the line L_1 from the origin o to the boundary points x_+ and x_-. Here, a planar section $Z \cap L$ of Z with a plane L, containing L_1, is shown.

It follows that

$$m(Z \cap L) = \frac{\pi^{(d-k)/2} \Gamma(k/2)}{\Gamma(d/2)} \int_{Z \cap L} |x|^{d-k} v_k(dx) \tag{8.25}$$

is an unbiased estimator of $\bar{V}_d(Z)$.

Example 8.4 (Local estimation of volume in $\mathbb{R}^3$). Let $d = 3$ and $k = 1$. It follows from (8.25) that for $L_1 \in \mathcal{L}_1^3$

$$m(Z \cap L_1) = 2\pi \int_{Z \cap L_1} |x|^2 v_1(dx) \tag{8.26}$$

is an unbiased estimator of the mean volume $\bar{V}_3(Z)$ of Z. This estimator is called the nucleator in the applied literature ([159]) and will here be denoted by $m_{cl_1}(Z \cap L_1)$ (the index cl_1 stands for *classical nucleator* based on observation along 1 line). If $o \in Z$, $Z \cap L_1 = [x_-, x_+]$ is a line segment containing the origin and (8.26) reduces to

$$m_{cl_1}(Z \cap L_1) = \frac{2\pi}{3}(|x_+|^3 + |x_-|^3),$$

cf. Figure 8.1. Note that if Z is a ball centred at the origin o with random radius, then $m_{cl_1}(Z \cap L_1)$ is identically equal to the volume of Z. The estimator based on observation along two perpendicular lines L_1 and L_1' through o,

$$m_{cl_2} = \tfrac{1}{2}[m_{cl_1}(Z \cap L_1) + m_{cl_1}(Z \cap L_1')],$$

is widely used and highly cited in the biosciences.

For $d = 3$ and $k = 2$, (8.25) leads to an unbiased estimator of $\bar{V}_3(Z)$ based on measurements in a plane L through o. The estimator is called the *integrated nucleator*, cf. [165], and takes the following form

$$m_{int}(Z \cap L) = 2 \int_{Z \cap L} |x| v_2(dx).$$

The reason why the estimator is called the integrated nucleator is the following result

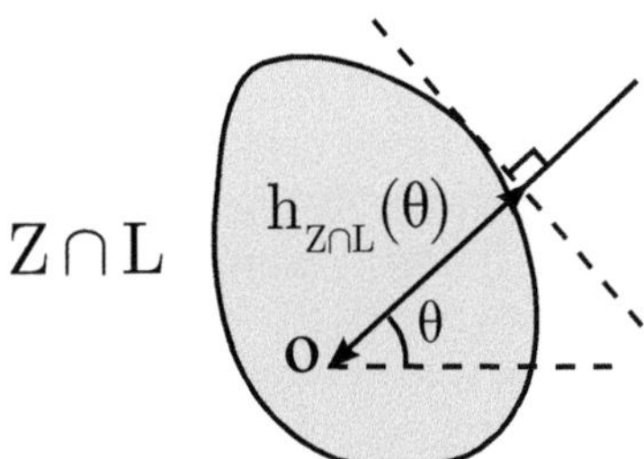

Fig. 8.2 For $\theta \in [0, 2\pi)$, the value $h_{Z \cap L}(\theta)$ of the support function is the distance from the origin o to the touching dashed line.

$$m_{int}(Z \cap L) = \int_{\mathcal{L}_1^L} m_{cl_1}(Z \cap L_1)\, dL_1,\tag{8.27}$$

that can be shown, using polar decomposition in the plane L. Recently, in [90], it has been shown that the integrated nucleator is identical to the so-called *wedge estimator*. A discretized version of $m_{int}(Z \cap L)$, called the isotropic *rotator*, was introduced already in [186] together with another local stereological estimator of volume, the so-called *vertical rotator*.

Example 8.5 (Local estimation of surface area in $\mathbb{R}^3$). Let $d = 3$ and consider estimators of the mean surface area $\bar{S}(Z) = 2\bar{V}_2(Z)$. Using (8.23) with $d = 3$, $k = 2$ and $j = 1$, we find that

$$\bar{V}_2(Z) = 2\pi \int_{\mathcal{E}_1^L} d(o, E)\bar{V}_0(Z \cap E)\, dE.$$

It follows that

$$m(Z \cap L) = 4\pi \int_{\mathcal{E}_1^L} d(o, E)\mathbf{1}\{Z \cap E \neq \emptyset\}\, dE$$

is an unbiased estimator of $\bar{S}(Z)$. If $o \in Z$, we can express $m(Z \cap L)$ in a simple way by means of the support function $h_{Z \cap L}$ of $Z \cap L$. (The definition of the support function is illustrated in Fig. 8.2.) We find

$$
\begin{aligned}
m(Z \cap L) &= 4\pi \int_0^\pi \int_{-h_{Z \cap L}(\theta + \pi)}^{h_{Z \cap L}(\theta)} |r|\, dr\, \frac{d\theta}{\pi} \\
&= 4 \int_0^\pi \left[\int_0^{h_{Z \cap L}(\theta)} r\, dr + \int_0^{h_{Z \cap L}(\theta + \pi)} r\, dr \right] d\theta \\
&= 2 \int_0^{2\pi} h_{Z \cap L}(\theta)^2\, d\theta.
\end{aligned}
$$

This representation shows that $m(Z \cap L)$ is equal to four times the area of the flower set associated with $Z \cap L$, defined by

$$F(Z \cap L) = \{r(\cos\theta, \sin\theta) \mid 0 \leq r \leq h_{Z \cap L}(\theta)\}.$$

This result was first published in [87]. The estimator has accordingly been called the *flower estimator*. In [88, 89], a discretization of $m(Z \cap L)$ based on measurement of the support function in both directions along two perpendicular lines is further discussed. The resulting discretized estimator is called the *pivotal estimator* and is very efficient, see [108]. An alternative representation of $m(Z \cap L)$ may be obtained by using the second result of Corollary 8.1 for $d = 3$ and $k = 2$ and the fact that

$$F_{-\frac{1}{2},-\frac{1}{2};\frac{1}{2}}\left(\sin^2 \angle(x,u)\right) = \cos \angle(x,u) + \angle(x,u)\sin \angle(x,u),$$

cf. e.g. [185, p. 146]. Using this, the close relation to another estimator of surface area, the *surfactor* ([184]), may be seen. For further details, see [188, Section 5.1.4].

Using the results presented in Sect. 8.1.3, we can derive local stereological estimators of Minkowski tensors. For simplicity, we will here focus on the case of volume tensors. A more comprehensive treatment is given in [188]. Combining (8.19) and (8.20), we find for $L \in \mathcal{L}_k^d$

$$\bar{\Phi}_{d,r,0}(Z) = \frac{\pi^{(d-k)/2}\Gamma(k/2)}{\Gamma(d/2)} \frac{1}{r!} \int_L x^r |x|^{d-k} \, \bar{\Phi}_k(Z \cap L, \mathrm{d}x). \tag{8.28}$$

For $r = 0$, the result reduces to (8.24). Using (8.28), we can construct local stereological estimators of centres of gravity ($r = 1$) and volume tensors of rank two ($r = 2$) that can be used to obtain information about orientation and shape of Z. Below, we only consider the case $r = 1$.

Example 8.6 (Local estimation of centre of gravity in $\mathbb{R}^3$). Let $d = 3$, $r = 1$ and $k = 1$. Then, we find, using (8.28),

$$\bar{\Phi}_{3,1,0}(Z) = 2\pi \int_L x |x|^2 \, \bar{\Phi}_1(Z \cap L, \mathrm{d}x).$$

It follows that

$$m(Z \cap L) = 2\pi \int_{Z \cap L} x |x|^2 \, v_1(\mathrm{d}x)$$

is an unbiased estimator of $\bar{\Phi}_{3,1,0}(Z)$. If $o \in Z$, then $Z \cap L$ is a line segment $[x_-, x_+]$, containing o. If e is a unit vector spanning L and pointing in the same direction as x_+, then

$$m(Z \cap L) = \frac{\pi}{2}(|x_+|^4 - |x_-|^4)e.$$

Note that if Z is centrally symmetric around o, then $m(Z \cap L) = o$, always.

The local stereological estimators can be used to analyze a particle population, using local sectional data, thereby providing information about the size, position, orientation and shape of the particles. Let us assume that the particles may be described by a stationary germ-grain model $\cup_{i=1}^{\infty}(x_i + Z_i)$, where $\{x_i\}$ is a stationary

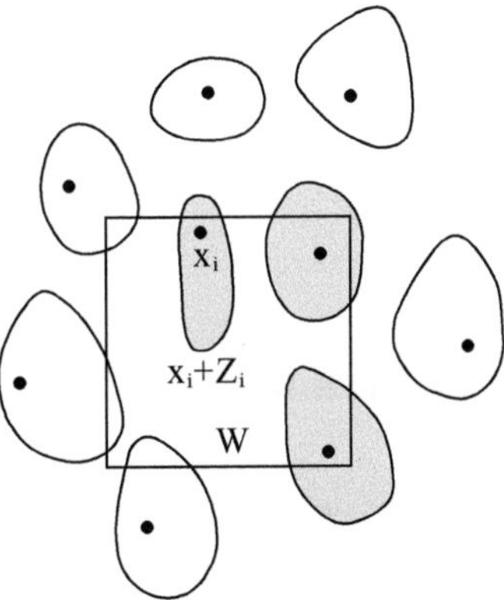

Fig. 8.3 Sampling of particles with reference point in W. Sampled particles are shown grey.

point process in $\mathbb{R}^d$ and $\{Z_i\}$ are i.i.d. nonempty, compact and convex random subsets of $\mathbb{R}^d$, independent of $\{x_i\}$. We let Z_0 be a random set with the common distribution of the Z_is, denoted by Q. We will assume that Q is invariant under rotations in $\mathbb{R}^d$.

Our aim is to estimate the distribution of $\beta(Z_0)$ from local sectional data where β may be an intrinsic volume or, more generally, a Minkowski tensor. Available for observation is a sample of particles $\{x_i + Z_i : x_i \in W\}$ collected in a d-dimensional sampling window W, see Fig. 8.3. We will focus on the situation in optical microscopy, where it is possible to perform measurements on any virtual section $Z_i \cap L$, $L \in \mathcal{L}_k^d$. If α is a rotation invariant functional, satisfying

$$\int_{\mathcal{L}_k^d} \alpha(K \cap L)\, \mathrm{d}L = \beta(K),$$

for any $K \in \mathcal{K}_{\mathrm{conv}}^d$, then

$$\bar{\alpha}(Z_0 \cap L) = \int_{\mathcal{L}_k^d} \bar{\alpha}(Z_0 \cap M)\, \mathrm{d}M$$

$$= \mathbf{E}\left(\int_{\mathcal{L}_k^d} \alpha(Z_0 \cap M)\, \mathrm{d}M\right)$$

$$= \mathbf{E}(\beta(Z_0)) = \bar{\beta}(Z_0).$$

If we let N be the number of particles sampled in W, then

$$\sum_{x_i \in W} \alpha(Z_i \cap L)/N$$

is a ratio-unbiased estimator of $\bar{\beta}(Z_0)$. If several sections are used per particle Z_i, one may estimate $\beta(Z_i)$ precisely and use the empirical distribution of $\{\hat{\beta}(Z_i) \mid x_i \in W\}$ as an estimate of the distribution of $\beta(Z_0)$.

Local stereology of spherical particles with non-centrally placed reference point has recently been studied in [395].

8.3 Variance Reduction Techniques

In the previous section, we have used rotational integral geometric identities to develop local stereological procedures for estimating quantitative properties of a spatial structure. The local stereological estimators can be applied without specific shape assumptions but may have a large variance. In this section, we will discuss procedures for reducing the variance of the estimators. Some of the procedures require the use of automatic image analysis. We will focus on the local stereological estimators presented in Examples 8.4 and 8.5 of the previous section.

So, in this section, Z will be an isotropic random convex body in $\mathbb{R}^3$. Let us first consider estimation of the mean volume $\bar{V}_3(Z)$ as described in Example 8.4. We let $L \in \mathcal{L}_2^3$ be a plane through the origin and $L_1, L_1' \in \mathcal{L}_1^L$ two perpendicular lines in L through the origin. Since $m_{cl_1}(Z \cap L_1)$ and $m_{cl_1}(Z \cap L_1')$ are identically distributed, we have

$$
\begin{aligned}
\mathbf{var}\, m_{cl_2} &= \mathbf{var}(\tfrac{1}{2}[m_{cl_1}(Z \cap L_1) + m_{cl_1}(Z \cap L_1')]) \\
&= \tfrac{1}{2}[\mathbf{var}\, m_{cl_1} + \mathbf{cov}(m_{cl_1}(Z \cap L_1), m_{cl_1}(Z \cap L_1'))] \\
&\leq \mathbf{var}\, m_{cl_1}.
\end{aligned}
$$

It follows that the variance of the classical nucleator based on observation along two perpendicular lines is smaller than or equal to the variance obtained when just observing along one line.

Because of (8.27), the integrated nucleator m_{int} may be regarded as a classical nucleator based on measurements along an infinite number of lines. As a consequence, the variance of m_{int} is expected to be smaller than or equal to the variance of m_{cl_1} (and m_{cl_2}). A formal argument for this result goes as follows. The identity (8.27) may be regarded as a conditional mean value result, viz.

$$
m_{int} = \mathbf{E}(m_{cl_1}(Z \cap L_1) \mid Z),
$$

where the mean value is conditional on Z and with respect to an isotropic line L_1 in the plane L through o, independent of Z. It follows that

$$
\begin{aligned}
\mathbf{var}\, m_{cl_1} &= \mathbf{var}(\mathbf{E}(m_{cl_1}(Z \cap L_1) \mid Z)) + \mathbf{E}(\mathbf{var}(m_{cl_1}(Z \cap L_1) \mid Z)) \\
&= \mathbf{var}\, m_{int} + \mathbf{E}(\mathbf{var}(m_{cl_1}(Z \cap L_1) \mid Z)) \\
&\geq \mathbf{var}\, m_{int}.
\end{aligned}
$$

Similarly, $\mathbf{var}\, m_{cl_2} \geq \mathbf{var}\, m_{int}$.

Because of these variance relations, it appears as an obvious idea to use m_{int} instead of m_{cl_1} or m_{cl_2}. In contrast to the two latter estimators that requires a few distance measurements by an expert, m_{int} needs automatic segmentation of $Z \cap L$. Let $\tilde{Z}_2$ be an estimate of the section $Z \cap L$, obtained by computerized image analysis. The automatic nucleator is now defined as

$$m_{aut} = m_{int}(\tilde{Z}_2).$$

Since the segmentation may not be precise in all cases, an intermediate version may be preferable. An expert supervises the process. If the segmentation is judged satisfactory, m_{aut} is used, otherwise the expert intervenes and determine m_{cl_2} manually. This estimator is called the semi-automatic nucleator and is denoted m_{semi}, cf. [165].

The estimators m_{cl_1}, m_{cl_2} and m_{int} are unbiased while m_{aut} and m_{semi} may be biased. In fact, m_{aut} may be heavily biased if the segmentation is generally unsatisfactory while the bias of m_{semi} is expected to be small because the segmented section $\tilde{Z}_2$ is only used when the segmentation is judged satisfactory by an expert. Also, m_{semi} is expected to be more precise (for instance, in terms of mean square error) than the best manual estimator m_{cl_2}, because m_{semi} only differs from m_{cl_2} when the segmentation is satisfactory and in these cases, the more precise estimator m_{int} is used.

In a concrete study of somastatin positive inhibitory interneurons from transgenic GFP-GAD mice hippocampi, cf. [165], it was found that m_{aut} had a bias of 32% while m_{semi} only 0.4%. The relative error ($\sqrt{\text{MSE}}$ / mean) was 0.58, 0.61 and 0.69 for m_{int}, m_{semi} and m_{cl_2}, respectively.

A similar comparative investigation has been performed in [108] for the local stereological estimators of mean surface area presented in Example 8.5. The estimator that requires automatic segmentation of the planar section $Z \cap L$ is here the flower estimator. Semi-automatic estimation based on two types of discretizations of the flower estimator, namely the pivotal estimator and the surfactor, has been investigated in [108]. For ellipsoidal particles, it is shown that the flower estimator is equal to the pivotal estimator based on support function measurements along four perpendicular rays. This makes the pivotal estimator a powerful approximation to the flower estimator. An important decrease in workload may be obtained by using the semi-automatic approach.

8.4 Computational Stereology

Stereology provides information about quantitative properties of spatial structures from observations in lower-dimensional sections of the spatial structure under study. A stereological procedure typically involves the following steps: (1) sampling of blocks to be analyzed, (2) generation of sections through the blocks and (3) analysis of the sampled sections ([17, Chap. 12]).

Until recently, stereology has mainly been a 'manual' discipline. Each of the three steps mentioned above has been performed manually by experts and technicians. However, during the last decades, computers have become an increasingly important tool in the stereological analysis of spatial structures, and a new sub-discipline of stereology, computational stereology, is emerging. Computational stereology may be defined as the sub-discipline of stereology that deals with the design of computational procedures that can substitute manual procedures in one of the three steps mentioned above ([326]).

It is expected that computational stereology will influence the practice in the laboratory. For instance, computational stereology may imply faster execution times compared to existing manual procedures or more efficient probes that require reliable automatic segmentation. Eventually, computational stereology may also affect the advance of theoretical stereology. Computational procedures thus open up the possibility for developing new stereological methods for estimating more complicated quantities than scalar quantities such as volume, surface area, length and number. One obvious example is the Minkowski tensors. On the other hand, a clear definition of computational stereology may make developers of computational image analysis tools realize the importance of 3D interpretations of 2D sections.

Before, images were typically recorded by systematically moving the microscope stage and taking photographs (micrographs) of the generated fields of view in the microscope. Subsequently, the analysis was performed manually on the generated micrographs. Nowadays, computers are used in the acquisition of the images to be analyzed by stereological methods. Such digital images are a prerequisite for any procedure in computational stereology.

The appearance of *whole slide scanners* has been a major advance for computational stereology. Here, the operator delineates the region of interest (ROI) of the sampled section and the whole slide scanner then generates a digital representation of the ROI. It is important that the scanning and storage of the digital images of the sampled sections can be performed without interference by the operator. As a consequence, the operator has the freedom to choose an appropriate time for analysis of the digital images, using developed software.

With digital representations of the sampled sections, it is possible to (1) extend the class of stereological estimators that can be implemented and (2) develop efficient subsampling of the sections under study.

An example of issue 1 relates to the disector that is used for estimating particle number ([230, p. 41 - 42]). It is here needed to identify particles in pairs of sections. The manual alignment of the pair of sections may be very time consuming. It is therefore an important advance that this alignment can be performed automatically on the basis of a digital representation of the two sections.

With a digital representation of the sampled sections, it is also possible to implement intelligent non-uniform sampling of fields of view within the section that may result in an important reduction in the variance of stereological estimators (issue 2). The standard procedure has until recently been to use systematic uniform random sampling of fields of view. If the particles (cells) of interest are distributed in an inhomogeneous pattern in the sampled section, this approach may, however, be rather inefficient. In such cases, many fields of view will contain no or very few cells, if systematic uniform random sampling is used. The idea is to use instead non-uniform sampling of fields of view with a probability of selecting a particular field of view that is roughly proportional to the number of cells seen in the field of view. Let the ith field of view contain y_i cells, $i = 1, \dots, N$, where N is the total number of fields of view. Let $S \subset \{1, \dots, N\}$ be the random sample of fields of view and p_i the probability that the ith field of view is included in the sample. Provided that $p_i > 0$ whenever $y_i > 0$,

$$\sum_{i \in S} y_i / p_i$$

is an unbiased estimator of the total number of the cells in the section. Typically, the sampling probability p_i is determined automatically from a scan of the section at low magnification, using a colour proportion that is roughly proportional to y_i while the actual counts in the sampled fields of view are determined by an operator at high magnification. If p_i is exactly proportional to y_i, this estimator always gives the right answer. In empirical studies, increase in efficiencies of a factor of 10 compared to ordinary systematic uniform random sampling has been found, see [164] and references therein. In the applied stereological literature, the estimator is called the *proportionator*.

Until recently, there have only been limited interactions between researchers in stereology and image analysis. One exception is [305]. This situation is very unfortunate, at least for the stereologists, because image analysis may actually be required for implementation of advanced stereological procedures. One example is the semi-automatic procedures described in the previous section of this chapter.

Acknowledgements This work has been supported by Centre for Stochastic Geometry and Advanced Bioimaging, funded by a grant from the Villum Foundation.

Chapter 9
An Introduction to Functional Data Analysis

Ulrich Stadtmüller and Marta Zampiceni

Abstract In this chapter we will give an introduction to the methods in functional data analysis. We will present the basics from principal component analysis for functional data together with the functional analytic background as well as the data analytic counterpart. As prerequisites we give an introduction to presentation techniques of functional data and some smoothing techniques. A few asymptotic results are presented as well.

9.1 Some Fundamental Tools

The aim of this contribution is to introduce into techniques which are helpful if one wants to work with models which use data consisting of functions or paths of stochastic processes. As further reading we suggest the book [324]. In particular the question of presenting this type of data and to do some descriptive statistics is well presented there and omitted in this short presentation.

9.1.1 Basic Statistics

Let $\{X(t), t \in \mathcal{T}\}$ be a stochastic process over a probability space $(\Omega, \mathcal{A}, \mathbf{P})$ and an index set being a compact interval $\mathcal{T} \subset \mathbb{R}$ (or $\mathcal{T} \subset \mathbb{R}^d$ for some $d > 1$). We

Ulrich Stadtmüller,
Ulm University, Department of Number Theory and Probability Theory, 89069 Ulm, Germany,
e-mail: ulrich.stadtmueller@uni-ulm.de

Marta Zampiceni,
Ulm University, Department of Number Theory and Probability Theory, 89069 Ulm, Germany,
e-mail: marta.zampiceni@uni-ulm.de

© Springer International Publishing Switzerland 2015

V. Schmidt (ed.), *Stochastic Geometry, Spatial Statistics and Random Fields*,
Lecture Notes in Mathematics 2120, DOI 10.1007/978-3-319-10064-7_9

call a realization of a stochastic process a *path*. In the following, we will focus on stochastic processes with continuous paths, satisfying the following basic condition:

$$\int_{\mathcal{T}} \mathbf{E}(X^2(t))\, dt < \infty, \tag{9.1}$$

which implies that almost all paths are square integrable functions.

Property (9.1) allows us to define mean-value, variance and covariance function of such stochastic processes. For $s, t \in \mathcal{T}$, we will use the following notation:

$$\begin{aligned}
\mu(t) &= \mu_X(t) = \mathbf{E}X(t), \\
\sigma^2(t) &= \sigma_X^2(t) = \mathbf{var}\,X(t), \\
C(s,t) &= \mathbf{cov}_X(s,t) = \mathbf{cov}(X(s), X(t)), \\
\mathbf{corr}_X(s,t) &= \frac{C(s,t)}{\sigma(s)\sigma(t)} = \frac{\mathbf{cov}_X(s,t)}{\sigma_X(s)\sigma_X(t)}.
\end{aligned}$$

These are the basic quantities we want to work with. To do statistical analysis we assume throughout this chapter that there are given n independent copies $X_1(t), \ldots, X_n(t),\, t \in \mathcal{T}$ of the process $\{X(t)\}$. Based on these, the previously introduced moment-functions can be estimated (for $s, t \in \mathcal{T}$) by the well-known standard estimators

$$\hat{\mu}(t) = \bar{X}(t) = \frac{1}{n} \sum_{i=1}^{n} X_i(t) \tag{9.2}$$

$$\widehat{\sigma^2}(t) = \frac{1}{n-1} \sum_{i=1}^{n} (X_i(t) - \bar{X}(t))^2 \tag{9.3}$$

$$\widehat{C}(s,t) = \frac{1}{n-1} \sum_{i=1}^{n} (X_i(s) - \bar{X}(s))(X_i(t) - \bar{X}(t)) \tag{9.4}$$

$$\widehat{\mathbf{corr}}(s,t) = \frac{\widehat{C}(s,t)}{\hat{\sigma}(s)\hat{\sigma}(t)}. \tag{9.5}$$

The first, obvious, problem with functional data is intrinsic in its nature: it is impossible to record the whole graph of a continuous function. So we do actually not observe the realization of $\{X_i(t)\}$, but rather a vector

$$X_i = (X_i(t_{i,1}), \ldots, X_i(t_{i,m_i}))^\top \in \mathbb{R}^{m_i} \tag{9.6}$$

of measurements of $\{X_i(t)\}$ taken at the "time points" $t_{i,1}, \ldots, t_{i,m_i}$.

If we are lucky, the recording times are the same for the whole sample of functional data (paths) $X_1(t), \ldots, X_n(t)$. In this case, one can compute the estimators considered in (9.2) - (9.5) pointwise at the given, common, time points and then reconstruct the whole functions by linear interpolation or some smoothing technique. We will briefly discuss some of these techniques in Sect. 9.1.2.

Often, the measurements can be affected by some additive noise. The recordings would then have the following form

$$Y_{i,j} = X_i(t_{i,j}) + \varepsilon_{i,j}, \qquad j = 1,\ldots,m_i \; i = 1,\ldots,n. \tag{9.7}$$

For $j = 1,\ldots,m_i$ and $i = 1,\ldots,n,$, we assume that the noise terms $\varepsilon_{i,j}$ are i.i.d. random variables, all independent of X_i, and such that $\mathbf{E}\,\varepsilon_{i,j} = 0$ and $\mathbf{var}\varepsilon_{i,j} = \sigma_\varepsilon^2$.

For the moment let us focus on the case with noise where all the n paths of a stochastic process are recorded at $t_1,\ldots,t_m$ and take a closer look at the (pointwise) estimators of mean and covariance. Of course, we will have somewhat different situations, depending on whether the measurements were altered by some error or not.

For the estimator of the mean, one obtains

$$\widehat{\mu}_X(t_j) = \frac{1}{n}\sum_{i=1}^{n} X_i(t_j), \qquad j = 1,\ldots,m,$$

$$\widehat{\mu}_Y(t_j) = \frac{1}{n}\sum_{i=1}^{n} Y_{i,j}, \qquad j = 1,\ldots,m.$$

in the cases considered in (9.6) and (9.7), respectively. Both estimators are consistent, since

$$\mathbf{E}\,\widehat{\mu}_X(t_j) = \mu_X(t_j), \qquad \mathbf{var}\,\widehat{\mu}_X(t_j) = \frac{1}{n}\sigma_X^2(t_j) \xrightarrow{n\to\infty} 0,$$

$$\mathbf{E}\,\widehat{\mu}_Y(t_j) = \mu_X(t_j), \qquad \mathbf{var}\,\widehat{\mu}_Y(t_j) = \frac{1}{n}(\sigma_X^2(t_j) + \sigma_\varepsilon^2) \xrightarrow{n\to\infty} 0.$$

The discrete estimator of the covariance function can be computed for $k,l = 1,\ldots,m$ as follows. Let

$$\widehat{\mathbf{cov}}_X(t_k,t_l) = \frac{1}{n-1}\sum_{i=1}^{n} (X_i(t_k) - \widehat{\mu}_X(t_k))\,(X_i(t_l) - \widehat{\mu}_X(t_l)),$$

$$\widehat{\mathbf{cov}}_Y(t_k,t_l) = \frac{1}{n-1}\sum_{i=1}^{n} (Y_{i,k} - \widehat{\mu}_Y(t_k))\,(Y_{i,l} - \widehat{\mu}_Y(t_l)).$$

Exercise 9.1. Show that $\mathbf{E}\,\widehat{\mathbf{cov}}_Y(t_k,t_l) = \mathbf{cov}_X(t_k,t_l) + \delta_{k,l}\sigma_\varepsilon^2 + O(1/n)$, where $\delta_{k,\ell}$ denotes Kronecker's delta.

The result of Exercise 9.1 implies that this estimator for the covariance function is asymptotically unbiased only in the first case.

Remark 9.1. A covariance function $C : \mathcal{T}^2 \to \mathbb{R}$ is a positive semidefinite kernel function (see Sect. 9.2.1 below). Therefore, for each $m \in \mathbb{N}$ and for all the points $t_1,\ldots,t_m \in \mathcal{T}$, the matrix $C = (C(t_k,t_l))_{k,l=1}^{m}$ is positive semidefinite as well. The same holds for the empirical covariance function, and for the matrix $\left(\widehat{C}(t_i,t_j)\right) \in \mathbb{R}^{m\times m}$. However, if we reconstruct a covariance function from the latter using some smoothing approach we are no longer sure that this property still holds.

9.1.2 Smoothing Techniques

A common problem in statistics is the following: Let us assume that the data $(y_1, t_1), \ldots, (y_k, t_k)$ are given, representing the values of an unknown, smooth function m at points t_i $(i = 1, \ldots, k)$. Furthermore, the values y_i contain some additive noise. We are then in the framework of a nonparametric regression model

$$Y_i = m(t_i) + \varepsilon_i, \qquad i = 1, \ldots, k. \tag{9.8}$$

The design points t_i may be *deterministic* (if the experimenter can control the design), in this case we assume that $\mathbf{E}\,\varepsilon_i = 0$ and $\mathbf{E}(\varepsilon_i \varepsilon_\ell) = \sigma^2 \delta_{i,\ell}$ in particular $\mathbf{E} Y_i = m(t_i)$, or *random*, then the regression curve $m(\cdot)$ is the conditional expectation $m(t) = \mathbf{E}(Y \mid X = t)$, with $\varepsilon = Y - \mathbf{E}(Y \mid X)$ and hence $\mathbf{E}(\varepsilon \mid X) = \mathbf{E}\varepsilon = 0$ and $\mathbf{var}(\varepsilon \mid X = t) = \sigma^2(t)$ and typically it is assumed that $(Y_i, X_i)_{i=1}^n$ are iid random variables and $\mathbf{E}(Y^2) < \infty$. In either case it is assumed that $m(\cdot)$ is a smooth function on some interval I and the task is to estimate $m(\cdot)$ from the data.

In functional data analysis, too, the first step is generally to reconstruct smooth functions (paths) from underlying discrete observations. We will now discuss some possible approaches to handle this situation.

9.1.2.1 Kernel Estimators

A first classical approach to estimate a function f or its derivatives $f^{(v)}$ is based on kernel approximation operators.

Definition 9.1. For any given $\mu, v \in \mathbb{N}$, $\mu > v$, the function $K : \mathbb{R} \to \mathbb{R}$ is called a *kernel function of order* (v, μ) if $K(x) \in L_1(\mathbb{R}) \cap L_2(\mathbb{R})$, $x^\mu K(x) \in L_1(\mathbb{R})$ and the following equations hold for $\ell \leq \mu$:

$$\int_{\mathbb{R}} x^\ell K(x) dx = \begin{cases} 0, & \text{if } 0 \leq \ell \neq v < \mu, \\ (-1)^v v!, & \text{if } \ell = v, \\ c_\mu \in \mathbb{R} \setminus \{0\}, & \text{if } \ell = \mu, \end{cases}$$

The index v refers to the derivative one is interested in, μ is connected with the smoothness of the function to be considered. Though there are some kernel functions defined on the whole real line, the most common ones are polynomials with compact support $[-1, 1]$. Let us consider some examples, see also Fig. 9.1.

1. Kernel functions of order $(0, 2)$

   a. $K(x) = \frac{1}{2} \mathbf{1}_{[-1,1]}(x)$ (uniform kernel);
   b. $K(x) = \frac{3}{4}(1 - x^2)\mathbf{1}_{[-1,1]}(x)$ (Epanechnikov kernel);
   c. $K(x) = \frac{15}{16}(1 - x^2)^2 \mathbf{1}_{[-1,1]}(x)$ (biweight kernel);
   d. $K(x) = \frac{1}{\sqrt{2\pi}} e^{-\frac{1}{2}x^2}$, the density function of the standard normal distribution.

2. Kernel function of order (1,3): a. $K(x) = \frac{15}{4}(x^3 - x)\mathbf{1}_{[-1,1]}(x)$

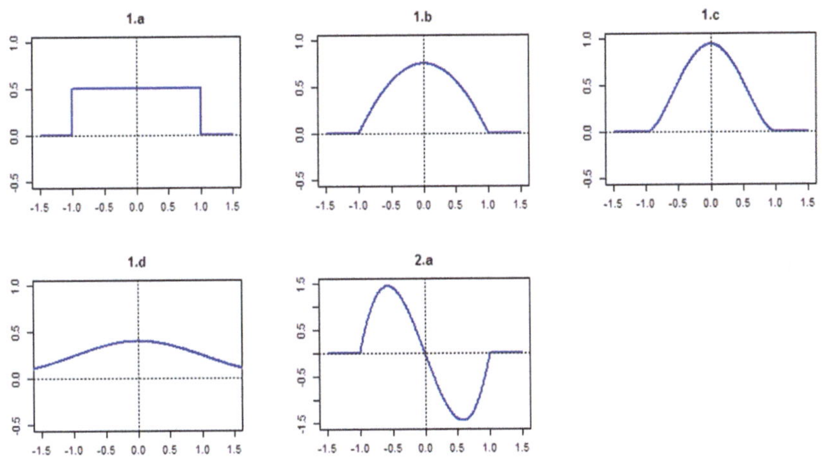

Fig. 9.1 Examples of kernel functions

Exercise 9.2. Consider the function $f(x) = \frac{15}{32}(3 + ax^2 + 7x^4)\mathbf{1}_{[-1,1]}(x)$. For which values of a is it a kernel function? What is its order (v,μ) then?

Given an appropriate real function $f \in L_2(\mathbb{R})$, the use of kernel functions allows to obtain an approximation for the function itself and its derivatives, sharing smoothness properties of the kernel function. Using a kernel function of order (v,μ) and some parameter $h > 0$ (the bandwidth) we consider the approximation

$$\bar{f}^{(v)}(x;h) = \frac{1}{h^{v+1}}\int_{\mathbb{R}} K\left(\frac{x-t}{h}\right) f(t)\, dt$$

for the v−th derivative $f^{(v)}$.

Example 9.1. Let $(X_1,\ldots,X_n)$ be a sample of size n for a distribution with density function f being μ−times differentiable. Then one can estimate the density function by $\hat{f}_n$ or its v−th derivative, using a kernel estimator of order $(0,\mu)$ or (v,μ) resp., in the following way:

$$\hat{f}_n^{(v)}(x;h) = \frac{1}{h^{v+1}}\int_{\mathbb{R}} K\left(\frac{x-t}{h}\right) d\hat{F}_n(t) = \frac{1}{nh^{v+1}}\sum_{j=1}^{n} K\left(\frac{x-X_j}{h}\right).$$

Lemma 9.1. *Let K be a kernel function of order (v,μ) with compact support and let $f \in C_\mu\left((x_0 - \varepsilon, x_0 + \varepsilon)\right)$ for some $\varepsilon > 0$ (i.e.: f has μ continuous derivatives in some ε-neighbourhood of x_0). Then, $\mathbf{E}\,\hat{f}_n^{(v)}(x;h) = \bar{f}^{(v)}(x;h)$ and, as $h \to 0+$,*

$$\bar{f}^{(v)}(x;h) = f^{(v)}(x) + c_\mu \frac{f^{(\mu)}(x)}{\mu!}(-1)^\mu h^{\mu-v} + o(h^{\mu-v}). \tag{9.9}$$

For details, see e.g. Sect. 4.1 in [290], or Chap. 3.3 in [369]. As a consequence, we see that the estimator $\hat{f}_n^{(v)}(x;h)$ is biased but asymptotically unbiased as $h \to 0+$. However, the variance is increasing with $h \to 0+$ since as $n \to \infty$ and h is depending on n such that $h = h_n$ with $h_n \to 0+$ but $nh^{2v+1} \to 0$, then $\mathbf{var}\hat{f}(x;h) \sim c/nh^{2v+1}$ with an explicit constant $c > 0$. In practice, the difficulty is to choose h appropriately.

Let us go back to the regression problem (9.8) stated at the beginning of this section. A possible way to deal with it is to use kernel-based estimators. These are linear estimators with respect to the data, having the form $\sum_{j=1}^n W_{j,n}(x)Y_j$, and the weights $W_{j,n}$ depend on some kernel function and a smoothing parameter. Let us focus on two examples.

The Gasser-Müller estimator. Suppose that $0 \le t_1 < \ldots < t_n \le 1$ are fixed time points, known a priori. Put $s_0 = 0$, $s_n = 1$ and $s_j = (t_{j+1}+t_j)/2$ for $j = 1, \ldots, n-1$. By choosing the weights

$$W_{j,n}(x;h) = \frac{1}{h} \int_{s_{j-1}}^{s_j} K\left(\frac{x-t}{h}\right) dt,$$

with a kernel function K of order $(0,\mu)$, one obtains the so-called *Gasser-Müller estimator* for the regression function m, i.e.

$$\hat{m}_n(x;h) = \sum_{j=1}^n \frac{1}{h} \int_{s_{j-1}}^{s_j} K\left(\frac{x-t}{h}\right) dt \cdot Y_j. \tag{9.10}$$

The Nadaraya-Watson estimator. We will explain this estimator in the case of a random design. Let the recording time points X_j be i.i.d., where. $X_1 \sim \mathrm{Unif}(0,1)$. Recall that, in this case, $m(x) = \mathbf{E}(Y \mid X = x)$. Then, taking the weights

$$W_{j,n} = \frac{K\left(\frac{x-X_j}{h}\right)}{\sum_{v=1}^n K\left(\frac{x-X_v}{h}\right)},$$

leads to the *Nadaraya-Watson estimator*

$$\hat{m}_n(x;h) = \frac{\sum_{j=1}^n K\left(\frac{x-X_j}{h}\right) Y_j}{\sum_{v=1}^n K\left(\frac{x-X_v}{h}\right)}, \tag{9.11}$$

see also Fig. 9.2.

For both estimators the bandwidth $h > 0$ has to be chosen appropriately.

Theorem 9.1 ([290]). *Let K be a kernel function of order $(0,2)$ and m a real-valued function in $C_2([0,1])$ (twice continuously differentiable on the compact interval $[0,1]$). Suppose to have n equidistant deterministic recording times on the inter-*

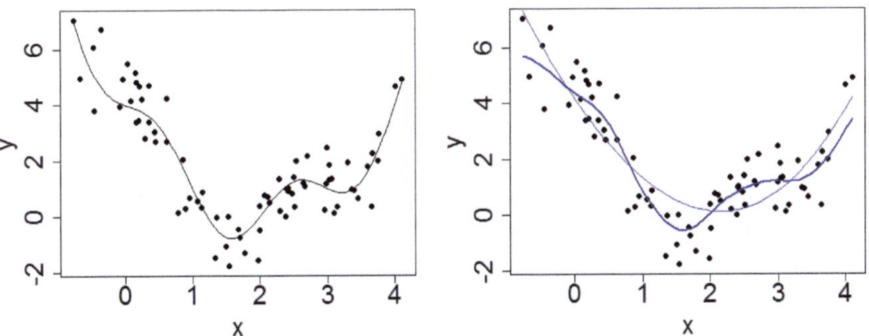

Fig. 9.2 Left panel: Original curve. Right panel: Nadaraya-Watson kernel estimator with Gaussian kernel and bandwidth $h = 0.75$ (thick line) and quadratic least squares regression line (thin)

val $[0, 1]$ and a positive null-sequence $h = (h_n)_{n \in \mathbb{N}}$ such that $nh_n \to \infty$ as $n \to \infty$. Let $\hat{m}_n$ be the Gasser-Müller estimator. Then, for $x \in (0, 1)$ and $n \to \infty$,

$$\mathbf{E}\,\hat{m}_n(x; h_n) = m(x) + \frac{1}{2}m''(x) \cdot c_2 \cdot h_n^2 + o\left(h_n^2\right) + O\left(\frac{1}{n}\right),$$

$$\mathbf{var}\,\hat{m}_n(x; h_n) = \sigma_\varepsilon^2 \frac{\|K\|_2^2}{nh_n} + o\left(\frac{1}{nh_n}\right).$$

It is also possible to obtain the corresponding expansion of the expectation and variance of the Nadaraya-Watson estimator.

Both estimators depend heavily on a suitable choice of the bandwidth. Furthermore, both estimators are in general not consistent at the boundary points $x \in \{0, 1\}$ and for a finite sample size one has to pay attention in a neighborhood of these points.

9.1.2.2 Local Polynomials

Linear regression is the simplest regression model. Then, the function m in problem (9.8) is assumed to be a straight line $\beta_0 + \beta_1 x$, and the unkwown coefficients β_0 and β_1 can be estimated for instance via the method of least squares. Since any smooth function looks locally almost like a straight line, we can use linear regression but just locally in almost every situation.

Suppose we have a random design, $X_1, \ldots, X_n$, with iid measuring points $X_i \sim f$ and $\mathrm{supp}(f) = I$. Then we can perform local linear regression by using a kernel K of order $(0, 2)$ with a bandwidth $h > 0$ to select and weight the local observations around each point $x \in I$, where we solve the minimization problem

$$\text{argmin}\left\{\beta_0,\beta_1 \in \mathbb{R}: \sum_{j=1}^{n}(Y_j-\beta_0-\beta_1(x-X_j))^2 K\left(\frac{x-X_j}{h}\right)\right\},$$

and receive the estimators $\hat{\beta}_0(x;h)$ and $\hat{\beta}_1(x;h)$ which, again, depend on the bandwidth h. For fixed $x \in I$ we have the estimates $\hat{m}(x;h) = \hat{\beta}_0(x;h)$ and $\widehat{m'}(x;h) = -\hat{\beta}_1(x;h)$.

Another possibility is to use local polynomials of higher order than one, see Fig. 9.3. Then, one solves the following problem

$$\text{argmin}\left\{\beta_0,\ldots,\beta_p \in \mathbb{R}: \sum_{j=1}^{n}(Y_j-\beta_0-\beta_1(x-X_j)^1-\ldots\right.$$
$$\left.-\beta_p(x-X_j)^p)^2 K\left(\frac{x-X_j}{h}\right)\right\},$$

obtaining approximations for the first p derivatives of m (if they exist), by

$$\widehat{m^{(v)}}(x;h) = (-1)^v v! \widehat{\beta}_v(x;h) \qquad v=0,\ldots,p.$$

For a fixed $x \in I$ the previous minimization problem is a weighted least-squared minimization problem with weight matrix

$$W_x = \text{diag}\left(K\left(\frac{x-X_1}{h}\right),\ldots,K\left(\frac{x-X_n}{h}\right)\right) \in \mathbb{R}^{n\times n}$$

and design matrix

$$A_x = \begin{pmatrix} 1 & (x-X_1) & \ldots & (x-X_1)^p \\ \vdots & \vdots & & \vdots \\ 1 & (x-X_n) & \ldots & (x-X_n)^p \end{pmatrix} \in \mathbb{R}^{n\times(p+1)}.$$

Minimizing $(Y-A_x\beta)^\top W_x(Y-A_x\beta)$ with respect to the parameter vector $\beta \in \mathbb{R}^{p+1}$, one obtains the estimator

$$\hat{\beta} = (A_x^\top W_x A_x)^{-1} A_x^\top W_x Y,$$

if $A_x^\top W_x A_x$ is invertible. Therefore, the estimators of the first p derivatives of m are again linear in Y, as $\widehat{m^{(v)}}(x;h) = (-1)^v e_v^\top v! \hat{\beta} = e_v^\top S_x Y$, $v=0,\ldots,p$ with the corresponding matrix S_x. Let us take a closer look at the two most common cases.

Exercise 9.3. Show that, if $p=0$, then $\hat{\beta}_0(x;h)$ is the Nadaraya-Watson estimator given in (9.11). *Hint.* Show first that

$$(A_x^\top W_x A_x)^{-1} = \left(\sum_{i=1}^{n}K\left(\frac{x-X_i}{h}\right)\right)^{-1} \quad \text{and} \quad A_x^\top W_x Y = \sum_{i=1}^{n}K\left(\frac{x-X_i}{h}\right)Y_i.$$

Exercise 9.4. Show that if $p = 1$, then

$$\hat{m}(x;h) = \frac{1}{nh} \sum_{j=1}^{n} \frac{\left(s_{2,n}(x;h) - s_{1,n}(x;h)(x - X_j)\right) K\left(\frac{x - X_j}{h}\right) Y_j}{s_{2,n}(x;h) s_{0,n}(x;h) - s_{1,n}^2(x;h)},$$

where $s_{v,n}(x;h) = \frac{1}{nh} \sum_{j=1}^{n} (x - X_j)^v K\left(\frac{x - X_j}{h}\right)$ is the so-called v−th local empirical moment.

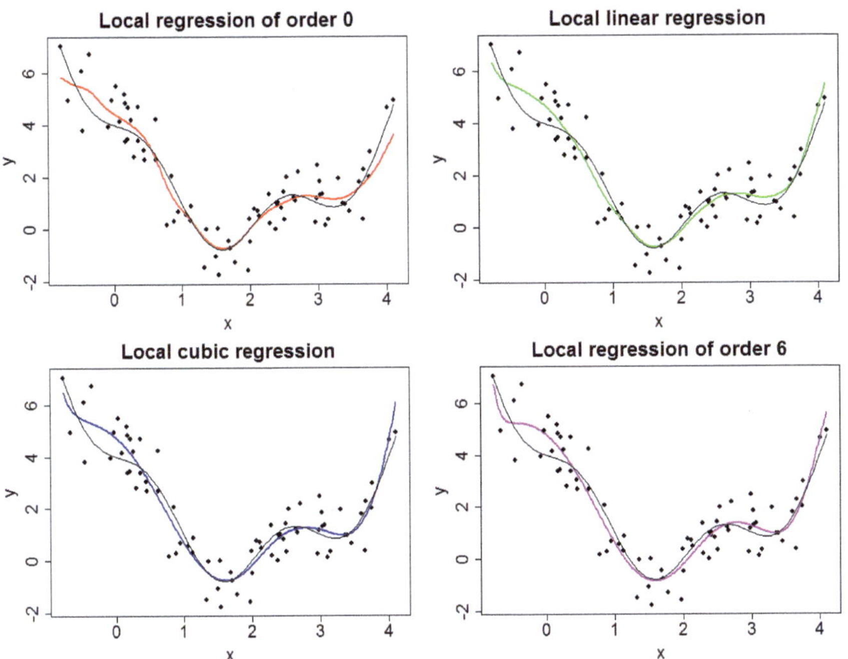

Fig. 9.3 Local polynomial regression line for different orders, compared with the actual function m (thin line). From top left to bottom right, the order of the polynomial regression is 0, 1, 3 and 6, while the bandwidth is $h = 0.2, 0.2, 0.4$ and 0.7. A Gaussian kernel was used for all the estimators.

Theorem 9.2 ([370]). *Let* $p = 1$, $f_{X,Y} \in C_2(\mathbb{R}^2)$, $\mathbf{E}\,|\,X\,|, \mathbf{E}\,|\,Y^2\,| < \infty$, *K be a kernel of order* $(0,2)$ *and* x_0 *such that* $f_X(x_0) > 0$. *Then with* h_n *as before it holds as* $n \to \infty$ *that*

$$\mathbf{E}(\hat{m}_n(x_0; h_n) \mid X_v = x_v, v = 1, \ldots, n) = m(x_0) + \frac{1}{2} m^{(2)}(x_0) \cdot c_2 \cdot h_n^2 + o\left(h_n^2\right),$$

$$\mathbf{var}(\hat{m}_n(x_0; h_n) \mid X_v = x_v, v = 1, \ldots, n) = \sigma^2(x_0) \frac{\|K\|_2^2}{nh_n f_X(x_0)} + o\left(\frac{1}{nh_n}\right).$$

Besides the simple idea they are based on, local polynomials present another big advantage: if the function f has a compact support I, the estimators they provide are consistent, even at the boundary points of I which is not the case for e.g. the kernel estimators presented above.

Note that the variance also depends on $f_X(x)^{-1}$, and so will the optimal value of a locally chosen optimal bandwidth h_n. The choice of the polynomial's degree p and more important of the bandwidth h are two main issues in the application of this approach.

Choice of p. A larger value of the polynomial order p allows for more flexibility, so that also strongly fluctuating (but smooth) functions can be well approximated. Moreover, it leads to better asymptotic results, provided h is chosen wisely, too. The estimators' bias is $O(h^{p+2})$ for an even p and $O(h^{p+1})$ if p is uneven. The constants involved grow as the degree p increases, this matters for finite sample size. Therefore, p is usually chosen as either 1, 2 or 3. Also, the bias for p uneven is better than the ones for p even. The choice of p (and of the kernel K) is, however, less important than the choice of the bandwidth h. As before, a large bandwidth leads to a large bias but smaller variance, and a small one to a small bias but larger variance.

Choice of h. This is the main practical challenge when working with local polynomials. The idea is to select a proper h by minimizing some error measure. Global measures of accuracy of $\hat{m}$ are, for instance, the mean integrated squared error, $\mathrm{MISE}(\hat{m}_n(\cdot;h) \mid X = x)$ or the asymptotic mean integrated squared error. Unfortunately, those depend also on the unknown function m and on its derivatives. That is why one uses some estimation techniques to approximate these quantities.

1. Cross-validation: $\hat{h}_n = \mathrm{argmin}_{h>0}\{CV(h)\}$, where

$$CV(h) = \frac{1}{n}\sum_{j=1}^{n}(Y_j - \hat{m}_{n,-j}(X_j;h))^2$$

   and $m_{n,-j}$ is the local polynomial estimator based on the observations without (X_j, Y_j). This procedure is time expensive, and it tends towards undersmoothing.
2. Plug-in estimation: instead of using the theoretical values in the $\mathrm{MISE}(h)$ or in the asymptotic $\mathrm{MISE}(h)$, one uses estimated values, see also Fig. 9.4. This procedure requires some caution, see Sect. 6.3 in [370] or Sect. 7.4 in [290] for a thorough discussion of the topic.

For more details, see, e.g., [116] or Chap. 5 in [370].

9.1.2.3 Orthogonal Series Expansion

Let us consider the Hilbert space $L_2(\mathcal{T})$ (with the usual scalar product) over a compact interval $\mathcal{T}$. A sequence $(\varphi_k)_{k=1}^{\infty} \in L_2(\mathcal{T})$ is called

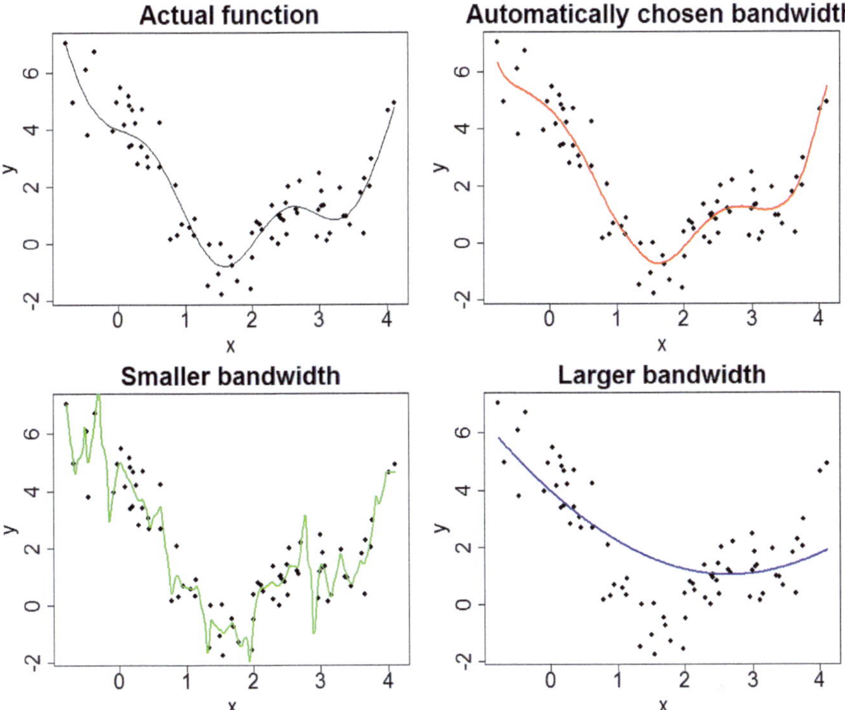

Fig. 9.4 Local linear regression with Gaussian kernel and different bandwidths. Top left: original function m. Top right: bandwidth selected by plug-in estimation ($h = 0.2141$). Bottom left: $h = 0.05$. Bottom right: $h = 1.50$

1. *complete* if each function $f \in L_2(\mathcal{T})$ can be approximated in the sense of the norm by a linear combination of the functions φ_k, and
2. *minimal* if no function from $(\varphi_k)_{k=1}^{\infty}$ can be approximated - in the sense of the norm - by the remaining ones, i.e, for every $m \in \mathbb{N}$,

$$\varphi_m \notin \overline{\mathrm{span}\{\varphi_k, k \neq m\}}.$$

Definition 9.2. A sequence $(\varphi_k)_{k=1}^{\infty} \subset L_2(\mathcal{T})$ is called a *Schauder basis* of $L_2(\mathcal{T})$ if, for every $f \in L_2(\mathcal{T})$, exists a unique sequence of scalars $(c_k)_{k=1}^{\infty}$ so that

$$\lim_{n \to \infty} \left\| f - \sum_{k=1}^{n} c_k \varphi_k \right\|_2 = 0.$$

In this case, we can write $f(x) = \sum_{k=1}^{\infty} c_k \varphi_k(x)$, but we note that in general we will only have L_2-convergence but not pointwise convergence. We can then approximate every function $f \in L_2(\mathcal{T})$ by "cutting" its Schauder basis expansion at some level ℓ, i.e.,

$$f(x) \simeq \sum_{k=1}^{\ell} c_k \varphi_k(x).$$

Note that every Schauder basis is complete and minimal, but not all complete and minimal sequences of functions in $L_2(\mathcal{T})$ are Schauder bases. Some examples of Schauder bases are

1. monomials on $\mathcal{T} = [a,b]$, $\varphi_k(t) = t^{k-1}, k \in \mathbb{N}$,
2. Fourier functions on $[0,T]$, $\varphi_k(t) = 1$, $\sin\left(\frac{2\pi kt}{T}\right)$, $\cos\left(\frac{2\pi kt}{T}\right)$, $k \in \mathbb{N}$,
3. spline bases,
4. wavelet bases.

More convenient are special classes of bases, so-called orthonormal ones, where for two functions φ_k, φ_ℓ we have that the inner product satisfies $\langle \varphi_k, \varphi_\ell \rangle = \delta_{k,\ell}$. Then, the following result holds.

Theorem 9.3. *Given an orthonormal basis* $(\varphi_k)_{k=1}^{\infty} \subset L_2(\mathcal{T})$, *we have for every function* $f \in L_2(\mathcal{T})$ *it holds that*

1. *if* $c_k = \langle f, \varphi_k \rangle$ *for* $k \in \mathbb{N}$, *then* $f = \sum_{k=1}^{\infty} c_k \varphi_k$ *in the* L_2-*sense,*
2. *for every* $\ell \in \mathbb{N}$

$$\left\| f - \sum_{k=1}^{\ell} c_k \varphi_k \right\|_2 \leq \left\| f - \sum_{k=1}^{\ell} d_k \varphi_k \right\|_2 \qquad \text{for all } d_1,\ldots,d_\ell \in \mathbb{R}.$$

Let us go back to the regression problem (9.8) stated at the beginning of Sect. 9.1.2. Given the data (y_i, t_i) for $i = 1,\ldots,n$, one would like to estimate the function m of the model $Y_i = m(t_i) + \varepsilon_i$, where ε_i is some noise with zero expectation and covariances $\mathbf{E}(\varepsilon_i \varepsilon_j) = \delta_{ij}\sigma_\varepsilon^2$. We can also suppose that the function m belongs to $L_2(\mathcal{T})$, so that, for a given orthonormal basis, $m(\cdot) = \sum_{k=1}^{\infty} c_k \varphi_k(\cdot)$ or, for large ℓ,

$$m(\cdot) \simeq \sum_{k=1}^{\ell} c_k \varphi_k(\cdot).$$

The orthonormal basis is given, as well as the value of ℓ (for the moment), while the coefficients $c = (c_1,\ldots,c_\ell)^\top$ are unknown. As stated in Theorem 9.3, the method of least squares is a clever way to determine c. We will consider the minimization problem

$$\operatorname{argmin} \left\{ c \in \mathbb{R}^\ell : \sum_{j=1}^{n} \left(Y_j - \sum_{k=1}^{\ell} c_k \varphi_k(t_j) \right)^2 = \|Y - \Phi c\|_2^2 \right\},$$

with $Y = (Y_1,\ldots,Y_n)^\top$ and the matrix $\Phi = (\varphi_k(t_j))_{j=1,k=1}^{n,\ell} \in \mathbb{R}^{n \times \ell}$, which plays the role of the design matrix in a typical regression problem. The solution, provided $\Phi^\top \Phi$ is regular, is then uniquely determined as

$$\hat{c} = (\Phi^\top \Phi)^{-1} \Phi^\top Y.$$

The vector $(\hat{m}(t_1), \ldots, \hat{m}(t_n))^\top$ can be represented as $\Phi\hat{c} = \Phi(\Phi^\top\Phi)^{-1}\Phi^\top Y =: SY$, where the matrix S is an orthogonal projection matrix with $\mathrm{trace}(S) = \ell$ expressing the degree of freedom in the estimator. The empirical residuals are $\hat{\varepsilon} = (I_n - S)Y$ and the residual variance can be estimated by $(1/n)\|\hat{\varepsilon}\|^2$. Furthermore we can estimate the whole function m by

$$\hat{m}(t) = \sum_{k=1}^{\ell} \hat{c}_k\, \varphi_k(t), \qquad t \in \mathcal{T}.$$

Suppose that the noises $\varepsilon_1, \ldots, \varepsilon_n$ are correlated and that we know the correlation matrix, $\mathbf{cov}\,\varepsilon = \Sigma_\varepsilon$. Then, if the matrix $\Phi^\top\Sigma_\varepsilon^{-1}\Phi$ is regular, the estimator $\hat{c} = (\Phi^\top\Sigma_\varepsilon\Phi)^{-1}\Phi^\top\Sigma_\varepsilon^{-1}Y$ is unbiased and has minimal variance among all linear estimators (which follows from the classical Gauss-Markov theorem). The matrix $\Phi(\Phi^\top\Sigma_\varepsilon\Phi)^{-1}\Phi^\top\Sigma_\varepsilon^{-1}$ is still an orthogonal projection but with respect to the appropriate inner product.

9.1.2.4 Method of Least-Squares with Penalty Term

If we would try to fit a smooth function trough a scatterplot minimizing the squared distance at the data points, i.e.

$$\sum_{i=1}^{n}(Y_i - m(t_i))^2,$$

we would get a solution interpolating each point. As the number of observations increases (and the distance between the observation times decreases), these interpolating functions will oscillate more and more, since they have to catch every observation point. To avoid this, one can discourage roughness by adding a penalty term to the original problem as, for instance, $\mathrm{Pen} = \int_T (m^{(2)}(t))^2 dt$. The new quantity to minimize is then

$$\mathrm{PENSSE}(\lambda) = \sum_{i=1}^{n}(Y_i - m(t_i))^2 + \lambda\,\mathrm{Pen}. \tag{9.12}$$

The parameter $\lambda > 0$ has to be chosen by the statistician. It controls the smoothness of the minimizing function m: if λ is close to zero, m is rougher, while it is smoother for larger values of λ. Hence λ is a smoothing parameter like the bandwidth in kernel estimators. The curve that minimizes the expression in (9.12) is a cubic spline. In practice one uses order four B-splines basis functions $(B_k(.))$ and minimizes with respect to the coefficients. Again we approximate by $f(t) = \sum_{k=1}^{\ell} c_k \varphi_k(t)$, with $\varphi_k(t) = B_k(t)$. Minimizing $\sum_i (Y_i - f(t_i))^2 + \lambda\,\mathrm{Pen}_2$ leads to the task to minimize

$$(Y - \Phi c)^\top (Y - \Phi c) + \lambda c^\top R c,$$

where $R_{ij} = \int_T B_i^{(2)}(t) B_j^{(2)}(t)\, dt$. The solution is given by

$$\hat{c} = (\Phi^\top \Phi + \lambda lR)^{-1} \Phi Y.$$

Furthermore we have

$$\hat{Y} = \Phi (\Phi^\top \Phi + \lambda R)^{-1} \Phi Y = S_R Y,$$

which is a linear operation again. But S_R is no longer an orthogonal projection since $S_R S_R \neq S_R$. For further details see e.g. [324] and the references therein.

9.2 Principal Components Analysis (PCA)

9.2.1 PCA for Multivariate Data

Let us consider the multivariate regression problem

$$Y_i = \langle (X_1^{(i)}, \dots, X_p^{(i)})^\top, \beta \rangle + \varepsilon_i \qquad i = 1 \dots n \tag{9.13}$$

with the unknown parameter vector $\beta \in \mathbb{R}^p$, the covariables $X^{(1)}, \dots, X^{(n)} \in \mathbb{R}^p$ building the design matrix $A \in \mathbb{R}^{n \times p}$ in the linear model $Y = A\beta + \varepsilon$ and with some noise vector $\varepsilon = (\varepsilon_1, \dots, \varepsilon_n)^\top$.

Suppose that the dimension of the covariable vectors $X^{(i)}$ is big in comparison to the sample size. A problem like that is hard to handle, in particular if there are more unknown parameters $\beta_1, \dots, \beta_p$ than sample values. In consequence, one has to reduce the dimension of $X^{(i)}$ in a clever way. If we simply would 'take less covariates', a lot of information could get lost. Therefore, we will look for linear combinations of the covariables containing as much information as possible or, in other terms, having maximal variance. This task can be seen as the following maximization problem

$$\begin{cases} \mathbf{var}(\rho^\top X) = \rho^\top (\mathbf{cov} X) \rho = \rho^\top C \rho \to \max \\ \|\rho\| = 1. \end{cases} \tag{9.14}$$

The constraint $\mid \rho \mid = 1$ ensures the existence of a solution, since we are maximizing a continuous function over the unit sphere $S_1(o) \in \mathbb{R}^p$, which is compact, and the solution is unique up to the sign. The maximizing vector, ρ_1, represents the coefficients of the linear combination of covariates with maximal variance.

By maximizing the same function, $\mathbf{var}(\rho^\top X)$, over the set $S_1(o) \cap \rho_1^\perp$, where $\rho_1^\perp = \{\rho \in \mathbb{R}^p : \rho \perp \rho_1\}$, we obtain the "second best" linear combination of covariates, the vector ρ_2. Furthermore, the random variable $\rho_2^\top X$ is, by construction, uncorrelated to $\rho_1^\top X$. In the next step we maximize over the set $S_1(o) \cap \text{span}\{\rho_1, \rho_2\}^\perp$ etc.. If we proceed we obtain for some $r \in \mathbb{N}$ that in step $r + 1 \leq p$ the maximum is zero for the first time. Then $(\text{span}(\rho_1, \dots, \rho_r))^\perp = \ker(C)$ and we have an orthonormal basis $\{\rho_1, \dots, \rho_r\}$ of $\ker(C)^\perp$. In the special case that C has full rank we end with $r = p$ and an orthonormal basis $\{\rho_1, \dots, \rho_p\}$ of $\mathbb{R}^p$.

This task is strictly connected with solving the eigenvalue problem

$$Cy = \lambda y, \quad \lambda \in \mathbb{C}, \, y \in \mathbb{C}^p,$$

where C is, as before, the covariance matrix of X, and therefore symmetric and positive semidefinite. Thus the eigenvalues and eigenvectors are real and the eigenvalues are non-negative, so that it is possible to order them as: $\lambda_1 \geq \ldots \geq \lambda_p \geq 0$. If C has rank r, the corresponding eigenvectors $y_1, \ldots, y_r$ are without any loss of generality an orthonormal basis of $\ker(C)^\perp$. In case the matrix C has full rank, i.e., $r = p$ we have an orthonormal basis of $\mathbb{R}^p$. Then we can decompose every vector $x \in \mathbb{R}^p$ as $x = \sum_{i=1}^p \langle x, y_i \rangle y_i = \sum_{i=1}^p c_i y_i$. Moreover, every unit vector $x \in \mathbb{R}^p$ has coefficients in this expansion with norm one, i.e., $\sum_{i=1}^p c_i^2 = 1$. Let us go back to the maximization problem (9.14), which is equivalent to

$$\begin{cases} \sum_{i,j=1}^p c_i y_i^\top C c_j y_j = \sum_{i=1}^p c_i^2 \lambda_i \to \max, \\ \sum_{i=1}^p c_i^2 = 1. \end{cases}$$

Since the eigenvalues of C are ordered from the biggest to the smallest, the maximum is attained for $|c_1| = 1, c_2 = \ldots = c_p = 0$. The maximizing vector ρ_1 is therefore equal (modulo the sign) to the first eigenvector y_1. The second maximization step would identify ρ_2 with y_2 and so on.

A random vector X taking values in $\mathbb{R}^p$, with $\mathbf{E}X = \mu$ and $\mathbf{cov}\,X = C$ can then be written as

$$X = \sum_{v=1}^p C_v \rho_v + \mu.$$

The coefficients $C_v = \langle X - \mu, \rho_v \rangle$ are called the *principal components* of the random variable X. They are random variables themselves, with the following moment property. For any $v, k = 1, \ldots, p$, it holds that

$$\begin{aligned} \mathbf{E}\,C_v &= 0, \\ \mathbf{var}\,C_v &= \mathbf{E}(\rho_v^\top (X - \mu)(X - \mu)^\top \rho_v) = \rho_v^\top C \rho_v = \lambda_v, \\ \mathbf{cov}(C_v, C_k) &= \mathbf{E}(\rho_v^\top (X - \mu)(X - \mu)^\top \rho_k) = \rho_v^\top C \rho_k = 0, \qquad v \neq k. \end{aligned}$$

By taking $\ell < p$ one obtains an approximation of X denoted by $X_\ell = \sum_{v=1}^\ell C_v \rho_v + \mu$. The error made by this approximation is then minimal among all orthonormal expansions of order ℓ, and it equals:

$$\mathbf{E}\left(\|X - X_\ell\|^2 \right) = \mathbf{E}\left(\left\| \sum_{v=\ell+1}^p C_v \rho_v \right\|^2 \right) = \sum_{v=\ell+1}^p \mathbf{E}(C_v^2) = \sum_{v=\ell+1}^p \lambda_v.$$

Usually, one is satisfied having covered the information up to a small percentage α as e.g. $\alpha = 5\%$ and then ℓ is chosen as the smallest value ℓ such that $\sum_{v=\ell+1}^p \lambda_v \leq 0.05 \sum_{v=1}^p \lambda_v$.

9.2.2 PCA for Functional Data

Let us move back to the framework anticipated in Sect. 9.1.1, where we considered a stochastic process $\{X(t), t \in \mathcal{T}\}$ on a compact set $\mathcal{T}$ such that $\int_\mathcal{T} \mathbf{E}(X(t)^2)dt = \kappa < \infty$. This means in particular that for $s, t \in \mathcal{T}$, the following moment functions exist:

$$\mu_X(t) = \mathbf{E}X(t),$$
$$C(s,t) = \mathbf{cov}_X(s,t).$$

We also suppose that these functions are continuous on their support. First, we will focus on the one dimensional case, where $\mathcal{T}$ is a compact interval. Later on in this section we will discuss the multidimensional case.

Let us consider the Hilbert space $L_2(\mathcal{T})$ (with the usual scalar product), and the operator $\mathcal{C}$ on $L_2(\mathcal{T})$ that maps f into $\int_\mathcal{T} C(\cdot,t)f(t)dt$, i.e., $\mathcal{C} \circ f(s) = \int_\mathcal{T} C(s,t)f(t)dt$ for $s \in \mathcal{T}$.

Theorem 9.4. *The operator $\mathcal{C}$ has the following properties.*

1. *$\mathcal{C}$ is a linear, self-adjoint, non negative and continuous operator from $L_2(\mathcal{T})$ onto itself.*
2. *As kernel operator, $\mathcal{C}$ is a Hilbert-Schmidt operator (and therefore compact).*

Proof. Without loss of generality we assume that $\mathbf{E}X(t) \equiv 0$. The Cauchy- Schwarz inequality implies that

$$\int_T \left(\int_T C(s,t)f(t)dt \right)^2 ds \leq \int_T \left(\int_T \sqrt{\mathbf{E}(X_s^2)\mathbf{E}(X_t^2)}f(t)dt \right)^2 ds$$
$$\leq \int_T \mathbf{E}(X_s^2)ds \int_T \mathbf{E}(X_t^2)dt \int_T f^2(v)dv$$
$$\leq \kappa^2 \int_T f^2(v)dv$$
$$\leq \kappa^2 \|f\|^2.$$

Thus, $\mathcal{C}$ is an endomorphism on $L_2(\mathcal{T})$, and it is bounded. Moreover, $\mathcal{C}$ is obviously linear and, being bounded, it is continuous. It is also self-adjoint (because of the symmetry of C), where

$$\langle g, \mathcal{C} \circ f \rangle = \int_T g(s) \int_T C(s,t)f(t)dt\,ds = \int_T \int_T g(s)C(t,s)f(t)dt\,ds$$
$$= \langle \mathcal{C} \circ g, f \rangle.$$

Furthermore, $\mathcal{C}$ is non-negative, as for all $f \in L_2(\mathcal{T})$,

$$\langle f, \mathcal{C} \circ f \rangle = \int_T f(s) \int_T C(s,t)f(t)dt\,ds = \mathbf{E}\left(\left(\int_T X(t)f(t)dt \right)^2 \right) \geq 0.$$

Since C is uniformly continuous on $\mathcal{T} \times \mathcal{T}$, for every $\varepsilon > 0$ there is a $\delta > 0$ such that for $s_1, s_2 \in \mathcal{T}$ with $|s_1 - s_2| \leq \delta$ it holds that $|C(s_1,t) - C(s_2,t)| \leq \varepsilon$ for all $t \in \mathcal{T}$. Let $F \subset L_2(\mathcal{T})$ be the family of functions f with $|f| = 1$. Then the Cauchy-Schwartz inequality impies that, for any $f \in F$ and $\varepsilon > 0$,

$$|\mathcal{C}(f)(s_1) - \mathcal{C}(f)(s_2)| \leq \int_T |(C(s_1,t) - C(s_2,t))f(t)| \, dt$$

$$\leq |f| \left(\int_T |C(s_1,t) - C(s_2,t)|^2 dt \right)^{1/2}$$

$$\leq k\varepsilon,$$

provided that $|s_1 - s_2| \leq \delta$, where k is some constant. Thus, $\mathcal{C}(F)$ is uniformly equicontinuous. The Arzelà-Ascoli theorem ensures then that $\mathcal{C}$ is a compact operator, i.e. for every bounded sequence $\{g_n\} \subset L_2(\mathcal{T})$, the sequence of the images $\{\mathcal{C} \circ g_n\}$ has a convergent sub-sequence. $\qquad\square$

Note that a compact operator between Hilbert spaces is such that for every sequence $\{f_n\}_{n \in \mathbb{N}}$ in the given space such that $\|f_n\| \leq 1$ for all $n \in \mathbb{N}$, the sequence of the images $\{\mathcal{C} \circ f_n\}$ has a convergent subsequence.

Theorem 9.5 (Spectral theorem, see e.g. [80]). *Let C be defined as above. Consider the eigenequation*

$$\mathcal{C} \circ f = \lambda f.$$

Then, the eigenvalues are real, discrete and non negative, (so it is possible to write $\lambda_1 \geq \lambda_2 \geq \dots \geq 0$). There are countably many eigenvalues but their sum is finite. Moreover, the eigenspaces corresponding to different eigenvalues are orthogonal one to another and, as long as the corresponding eigenvalues are not equal to zero, the eigenspaces are finite-dimensional. Therefore, there is a basis $\{\varphi_n\}$ of $ker(\mathcal{C})^\perp$ consisting of orthonormal eigenfunctions corresponding to the positive eigenvalues of C.

Let f be in $L_2(\mathcal{T})$. Then, if c_v denotes the projection of f onto the eigenspace generated by φ_v, we have that

$$(\mathcal{C} \circ f)(s) \overset{L_2(\mathcal{T})}{=} \sum_{v=1}^{\infty} \lambda_v c_v \varphi_v(s). \qquad (9.15)$$

If C is such that $\int_T C^2(s,t)dt = M < \infty$ on $\mathcal{T}$, we also have pointwise and uniform convergence on the right-hand side of (9.15). Moreover, every function $f \in \ker(\mathcal{C})^\perp$ can be decomposed (again, in the $L_2(\mathcal{T})$ sense) as $f(\cdot) = \sum_{v=1}^{\infty} c_v \varphi_v(\cdot)$.

A stochastic process $\{X(t), t \in \mathcal{T}\}$ as considered at the beginning of this section with moments $\mu(t) = \mathbf{E}X(t)$ and $C(s,t) = \mathbf{cov}(X(s), X(t))$ can be decomposed according to the eigenfunctions of its covariance function, i.e.,

$$X(t) = \mu(t) + \sum_{v=1}^{\infty} \xi_v \varphi_v(t). \qquad (9.16)$$

The coefficients ξ_v in this expansion are computed, as usual, as $\xi_v = \langle X - \mu, \varphi_v \rangle_{L_2(\mathcal{T})}$. Note that the ξ_v are random variables with $\mathbf{E}\xi_v = 0$. As for their covariance structure, we have

$$
\begin{aligned}
\mathbf{cov}(\xi_v, \xi_\mu) = \mathbf{E}(\xi_v \xi_\mu) &= \mathbf{E}\left(\int_T \int_T (X(t) - \mu(t))(X(s) - \mu(s))\varphi_v(t)\varphi_\mu(s)\,dt\,ds \right) \\
&= \int_T \int_T C(s,t)\varphi_v(t)\varphi_\mu(s)\,dt\,ds \\
&= \int_T \lambda_v \varphi_v(t)\varphi_\mu(t)\,dt = \lambda_v \delta_{v,\mu}.
\end{aligned}
$$

Approximating the process $\{X(t)\}$ by its ℓ-truncated expansion, we get that

$$
X(t) \simeq \mu(t) + \sum_{v=1}^{\ell} \xi_v \varphi_v(t) =: X_\ell(t).
$$

The error $\|X - X_\ell\|$ is minimal over all the orthonormal basis expansions truncated at the level ℓ.

Exercise 9.5. Show that

$$
\mathbf{E}\left(\|X(t) - X_\ell(t)\|^2 \right) = \sum_{v=\ell+1}^{\infty} \lambda_v.
$$

Since the result of Exercise 9.5 holds for $d = 0$ as well, we also know that

$$
\mathbf{E}\left(\|X(t) - \mu(t)\|^2 \right) = \int_T \mathbf{var}\, X(t)\,dt = \sum_{v=1}^{\infty} \lambda_v.
$$

Note that, as in the multivariate case, the eigenfunctions φ_v are defined only up to the sign.

Example 9.2. Let $\{W(t), t \in [0,1]\}$ be a Wiener process. Then, for $s,t \in [0,1]$, $\mathbf{E}W(t) \equiv 0$, while $\mathbf{cov}(W(s), W(t)) = \min\{s,t\}$. In this very special case, it is possible to explicitly compute eigenfunctions and eigenvalues of the operator $C(s,t)$. The eigenequation is given by

$$
\int_0^1 \min\{s,t\}\varphi(s)\,ds = \lambda \varphi(t), \qquad t \in [0,1],
$$

and if we differentiate this equation twice we obtain the following differential equation of second order with constant coefficients:

$$
\lambda \varphi^{(2)}(t) = -\varphi(t), \tag{9.17}
$$

whose general solution is well known. Namely, $\varphi(t) = c_1 \sin \frac{t}{\sqrt{\lambda}} + c_2 \cos \frac{t}{\sqrt{\lambda}}$.

Exercise 9.6. Show that the solutions of (9.17) statisfying the initial conditions and the fact that we are looking for eigenfunctions with unitary norm (see Fig. 9.5) are given by

$$\lambda_v = \frac{4}{\pi^2 (2v-1)^2},$$

$$\varphi_v(t) = \sqrt{2}\sin\left((2v-1)\frac{\pi}{2}t\right), \qquad v = 1,2,\ldots$$

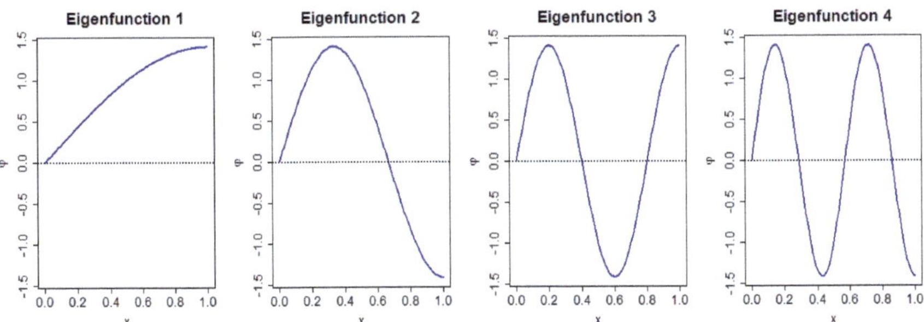

Fig. 9.5 First four eigenfunctions of $\mathcal{C}_W(s,t)$

Note that in Exercise 9.6 and in Fig. 9.5, as well, we have picked the eigenfunctions φ_v so that they start positive. Thus, the Wiener process can be defined as the following infinite sum (pointwise a.s. convergence follows from Kolmogorov's citerion, see e.g. [160], Thm.6.5.2, but even uniform convergence in $[0,1]$ holds true, see [181])

$$W(t) = \sum_{v=1}^{\infty} \xi_v \sqrt{2}\sin\left((2v-1)\frac{\pi}{2}t\right).$$

Usually, one only knows that the scores ξ_v, $v \in \mathbb{N}$, have expectation equal to zero, variance equal to λ_v (the v-th eigenvalue of the covariance matrix), and are uncorrelated. In this special case, we also know that they are normally distributed as

$$\xi_v \sim N(0, \lambda_v) \qquad v = 1,2,\ldots$$

and independent. This observation is useful, for instance, to simulate realizations of a Wiener process. Then, chosen a certain truncation level ℓ for the approximation, one just has to generate for $v = 1,\ldots\ell$ the random variables ξ_v (which have a known distribution) to obtain the weights of the corresponding eigenfunctions (see Fig 9.6). Note that this is exactly the reverse of what we usually do. The typical situation is: given n independent realizations of a stochastic process X, which we do not know a priori, we look for the eigenfunctions and eigenvalues of the (estimated) covariance function of X. This example is peculiar also for another reason: the paths of a Wiener process are nowhere differentiable, and we approximate them using very smooth functions. That is why we need so many eigenfunctions to obtain something which "looks like a Wiener process". We will soon see how to extend this procedure to the

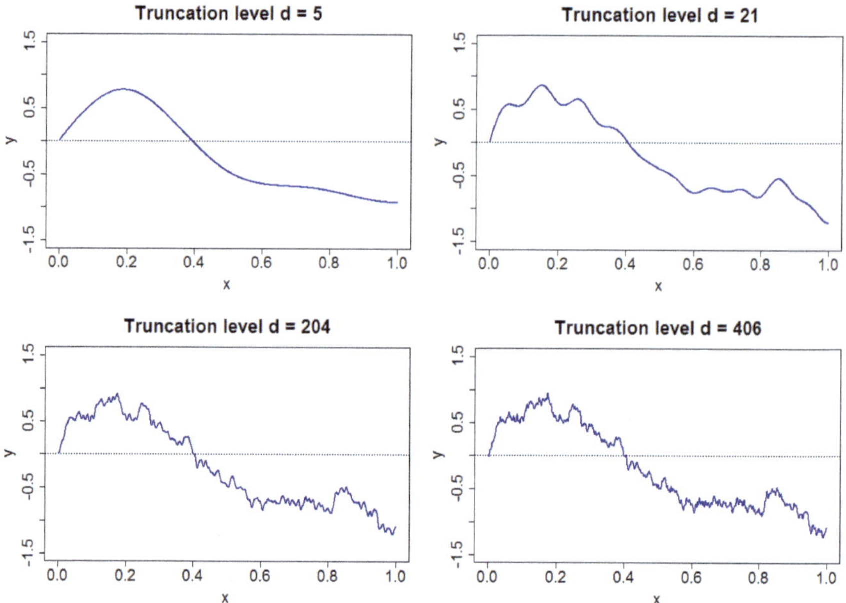

Fig. 9.6 Simulations of the paths W_ℓ, representing the truncation at level ℓ of a Wiener process realization for $\ell = 5, 21, 204, 406$, respectively.

$d-$ dimensional case, and more in detail, to the $2-$dimensional case, see Example 9.3 below.

Since we know both the mean integrated squared error for a truncation X_ℓ and the mean variance of X, we obtain by $\frac{\sum_{\nu=\ell+1}^\infty \lambda_\nu}{\sum_{\mu=1}^\infty \lambda_\mu}$ a measure of the "relative" error committed by truncating to the level ℓ. Moreover, $\lambda_\nu / \sum_{\mu=1}^\infty \lambda_\mu$ tells us which percentage of the expected variance the $\nu-$th eigenfunction could explain. For instance, the eigenfunctions from Fig. 9.5 account for 81.057%, 9.006%, 3.242% and 1.654% of the total variation, respectively.

On the other hand, the "relative" mean squared integrated errors for the paths in Fig. 9.6 are approssimately: 4.040% ($< 5\%$), 0.965% ($< 1\%$), 0.099% ($< 0.1\%$) and 0.0499% ($< 0.05\%$).

We can apply exactly the same reasoning concerning PCA in a higher-dimensional setting. Suppose that the process $\{X(t)\}$ is indexed by a compact set $\mathcal{T}$ in $\mathbb{R}^d$, with $d > 1$. It is still possible to consider the Hilbert space $L_2(\mathcal{T})$ with the usual scalar product $\langle f, g \rangle = \int_{\mathcal{T}} f(t)g(t)\,dt$, given by the d-dimensional integral.

Condition (9.1) ensures the existence of the covariance function of $\{X(t)\}$, so that it is still possible to define the covariance operator $\mathcal{C}$, which has all the properties states in Theorem 9.4. Then, the spectral theorem (see Theorem 9.5) applies, and we can obtain eigenvalues and eigenspaces of $\mathcal{C}$

The basic idea thus does not change, but as the dimension d increases, the eigenequation becomes more and more difficult to solve. Let us go back to Example 9.2, and see what happens if $d = 2$.

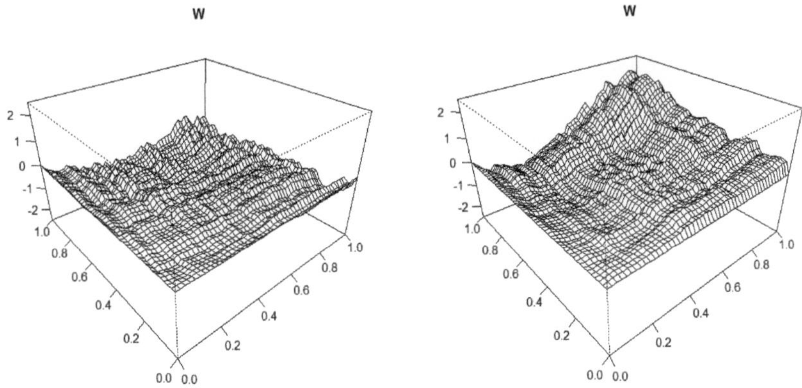

Fig. 9.7 Two realizations of a Wiener sheet

Example 9.3. Let $\{W(t), t \in [0,1]^2\}$ be a Wiener sheet with $\mathbf{E}W(t) \equiv 0$ and

$$\mathbf{cov}(W(s), W(t)) = \min\{s_1, t_1\} \cdot \min\{s_2, t_2\}$$

for $s = (s_1, s_2)^\top, t = (t_1, t_2)^\top \in [0,1]^2$. It is still possible to write down the eigenequation

$$\int_0^1 \int_0^1 \min\{s_1, t_1\} \min\{s_2, t_2\} \varphi(s_1, s_2) ds_1 ds_2 = \lambda \varphi(t_1, t_2), \qquad (9.18)$$

for $t \in [0,1]^2$. Equation (9.18) can be solved, with a little effort, and the resultig eigenvalues and corresponding eigenfunctions are analogous to the ones from Example 9.2, namely

$$\lambda_n = \prod_{k=1,2} \left(\frac{2}{\pi(2n_k - 1)} \right)^2,$$

$$\varphi_n(t) = 2 \prod_{k=1,2} \sin\left((2n_k - 1)\frac{\pi}{2}t\right), \qquad n = (n_1, n_2) \in \mathbb{N} \times \mathbb{N}.$$

As we see, the eigenspaces and eigenvalues are given by all possible compositions of the one-dimensional eigenvalues and eigenfunctions. This time, though, one has to pay more attention, since many eigenspaces have dimension higher than one. For instance $\lambda_{(1,2)} = \lambda_{(2,1)} = \frac{16}{9\pi^4}$. This means that the eigenspace corresponding to the eigenvalue $\frac{16}{9\pi^4}$ is generated by the two functions $\varphi_{(1,2)}$ and $\varphi_{(2,1)}$, i.e., $\mathcal{L}\{\varphi_{(1,2)}(t), \varphi_{(2,1)}(t)\}$. A clever way to rewrite the solution is the following: If we

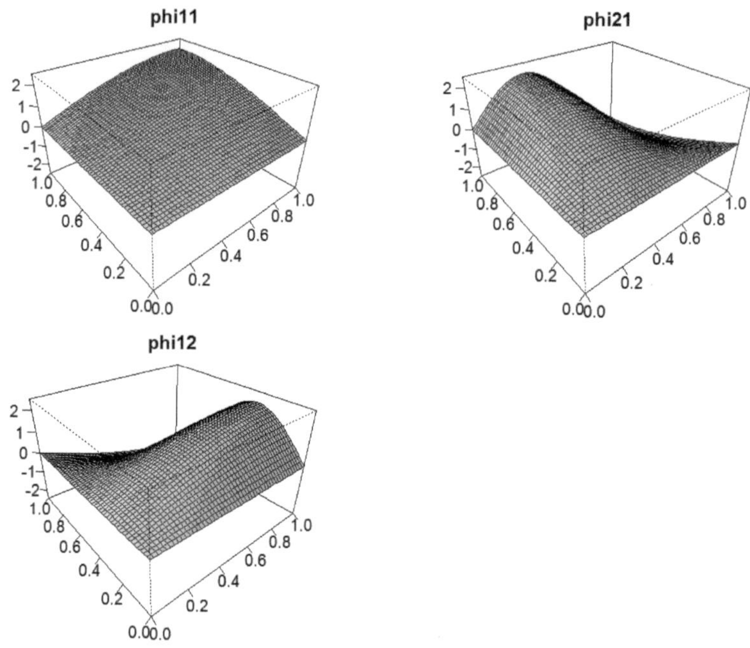

Fig. 9.8 Top left: Eigenspace corresponding to the largest eigenvalue, $\lambda = \frac{16}{\pi^4}$. Top right and bottom left: The two eigenfunctions corresponding to the second largest eigenvalue, $\lambda = \frac{16}{9\pi^4}$.

take $M = (2n_1 - 1)^2(2n_2 - 1)^2$ for $n_1, n_2 \in \mathbb{N}$, then we have

$$\lambda_M = \frac{16}{\pi^4 M}$$

$$E_M(t) = \mathcal{L}\{\varphi_{(n_1,n_2)}(t) : (2n_1 - 1)^2(2n_2 - 1)^2 = M\},$$

and we can recognize eigenspaces of higher dimension easier.

Exercise 9.7. Summarize the features of the eigenspaces corresponding to the largest six eigenvalues: fill in the table for the third up to the sixth eigenspace.

Eigenspace no.	M	λ_M	dim(E_M)	n
1	1	$\frac{16}{\pi^4}$	1	(1,1)
2	9	$\frac{16}{9\pi^4}$	2	(1,2), (2,1)

Hint. See Sect. 3.2 in [3] for further examples.

9.3 Statistical Models for Functional Data

9.3.1 Linear Regression

Let us briefly recall the typical setup in multivariate linear regression. Given a scalar response variable Y and a random vector $X \in \mathbb{R}^p$ of explanatory variables, one assumes that there is a linear relationship between Y and X. Namely,

$$Y = \tilde{\alpha} + \langle X, \beta \rangle + \varepsilon = \alpha + \langle X - \mu, \beta \rangle + \varepsilon, \tag{9.19}$$

with the usual scalar product in $\mathbb{R}^p$. The term α allows us to assume that the covariates have expectation equal to zero. The random variable ε is assumed to be independent of X, and such that $\mathbf{E}\varepsilon = 0$ and $\mathbf{var}\,\varepsilon = \sigma^2$, while the vector $\beta \in \mathbb{R}^p$, the parameter vector, contains the unkown coefficients for the explanatory variables. If there are n observations, (Y_i, X_i), one usually writes the n equations of the form (9.19) into a cumulative vector equation

$$Y = \tilde{\alpha} + X\beta + \varepsilon,$$

where X is now a $n \times p$ matrix, α a vector in $\mathbb{R}^n$, and the random vector ε is such that $\mathbf{E}\varepsilon = o$, $\mathbf{cov}\,\varepsilon = \sigma^2 I_n$, with I_n denoting the identity matrix in $\mathbb{R}^{n \times n}$. Provided that $(X^\top X)^{-1}$ exists, we can solve the regression problem and obtain

$$\hat{\alpha} = \bar{Y},$$
$$\hat{\mu} = \bar{X},$$
$$\hat{\beta} = (X^\top X)^{-1} X^\top Y,$$
$$\hat{\tilde{\alpha}} = \hat{\alpha} - \hat{\mu}^\top \hat{\beta}.$$

We can extend this setup to functional data. Instead of having a random vector X as explanatory variable, we will consider a stochastic process $\{X(t), t \in \mathcal{T}\}$. In the following, we suppose that $\{X(t)\}$ has the properties stated in Sect. 9.2.2.

The most natural idea is to reduce this problem to a known one and, more precisely, to multivariate linear regression. Of course, the functional covariate $X(t)$ can be discretized, and thus transformed in a random vector $X = (X(t_1), \dots, X(t_p))^\top$. This approach does, however, not use the continuity of the processes.

Another option is to use the fact that $L_2(\mathcal{T})$, the space of the paths of the stochastic process $\{X(t)\}$, has a canonical scalar product. Then, we can consider a functional linear model which works similarly to the multivariate case stated in (9.19):

$$Y = \alpha + \langle X(\cdot) - \mu(\cdot), \beta(\cdot) \rangle + \varepsilon. \tag{9.20}$$

The unknown parameter function $\beta \in L_2(\mathcal{T})$ gives the weights for the stochastic process $\{X(t)\}$ at every time point $t \in \mathcal{T}$.

Suppose that we know $C(s,t)$, the covariance function of $\{X(t)\}$, or its estimator $\widehat{C}(s,t)$ (see Sect. 9.3.2 below). Then we can also compute its eigenvalues and eigenfunctions, λ_k and φ_k for $k \in \mathbb{N}$, and write the stochastic process $\{X(t)\}$ as

$$X(t) = \mu(t) + \sum_{k=1}^{\infty} \xi_k \varphi_k(t).$$

Even if $\beta(t)$ itself does not belong to $\ker(\mathcal{C})^{\perp} \subset L_2(\mathcal{T})$, it is decomposable as $\beta(t) = \beta_0(t) + \beta_1(t)$, with $\beta_0 \in \mathrm{kern}(\mathcal{C})$ and $\beta_1(t) \in \mathrm{kern}(\mathcal{C})^{\perp}$. Then,

$$\mathbf{E}(\langle \beta_0(\cdot), X(\cdot) \rangle^2) = \mathbf{E}\left(\int_T \int_T \beta_0(s) X(s) X(t) \beta_0(t) dt dt \right)$$
$$= \int_T \int_T \beta_0(s) C(s,t) \beta_0(t) dt dt = 0.$$

Thus, the term $\beta_0(t)$ term is irrelevant and we can assume, without loss of generality, that for the parameter function β it holds that $\beta \in \ker(\mathcal{C})^{\perp}$. Then $\beta(t) = \sum_{k=1}^{\infty} \langle \beta, \varphi_k \rangle \varphi_k(t) = \sum_{k=1}^{\infty} \beta_k \varphi_k(t)$.

With these decompositions, we get that

$$\langle X(\cdot) - \mu(\cdot), \beta(\cdot) \rangle = \int_T X(t) \beta(t) dt = \int_T \sum_{k=1}^{\infty} \xi_k \varphi_k(t) \sum_{j=1}^{\infty} \beta_j \varphi_j(t) dt$$
$$= \sum_{k,j=1}^{\infty} \xi_k \beta_j \langle \varphi_k, \varphi_j \rangle = \sum_{k=1}^{\infty} \xi_k \beta_k.$$

This relationship allows us to reformulate the functional regression problem (9.20) as

$$Y = \alpha + \sum_{k=1}^{\infty} \xi_k \beta_k + \varepsilon. \tag{9.21}$$

We want to evaluate the parameters α and $\{\beta_k\}_{k \in \mathbb{N}}$.

Given n observations $(Y_i, X_i(t))$, $i = 1, \ldots, n$, it is possible to proceed in an analogous way as for the multivariate case, i.e., we can summarize the n equations

$$Y_i = \alpha + \sum_{k=1}^{\infty} \xi_k^{(i)} \beta_k + \varepsilon_i, \qquad i = 1, \ldots, n$$

by

$$Y = \alpha + A\beta + \varepsilon,$$

with some design matrix $A \in \mathbb{R}^{n \times \infty}$ and parameter sequence $\beta \in \ell^2$ (i.e $\beta = \{\beta_k\}_{k \in \mathbb{N}}$ is a sequence such that $\sum_{k=1}^{\infty} |\beta_k|^2 < \infty$). We have reduced the problem's dimension from an uncountable to a countable (but still infinite) dimension. The problem considered in (9.21) can be approximated and regularized by fixing a truncation level $p \in \mathbb{N}$. One typically chooses $p \ll n$. Then,

$$Y_i \simeq \alpha + \sum_{k=1}^{p} \xi_k^{(i)} \beta_k + \varepsilon, \qquad i = 1,\ldots,n$$

which becomes $Y_i \simeq \sum_{k=0}^{p} \xi_k^{(i)} \beta_k + \varepsilon$ if one puts $\xi_0^{(i)} = 1$ and $\beta_0 = \alpha$. We thus obtain a finite-dimensional problem of the form

$$Y = \begin{pmatrix} Y_1 \\ \vdots \\ Y_n \end{pmatrix} \simeq \underbrace{\begin{pmatrix} 1 & \xi_1^{(1)} & \cdots & \xi_p^{(1)} \\ \vdots & \vdots & \vdots & \vdots \\ 1 & \xi_1^{(n)} & \cdots & \xi_p^{(n)} \end{pmatrix}}_{A \in \mathbb{R}^{n \times (p+1)}} \cdot \underbrace{\begin{pmatrix} \alpha \\ \beta_1 \\ \vdots \\ \beta_p \end{pmatrix}}_{\beta \in \mathbb{R}^{p+1}} + \begin{pmatrix} \varepsilon_1 \\ \vdots \\ \varepsilon_n \end{pmatrix}.$$

The theory of linear models assures that the estimators $\hat{\alpha}, \hat{\beta}_1, \ldots, \hat{\beta}_p$ are consistent. The final regression model is given by

$$\widehat{Y} = A \cdot \widehat{\beta} = A(A^\top A)^{-1} A^\top Y = SY$$

The parameter function $\hat{\beta}(t)$ can be reconstructed via $\hat{\beta}(t) = \sum_{k=1}^{p} \hat{\beta}_k \varphi_k(t)$. We can apply standard tools from linear regression to evaluate the goodness of our model, for instance:

1. We can study the residuals, $\hat{\varepsilon}_i = Y_i - \widehat{Y}_i$.
2. We can look at the R^2–statistic, which is defined as

$$R^2 = 1 - \frac{\sum_{i=1}^{n} (\hat{Y}_i - Y_i)^2}{\sum_{i=1}^{n} (Y_i - \bar{Y})^2} \in [0,1].$$

An R^2–value close to 1 means that the regression model is meaningful, in the sense that it contains much information and can describe the object Y well.

An interesting point is, as usually, the choice of p. That can be done, for instance, via generalized cross-validation if we denote the matrix in equation (9.22) by S_p, and by $\widehat{Y}_i^{(p)}$ the fitted values for Y. We want to find the value of p which minimizes the quantity

$$\mathrm{GCV}(p) = \sum_{i=1}^{n} \left(\frac{Y_i - \hat{Y}_i^{(p)}}{1 - S_p} \right).$$

9.3.2 Estimation, a Second Look

Let $\{X(t), t \in \mathcal{T}\}$ be a stochastic process as in Sect. 9.2.1, and let $\mu(t)$, $C(s,t)$ be its mean and covariance functions (for $s,t \in \mathcal{T}$). Without loss of generality, we can take $\mathcal{T} = [0,1]$. As stated in the introduction of this chapter, there are three possibilities (three different stages):

1. One could observe the whole paths in continues time, i.e.,

$$\{X_i(t), t \in \mathcal{T}\} \qquad i = 1, \ldots, n. \tag{9.22}$$

2. Each realization is measured at discrete time points, i.e.,

$$X_i(t_{i1}), \ldots, X_i(t_{im_i}) \qquad i = 1, \ldots, n. \tag{9.23}$$

3. Each realization is recorded at discrete time points and contains independent measuring errors,i.e.,

$$X_{ij} = X_i(t_{ij}) + \eta_{ij} \qquad i = 1, \ldots, n \text{ and } j = 1, \ldots, m_i, \tag{9.24}$$

where the η_{ij} are i.i.d. random variables, with $\mathbf{E}\eta_{ij} = 0$ and $\mathbf{E}(\eta_{ij}^2) = \sigma_\eta^2$.

Let us start with the easiest case, which is given in (9.22). The estimators for mean and covariance functions are then exactly the ones from Sect. 9.1.1. Namely,

$$\hat{\mu}(t) = \frac{1}{n}\sum_{i=1}^n X_i(t) = \bar{X}(t),$$

$$\widehat{C}(s,t) = \frac{1}{n-1}\sum_{i=1}^n (X_i(s) - \bar{X}(s))(X_i(t) - \bar{X}(t)).$$

The estimator $\hat{\mu}(t)$ is unbiased and its variance is $\mathbf{var}\hat{\mu}(t) = C(t,t)/n$ for $t \in \mathcal{T}$. Note that $\hat{\mu}(t)$ is also a consistent estimator, in the sense that

$$\mathbf{E}\left(\int_T (\hat{\mu}(t) - \mu(t))^2 \, dt\right) = \int_T \mathbf{E}(\hat{\mu}(t) - \mu(t))^2 \, dt =$$

$$= \frac{1}{n}\underbrace{\int_T C(t,t)dt}_{\sigma^2} = \frac{1}{n}\int_T \mathbf{var}X(t)dt = \frac{\sigma^2}{n},$$

i.e., the mean integrated squared error has order $O(1/n)$. The quantity σ^2 allows us to obtain pointwise confidence intervals. Furthermore, since $\sqrt{n}(\hat{\mu}(t) - \mu(t)) \xrightarrow{D} N(0, C(t,t))$, a global asymptotic confidence interval can be obtained from

$$n\|\hat{\mu}(t) - \mu(t)\|_2^2 \xrightarrow{D} \int_t G^2(t)dt,$$

where $\{G(t)\}$ is a Gaussian process with mean zero and $\mathbf{cov}(G(s), G(t)) = C(s,t)$. Note that $\widehat{C}(s,t)$ is unbiased, too, and its variance is of order $O(1/n)$, provided that $\mathbf{E}X^4(t) < \infty$.

The previous situation is rather uncommon, because usually the measurements are taken at discrete time points. Recall from the introduction: if the recording times are the same for the whole sample, and if they are dense enough, then the best way to approximate the mean and covariance functions is to find their point estimators (at the given time points $t_1,\ldots,t_m$) and to interpolate. However, in this case, the estimated covariance function is not necessarily positive semidefinite.

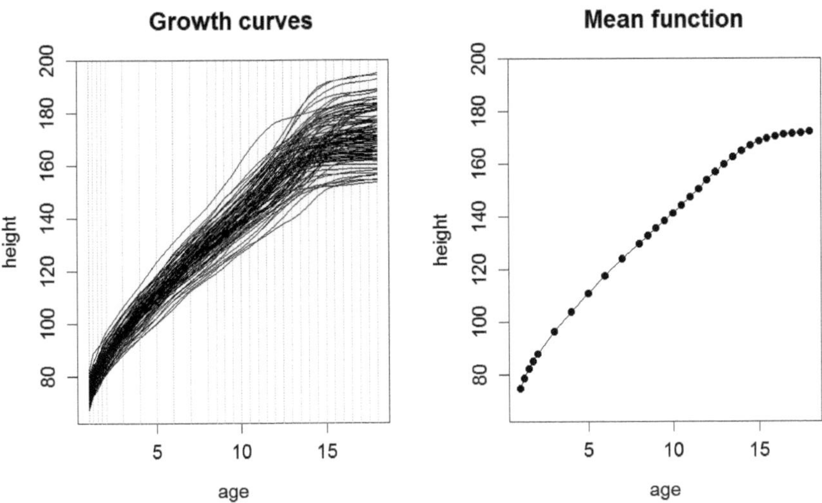

Fig. 9.9 Left: Growth curves of 93 children from Berkeley Growth Study data. The dotted vertical lines indicate the measuring timepoints. Right: The thick points are the computed mean values, the line represents the simplest possible interpolation: piecewise linear.

Example 9.4 (Growth curve data). Consider the data set shown in Fig. 9.9, obtained from the Berkeley Growth Study data (see [397]). In this study, the heights of 93 children were measured at 31 stages, from 1 to 18 years. The measuring times are common to all the observations, the measurement errors are negligible. By computing individually mean and covariance for all the known recording times $t_1,\ldots,t_{31}$, we obtain a vector $\widehat{\mu} = (\widehat{\mu}(t_1),\ldots,\widehat{\mu}(t_{31}))^\top \in \mathbb{R}^{31}$ and a matrix $\widehat{C} = (\widehat{C}(t_i,t_j)) \in \mathbb{R}^{31\times 31}$. In this case, as the recordings are actually very smooth per se, we do not need any further smoothing. As shown in Fig. 9.9 and in Fig. 9.10, already local linear interpolation provides satisfactory results. The scores of the eigenfunctions (see Fig. 9.12) provide useful information, as well. For instance, one could have realizations of two (or more) different stochastic processes in the same dataset. Then the scores will probably be different, depending on which realization of which process is considered. In this case, for instance, plotting $\xi_1^{(i)}$ against $\xi_2^{(i)}$ for all the realizations ($i = 1,\ldots,n = 93$) as in Fig. 9.12 suggests that the recordings could be separated into two groups. This seems to be reasonable, as the study contains growth data from girls and from boys. Thus, this is a possible tool for the classification of such stochastic processes (see [72], or [183] for more details and more advanced procedures).

Covariance function

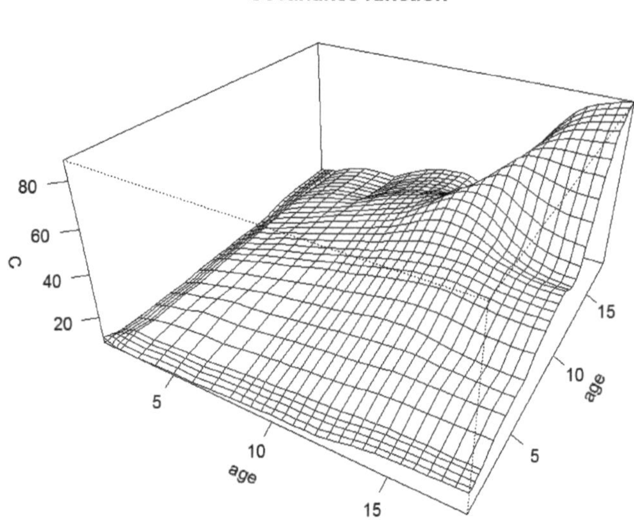

Fig. 9.10 Estimation of covariance function for the growth data. Since the recordings are very smooth it is sufficient to compute the covariance matrix $\widehat{C} \in \mathbb{R}^{31 \times 31}$ and estimate the covariance function via piecewise linear interpolation.

If the observation points are sparse (low frequency data), if they are not the same for all the paths, if there are multiple measurements, or in presence of errors, a smoothing method is required. For instance, if one is interested in estimating $\mu(t)$ the following local least squares procedure would work, considering

$$\operatorname*{argmin}_{\alpha,\beta} \left\{ \sum_{i,j=1}^{n,m_i} \left(X_i(t_{ij}) - \alpha - \beta(t - t_{ij}) \right)^2 K\left(\frac{t - t_{ij}}{h_1} \right) \right\}.$$

The solution $(\alpha^*(t), \beta^*(t))$ contains an estimation of the mean-value function, as $\hat{\mu}(t) = \alpha^*(t)$ for all $t \in \mathcal{T}$. Estimating the covariance function is, of course, more laborious. Let us define an auxiliary quantity $\widetilde{C}_i(t_{ij}, t_{ik})$ by

$$\widetilde{C}_i(t_{ij}, t_{ik}) = (X_i(t_{ij}) - \hat{\mu}(t_{ij}))(X_i(t_{ik}) - \hat{\mu}(t_{ik})).$$

Then one can solve the minimization problem

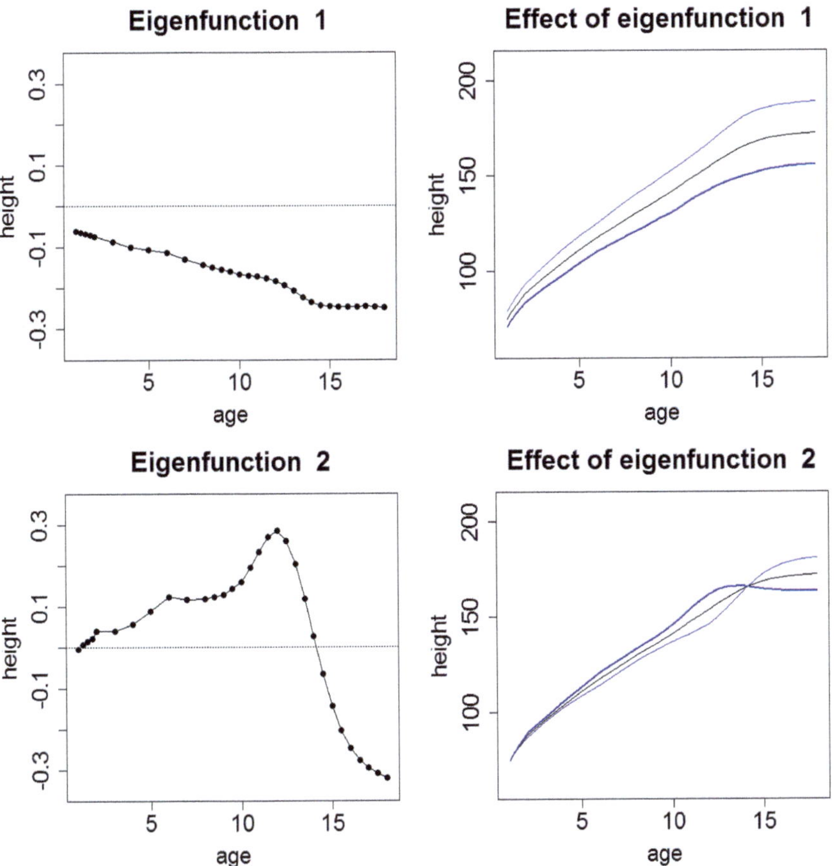

Fig. 9.11 First two eigenfunctions for the growth data: they explain 80.36% and 14.06% of the total variation, respectively. A common procedure to show the effects of an eigenfunction consists in adding (and subtracting) a suitable multiple of it to the mean function. In this case, we took $2\sqrt{\lambda_i}$ as a factor.

$$\mathrm{argmin}_{\alpha,\beta,\gamma}\left\{\sum_{i,j,k=1}^{n,m_i,m_i}\left(\widetilde{C}_i(t_{ij},t_{ik})-\alpha-\beta(t-t_{ij})-\gamma(s-t_{ik})\right)^2 K\left(\frac{t-t_{ij}}{h_2}\right)\right.$$
$$\left. K\left(\frac{s-t_{ik}}{h_2}\right)\right\}$$

in the error-free case (9.23). Otherwise, see (9.24), the minimization problem

 Ulrich Stadtmüller and Marta Zampiceni

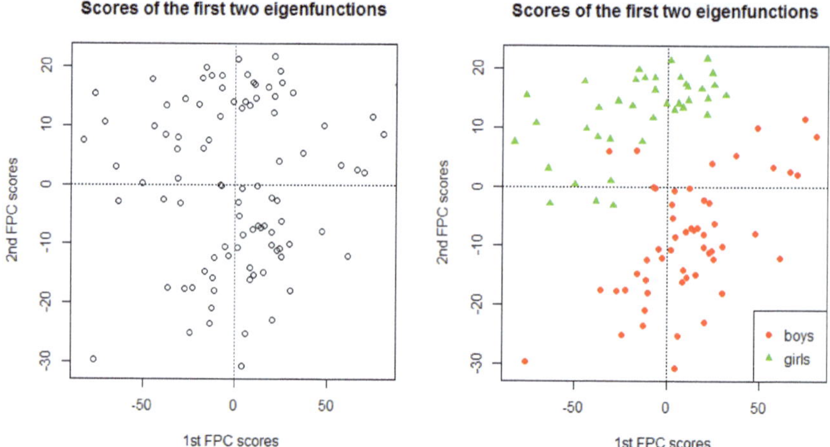

Fig. 9.12 Left: Scatterplot of the first two functional principal components scores. Right: same scatterplot with gender indications.

$$\operatorname{argmin}_{\alpha,\beta,\gamma}\left\{\sum_{\substack{i,j,k=1\\j\neq k}}^{n,m_i,m_i}\left(\widetilde{C}_i(t_{ij},t_{ik})-\alpha-\beta(t-t_{ij})-\gamma(s-t_{ik})\right)^2 K\left(\frac{t-t_{ij}}{h_2}\right)\right.$$

$$\left. K\left(\frac{s-t_{ik}}{h_2}\right)\right\}$$

can be considered. In both cases, the optimal value of α, $\alpha^*(s,t)$, is the sought estimator of the covariance function at the point $(s,t) \in \mathcal{T}^2$.

Theorem 9.6. *In the case considered in (9.22), under appropriate moment conditions, the following results hold:*

$$\sup_{t\in\mathcal{T}}|\hat{\mu}(t)-\mu(t)| = O_p\left(\sqrt{\frac{\log n}{n}}\right),$$

$$\sup_{s,t\in\mathcal{T}}|\hat{C}(s,t)-C(s,t)| = O_p\left(\frac{\log n}{\sqrt{n}}\right).$$

In the cases considered in (9.23) and (9.24), under appropriate conditions (see [424]), the following results hold:

$$\sup_{t\in\mathcal{T}}|\hat{\mu}(t)-\mu(t)| = O_p\left(\frac{1}{\sqrt{n}h_1}\right),$$

$$\sup_{s,t\in\mathcal{T}}|\hat{C}(s,t)-C(s,t)| = O_p\left(\frac{1}{\sqrt{n}h_2^2}\right).$$

Once we have $\hat{C}$, we can plug it into the equation

$$\int_T \hat{C}(s,t)\varphi(t)dt = \lambda\varphi(s), \qquad s \in T, \tag{9.25}$$

and obtain estimators of the eigenvalues and eigenfunctions. Let us study the numerical aspects of this procedure in more detail. First, we will make a discretization step, by taking a sufficient large number of evenly spaced points in T, $\tau_1,\ldots,\tau_{N_n}$, having (constant) distance $\Delta = \Delta_{N_n}$, and focus on the estimation of the covariance function restricted to the chosen discretization points, i.e., $\tilde{C} = \left(\tilde{C}(i,j)\right) = \left(\hat{C}(\tau_i,\tau_j)\right)$. We can then discretize the equation (9.25) by

$$\tilde{C}\Delta\tilde{\varphi} = \tilde{\lambda}\tilde{\varphi} \qquad \text{resp.} \qquad \tilde{C}\tilde{\varphi} = \frac{\tilde{\lambda}}{\Delta}\tilde{\varphi},$$

solve it, and obtain, for $i = 1,\ldots,N_n$, the eigenvalues $\tilde{\lambda}_i/\Delta$ and the corresponding eigenvectors $\tilde{\varphi}_i \in \mathbb{R}^{N_n}$. The i−th eigenvector is the discrete approximation of the i−th eigenfunction, evaluated at $\tau_1\ldots,\tau_{N_n}$ (with an appropriate scaling).

Exercise 9.8. Find out which scaling would be appropriate.

Note that the eigenvectors are orthogonal but the smoothed or interpolated eigenfunctions are not necessarily orthogonal to each other. Since we can choose a relatively fine discretization grid, the discretization error is not particularly relevant. The approximation error made by solving (9.25) instead of the original problem is more relevant.

Theorem 9.7. *If, for $p \in \mathbb{N}$, $\lambda_1 > \ldots > \lambda_p$ and*

$$\|\hat{C} - C\|_2 = O_p(a_n), \qquad \sup_{s,t \in T} |\hat{C}(s,t) - C(s,t)| = O_p(b_n)$$

for two zero-sequences (a_n) and (b_n), then, for $k = 1,\ldots,p$,

$$|\hat{\lambda}_k - \lambda_k| = O_p(a_n), \qquad \|\hat{\varphi}_k - \varphi_k\|_2 = O_p(a_n),$$
$$\|\hat{\varphi}_k - \varphi_k\|_\infty = O_p(b_n).$$

The proof is based on the following well-known result from functional analysis (see Chapt. X of [171]):

Theorem 9.8.

1. *If C is a kernel operator with kernel $K \in L_2(T \times T)$ (without sign restrictions), then*

$$\|\mathcal{C}\| \leq \|\mathcal{C}\|_H = \|K\|_2,$$

   and,

$$\|\mathcal{C}\| = |\lambda|_{\max},$$

   where $|\lambda|_{\max}$ is the largest absolute value of an eigenvalue of C Furthermore, for every eigenvalue λ of C, it holds that

$$-\|\mathcal{C}\|_H \leq -\|\mathcal{C}\| \leq \lambda \leq \|\mathcal{C}\| \leq \|\mathcal{C}\|_H.$$

2 (Weyl's inequality). *If $\mathcal{C}_g$, $\mathcal{C}_1$ and $\mathcal{C}_2$ are kernel operators as above such that $\mathcal{C}_g \leq \mathcal{C}_1 + \mathcal{C}_2$, then, if $\left\{\lambda_k^{(g,+)}\right\}$, $\left\{\lambda_k^{(1,+)}\right\}$ and $\left\{\lambda_k^{(2,+)}\right\}$ are their positive (ordered) eigenvalues, the inequality*

$$\lambda_k^{(g,+)} \leq \lambda_k^{(1,+)} + \lambda_k^{(2,+)}$$

holds for every $k \in \mathbb{N}$.

9.3.3 More General Regression Models

9.3.3.1 Linear Regression with Functional Response

Suppose now that not only the explanatory variable X, but also the response variable Y are stochastic processes (over some compact intervals $\mathcal{T}$ and $\mathcal{S}$ respectively). We will call C_X the covariance function of $\{X(t)\}$, C_Y the covariance function of $\{Y(s)\}$. Furthermore, we consider the cross-covariance function $C_{XY}(t,s) = \mathbf{cov}(X(t), Y(s)) = C_{YX}(s,t)$ for $t \in \mathcal{T}$ and $s \in \mathcal{S}$. In this setup it is possible to define two different models.

1. If the whole explanatory variable has to be included in the model, we will have

$$Y(s) = \alpha(s) + \int_T (X(t) - \mathbf{E}X(t)) \beta(s,t) \, dt + \varepsilon(s), \qquad s \in \mathcal{S},$$

   with a residual process $\{\varepsilon(s)\}$ such that $\mathbf{E}(\varepsilon(s)|X(t), t \in \mathcal{T}) \equiv 0$.
2. If only the current value of the covariate is of interest, then

$$Y(s) = \alpha(s) + X(s)\beta(s) + \varepsilon(s), \qquad s \in \mathcal{S},$$

   with a residual process such that $\mathbf{E}(\varepsilon(t)|X(t)) \equiv 0$.

We will focus on the first case. We have two covariance operators, $\mathcal{C}_Y$ and $\mathcal{C}_X$, with their eigenvalues $\{v_j\}_{j \in \mathbb{N}}$ and $\{\lambda_k\}_{k \in \mathbb{N}}$ and eigenfunctions $\{\psi_j\}_{j \in \mathbb{N}}$ and $\{\varphi_k\}_{k \in \mathbb{N}}$. If we define $\mu_X(t) = \mathbf{E}X(t)$ for $t \in \mathcal{T}$ and $\mu_Y(s) = \mathbf{E}Y(s)$ for $s \in \mathcal{S}$, then we can put

$$\zeta_j = \langle Y - \mu_Y, \psi_j \rangle, \qquad \xi_k = \langle X - \mu_X, \varphi_k \rangle,$$

and decompose Y and X into

$$Y(s) = \mu_Y(s) + \sum_{j=1}^{\infty} \zeta_j \psi_j(s), \qquad s \in \mathcal{S},$$

$$X(t) = \mu_X(t) + \sum_{k=1}^{\infty} \xi_k \varphi_k(t), \qquad t \in \mathcal{T}.$$

Furthermore, we can expand the functions $\alpha(s)$ and $\beta(t,s)$ from equation (9.26) with respect to these orthonormal bases, too. Namely,

$$\alpha(s) = \sum_{j=1}^{\infty} \alpha_j \psi_j(s), \qquad s \in \mathcal{S},$$

$$\beta(s,t) = \sum_{j,k=1}^{\infty} \beta_{j,k} \psi_j(s) \varphi_k(t), \qquad (s,t) \in \mathcal{S} \times \mathcal{T},$$

where $\alpha_j = \langle \alpha, \psi_j \rangle$ and $\beta_{j,k} = \int_\mathcal{S} \int_\mathcal{T} \beta(s,t) \psi_j(s) \varphi_k(t) dt ds$. We can apply the least-squares method to find approximations for α and β, where we look for functions which minimize

$$\mathbf{E}\left(\int_\mathcal{S} \left(Y(s) - \alpha(s) - \int_\mathcal{T} (X(t) - \mu_X(t)) \beta(s,t) dt \right)^2 ds \right). \qquad (9.26)$$

We can exploit the previous decompositions and rewrite it in a more suitable way. Since

$$\int_\mathcal{T} (X(t) - \mu_X(t)) \beta(s,t) dt = \sum_{k=1}^{\infty} \xi_k \sum_{j,l=1}^{\infty} \beta_{j,l} \psi_j(s) \underbrace{\int_\mathcal{T} \varphi_k(t) \varphi_l(t) dt}_{\delta_{k,l}}$$

$$= \sum_{j,k=1}^{\infty} \xi_k \beta_{j,k} \psi_j(s),$$

the expression in (9.26) is equal to

$$\mathbf{E}\left(\int_\mathcal{S} \left(\sum_{j=1}^{\infty} \left(\zeta_j - \alpha_j - \sum_{k=1}^{\infty} \xi_k \beta_{j,k} \right) \psi_j(s) \right)^2 ds \right),$$

and it is minimal whenever

$$\mathbf{E}\left(\sum_{j=1}^{\infty} \left(\zeta_j - \alpha_j - \sum_{k=1}^{\infty} \xi_k \beta_{j,k} \right)^2 \right)$$

is minimal.

Suppose that we have n observations $\{Y_i(s), X_i(t), s \in \mathcal{S}, t \in \mathcal{T}\}$. Then we have to minimize the empirical counterpart of (9.26) with respect to α and β, i.e. we consider the expression

$$\sum_{i=1}^{n} \left(\int_\mathcal{S} \left(Y_i(s) - \alpha(s) - \int_\mathcal{T} (X_i(t) - \bar{X}(t)) \beta(s,t) dt \right)^2 ds \right).$$

We can still plug in the previous expansions of X, Y, α and β and, if we take truncations to some levels p and q instead of the whole expansions, we obtain the following minimization problem:

$$\sum_{i=1}^{n} \left(\sum_{j=1}^{p} \left(\zeta_j^{(i)} - \alpha_j - \sum_{k=1}^{q} \xi_k^{(i)} \beta_{j,k} \right)^2 \right) \longrightarrow \min_{\alpha_j, \beta_{j,k}} \qquad (9.27)$$

and, equivalently,

$$\sum_{i=1}^{n} \left(\zeta_j^{(i)} - \alpha_j - \sum_{k=1}^{q} \xi_k^{(i)} \beta_{j,k} \right)^2 \longrightarrow \min_{\alpha_j, \beta_{j,1} \ldots \beta_{j,q}} \qquad j = 1, \ldots, p,$$

which is a standard least squares problem for every fixed j. Its solutions, $\widehat{\alpha}_j$ and $\widehat{\beta}_{jk}$ (for $j = 1 \ldots p$ and $k = 1 \ldots q$) allow us to estimate the functions α and β by

$$\widehat{\alpha}(s) = \sum_{j=1}^{p} \widehat{\alpha}_j \psi_j(s), \qquad \widehat{\beta}(s,t) = \sum_{j=1}^{p} \sum_{k=1}^{q} \widehat{\beta}_{jk} \psi_j(s) \varphi_k(t).$$

In practice, the eigenfunctions $\{\psi_j\}$ and $\{\varphi_k\}$ have to be estimated as discussed in the previous sections, using the estimators of the autocovariance functions. In the end, we obtain predictions in the form

$$\widehat{Y}_i(s) = \widehat{\alpha}(s) + \sum_{j,k=1}^{p,q} \xi_k^{(i)} \widehat{\beta}_{jk} \widehat{\psi}_j(s) \qquad i = 1, \ldots, n.$$

Remark 9.2.   1. A first model check involves the residual processes $\{Y_i(s) - \widehat{Y}_i(s), s \in \mathcal{S}\}$.
2. The choice of p, q is, as usual, difficult. Cross-validation techniques could prove useful, but they are very laborious, too. Otherwise, AIC-like procedures, as minimizing the expression in (9.27) plus a term like $pq + 1$ could help.
3. Let $\{\lambda_k\}$ denote the eigenvalues of the covariance function $\mathcal{C}_X$. Then, since

$$\sum_{j=1}^{\infty} \zeta_j \psi_j(s) = \alpha(s) + \sum_{j,k=1}^{\infty} \xi_k \beta_{jk} \psi_j(s) + \varepsilon(s),$$

we get that

$$\mathbf{E}\left(\int_T \xi_{k_0} \psi_{j_0}(s) ds \sum_{j=1}^{\infty} \zeta_j \psi_j(s) \right)$$

$$\mathbf{E}\left(\int_T \xi_{k_0} \psi_{j_0}(s) ds \left(\alpha(s) + \sum_{j,k=1}^{\infty} \xi_k \beta_{jk} \psi_j(s) + \varepsilon(s) \right) \right),$$

which is equivalent with

$$E(\xi_{k_0}\zeta_{j_0}) = 0 + \beta_{j_0,k_0}\,\mathbf{E}(\xi_{k_0}^2) + 0 = \beta_{j_0,k_0}\lambda_{k_0}$$

and

$$\beta_{j_0 k_0} = \frac{\mathbf{E}(\zeta_{j_0}\xi_{k_0})}{\lambda_{k_0}},$$

respectively. Thus, we can estimate the coefficients β_{jk} using the estimated correlation between the random variables ζ_j and ξ_k. More precisely,

$$\widehat{\mathbf{E}(\zeta_j\xi_k)} = \frac{\sum_{i=1}^{n}(\zeta_j^{(i)} - \bar{\zeta}_j)(\xi_k^{(i)} - \bar{\xi}_k)}{\sqrt{\sum_{i=1}^{n}\left(\zeta_j^{(i)} - \bar{\zeta}_j\right)^2 \sum_{l=1}^{n}\left(\xi_k^{(l)} - \bar{\xi}_k\right)^2}},$$

$$\widehat{\beta}_{jk} = \frac{\widehat{\mathbf{E}(\zeta_j\xi_k)}}{\widehat{\lambda}_k} \qquad\qquad j,k = 1,2,\dots$$

9.3.3.2 Generalized Linear Model

Suppose that the covariate still has a functional form, while the response variable Y is dichotomous. If we have n samples, the generalized linear model will be

$$Y_i \sim \mathrm{Binom}(1,p_i), \qquad \text{where } p_i = \mathbf{E}Y_i = g(\eta_i) = \mu_i, \qquad i = 1,\dots,n,$$

with a linear predictor η_i and logit- link- function g given by

$$g(x) = \frac{e^x}{1 - e^x}.$$

The random variables Y_i have variance $\mathbf{var}Y_i = p_i(1-p_i) = \mu_i(1-\mu_i) = \sigma^2(\mu_i)$. Since we are working with a functional covariate, the linear predictors have the usual form

$$\eta_i = \alpha + \langle X_i(\cdot) - \mu(\cdot), \beta(\cdot)\rangle = \alpha + \sum_{k=1}^{\infty}\xi_k^{(i)}\beta_k, \tag{9.28}$$

approximated by the truncated expansion $\alpha + \sum_{k=1}^{p}\xi_k^{(i)}\beta_k$. We can find the vector $\beta \in \mathbb{R}^p$ by the maximum-likelihood method. In fact, the probability mass function of the vector Y is given by

$$\prod_{i=1}^{n}p_i^{Y_i}(1-p_i)^{1-Y_i}.$$

Then the log-likelihood is given by

$$l(Y;\beta) = \sum_{i=1}^{n}\left(Y_i\log\frac{p_i}{1-p_i} + \log(1-p_i)\right)$$

$$= \sum_{i=1}^{n}\left(Y_i\eta_i - \log(1+e^{\eta_i})\right).$$

Recall that we are approximating η_i by $\alpha + \sum_{k=1}^{p} \xi_k^{(i)} \beta_k$. At the point of $l(Y;\beta)$'s maximum (with respect to the vector β), the gradient has to vanish, i.e.

$$
\begin{aligned}
\nabla_\beta l(Y;\beta) &= \left(\sum_{i=1}^{n} Y_i \xi_k^{(i)} - \frac{1}{1+e^{\eta_i}} e^{\eta_i} \xi_k^{(i)} \right)_{k=0}^{p} \\
&= \left(\sum_{i=1}^{n} Y_i \xi_k^{(i)} - p_i \xi_k^{(i)} \right)_{k=0}^{p} \\
&= A^\top (Y - \mu) \overset{!}{=} 0,
\end{aligned}
\tag{9.29}
$$

provided that the matrix $A \in \mathbb{R}^{n \times (p+1)}$ is defined as usually, by

$$
A = \left(1, \xi_1^{(k)}, \ldots, \xi_p^{(k)} \right)_{k=1}^{n}.
$$

Note that the system of equations (9.29) is not linear in β, since the vector μ depends on it, as well. The system (9.29) is convex, though, i.e., it can be solved using the Newton algorithm. Using the resulting solution $\hat{\beta}$, one can obtain the linear predictors η_i and the probabilities p_i of Y_i (and expectations, as well) for $i = 1, \ldots, n$. If we want to test whether the covariate has an influence on Y, the null-hypothesis is given by

$$
H_0 : \beta = \beta_0 = 0.
$$

In this setup, the matrix $A^\top A / n$ is a random one. We know that, under the null hypothesis, $A^\top A \overset{\mathbf{P}}{\to} \Gamma_{p+1}$, and, if $p_n \to \infty$ slowly, we get that

$$
\frac{\hat{\beta}_n^\top \Gamma_{p_n} \hat{\beta}_n - (p_n + 1)}{\sqrt{2(p_n + 1)}} \overset{D}{\to} N(0,1)
$$

under H_0. For more details, see [291].

Exercise 9.9. Suppose that the response variable Y is Poisson distributed. In this case, if we have n samples, the GLM is given by

$$
Y_i \sim \text{Poi}(\lambda_i), \qquad \text{with } \lambda_i = \mathbf{E} Y_i = \mu_i = g(\eta_i),
$$

where η_i is a linear predictor as in (9.28) and $g(x) = e^x$. Compute the log-likelihood function $l(Y;\beta)$ and show that, also in this case,

$$
\nabla_\beta l(Y;\beta) = A^\top (Y - \mu), \qquad \text{with } A = \left(1, \xi_1^{(k)}, \ldots, \xi_p^{(k)} \right)_{k=1}^{n}.
$$

Chapter 10
Some Statistical Methods in Genetics

Alexander Bulinski

Abstract A challenging problem in modern genetics is to identify the collection of factors responsible for increasing the risk of specified complex diseases. The progress in the human genome reading permitted to collect the genetic datasets for analysis by means of various complementary statistical tools. The intensive studies in this research domain are carried out in leading research centers all over the world. One has to operate with data of huge dimensions and this is one of the main difficulties in detection of genetic susceptibility to common diseases such as hypertension, myocardial infarction and others. In this chapter, we concentrate on the multifactor dimensionality reduction method, and we also discuss its modifications and extensions. Our recent results on the central limit theorem related to this method are provided as well. Moreover, we explain the main features of the logic regression where we tackle the simulated annealing for stochastic minimization of functions defined on a graph with forests as vertices. Finally, we mention several important research directions which are out of the scope of the present chapter.

10.1 Introduction

There is a number of deep mathematical results related to problems in genetics. We mention the fundamental contributions by J.B.S. Haldane (1892-1964), R.H. Fisher (1890-1962) and S.G. Wright (1889-1988) made in the first half of the 20-th century. One can say that these scientists created the domain of population genetics, which deals with the complex issues of genotype involving many alleles, over long periods of time and under different modes of mating (for basic concepts in genetics we refer, e.g., to [23] and [48]). Further development of this branch of investigations is considered, e.g., in [39]. The discovery of the DNA structure in 1953 by

Alexander Bulinski
Faculty of Mathematics and Mechanics, Lomonosov Moscow State University, Moscow 119991, Russia, e-mail: bulinski@yandex.ru

© Springer International Publishing Switzerland 2015

V. Schmidt (ed.), *Stochastic Geometry, Spatial Statistics and Random Fields*, Lecture Notes in Mathematics 2120, DOI 10.1007/978-3-319-10064-7_10

F.H.C. Crick (1916-2004) and J.D. Watson was a milestone determining many features of modern molecular biology. Moreover, the situation has radically changed at the beginning of the 21-st century when the laboratory methods (especially the microchip technology) permitted to obtain data concerning the individual genetic code.

According to the information provided by the World Health Organization (see http://www.who.int/topics), in 2030 more than 23 million people will die due to cardiovascular diseases and more than 13 million will be the victims of the oncological ones. Thus it is not surprising that in research centers all over the world intensive studies are devoted to detection of genetic susceptibility of individuals to complex diseases, mentioned above as well as others. This research direction called the *genome-wide association study* (GWAS) involves specialists in medicine, biology, chemistry, informatics and mathematical statistics. The successes in the human genome reading led to the creation of data sets in the framework of several international projects, see, e.g., http://www.gwascentral.org/. Note that GWAS is of great importance as the diagnostics, prophylaxis and therapy of complex diseases could be improved. The ultimate goals are to develop new drugs for their treatment and create a personalized medicine. A review of investigations in GWAS during the last five years is provided in [403].

One uses the term *simple (or Mendelian) disease* when it is related to one mutation in a specified segment (locus) of the DNA molecule. A classical example of such disease is sicklemia. In contrast many other diseases, e.g., Altzheimer's disease or schizophrenia, are provoked by mutations in different parts (loci) of the DNA molecule which are engaged in the formation of certain proteins. Diseases of the latter type are called *complex.*

In genetic data analysis the *single nucleotide polymorphism* (SNP) plays an important role among other markers. SNP means a small variation in the genetic code, namely, the change of a specific nucleotide (denoted as A, T, C and G, that is Adenine, Thymine, Cytosine and Guanine) in a strand of the double helix of DNA. More exactly such changes or "damages" must occur in a certain percent of population (not less than 5% in general). The problem of detection of the collections of SNP which have the impact on specified phenotype properties is a challenging one as here one needs to create new biological approaches, in combination with new tools in statistics and informatics, to the analysis of data having huge dimensionality. Recall that the human genom contains about 6 milliard letters (nucleotides) A,T,C and G.

In this chapter, we will discuss various statistical methods of genetic and non-genetic (environmental) data analysis. The corresponding papers published in such journals as Advances in Genetics, American Journal of Human Genetics, Biometrika, Bioinformatics, Biostatistics, Briefings in Bioinformatics, Human Heredity and others demonstrate that modern investigations in GWAS include the construction of algorithms for data analysis, their computer implementation as well as applications to simulated and real data. The statisticians employ sophisticated models described, e.g., by means of hidden Markov processes (or fields) and spatial point processes. Various methods are used, for instance, simulated annealing, Markov chain Monte

Carlo (MCMC) as well as Bayesian inference and machine learning techniques. However, this scenario does not imply that each time rigorous proofs of mathematical results are provided.

The present chapter is organized as follows. After the brief introduction in Sect. 10.2, we consider the problem of identification of the most significant factors which could increase the risk of complex diseases. To this end we treat the multifactor dimensionality reduction (MDR) method and its modifications and extensions. In Sect. 10.3 we dwell on logistic and logic regression, as well as on stochastic optimization techniques. Section 10.4 is devoted to some models involving random fields. Sect. 10.5 contains some concluding remarks. We also tackle some other research directions which deserve special attention.

10.2 The MDR method and its Modifications

In this section we concentrate on the development of the *multifactor dimensionality reduction* (MDR) method. It has been introduced in [332]. One can say that this method is a constructive induction that seeks to identify combinations of multi-locus genotypes that are associated with either high or low risk of disease. The constructive induction described in [273] is a general process of defining a new attribute as a function of two or more other attributes. A comprehensive survey concerning the MDR method is provided in [333]. Some new results which are not in the scope of the latter paper are tackled as well. We are interested in the "dimensionality reduction" of the whole collection of factors describing the response variable. Therefore, we use the term "MDR method". However instead of considering the contingency tables (to specify zones of low and high risk) presented in [332] and various subsequent works, we choose another way.

10.2.1 Implementation of the MDR Method

Assume that all random variables under consideration are defined on a probability space $(\Omega, \mathcal{F}, \mathbf{P})$. Let $X_i : \Omega \to \{0, 1, \ldots, q\}$ be a random variable where $i = 1, \ldots, n$ (q and n are some positive integers), and let $X = (X_1, \ldots, X_n)$. Hence the random vector X with components X_i takes values in $\mathbb{X} = \{0, 1, \ldots, q\}^n$. Consider a random (*response*) variable $Y : \Omega \to \{-1, 1\}$ depending on the *factors* (explanatory variables) $X_1, \ldots, X_n$. For example in medical studies such response variable Y can describe the health state, e.g., $Y = 1$ or $Y = -1$ means "sick" or "healthy" (one says also "case" or "control"), and $X_1, \ldots, X_m$ and $X_{m+1}, \ldots, X_n$ are genetic and non-genetic factors, respectively. Usually X_i ($1 \leq i \leq m$) characterizes a *single nucleotide polymorphism* (SNP) in a specified locus of the DNA molecule. In this case one considers such X_i having three values, for instance, 0, 1 and 2 according to genotypes aa, aA and AA, see, e.g., [54]. It is convenient to suppose that the other X_i ($m + 1 \leq i \leq n$) also take values in $\{0, 1, 2\}$. For instance the domain of blood pressure can be partitioned in zones of low, normal and high values. However, in the following, we will suppose that all factors take values in an arbitrary finite set. Assume that there are N individuals. Let $X^j = (X_1^j, \ldots, X_n^j)$ and Y^j be (random) factors and the response variable for the j-th individual ($j = 1, \ldots, N$). As usual we write small letters for realizations of random variables (or vectors).

First of all we recall the details of the original MDR method [332]. Its implementation involved balanced case-control studies. There are six steps inherent to this method. At *step one*, the dataset consisting of N individuals is (randomly) divided into K pair-wise disjoint parts (groups) $G_1, \ldots, G_K$ for cross-validation to avoid over-fitting. Usually $K = 10$. In the case of ten-fold cross-validation, the training set is given by 90 percent of the data, whereas the testing set comprises the remaining 10 percent of the data. For cross-validation see, e.g., [6]; further we explain how to modify the procedure if N is not divisible by K. We start to deal with group G_1 as testing set whereas the training set $V_1 = \bigcup_{t=2}^{K} G_t$. In *step two*, for a given integer r ($1 \leq r < n$), a set of genetic and/or environmental factors $X_{m_1}, \ldots, X_{m_r}$ is selected among $X_1, \ldots, X_n$. Clearly, there exist $\binom{n}{r}$ subsets $\{m_1, \ldots, m_r\} \subset \{1, \ldots, n\}$. In *step three* one introduces cells (multifactor classes) in $\mathbb{R}^r$ for all possible values of the vector $(X_{m_1}, \ldots, X_{m_r})$. For example, for two loci with three genotypes each, there are nine possible twolocus-genotype combinations. Then for each individual j belonging to the training dataset we consider the vector $(x_{m_1}^j, \ldots, x_{m_r}^j)$, choose the corresponding cell and write down into it the value y^j of Y^j. Then, for each cell we find the number of controls and cases. In other words one constructs a contingency table. The ratio of the number of cases to the number of controls is calculated within each cell. We formally put $0/0 = 0$ and $C/0 = \infty$ for $C > 0$ (another possibility is to introduce empty cells). After that, in *step four* each cell is labeled as "high risk" if the cases to controls ratio exceeds some threshold (say, 1) and as "low risk" otherwise. Thus we get a reduction of the multidimensional space in which the vector $(X_{m_1}, \ldots, X_{m_r})$ takes values to the simple "one dimensional" case of two labels "high" or "low". In Fig.1 of [332] the dark-shaded cells represent high-risk genotype combinations, whereas the light-shaded cells correspond to low-risk genotype combinations and the white cells indicate empty cells without observations. We make predictions about the disease status of each individual belonging to the testing group G_1. If the value of the observation vector with components labeled by $m_1, \ldots, m_r$ belongs to "high risk" cells we predict the value 1 for the response variable and -1 otherwise. The proportion of individuals for which an incorrect prediction was made is an estimator of the prediction error. Then, in *step five*, steps three and four are repeated for each possible cross-validation interval. Namely, each time we take the group G_t as testing set and the union of other groups as training set, $t = 1, \ldots, K$. Thus the K-fold cross-validation is repeated K times and the prediction errors are averaged. Finally, in *step six*, we get a list of the averaged prediction errors for all possible combinations $X_{m_1}, \ldots, X_{m_r}$ and choose among them one or more combinations having the minimum value of this error.

The procedure described above is summarized in Table 10.1 where for $x = (x_1, \ldots, x_n)$ and $\Lambda = \{m_1, \ldots, m_r\} \subset \{1, \ldots, n\}$ we put $x_\Lambda = (x_{m_1}, \ldots, x_{m_r})$. For $A \subset D$ we consider the *indicator function*

$$1\{A\}(x) = \begin{cases} 1, & \text{if } x \in A, \\ 0, & \text{if } x \in D \setminus A, \end{cases}$$

where we write $1\{A\}$ instead of $1_D\{A\}$ when the choice of D is clear. Further $|G|$ stands for the cardinality of a finite set G.

Table 10.1 Implementation of the MDR method

Step 1	Step 2
Fix $K \in \mathbb{N}$, $r \in \{1,\dots,n-1\}$. Let $\{1,\dots,N\} = \cup_{t=1}^{K} G_t$, $G_s \cap G_t = \varnothing, s \neq t$, $V_t = \{1,\dots,N\} \setminus G_t$.	Choose $\Lambda = \{m_1,\dots,m_r\} \subset \{1,\dots,n\}$ and take $t = 1$.
Step 3	**Step 4**
For $z \in \{0,1,2\}^r$, $t \in \{1,\dots,K\}$ put $$x_\Lambda^j = (x_{m_1}^j,\dots,x_{m_r}^j),$$ $$M_t^+(z) = \sum_{j \in V_t : x_\Lambda^j = z} 1\{y^j = 1\},$$ $$M_t^-(z) = \sum_{j \in V_t : x_\Lambda^j = z} 1\{y^j = -1\},$$ $$R_t(z) = M_t^+(z)/M_t^-(z).$$	Let $F_t(z) = 1\{R_t(z) > 1\} - 1\{R_t(z) \leq 1\}$, $z \mapsto$ high risk if $F_t(z) = 1$, $z \mapsto$ low risk if $F_t(z) = -1$. For G_t calculate the prediction error $$E_t(\Lambda) = \tfrac{1}{|G_t|} \sum_{j \in G_t} 1\{y^j \neq F_t(x_\Lambda^j)\}.$$
Step 5	**Step 6**
Make cross-validation, i.e. take $t+1$ instead of t, repeat steps 3 and 4, calculate $E_{t+1}(\Lambda)$ until $t+1 \leq K$. Find average error $AE_K(\Lambda) = \tfrac{1}{K}\sum_{t=1}^{K} E_t(\Lambda)$.	Repeat steps 2 to 5 for each $\Lambda = \{m_1,\dots,m_r\}$. Choose Λ corresponding to minimal $AE_K(\Lambda)$

Note that in step four one could compare the ratio $R_t(z)$ (see Table 10.1) not with 1 but with the global ratio of cases over controls in the particular genotype combination being evaluated. Also there are several characteristics which can be derived from the number of *true positives* (TP), *true negatives* (TN), *false positives* (FP) and *false negatives* (FN) in the test set, describing the performance of classification model. For example the notion of *accuracy* (Acc), *specificity* (SP) and *true positive rate* or *sensitivity* (TPR) are defined as follows:

$$Acc = \frac{\text{TP}+\text{TN}}{\text{TP}+\text{TN}+\text{FP}+\text{FN}}, \quad SP = \frac{\text{TN}}{\text{FP}+\text{TN}}, \quad \text{TPR} = \frac{\text{TP}}{\text{TP}+\text{FN}}.$$

Note that we do not tackle the problem of balanced or unbalanced samples (see, e.g., [402]), since below we do not need to distinguish these cases. Later on we return to the problem of finding r factors among n used for further detailed analysis and the estimation of the disease risk.

10.2.2 Prediction Algorithm

Now we are interested in the approximation of Y by means of $f(X)$ where $f: \mathbb{X} \to \{-1,1\}$ is a non-random function. In this way one can justify the choice of a reduced collection of factors to describe the response variable. Using a penalty function $\psi: \{-1,1\} \to \mathbb{R}_+$ (the trivial case $\psi \equiv 0$ is excluded) the quality of such approximation is defined by

$$\text{Err}(f) = \mathbf{E}\left(|Y - f(X)|\psi(Y)\right). \tag{10.1}$$

In other words we ascribe, in general, different weights to the approximation of values 1 and -1 taken by Y. Introduce $M = \{x \in \mathbb{X} : \mathbf{P}(X = x) > 0\}$ and

$$F(x) = \psi(-1)\mathbf{P}(Y = -1 \mid X = x) - \psi(1)\mathbf{P}(Y = 1 \mid X = x), \quad x \in M.$$

Exercise 10.1. Prove that solutions of the problem $\text{Err}(f) \to \inf$ have the form

$$f = \mathbf{1}\{A\} - \mathbf{1}\{\overline{A}\}, \quad A \in \mathcal{A}, \tag{10.2}$$

where $\mathbf{1}\{\varnothing\} = 0$ and $\mathcal{A}$ consists of the sets $A = \{x \in M : F(x) < 0\} \cup B \cup C$. Here B is an arbitrary subset of $\{x \in M : F(x) = 0\}$ and C is any subset of $\mathbb{X} \setminus M$.

We say that f is an *optimal function* when it belongs to the class of functions described in (10.2).

If we choose $A^* = \{x \in M : F(x) < 0\}$, then A^* has the minimal cardinality among all subsets of $\mathcal{A}$. Taking into account that $\psi(-1) + \psi(1) \neq 0$ we can write

$$A^* = \{x \in M : \mathbf{P}(Y = 1 \mid X = x) > \gamma(\psi)\}, \quad \gamma(\psi) = \frac{\psi(-1)}{\psi(-1) + \psi(1)}. \tag{10.3}$$

Note that we can rewrite (10.1) as

$$\text{Err}(f) = 2 \sum_{y \in \{-1,1\}} \psi(y)\mathbf{P}(Y = y, f(X) \neq y). \tag{10.4}$$

The distribution of the random vector (X,Y) is unknown and we cannot calculate the value $\text{Err}(f)$. Therefore statistical inference on the quality of approximation of Y by means of $f(X)$ is based on an estimator of $\text{Err}(f)$ involving i.i.d. random vectors $\xi^1, \xi^2, \ldots$ with the same distribution as that of vector (X,Y). We consider all vectors as column ones and to simplify notation we write (X, Y) instead of $(X^\top, Y)^\top$ where $\top$ stands for transposition.

To approximate $\mathrm{Err}(f)$ as $N \to \infty$ we use a *prediction algorithm*. It involves a function $f_{\mathrm{PA}} = f_{\mathrm{PA}}(x, \xi_N)$ with values $\{-1, 1\}$ which is defined for $x \in \mathbb{X}$ and $\xi_N = (\xi^1, \ldots, \xi^N)$. More exactly we use a *family* of functions $f_{\mathrm{PA}}^m(x, v_m)$ defined for $x \in \mathbb{X}$ and $v_m \in \mathbb{V}_m$ where $\mathbb{V}_m = (\mathbb{X} \times \{-1, 1\})^m$, $m \in \mathbb{N}$, $m \leq N$. To simplify the notation we write $f_{\mathrm{PA}}(x, v_m)$ instead of $f_{\mathrm{PA}}^m(x, v_m)$.

For $S = \{j_1, \ldots, j_s\}$ where $1 \leq j_1 < \ldots < j_s \leq N$ put $\xi_N(S) = (\xi^{j_1}, \ldots, \xi^{j_s})$ and let $\overline{S} = \{1, \ldots, N\} \setminus \{j_1, \ldots, j_s\}$. For $K \in \mathbb{N}$ $(K > 1)$ consider a partition of $\{1, \ldots, N\}$ formed by the sets

$$S_k(N) = \{(k-1)[N/K] + 1, \ldots, k[N/K]\mathbf{1}\{k < K\} + N\mathbf{1}\{k = K\}\} \tag{10.5}$$

where $k = 1, \ldots, K$ and $[a]$ denotes the integer part of $a \in \mathbb{R}$. Obviously, for the cardinality $|S_k(N)|$ of the finite set $S_k(N)$ it holds that $|S_k(N)| = [N/K]$ for index $k = 1, \ldots, K-1$, and $[N/K] \leq |S_K(N)| < [N/K] + K$.

10.2.3 Estimated Prediction Error

Generalizing [54] we can construct an estimator of $\mathrm{Err}(f)$ using a sample ξ_N, a prediction algorithm with f_{PA} and a K-fold cross-validation where $K \in \mathbb{N}$, $K > 1$. Namely, let

$$\widehat{\mathrm{Err}}_K(f_{\mathrm{PA}}, \xi_N) =$$

$$2 \sum_{y \in \{-1,1\}} \frac{1}{K} \sum_{k=1}^{K} \sum_{j \in S_k(N)} \frac{\widehat{\psi}(y, S_k(N))\mathbf{1}\{Y^j = y, f_{\mathrm{PA}}(X^j, \xi_N(\overline{S_k(N)})) \neq y\}}{|S_k(N)|}, \tag{10.6}$$

where for each $k = 1, \ldots, K$ the random variables $\widehat{\psi}(y, S_k(N))$ are strongly consistent estimators (as $N \to \infty$) of $\psi(y)$, $y \in \{-1, 1\}$, constructed by the data $\{Y^j, j \in S_k(N)\}$. We call $\widehat{\mathrm{Err}}_K(f_{\mathrm{PA}}, \xi_N)$ an *estimated prediction error*.

We now formulate a result which is important for statistical applications.

Theorem 10.1 ([58]). *Let f_{PA} define a prediction algorithm for $f : \mathbb{X} \to \{-1, 1\}$. Assume the existence of a set $U \subset \mathbb{X}$ such that for each $x \in U$ and any $k = 1, \ldots, K$ it holds that*

$$f_{\mathrm{PA}}(x, \xi_N(\overline{S_k(N)})) \to f(x) \text{ a.s., for } N \to \infty. \tag{10.7}$$

Then

$$\widehat{\mathrm{Err}}_K(f_{\mathrm{PA}}, \xi_N) \to \mathrm{Err}(f) \text{ a.s., for } N \to \infty, \tag{10.8}$$

if and only if, for $N \to \infty$, the following relation holds a.s.

$$\sum_{k=1}^{K} \left(\sum_{x \in \mathbb{X}^+} \mathbf{1}\{f_{\mathrm{PA}}(x, \xi_N(\overline{S_k(N)})) = -1\}L(x) - \sum_{x \in \mathbb{X}^-} \mathbf{1}\{f_{\mathrm{PA}}(x, \xi_N(\overline{S_k(N)})) = 1\}L(x) \right) \to 0,$$

$$\tag{10.9}$$

where $\mathbb{X}^+ = (\mathbb{X} \setminus U) \cap \{x \in M : f(x) = 1\}$, $\mathbb{X}^- = (\mathbb{X} \setminus U) \cap \{x \in M : f(x) = -1\}$
and

$$L(x) = \psi(1)\mathbf{P}(X = x, Y = 1) - \psi(-1)\mathbf{P}(X = x, Y = -1), \quad x \in \mathbb{X}. \qquad (10.10)$$

The above criterion for the validity of (10.8) shows that condition (10.9) plays the key role (when U does not coincide with $\mathbb{X}$) to ensure that (10.8) holds. As usually, a sum over the empty set is put equal to 0.

We will use still another statement with an easily verifiable condition.

Corollary 10.1 ([58]). *Let* f_{PA} *define a prediction algorithm for* $f : \mathbb{X} \to \{-1, 1\}$. *Suppose that there exists a set* $U \subset \mathbb{X}$ *such that for each* $x \in U$ *and any* $k = 1, \ldots, K$ *relation (10.7) is true. If*

$$L(x) = 0 \quad \text{for } x \in (\mathbb{X} \setminus U) \cap M \qquad (10.11)$$

then (10.8) is satisfied.

Note also that Remark 4 from [58] explains why the natural choice of a penalty function is the one proposed in [402], i.e.,

$$\psi(y) = c(\mathbf{P}(Y = y))^{-1}, \quad y \in \{-1, 1\}, \ c > 0. \qquad (10.12)$$

Further discussion and examples can be found in [58].

10.2.4 Dimensionality Reduction

We now turn to the foundations of the *multifactor dimensionality reduction* (MDR) method. In many situations it is reasonable to suppose that the response variable Y depends only on some subcollection $X_{k_1}, \ldots, X_{k_r}$ of the explanatory variables $X_1, \ldots, X_n$, where $\{k_1, \ldots, k_r\} \subset \{1, \ldots, n\}$ $(1 \leq r < n)$. This means that for any $x \in M$

$$\mathbf{P}(Y = 1 \mid X_1 = x_1, \ldots, X_n = x_n) = \mathbf{P}(Y = 1 \mid X_{k_1} = x_{k_1}, \ldots, X_{k_r} = x_{k_r}). \qquad (10.13)$$

In the framework of complex disease analysis it is natural to assume that only a part of the risk factors could provoke that disease and the impact of the others can be neglected. Relation (10.13) can arise in many other situations, e.g., pertinent to pharmacology (in this context the response variable describes the efficiency or non-efficiency of a drug).

Any collection $\{k_1, \ldots, k_r\}$ implying (10.13) is called *significant*. If $\{k_1, \ldots, k_r\}$ is significant then any collection $\{m_1, \ldots, m_i\}$ such that $\{k_1, \ldots, k_r\} \subset \{m_1, \ldots, m_i\}$ is significant as well. For a set $D \subset \mathbb{X}$ we define its "projection"

$$\pi_{k_1, \ldots, k_r} D = \{u = (x_{k_1}, \ldots, x_{k_r}) : x = (x_1, \ldots, x_n) \in D\}.$$

For $B \subset \mathbb{X}_r$ where $\mathbb{X}_r = \{0, 1, \ldots, q\}^r$, let us introduce in $\mathbb{X} = \mathbb{X}_n$ the cylinder

$$C_{k_1,\ldots,k_r}(B) = \{x = (x_1,\ldots,x_n) \in \mathbb{X} : (x_{k_1},\ldots,x_{k_r}) \in B\}.$$

If $B = \{u\}$ where $u = (u_1,\ldots,u_r) \in \mathbb{X}_r$, we write $C_{k_1,\ldots,k_r}(u)$ instead of $C_{k_1,\ldots,k_r}(\{u\})$. Clearly, for

$$u = \pi_{k_1,\ldots,k_r}\{x\}, \quad \text{i.e. } u_i = x_{k_i}, \quad \text{for } i = 1,\ldots,r, \tag{10.14}$$

we have

$$\mathbf{P}(Y = 1 \mid X_{k_1} = x_{k_1},\ldots,X_{k_r} = x_{k_r}) \equiv \mathbf{P}(Y = 1 \mid X \in C_{k_1,\ldots,k_r}(u)). \tag{10.15}$$

If (10.13) holds then the optimal function f^* introduced in (10.2) with $A = A^*$ defined in (10.3) has the form

$$f^{k_1,\ldots,k_r}(x) = \begin{cases} 1, & \text{if } \mathbf{P}(Y = 1 \mid X \in C_{k_1,\ldots,k_r}(u)) > \gamma(\psi), \ x \in M, \\ -1, & \text{otherwise,} \end{cases} \tag{10.16}$$

where u and x satisfy (10.14) because $\mathbf{P}(X \in C_{k_1,\ldots,k_r}(u)) \geq \mathbf{P}(X = x) > 0$ for $x \in M$. Hence, for each significant $\{k_1,\ldots,k_r\} \subset \{1,\ldots,n\}$ and any collection $\{m_1,\ldots,m_r\} \subset \{1,\ldots,n\}$ we have

$$\mathrm{Err}(f^{k_1,\ldots,k_r}) \leq \mathrm{Err}(f^{m_1,\ldots,m_r}). \tag{10.17}$$

For $C \subset \mathbb{X}$, $N \in \mathbb{N}$ and $W_N \subset \{1,\ldots,N\}$, we put

$$\widehat{\mathbf{P}}_{W_N}(Y = 1 \mid X \in C) = \frac{\sum_{j \in W_N} \mathbf{1}\{Y^j = 1, X^j \in C\}}{\sum_{j \in W_N} \mathbf{1}\{X^j \in C\}}. \tag{10.18}$$

Furthermore, we formally put $0/0 = 0$, and write $\widehat{\mathbf{P}}_{W_N}(Y = 1)$ when $C = \mathbb{X}$ in (10.18). According to the *strong law of large numbers for arrays* (SLLNA), see, e.g., [388], for any $C \subset \mathbb{X}$ with $\mathbf{P}(X \in C) > 0$ one can claim that

$$\widehat{\mathbf{P}}_{W_N}(Y = 1 \mid X \in C) \to \mathbf{P}(Y = 1 \mid X \in C) \text{ a.s., if } |W_N| \to \infty \text{ as } N \to \infty. \tag{10.19}$$

Let $\widehat{\gamma}_{W_N}(\psi)$ be a strongly consistent estimator of $\gamma(\psi)$ constructed by means of $\xi_N(W_N)$. For arbitrary $\{m_1,\ldots,m_r\} \subset \{1,\ldots,n\}$, $x \in \mathbb{X}$, $v = \pi_{m_1,\ldots,m_r}\{x\}$ and a penalty function ψ we consider the prediction algorithm with a function $\widehat{f}_{\mathrm{PA}}^{m_1,\ldots,m_r}$ such that

$$\widehat{f}_{\mathrm{PA}}^{m_1,\ldots,m_r}(x, \xi_N(W_N)) = \begin{cases} 1, & \text{if } \widehat{\mathbf{P}}_{W_N}(Y = 1 \mid X \in C_{m_1,\ldots,m_r}(v)) > \widehat{\gamma}_{W_N}(\psi), \ x \in M, \\ -1, & \text{otherwise.} \end{cases} \tag{10.20}$$

Let

$$U = \{x \in M : \mathbf{P}(Y = 1 \mid X \in C_{m_1,\ldots,m_r}(v)) \neq \gamma(\psi)\}. \tag{10.21}$$

Then, Corollary 10.1 (with Examples 1 and 2 of [58]) yields that

$$\widehat{\mathrm{Err}}_K(\widehat{f}_{\mathrm{PA}}^{m_1,\ldots,m_r}, \xi_N) \to \mathrm{Err}(f^{m_1,\ldots,m_r}) \quad \text{a.s., as } N \to \infty. \tag{10.22}$$

Therefore relations (10.17) and (10.22) imply that, for each $\delta > 0$, any significant collection $\{k_1,\ldots,k_r\} \subset \{1,\ldots,n\}$ and an arbitrary set $\{m_1,\ldots,m_r\} \subset \{1,\ldots,n\}$, it holds that

$$\widehat{\mathrm{Err}}_K(\widehat{f}_{\mathrm{PA}}^{k_1,\ldots,k_r}, \xi_N) \leq \widehat{\mathrm{Err}}_K(\widehat{f}_{\mathrm{PA}}^{m_1,\ldots,m_r}, \xi_N) + \delta \quad \text{a.s.} \tag{10.23}$$

when N is large enough.

Thus, according to results established in [58], we come to the following conclusion. For a given $r = 1,\ldots,n-1$, in view of (10.23) it is natural to choose a collection $X_{k_1},\ldots,X_{k_r}$ among the factors $X_1,\ldots,X_n$ leading to the smallest estimated prediction error $\widehat{\mathrm{Err}}_K(\widehat{f}_{\mathrm{PA}}^{k_1,\ldots,k_r}, \xi_N)$. After that it is desirable to employ the permutation tests (see, e.g., [54] and [145]) to validate the prediction power of the selected factors. We do not tackle the choice of r (usually r is significantly smaller than n), some recommendations can be found in [333].

Remark 10.1. It is worth to emphasize that for each $\{m_1,\ldots,m_r\} \subset \{1,\ldots,n\}$ we have constructed strongly consistent estimators of $\mathrm{Err}(f^{m_1,\ldots,m_r})$ and we can compare these estimators on an event having probability one. If we had only the convergence in probability instead of a.s. convergence in (10.22) then to compare $\widehat{\mathrm{Err}}_K(\widehat{f}_{\mathrm{PA}}^{m_1,\ldots,m_r}, \xi_N)$ for different subsets $\{m_1,\ldots,m_r\}$ of $\{1,\ldots,n\}$ one must take into account the Bonferroni corrections. This is not reasonable when the sets $\{m_1,\ldots,m_r\}$ form a large class.

10.2.5 Central Limit Theorem

The next natural problem is to estimate the convergence rate in (10.22). We provide a quite recent result proven in [59]. For this purpose let us further consider a function ψ having the form (10.12). In this case $\gamma(\psi) = \mathbf{P}(Y = 1)$. Then due to (10.3) without loss of generality we can put $c = 1$ in (10.12). Define the events

$$A_{N,k}(y) = \{Y^j = -y, \ j \in S_k(N)\}$$

and the random variables

$$\widehat{\psi}(y, S_k(N)) = \frac{\mathbf{1}\{\overline{A_{N,k}(y)}\}}{\widehat{\mathbf{P}}_{S_k(N)}(Y = y)}, \quad \text{for } N \in \mathbb{N}, \ k = 1,\ldots,K, \ y \in \{-1,1\}, \tag{10.24}$$

where the trivial cases $\mathbf{P}(Y = y) \in \{0,1\}$ are excluded (recall that $0/0 = 0$).

For $\{m_1,\ldots,m_r\} \subset \{1,\ldots,n\}$ we define functions which can be viewed as the *regularized versions* of the estimators $\widehat{f}_{\mathrm{PA}}^{m_1,\ldots,m_r}$ of $f^{m_1,\ldots,m_r}$ (see (10.20) and (10.16)). Namely, for $x \in \mathbb{X}$, $v = \pi_{m_1,\ldots,m_r}x$, $N \in \mathbb{N}$, $W_N \subset \{1,\ldots,N\}$ and $\varepsilon = (\varepsilon_N)_{N \in \mathbb{N}}$ where the non-random $\varepsilon_N \to 0$ as $N \to \infty$, we put

$$\widehat{f}_{\mathrm{PA},\varepsilon}^{m_1,\ldots,m_r}(x,\xi_N(W_N)) = \begin{cases} 1, & \text{if } \widehat{\mathbf{P}}_{W_N}(Y=1|X\in C_{m_1,\ldots,m_r}(v)) > \widehat{\gamma}_{W_N}(\psi)+\varepsilon_N,\, x\in M, \\ -1, & \text{otherwise.} \end{cases}$$

$$(10.25)$$

In other words we use $\widehat{\gamma}_{W_N}(\psi) + \varepsilon_N$ instead of the threshold $\widehat{\gamma}_{W_N}(\psi)$ in (10.20).

Take now U as defined in (10.21). Applying Corollary 10.1 once again (and Examples 1 and 2 of [58]) we can claim that statements analogous to (10.22) and (10.23) are valid for the estimators introduced in (10.25). However in this case we have the following more precise result.

Theorem 10.2 ([59]). *Let $\varepsilon_N \to 0$ and $N^{1/2}\varepsilon_N \to \infty$ as $N \to \infty$. Then, for each $K \in \mathbb{N}$, any subset $\{m_1,\ldots m_r\}$ of $\{1,\ldots,n\}$ $(r \in \mathbb{N})$, $f = f^{m_1,\ldots,m_r}$ and a prediction algorithm defined by a function $f_{\mathrm{PA}} = \widehat{f}_{\mathrm{PA},\varepsilon}^{m_1,\ldots,m_r}$ the following central limit theorem holds:*

$$\sqrt{N}(\widehat{Err}_K(f_{\mathrm{PA}},\xi_N) - Err(f)) \xrightarrow{D} Z \sim N(0,\sigma^2), \quad \text{as } N \to \infty. \tag{10.26}$$

Here the estimators considered in (10.24) are those used in the construction of $\widehat{Err}_K(f_{\mathrm{PA}},\xi_N)$, and $\sigma^2 = \mathbf{var}\,V$ where

$$V = 2\sum_{y\in\{-1,1\}} \frac{\mathbf{1}\{Y=y\}}{\mathbf{P}(Y=y)}\left(\mathbf{1}\{f(X)\neq y\} - \mathbf{P}(f(X)\neq y \mid Y=y)\right). \tag{10.27}$$

Recall that for a sequence of random variables $(\eta_N)_{N\in\mathbb{N}}$ and a sequence $(a_N)_{N\in\mathbb{N}}$ of positive numbers one writes $\eta_N = o_{\mathbf{P}}(a_N)$ if $\eta_N/a_N \xrightarrow{\mathbf{P}} 0$, $N \to \infty$.

Remark 10.2. Clearly one can interpret the central limit theorem (CLT) as a result describing the rate of approximation for the random variables under consideration. Thus Theorem 10.2 implies that

$$\widehat{Err}_K(f_{\mathrm{PA}},\xi_N) - Err(f) = o_{\mathbf{P}}(a_N), \quad N \to \infty, \tag{10.28}$$

where $a_N = o(N^{-1/2})$, and this is an optimal estimator if $\sigma^2 \neq 0$, i.e., one cannot take $a_N = O(N^{-1/2})$ in (10.28).

Exercise 10.2. Let $\{\zeta_N\}_{N\in\mathbb{N}}$ be a sequence of random variables. Assume that for some sequence of real numbers $\{b_N\}_{N\in\mathbb{N}}$ it holds that $b_N\zeta_N \xrightarrow{D} \zeta$, as $N \to \infty$, where ζ is a random variable with continuous distribution function. Prove that $b_N \neq 0$ for all N large enough and $\zeta_N = o_{\mathbf{P}}(c_N)$ as $N \to \infty$, for any $c_N = o(1/b_N)$. Show that if one takes $c_N = O(1/b_N)$, as $N \to \infty$, then the relation $\zeta_N = o_{\mathbf{P}}(c_N)$ is not true.

Proofs and further discussion including a multidimensional CLT and its statistical version with self-normalization can be found in [59]. The generalizations for a non-binary response variable Y are obtained in [55].

10.2.6 Modifications of the MDR Method

Now we consider modifications of the MDR-method. In [54] a new version of the MDR method has been introduced and called the *MDR method with independent rule*. The idea is to combine the MDR method with an approach proposed in [310] to classify a large array of binary data. Namely, for certain $C = C_{m_1,\ldots,m_r}(v)$, the use of the estimator $\widehat{\mathbf{P}}_{W_N}(Y = 1 \mid X \in C)$ introduced in (10.18) could be unreasonable. In fact the number of non-zero summands in the nominator and denominator of the fraction appearing in (10.18) could be rather small (or zero) for some cells, even for large N and large sample W_N. Due to Bayes' formula we have

$$\mathbf{P}(Y = 1 \mid X \in C) = \frac{\mathbf{P}(X \in C \mid Y = 1)\mathbf{P}(Y = 1)}{\mathbf{P}(X \in C \mid Y = 1)\mathbf{P}(Y = 1) + \mathbf{P}(X \in C \mid Y = -1)\mathbf{P}(Y = -1)}.$$

In [310] it was shown that for some class of models, $\mathbf{P}(X \in C_{m_1,\ldots,m_r}(v) \mid Y = y)$ admits the estimate

$$\widehat{\mathbf{P}}_{W_N}(X \in C_{m_1,\ldots,m_r}(v)|Y = y) = \prod_{i=1}^{r} \frac{\sum_{j \in W_N} \mathbf{1}\{X_{m_i}^j = x_{m_i}, Y^j = y\}}{\sum_{j \in W_N} \mathbf{1}\{Y^j = y\}}, \quad y \in \{-1, 1\}.$$

Obviously, the sum $\sum_{j \in W_N} \mathbf{1}\{Y^j = y\}$ has regular behavior for large W_N.

Note that to reduce the time for searching significant combinations of factors, two-step procedures are used. At the first step one chooses the subset of all factors for further detailed analysis performed at the second step. For this purpose a certain function is introduced to characterize the dependence between Y and a collection of variables $X_{m_1}, \ldots, X_{m_r}$ where $\{m_1, \ldots, m_r\} \subset \{1, \ldots, n\}$. The collections with specified values of this function are rejected. After that other collections are considered.

In [288] the following approach based on the Shannon entropy was employed. For response variable Y and each factor X_k introduce the *information*

$$I(X_k) = H(X_k) + H(Y) - H(X_k, Y), \tag{10.29}$$

where, for a random variable (or vector) $Z : \Omega \to \{z_1, \ldots, z_m\}$ such that the probability $\mathbf{P}(Z = z_k)$ is positive, $k = 1, \ldots, m$, the *Shannon entropy* is defined by

$$H(Z) = -\sum_{k=1}^{m} \mathbf{P}(Z = z_k) \log_2 \mathbf{P}(Z = z_k). \tag{10.30}$$

The value of $I(X_k)$ characterizes the interrelation between Y and X_k. If Y and X_k are independent then $I(X_k) = 0$. One can say that if $I(X_k) > I(X_q)$ then the interrelation between X_k and Y is stronger than that between Y and X_q. Furthermore, one can characterize the relation between Y and the pairs (X_k, X_q) where $k, q = 1, \ldots, n$. Namely, put

$$I(X_k, X_q) = -H(X_k) - H(X_q) - H(Y)$$
$$+H(X_k, X_q) + H(X_k, Y) + H(X_q, Y) - H(X_k, X_q, Y). \quad (10.31)$$

Now, similar to the MDR method, one uses estimators of $I(X_k)$ and $I(X_k, X_q)$ for inference. Thus, for $H(Z)$ appearing in (10.30) take

$$\widehat{H}(Z) = -\sum_{k=1}^{m} \widehat{\mathbf{P}}(Z = z_k) \log_2 \mathbf{P}(Z = z_k). \quad (10.32)$$

where

$$\widehat{\mathbf{P}}(Z = z_k) = \frac{1}{N} \sum_{j=1}^{N} \mathbf{1}\{Z^j = z_k\}, \quad k = 1, \ldots, m,$$

and $Z^1, \ldots, Z^N$ are i.i.d. copies of Z (in other words we employ $X_1^j, \ldots, X_n^j$ and Y^j with $j = 1, \ldots, N$). For a given $s \in \mathbb{N}$ we can take for further analysis the factors X_i with i belonging to the set of indices k which correspond to the s largest values of $\widehat{I}(X_k) = \widehat{H}(X_k) + \widehat{H}(Y) - \widehat{H}(X_k, Y)$ or to the set of indices k, q which describe the s largest values of $\widehat{I}(X_k, X_q)$ (one uses (10.32) to estimate the right-hand side of (10.31)). However, the first step with the choice based only on the s largest values of $\widehat{I}(X_k)$ would not be convincing as the liaison between Y and single factors could be small (the absence of the main effect) whereas the dependence between Y and a certain combination of factors could be rather strong.

Measures of dependence for a collection of random variables more general than those defined in (10.29) and (10.31) are considered in [66]. For random variables X_i, $i = 1, \ldots, n$, and $X = (X_1, \ldots, X_n)$ write $X_J = (X_{j_1}, \ldots, X_{j_d})$ where $J = \{j_1, \ldots, j_d\}$ is a subset of $G = \{1, \ldots, n\}$. Put

$$\mu(X) = \sum_{J \subset G} (-1)^{|G \setminus J|} H(X_J)$$

where $H(X_\varnothing) = 0$. Furthermore, define

$$v(X) = \sum_{i=1}^{n} H(X_i) - H(X), \quad \tau(X) = v(X, Y) - v(X).$$

According to [66] the functions μ, v and τ are called n-*way interaction information* (nWII), *total correlation information* (TCI) and *phenotype-associated information* (PAI), respectively. The goals of [66] were 1) to develop a novel metric PAI that is robust to the confounding effects of factors, 2) demonstrate that the PAI is a useful information-theoretic metric for effectively screening gene-gene and gene-environment interactions, and 3) develop the algorithm AMBIENCE that employs the PAI metric to identify the variables involved in the strongest interactions.

The *generalized* MDR (GMDR) method employs the framework of generalized linear models for scoring in conjunction with MDR [257]. GMDR enables inclusion of covariants and handles both discrete and continuous traits in population-based study designs. GMDR utilizes the same riskpooling, dimensionality reduction strategy as MDR and yields the original MDR as a special case. However, despite availability of a more efficient implementation based on parallel computing (see [62]), MDR and its variants, including GMDR, are computationally intensive. We also mention the *model-based MDR* (MB-MDR) method proposed in [63] which allows a more flexible definition of risk cells than the application of MDR techniques. Furthermore, we refer to the *MDR pedigree disequilibrium test* (MDR-PDT) considered in [111] and *MDR in structured populations* (MDR-SP) proposed in [300].

In [303] the following *Gene-MDR* method has been introduced. First, MDR analysis is applied for combining multiple SNPs from the same gene. Second, between-gene MDR analysis then performs interaction exploration using the summarized gene effects from within-gene MDR analysis. This method was applied to bipolar disorder (BD) GWAS data from Welcome Trust Case Control Consortium (WTCCC). The results demonstrate that Gene-MDR is capable of detecting high order gene-gene interactions associated with BD. Thus reducing the dimension of genome-wide data from SNP level to gene level, Gene-MDR efficiently identifies high order gene-gene interactions. Therefore, according to [303] Gene-MDR can provide the key to understand complex disease etiology.

A *robust MDR method* (RMDR) has been introduced in [158] to perform constructive induction using Fisher's exact Test rather than a predetermined threshold. According to [158] the advantage of this approach is that one considers in the MDR analysis only those genotype combinations which are determined to be statistically significant. The RMDR method is applied to the detection of gene-gene interactions in genotype data from a population-based study of bladder cancer in New Hampshire.

To complete this section we mention that a deep problem is to choose and study the measure characterizing the importance of certain collections of factors. In this regard we refer to [360].

10.3 Logic Regression and Simulated Annealing

Consider now a model in which the factors $X_1, \ldots, X_n$ are binary random variables taking values 0 or 1. In the framework of SNP studies we can assume, e.g., that SNP at locus labeled by i is characterized by the random variable X_i where $X_i = 0$ for recessive genotype (aa) and $X_i = 1$ for dominant one (AA or Aa). It is natural to assume that certain logic combinations of these factors can determine the value of the response variable Y showing the state of the health of the individual under consideration.

10.3.1 Logic Trees

We use the operators $\wedge$ (AND), $\vee$ (OR), ^{c} (NOT). Thus $X_i \wedge X_j = \min\{X_i, X_j\}$, $X_i \vee X_j = \max\{X_i, X_j\}$ and X_i^c (the *conjugate of* X_i) means "not X_i", that is $X_i^c = 1$ whenever $X_i = 0$ and $X_i^c = 0$ whenever $X_i = 1$. In other words, $X_i^c = 1 - X_i$. For the functions $X_i = X_i(\omega)$ these operations are carried out at each point $\omega \in \Omega$. As usual one can consider the values 1 and 0 corresponding to the logic statements "true" and "false". With the variables $X_1, \ldots, X_n$ and the operators mentioned above we can construct a Boolean function (or expression) L. For example,

$$L = [(X_1 \vee X_2) \wedge X_5^c] \wedge (X_7 \vee X_9). \tag{10.33}$$

Note that any Boolean function can be written in the "disjunctive normal form", that is as $L_1 \vee \ldots \vee L_m$ where $L_k = Z_{i_1(k)} \wedge \ldots \wedge Z_{i_{j(k)}(k)}$ and Z_i is equal to X_i or to X_i^c for $i = 1, \ldots, n$, here $\{i_1(k), \ldots, i_{j(k)}(k)\} \subset \{1, \ldots, n\}$, $k = 1, \ldots, m$, $m \in \mathbb{N}$.

Exercise 10.3. Show that expression (10.33) can be written in the disjunctive normal form

$$L = (X_1 \wedge X_5^c \wedge X_7) \vee (X_2 \wedge X_5^c \wedge X_7) \vee (X_1 \wedge X_5^c \wedge X_9) \vee (X_2 \wedge X_5^c \wedge X_9).$$

Thus the same logic function admits different representations. Clearly any logic function introduced above can be represented by means of a certain tree (also non-uniquely). We mention in passing that logic trees as well as *classification and regression trees* (CART) share some similarities, although they are different.

Below we list the standard terminology for logic trees under consideration (see [359]).

1. At each *knot* of the tree there is a single element (X_i or X_i^c, operators $\wedge$ or $\vee$).
2. Every knot has either zero or two subknots (*children* of the *parent* knot).
3. The subknots are each other's *siblings*.
4. A knot devoid of parent knot is called a *root*.
5. The knots having no children are called *leaves*.
6. Leaves can only be occupied by variables, all other knots by operators.

The expression L defined in (10.33) can be represented, e.g., by the tree in Fig. 10.1.

10.3.2 Logic Regression

Let $X = (X_1, \ldots, X_n)$ be a predictor vector with binary components. For a response variable Y and specified *link function* g we consider the model

$$g(\mathbf{E}(Y \mid X)) = \beta_0 + \sum_{j=1}^{s} \beta_j L_j(X), \tag{10.34}$$

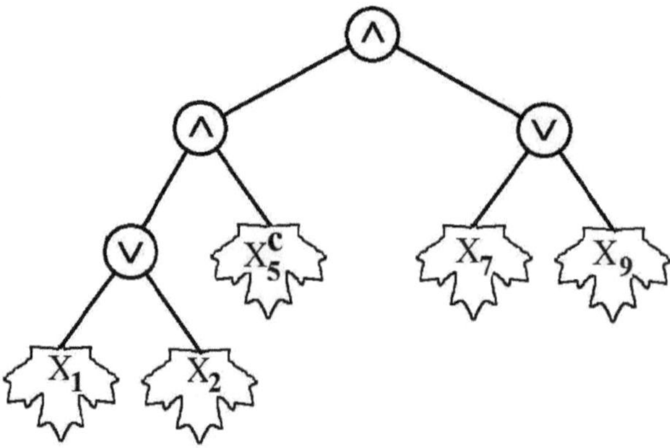

Fig. 10.1 Example of logic tree

where L_j $(j = 1,\dots,s)$ is a Boolean expression in the predictors $X_1,\dots,X_n$ and $\beta = (\beta_0,\dots,\beta_s)$ is a vector of real-valued coefficients. Following [335] the model introduced by (10.34) is called the *logic regression*. The above framework includes, for example, *linear regression*, when $g(x) = x$, $x \in \mathbb{R}$, and $L_j(X) = X_j$, $j = 1,\dots,n$ $(s = n)$. For the latter choice of L_j $(j = 1,\dots,n)$ we come to the definition of *logistic regression* when Y is a binary variable with $\mathbf{P}(Y = 1 \mid X) = p$, $p = p(X)$, and $g(x) = \lambda(x) = \log(x/(1-x))$, $x \in (0,1)$. Namely, in this case $\mathbf{E}(Y \mid X) = p$ and we have the formula

$$\log\left(\frac{p}{1-p}\right) = \beta_0 + \sum_{j=1}^{n} \beta_j X_j. \tag{10.35}$$

Note that one can rewrite (10.35) as

$$p = \Lambda\left(\beta_0 + \sum_{j=1}^{n} \beta_j X_j\right),$$

where the *logistic function* Λ is given by

$$\Lambda(x) = \frac{e^x}{1+e^x}, \quad x \in \mathbb{R}.$$

For every model type we define a score function which reflects the "quality" of the studied model. For example, for linear regression the score could be the residual sum of squares and for logistic regression the score could be the binomial deviance. Assume that (10.34) holds with $g = \lambda$. Let (X^j, Y^j) be i.i.d. copies of (X, Y), $j = 1,\dots,N$ and $\xi_N = (X^1, Y^1,\dots,X^N, Y^N)$. Introduce the *normalized logarithmic likelihood function*

$$\mathcal{L}(h,\xi_N) = \frac{1}{N}\sum_{j=1}^{N} \mathbf{1}\{Y^j = 1\}\log(\Lambda(h(X_1^j,\ldots,X_n^j))$$

$$+\frac{1}{N}\sum_{j=1}^{N} \mathbf{1}\{Y^j = -1\}\log(1 - \Lambda(h(X_1^j,\ldots,X_n^j))), \qquad (10.36)$$

where h belongs to a certain class of functions $\mathcal{M}$. We will search for a function $\widehat{h}_{\mathcal{M}}(N) = \mathrm{argmax}_{h\in\mathcal{M}}\mathcal{L}(h,\xi_N)$.

Due to the SLLN for i.i.d. random vectors one has that a.s.

$$\mathcal{L}(h,\xi_N) \to \mathcal{L}(h) = \mathbf{E}\left(\mathbf{1}\{Y = 1\}\right)\log\Lambda\left(h(X)\right) + \mathbf{E}\left(\mathbf{1}\{Y = -1\}\right)\log(1 - \Lambda\left(h(X)\right)),$$

as $N \to \infty$, whenever $\mathbf{E}\log\Lambda\left(h(X)\right)$ and $\mathbf{E}\log(1 - \Lambda\left(h(X)\right))$ exist.

Let μ and ν be two probability measures on a measurable space $(S,\mathcal{B})$. Introduce the *Kullback–Leibler divergence* or *relative entropy* $D(\mu\|\nu)$ of ν with respect to μ as follows:

$$D(\mu\|\nu) = \begin{cases} \mathbf{E}_\mu\left(\frac{d\nu}{d\mu}\right), & \text{if } \nu << \mu, \\ \infty, & \text{otherwise.} \end{cases}$$

Here $\mathbf{E}_\mu$ stands for the expectation with respect to the probability measure μ on $(S,\mathcal{B})$, $\nu << \mu$ means that ν is absolutely continuous with respect to μ, and, in this case, $d\nu/d\mu$ is the corresponding Radon–Nykodym derivative.

Exercise 10.4. For any probability measures μ and ν on $(S,\mathcal{B})$ show that the *Gibbs inequality* holds, i.e. $D(\mu\|\nu) \geq 0$, with equality if and only if $\mu = \nu$.

In view of the Gibbs inequality the function $\mathcal{L}$ attains its maximum at $h_0(x) = \lambda(\mathbf{P}(Y = 1 \mid X = x))$, $x \in \mathbb{X}$, where λ is the inverse function to Λ. Thus, in this way, one can justify the use of the score-functions introduced in (10.36).

There are various modifications and extensions of the logic regression (LR). We note for example *multinomial LR*, *trio LR*, *Monte Carlo LR* and *logic FS* (feature selection). In this regard we refer to [359] (part III) and references therein. In [54] the *ternary logic regression* was introduced and employed.

10.3.3 Operations on Logic Trees

In the framework of logic regression it is practically impossible to use in (10.34) arbitrary combinations of logic expressions. We have to restrict our collection. We will interpret the logic expression L_j from (10.34) as a tree . Thus the right-hand side of formula (10.34) is defined by a vector of coefficients β and a *forest* $F = \{L_1,\ldots,L_s\}$ (any forest is a collection of trees). Let us introduce some notion of *complexity* $C(T)$ of a tree T and let the *complexity* $C(F)$ of a forest be given by $C(F) = \max_{k=1,\ldots,s} C(T_k)$ where T_k is a tree corresponding to L_k. For instance one can define $C(T)$ as the number of leaves in the tree (another possibility is to consider the height of the tree).

We try to find the maximum of the function $\mathcal{L}(h, \xi_N)$ given in (10.36) over all β and F appearing in (10.34) (such that $C(F) \leq q$ for a given positive q). In general it is an easy problem to find the maximum over all $\beta \in \mathbb{R}^n$. However, the subsequent search of the maximum over all F in a certain class is very hard when the cardinality of this set is huge. Instead of exhaustive search one uses the stochastic maximization. To employ this procedure we have to indicate for each point of a set S its neighbors (or in an equivalent way to define a graph (S, E)).

Following [335] we say that a tree T_2 is a *neighbor* of a tree T_1 if T_2 can be obtained from T_1 by means of one of the following six operations.

1. *Alternating a leaf.* We pick a leaf with some X_i inside, change it by X_j or X_j^c. However to avoid tautologies one cannot have X_j or X_j^c in the leaf being sibling of that where we changed the variable. For example, in Fig. 10.2 (A), X_1 in the leaf of Fig. 10.1 was replaced with X_3.

2. *Changing operators.* Any "$\wedge$" can be replaced by a "$\vee$", and vice versa (e.g., the operator at the root of the initial tree in Fig. 10.1 has been changed in Fig. 10.2 (B)).

3. *Growing.* At any knot that is not a leaf, one allows a new branch to grow. This is done by declaring the subtree starting at this knot to be the right side branch of the new subtree at this position, and the left side branch to be a leaf representing any predictor X_i. These two side trees are connected by a "$\wedge$" or "$\vee$" at the location of the knot. An example of this operation is given in Fig. 10.2 (C).

4. *Pruning.* A leaf is trimmed from the existing tree, and the subtree starting at the sibling of the trimmed leaf is "shifted" up to start at the parent of the trimmed leaf. This is illustrated in Fig. 10.2 (D).

5. *Splitting.* Any leaf can be split by creating a sibling, and determining a parent for those two leaves. For example, in Fig. 10.2 (E) the leaf containing X_9 from the initial tree (see Fig. 10.1) has been split, with leaf containing X_6 as its new sibling.

6. *Deleting.* One can delete a leaf in a pair of siblings which are both leaves, see Fig. 10.2 (F), where X_1 has been deleted from the initial tree.

Clearly these operations can be inverted. Thus the relation of neighboring is symmetric.

Exercise 10.5. Show that one can manage with only four operations instead of six.

However, a wider class of possibilities "to move" from one tree to another could provide more rapid convergence for the stochastic optimization algorithm discussed in Sect. 10.3.4 below. Note that in some situations we do not use all possible moves introduced above. For instance, if we impose a restriction on the model size, i.e. the total number of binary variables in the logic trees, then any move that increases the model size (splitting a leaf, growing a branch) could not be permitted.

One says that the forests F_1 and F_2 are neighbors if they admit the representation $F_1 = \{T_1, T_2, \dots, T_s\}$ and $F_2 = \{\widetilde{T}_1, T_2, \dots, T_s\}$ where the trees T_1 and $\widetilde{T}_1$ are neighbors. This definition is convenient as the neighboring of two forests is determined only by the two trees.

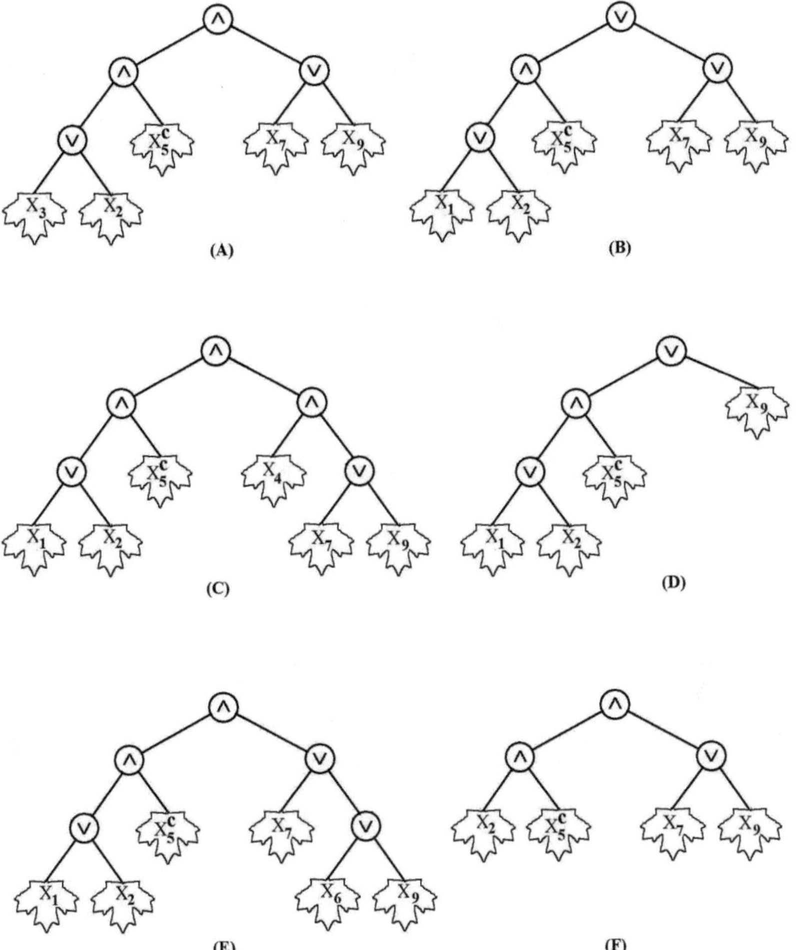

Fig. 10.2 Operations with logic trees

Before considering the optimization problem related to logic regression we mention that there exists free software which can be downloaded as package LogicReg from the The Comprehensive R Archive Network (http://cran.r-project.org/).

10.3.4 Simulated Annealing

Let J be a real-valued *objective* (or cost) *function* defined on a finite set S. We write $S^* = \arg\min J$, i.e., a subset of S consisting of the points where J attains its minimum value on S. Clearly the problem to find the set $\widetilde{S} = \arg\max J$ is reduced to the search of the points where the function $-J$ attains its minimum. Assume that $S^* \neq S$, that is, J does not take a constant value all over S. For each $i \in S$ we introduce a neighborhood $S(i) \subset S \setminus \{i\}$ in such a way that $j \in S(i)$ if and only if $i \in S(j)$. An equivalent approach is to define a graph $G = (S, E)$ and suppose that there exists an edge connecting the vertices i and j if and only if $j \in S(i)$.

Consider a matrix $Q = (q_{ij})$ with nonnegative entries such that for any $i \in S$

$$\sum_{j \in S} q_{ij} = 1.$$

We also define the *temperature* T as a function $T : \mathbb{Z}_+ \to (0, 1)$. Here $T(t)$ is interpreted as the temperature at time $t \in \mathbb{Z}_+$, where $\mathbb{Z}_+ = \{0, 1, \dots\}$.

Now we describe the classical *simulated annealing algorithm* which gives the possibility to find the set S^* with probability close to one. For this purpose we introduce the following Markov chain $U = \{U(t), t \in \mathbb{Z}_+\}$. Let $U(0) = x_0$ where $x_0 \in S$ is called the *initial state*. Let $U(t) = i$ for $t \in \mathbb{Z}_+$ and $i \in S$. Then we take $j \in S(i)$ with probability q_{ij}. The state of the Markov chain at time $t + 1$ now is defined as follows. If $J(j) \leq J(i)$ we put $U(t+1) = j$. If $J(j) > J(i)$ we put $U(t+1) = j$ with probability $\exp\{-(J(j) - J(i))/T(t)\}$, and $U(t+1) = i$ with probability $1 - \exp\left(-(J(j) - J(i))/T(t)\right)$. Namely,

$$\mathbf{P}(U(t+1){=}j|U(t){=}i){=}\begin{cases} q_{i,j}\exp\left(-T(t)^{-1}\max\{J(j)-J(i),0\}\right), & \text{if } j \in S(i), \\ 0, & \text{if } j \notin S(i) \cup \{i\}. \end{cases}$$

Put

$$\mathbf{P}(U(t+1) = i \mid U(t) = i) = 1 - \sum_{j \neq i} \mathbf{P}(U(t+1) = j \mid U(t) = i).$$

Recall that a (homogeneous) Markov chain $X = \{X_n, n \in \mathbb{Z}_+\}$ with transition matrix $P = (p_{ij})$ is called *irreducible* if for any $i, j \in S$ (the state space) there exists $n = n(i, j) \in \mathbb{N}$ such that $p_{i,j}(n) > 0$. Here $p_{i,j}(n)$ is an element of the matrix P^n, that is probability of transition from i to j in n steps. Let $D_i = \{n \in \mathbb{N} : p_{i,i}(n) > 0\}$ for each $i \in S$. Denote by $\gcd(D_i)$ the greatest common divisor of elements belonging to D_i. One says that a Markov chain is *aperiodic* if $\gcd(D_i) = 1$ for each $i \in S$.

Let us explain the idea of using the Markov chain U. Consider a special case of a homogeneous Markov chain U_T defined above where $T(t) = T$ for each $t \in \mathbb{Z}_+$. If U_T is irreducible and aperiodic and if it holds that

$$q_{i,j} = q_{j,i}, \quad i, j \in S, \tag{10.37}$$

then there exists (even under more general conditions) a unique invariant (initial) distribution $\mathbf{P}_T$ of $U_T(0)$ which is given by

$$\mathbf{P}_T(\{i\}) = \frac{1}{Z_T} \exp\left(-\frac{J(i)}{T}\right), \quad i \in S, \tag{10.38}$$

where $Z_T = \sum_{i \in S} \exp\left(-J(i)/T\right)$ is a normalizing constant, see, e.g. [1], Chap. 3.

It is not difficult to show that the probability measure $\mathbf{P}_T$ will eventually be concentrated on S^* as $T \downarrow 0$. In fact, take $J^* = J(j)$ where $j \in S^*$. Obviously,

$$\lim_{u \to 0+} \exp\left(a/u\right) = \begin{cases} 0, & \text{if } a < 0, \\ 1, & \text{if } a = 0, \\ \infty, & \text{if } a > 0. \end{cases}$$

Therefore, for each $i \in S$ it holds that

$$\begin{aligned}
\lim_{T \downarrow 0} \mathbf{P}_T(\{i\}) &= \lim_{T \downarrow 0} \frac{\exp\left((J^* - J(i))/T\right)}{\sum_{j \in S} \exp\left((J^* - J(j))/T\right)} \\
&= \lim_{T \downarrow 0} \frac{\exp\left((J^* - J(i))/T\right)}{\sum_{j \in S} \exp\left((J^* - J(j))/T\right)} \mathbf{1}\{i \in S^*\} \\
&\quad + \lim_{T \downarrow 0} \frac{\exp\left((J^* - J(i))/T\right)}{\sum_{j \in S} \exp\left((J^* - J(j))/T\right)} \mathbf{1}\{i \in S \setminus S^*\} \\
&= \frac{1}{|S^*|} \mathbf{1}\{i \in S^*\} + \frac{0}{|S^*|} \mathbf{1}\{i \in S \setminus S^*\} \\
&= \frac{1}{|S^*|} \mathbf{1}\{i \in S^*\}.
\end{aligned}$$

We have briefly described the classical *Metropolis algorithm* (proposed initially to study physical systems) permitting to find (asymptotically) the *Gibbs distribution* (10.38). Further improvement of this algorithm is due to W.K.Hastings. Also note that in [233] the simulated annealing has been developed in the framework of optimization problems. As to our minimization problem we will choose the set S^* with high probability if the choice is done according to the distribution $\mathbf{P}_T$ given in (10.38) with small positive T.

It is worth to mentioned that to perform the algorithm in a logic regression study we need a data set, an annealing cooling scheme, a link function g appearing in (10.34), an objective function J, the maximum complexity of forests and the number of iterations used in the annealing.

To formulate the general result of this section we need some more notation. A *path* from vertex i_1 to vertex i_n (where an integer $n \geq 2$) in S is a (directed) collection of vertices $i_1, \ldots, i_n$ such that i_k and i_{k+1} are connected by an edge (from E) for each $k = 1, \ldots, n-1$. One says that a vertex i communicates with S^* by a *path of the height $h \geq 0$* if for some $n \in \mathbb{N}$ there is a path $i_1, \ldots, i_n$ with the following property: $i_1 = i$, $i_n \in S^*$ and

$$\max_{k=1,\dots,n} J(i_k) \leq J(i) + h.$$

Let d^* be the smallest number such that each $i \in S$ communicates with S^* by a path of height d^* (see Fig. 10.3).

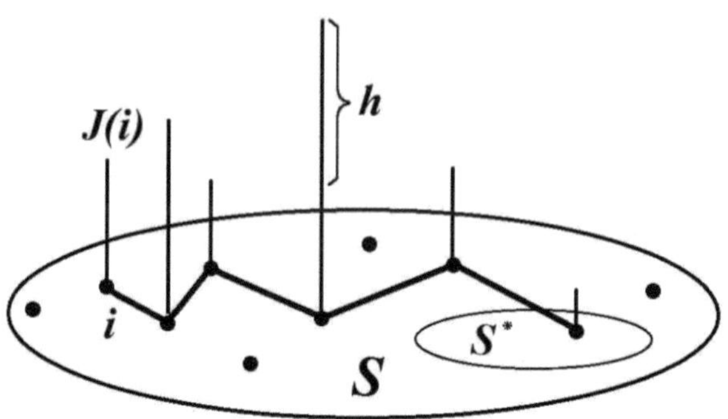

Fig. 10.3 Path of height h

Theorem 10.3 ([162]). *The relation*

$$\lim_{t\to\infty} \mathbf{P}(U(t) \in S^*) = 1 \tag{10.39}$$

holds if and only if $\lim_{t\to\infty} T(t) = 0$ *and*

$$\sum_{t=1}^{\infty} \exp\left(-\frac{d^*}{T(t)}\right) = \infty. \tag{10.40}$$

According to (10.40) one chooses $d > d^*$ and $T(t) = d/\log t$ for $t > 1$. The number d^* can be viewed as a characteristic of complexity for the Markov chain U to reach S^* escaping the local minimum of J. In [359] a typical cooling scheme starts at $T = 10^{z_{start}}$ and then the temperature is lowered to $T = T^{z_{end}}$ where $z_{end} < z_{start}$, in equal decrements on the $\log_{10}$-scale. Note also that there are various modifications of the simulated annealing algorithm. For example, to study the function defined on a collection of trees (i.e. the vertices of a graph were trees) the following algorithm was employed in [54]. If $U(t) = i$, then take a random arrangement of vertices belonging to the set $S(i)$ and obtain $\{j_1,\dots,j_{R(i)}\}$, here $R(i) = |S(i)|$. Furthermore calculate $J(j_q)$ for $q = 1,\dots,K(i)$ where $K(i) = [R(i)/e]$ and take $k \in \{K(i)+1,\dots,R(i)\}$ at random. If

$$J(j_k) < J(i) \wedge \min_{1\leq q\leq K(i)} J(j_q),$$

put $U(t+1) = j_k$ (as usual $a \wedge b = \min\{a,b\}$ for $a,b \in \mathbb{R}$). In the opposite case introduce $D(i,k) = \{j_1,\ldots,j_{K(i)}\} \cup \{j_k\} \cup \{i\}$ and perform at time $t+1$ the transition to the state j belonging to $D(i,k)$ with probability $\frac{1}{Z_{i,k}} \exp(-J(j)/T(t))$ where $Z_{i,k} = \sum_{v \in D(i,k)} \exp(-J(v)/T(t))$. Simulation experiments showed that this algorithm was efficient when the vertices of the graph were trees and the corresponding concept of neighbors involved the operations $1. - 6.$ introduced in Sect. 10.3.1.

Note in passing that various versions of simulated annealing have been developed, for example *fast annealing* and *very fast simulated reannealing* (VFSR) or *adaptive simulated annealing* (ASA). See also [76] for simulated annealing and genetic algorithms.

10.4 Models Involving Random Fields

Among the new statistical methods we also mention kernel-machine-based models, known as SKAT (see [419, 420]) and similarity regression (SIMreg) [399]. SKAT is developed in the framework of a random effect model and SIMreg directly describes the trait similarity as a function of genetic similarity. Recently, in [168] a genetic random field model (GenRF) has been proposed to test the joint association of multiple genetic variants. This approach is motivated by the progress of spatial statistics. GenRF, in contrast to SKAT and SIMreg, regresses the response of one subject on responses of all other subjects.

Let the trait or phenotype Y_i of the i-th individual ($i = 1,\ldots,N$) of a given region be described by the vectors of covariates G_i (characterizing the genotype by means of p variables, e.g., SNPs) and X_i (consisting of q nongenetic variables such as age, gender and others). Note that SKAT is a semiparametric linear model of the form

$$Y_i = \alpha^\top X_i + h(G_i) + \varepsilon_i, \quad i = 1,\ldots,N, \tag{10.41}$$

where Y_i is a continuous variable, $\alpha \in \mathbb{R}^q$, h is a real-valued nonparametric function belonging to the functional space generated by a positive semidefinite kernel function $K(\cdot,\cdot)$. Then, for this model, testing for the joint association is equivalent to testing the hypothesis $H_0 : h(G_i) = 0$, $i = 1,\ldots,N$, see, e.g., [256]. As explained in [420], the kernel function $K(G_i,G_j)$ can be interpreted as a measure for genetic similarity in the region of interest between the i-th and j-th subjects, where the kernel function better capturing the similarity between individuals and the causal variant effects can increase the power of the test. SIMreg explicitly defines a measure for trait similarity and directly regresses the trait similarity between each pair of subjects on genetic similarity. Note that SKAT and SIMreg lead to analogous test statistics, see [399].

The key idea of the GenRF studied in [168] is the following. If the genetic variants are jointly associated with a trait, then the genetic similarity across subjects will contribute to the trait similarity. Namely, for centered random variables Y_i the conditional distribution of Y_i given all other responses has the following form

$$Y_i | Y_{-i} \sim \gamma \sum_{j \neq i} s(G_i, G_j) Y_j + \varepsilon_i, \quad i = 1, \ldots, N. \tag{10.42}$$

Here Y_{-i} stands for the vector of all Y_j's without Y_i, whereas $s(\cdot, \cdot)$ is a known weight function, γ is a nonnegative parameter and the ε_i's are i.i.d. random errors. One can test the joint association of genetic variants with the trait by testing the null hypothesis $H_0 : \gamma = 0$. Thus, in the framework of (10.42), we can view responses as realization of a random field defined on a p-dimensional space of genetic variants (one can write $Y_{G_{i1}, \ldots, G_{ip}}$). The responses from locations that are "closer" in the genetic space are expected to be more similar if the genetic association exists. Models like (10.42) have been introduced in [37] for random fields. The GenRF model is closely related to the conditional auto-regressive model in spatial statistics, see [86].

Suppose that each component of G_i records the number of minor alleles in a single locus and takes values in the set $\{0, 1, 2\}$ corresponding to three possibilities $\{AA, Aa, aa\}$. In [419] the so-called identity-by-state (IBS) measure of similarity has been defined:

$$s(G_i, G_j) = \sum_{k=1}^{p} w_k (2 - |G_{ik} - G_{jk}|)$$

where the w_k's are some weights. Another choice of the function s is discussed in [252]. Note that in [419] the following generalization of (10.42) has been considered

$$Y_i | Y_{-i}, X_i \sim \beta^\top X_i + \gamma \sum_{j \neq i} s(G_i, G_j)(Y_j - \beta^\top X_j) + \varepsilon_i, \quad i = 1, \ldots, N, \tag{10.43}$$

where the Y_j's are not assumed to be centered, $\beta \in \mathbb{R}^q$.

For the Gaussian scenario when $\varepsilon_i \sim N(0, \sigma^2)$, $i = 1, \ldots, N$, the model given in (10.43) means that the conditional distribution of Y_i, given the responses from all other subjects and the covariates $X = (X_1^\top, \ldots, X_N^\top)^\top$, is normal with mean $\beta^\top X_i + \gamma \sum_{j \neq i} s(G_i, G_j)(Y_j - \beta^\top X_j)$ and variance σ^2. Thus, assuming that β is known, one can find the maximum pseudo-likelihood estimator for γ, namely,

$$\widetilde{\gamma} = \frac{(Y - X\beta)^\top S (Y - X\beta)}{(Y - X\beta)^\top S^2 (Y - X\beta)}.$$

Here $Y = (Y_1, \ldots, Y_N)^\top$ and the $N \times N$ matrix S has zero diagonal elements and $S_{ij} = s(G_i, G_j)$ for $i \neq j$. Intuitively one expects that a large value of $\widetilde{\gamma}$ would lead to rejecting the null hypothesis $H_0 : \gamma = 0$. However, in practice β is unknown and, therefore, the authors of [168] propose to replace β by the least squares estimator $\widehat{\beta} = (X^\top X)^{-1} X^\top Y$. Then we obtain the test statistic

$$\widehat{\gamma} = \frac{Y^\top B S B Y}{Y^\top B S^2 B Y}$$

where $B = I - X(X^\top X)^{-1} X^\top$ and I stands for the identity matrix. Taking into account that BS^2B is positive-definite we can write, for any real t, the following equality

$$\mathbf{P}_{H_0}(\hat{\gamma}>t)=\mathbf{P}_{H_0}((BY)^{\top}(S-tS^2)BY>0)=\mathbf{P}(Z^{\top}(S-tS^2)Z>0)=\mathbf{P}\left(\sum_{i=1}^{N}\lambda_i\Phi_i>0\right)$$

where $Z \sim N(0,B^2)$, the Φ_i's are independent and χ_1^2-distributed random variables and the λ_i's are the eigenvalues of $B(S-tS^2)B$. The final p-value can be found in a way similar to calculations used in [420] for SKAT statistics. Note that asymptotically the proposed test is robust with respect to distributions others than normal. Indeed, in [148] an upper bound has been provided for

$$\sup_{x\in\mathbb{R}}\left|\mathbf{P}_{H_0}((BY)^{\top}ABY<x)-\mathbf{P}_{H_0}(Z^{\top}AZ<x)\right|$$

where $A = S - tS^2$. This leads to the robustness of the GenRF test as long as BY has zero mean under null hypothesis which is true as the least squares estimator $X(X^{\top}X)^{-1}X^{\top}Y$ is unbiased for the mean of Y within the GenRF model when $\gamma = 0$. In [168] a simulation study is considered and applications are given as well.

Now we briefly discuss the hidden Markov random field model for GWAS. Assume that we have m cases and n controls that are genotyped over a set $S = \{1,\ldots,p\}$ of p SNPs. The problem is to determine which SNPs in S are associated with the studied disease. Let $Y = (Y_1,\ldots,Y_p)$ be the observed genotype data and $Y_s = (y_{s,1},\ldots,y_{s,m+n})$ with $y_{s,i}$ being the observed genotype for the i-th individual at the s-th SNP. In [252] it is proposed to develop a Markov random field (MRF) model to take into account the linkage disequilibrium (LD) information in identifying the disease-associated SNPs. Regarding LD see, e.g., [240, Sect. 5.4.]. First of all a weighted undirected LD graph G is constructed based on pairwise LD information derived from the data (or from the HapMap project). An edge between SNPs s and s' is drawn with weight $w_{s,s'} = \mathbf{1}\{r_{s,s'}^2 > \tau\}r_{s,s'}^2$ if $w_{s,s'} \neq 0$. Here $r_{s,s'}^2$ measures the LD between the SNPs s and s' (r is the Pearson coefficient of correlation for allele frequencies), τ is a fixed threshold (e.g. $\tau = 0.4$ as in [252]). For $s = 1,\ldots,p$, we introduce the random variable

$$X_s = \begin{cases} 1 & \text{if SNP } s \text{ is associated with the disease,} \\ 0 & \text{otherwise.} \end{cases}$$

One expects that X_s and $X_{s'}$ are dependent if s and s' are linked on the LD graph. The joint probability function of $X = (X_1,\ldots,X_p)$ is defined as for discrete Markov random fields (see, e.g., [373, Sect. 9.3]). Put $\Phi = (\gamma,\beta) \in \mathbb{R}_+^2$ and

$$\mathbf{P}(X=x;\Phi)=Z^{-1}\exp\left\{\gamma\sum_{s=1}^{p}x_s+\beta\sum_{s\sim s'}w_{s,s'}\mathbb{I}\{x_s=x_{s'}\}\right\} \qquad (10.44)$$

where $s \sim s'$ means that there exists an edge between s and s' in the LD graph and Z is a normalizing factor. In [252] it is supposed that, given any realization of X, the random variables $Y_1,\ldots,Y_p$ are conditionally independent. Moreover, involving a Dirichlet prior with parameter $\alpha = (\alpha_1,\alpha_2,\alpha_3)$ for genotype frequencies at the *s*-th SNP (for genotype values 0, 1 and 2) in the case population and the control population, the following relationships have been obtained in [252]:

$$\mathbf{P}(Y_s|X_s=0) = \frac{\Gamma(\sum_{j=1}^3 \alpha_j)\prod_{j=1}^3 \Gamma(\alpha_j+y_{s+,j}+y_{s-,j})}{\prod_{j=1}^3 \Gamma(\alpha_j)\Gamma(\sum_{j=1}^3(\alpha_j+y_{s+,j}+y_{s-,j})}$$ (10.45)

and

$$\mathbf{P}(Y_s|X_s=1) = \frac{\Gamma(\sum_{j=1}^3 \alpha_j)\prod_{j=1}^3 \Gamma(\alpha_j+y_{s+,j})}{\prod_{j=1}^3 \Gamma(\alpha_j)\Gamma(\sum_{j=1}^3(\alpha_j+y_{s+,j}))} \cdot \frac{\Gamma(\sum_{j=1}^3 \alpha_j)\prod_{j=1}^3 \Gamma(\alpha_j+y_{s-,j})}{\prod_{j=1}^3 \Gamma(\alpha_j)\Gamma(\sum_{j=1}^3(\alpha_j+y_{s-,j}))}.$$ (10.46)

Here $y_{s+} = (y_{s+,1}, y_{s+,2}, y_{s+,3})$ denotes, for SNP s, the observed genotype for data in m cases and the vector y_{s-} has an analogous meaning for n controls. The combination of the probability models given in (10.44)–(10.46) defines a hidden Markov random field (HMRF) model where X is the vector of the hidden states that follows a discrete MRF. An efficient iterative conditional mode (ICM) algorithm is provided to estimate the parameters and a Gibbs sampling approach is employed for estimating the posterior probabilities. The latter probabilities can be used to define a false discovery rate (FDR) controlling procedure in order to select the relevant SNPs. The developed approach is applied in [252] to the analysis of case-control neuroblastoma (NB) data set.

In [168] a genetic random field model has been applied to the analysis of the Multi-Ethnic Study of Atherosclerosis (MESA). A novel statistical approach has been proposed in [168], called the longitudinal genetic random field (LGRF) model, to test the joint association between a set of genetic variants and a phenotype measured repeatedly during the course of an observational study. Moreover, an improved version of LGRF (robust-LGRF) is developed to enhance the robustness to misspecification of the within-subject correlation and hence to avoid type-I error inflation. A fast version of LGRF (fast-LGRF), as a further extension of robust-LGRF, is considered to improve the computational efficiency when the total number of repeated measurements in the sample is large.

According to [411] an important problem is to identify genes that show different expression profiles over time and pathways that are perturbed during a given biological process. In [168] a hidden spatial-temporal Markov random field (hstMRF)-based method has been developed for identifying genes and subnetworks that are related to biological processes, where the dependency of the differential expression patterns of genes on the networks are modeled over time and over the network of pathways.

Important applications of HMRF can be found also in spatial population genetics, see, e.g., [125], where the natural neighborhood structure obtained from the Dirichlet tiling is used. Let $\{s_1, \ldots, s_n\}$ represent the set of observation sites (in $\mathbb{R}^2$) for n individuals. For each s_i, the Dirichlet cell (also called Voronoi cell, see e.g. Definition 3.13) consists of all points that are closer to s_i than to any other sampling site. Two sampling sites are neighbors if their cells have a common edge. The Potts-Dirichlet model is the Potts model (see, e.g., [151, Sect. 7.3]) built on the Dirichlet tiling generated by sampling sites. Let c_i be the cluster from which the individual i originates. We assume that there exist at most K_{max} clusters. The Potts model is described by the probability distribution on the set of cluster configurations. Namely, given n observation sites, we consider the probabilities

$$\pi(c) = \frac{e^{\psi U(c)}}{Z(\psi, K_{max})}, \quad c \in \{1, \ldots, K_{max}\}^n,$$

where ψ is a nonnegative (interaction) parameter, $U(c) = \sum_{i \sim j} \delta_{c_i, c_j}$, $i \sim j$ means that i and j are neighbors, δ_{c_i, c_j} is the Kronecker symbol, $Z(\psi, K_{max})$ is a normalizing constant called the partition function. Clearly, this model possesses the Markov property.

In [125] the hierarchical Bayes model based on an HMRF is studied. Moreover, in [125] deviations from the Hardy-Weinberg (HW) equilibrium are considered which are caused by inbreeding. Let L be the number of loci, J_l be the number of alleles at locus l, and z be the collection of all genotypes (the data). Given that the individual i originates from the cluster $c_i = k$ and given the allele frequencies $f_{k \cdot}$ in this cluster, the conditional probability of observing the genotype $z_i^l = (a_i^l, b_i^l)$ at locus l is $\mathcal{L}_k(f_{kla_i^l}, f_{klb_i^l})$ where $\mathcal{L}_k(f, f) = f^2 + \varphi_k f$ and $\mathcal{L}_k(f, g) = 2fg(1 - \varphi_k)$ for $f \neq g$. Diploidy is also assumed. The set of parameters $\theta = (\psi, c, f, \varphi)$ where ψ is the interaction parameter, c is the cluster configuration, $f = (f_{klj})$, $k = 1, \ldots, K_{max}$, $l = 1, \ldots, L$, $j = 1, \ldots, J_l$, are the allele frequencies and $\varphi = (\varphi_1, \ldots, \varphi_{K_{max}})$ gives the inbreeding coefficients in each subpopulation. The priors on allele frequencies are Dirichlet distributions $D(\alpha, \ldots, \alpha)$. The prior distributions on the φ_k's are beta $B(\lambda, \mu)$ distributions. The hierarchy of the model is reflected by

$$\pi(\theta) = \pi(\varphi)\pi(\psi|\varphi)\pi(c|\varphi, \psi)\pi(f|c, \psi, \varphi) = \pi(\varphi)\pi(\psi)\pi(c|\psi)\pi(f|c).$$

Assuming linkage equilibrium between loci, the likelihood is defined as follows:

$$\pi(z|\theta) = \prod_{i=1}^{n} \prod_{l=1}^{L} \mathcal{L}_{c_i}(f_{c_i l a_i^l}, f_{c_i l b_i^l}).$$

Inference on θ is carried out by simulating the posterior distribution $\pi(\theta|z)$ through a Markov chain Monte Carlo (MCMC) sampling algorithm. In contrast to previous work, in [125] the so-called regularization approach is employed to estimate the number of clusters. Some real data analysis (e.g., for the Scandinavian brown bear population) is provided as well.

10.5 Concluding Remarks

In this chapter we mainly discussed statistical problems concerning the identification of significant factors (among $X_1, \ldots, X_n$) having an essential impact on a binary random response variable Y. Such analysis is important for various applications, e.g., in medicine and biology. Note that spatial problems arise here as we have to consider random functions (random fields) defined on graphs with vertices representing some trees. So, we provided basic information concerning stochastic optimization problems. We considered and applied the simulated annealing techniques. Special attention was paid to a new version of the MDR method using a certain algorithm to estimate the prediction error (for approximation of Y) involving a penalty function and a cross-validation procedure. Here we mention our criterion of strong consistency of the estimators under consideration and the CLT for regularized versions of such estimators. We concentrated on rigorous results established for random functions.

To complete the chapter we mention several vast research directions related to model selection, having applications to the analysis of factors provoking complex diseases, which deserve special books or surveys. We draw the attention to classification theory and machine learning, Bayesian networks, graphical models including (hidden) Markov random fields and conditional random fields, see., e.g., [41], [50], [56], [167], [295], [336].

Furthermore, it is reasonable to compare the employment of different modern techniques for data analysis. In this respect we refer, e.g., to [54] where genetic and environmental risk factors for coronary heart disease and myocardial infarction were considered.

Acknowledgements This work is partially supported by RFBR grant 13-01-00612.

Chapter 11
Extrapolation of Stationary Random Fields

Evgeny Spodarev, Elena Shmileva and Stefan Roth

Abstract We introduce basic statistical methods for the extrapolation of stationary random fields. For square integrable fields, we consider kriging extrapolation techniques. For (non–Gaussian) stable fields, which are known to be heavy-tailed, we describe further extrapolation methods and discuss their properties. Two of them can be seen as direct generalizations of kriging.

11.1 Introduction

In this chapter, we consider the problem of extrapolation (prediction) of random fields arising mainly in geosciences, mining, oil exploration, hydrosciences, insurance, etc. The techniques to solve this problem are one of the fundamental tools in geostatistics that provide statistical inference for spatially referenced variables of interest. Examples of such quantities are the amount of rainfall, concentration of minerals and vegetation, soil texture, population density, economic wealth, storm insurance claim amounts, etc.

The origins of geostatistics as a mathematical science can be traced back to the works of L. Gandin (1963) [135], B. Matérn (1960) [262], and G. Matheron (1962-63) [263, 264]. However, the mathematical foundations were already laid by A.N.Kolmogorov (1941) [234] as well as by N.Wiener (1949) [414], where the ex-

Evgeny Spodarev
Ulm University, Institute of Stochastics, 89069 Ulm, Germany,
e-mail: evgeny.spodarev@uni-ulm.de

Elena Shmileva
St.Petersburg State University, Chebyshev Laboratory, St. Petersburg 199178, Russia,
e-mail: elena.shmileva@gmail.com

Stefan Roth
Ulm University, Institute of Stochastics, 89069 Ulm, Germany,
e-mail: stefan.roth@uni-ulm.de

© Springer International Publishing Switzerland 2015
V. Schmidt (ed.), *Stochastic Geometry, Spatial Statistics and Random Fields,*
Lecture Notes in Mathematics 2120, DOI 10.1007/978-3-319-10064-7_11

trapolation of stationary time series was studied, whereas their practical application is known since 1951 due to mining engineer D. G. Krige [236]. Typical practical problems to solve are e.g. plotting the contour concentration map of minerals (interpolation), inference of the mean areal precipitation and evaluation of the accuracy of estimators from spatial measurements (averaging), selection of locations of new monitoring points so that e.g. the concentration of minerals can be evaluated with sufficient accuracy (monitoring network design).

The remainder of this chapter is divided into three sections. Sect. 11.2 contains preliminaries about distributional invariance properties and the dependence structure of random fields. In Sect. 11.3, we concentrate on kriging, which is a widely used probabilistic extrapolation technique for fields with finite second moment. Sect. 11.4 contains recent results on the extrapolation of heavy tailed random fields with infinite variance, namely of stable random fields.

In Sect. 11.2 and 11.3 we mainly follow the books [71, 86, 373, 408]. Sect. 11.4 is based on [221], it also contains some new results for stable fields with infinite first moment, see Sect. 11.4.4.

11.2 Basics of Random Fields

Let $(\Omega, \mathcal{F}, \mathbf{P})$ be an arbitrary probability space.

Definition 11.1. A *random field* $X = \{X(t), \ t \in \mathbb{R}^d\}$ is a random function on $(\Omega, \mathcal{F}, \mathbf{P})$ indexed by the elements of $\mathbb{R}^d$ or of some subset $I \subset \mathbb{R}^d$, where $d \in \mathbb{N}$ is an arbitrary integer, i.e., X is a measurable mapping $X : \Omega \times \mathbb{R}^d \to \mathbb{R}$.

For an introduction into the theory of random functions see e.g. [373, Chap. 9].

11.2.1 Random Fields with Invariance Properties

A random field whose finite-dimensional distributions are invariant with respect to the action of a group G of transformations of $\mathbb{R}^d$ is called *G-invariant in the strict sense*. In case if this invariance is given only for the first two moments of the field, which are assumed to be finite we speak about the *G–invariance in wide sense*. Thus, if G is the group of all translations of $\mathbb{R}^d$, then one calls such random fields *stationary* (in the respective sense). For G being the group of rotations SO_d one claims the random field to be *isotropic*. If G is the group of all rigid motions in $\mathbb{R}^d$ then such field is called *motion invariant*. The same notions of invariance can be transferred to the increments of random fields. In this case, the property of stationarity is often called *intrinsic*. The intrinsic stationarity in the wide sense is called *intrinsic stationarity of order two*. For more details on invariance properties confer [373, Sect. 9.5].

Exercise 11.1. Show that the mean value function (if it exists) of any process ($d = 1$) with stationary increments is a linear function, i.e., $\mathbf{E}X(t) = at + c$ for all $t \in \mathbb{R}$, where $a \in \mathbb{R}$ and $c \in \mathbb{R}$ are some constants.

A popular class of random fields are *Gaussian fields*.

Definition 11.2. A random field $X = \{X(t),\, t \in \mathbb{R}^d\}$ is called *Gaussian* if its finite dimensional distributions are Gaussian.

The use of Gaussian fields for modelling purposes in applications can be explained mainly by the simplicity of their construction and analytic tractability combined with the normal distributions of marginals which describes many real phenomena due to the central limit theorem.

By Kolmogorov's theorem, the probability law of a Gaussian random field is uniquely defined by its mean value and covariance functions; see [373, Sect. 9.2.2] for more details. If the mean value function $\mathbf{E}X(t)$, $t \in \mathbb{R}^d$ is identically zero we call X to be *centered*. Without loss of generality we tacitly assume all random fields considered in this chapter to be centered.

Exercise 11.2. Show that for Gaussian random fields stationarity (isotropy, motion invariance) in the strict sense and stationarity (isotropy, motion invariance) in the wide sense are equivalent. In this case we call a Gaussian field just *stationary* (*isotropic, motion invariant*).

11.2.1.1 Examples: Gaussian Random Fields

In this section we briefly discuss several examples of Gaussian processes and Gaussian random fields.

1. Ornstein-Uhlenbeck Process. A centered Gaussian process $X = \{X(t),\, t \in \mathbb{R}\}$ with the covariance function $\mathbf{E}\left(X(s)X(t)\right) = e^{-|s-t|/2}$, $s,t \in \mathbb{R}$ is called an *Ornstein-Uhlenbeck process*. It has been shown in [49, p. 350] that X is the only stationary Markov Gaussian process which is stochastically continuous. Additionally, it has short memory, i.e.,

$$X(t) \overset{\mathrm{D}}{=} e^{-t/2}X(0) + V(t), \quad t > 0,$$

where $V(t)$ is independent of the past $\{X(s), s \le 0\}$, cf. [254, Example 2.6, p.11]. Defined on $\mathbb{R}_+$, $X = \{X(t),\, t \ge 0\}$ is the strong solution of the Langevin stochastic differential equation

$$dX(t) = -1/2X(t)dt + dW(t)$$

with initial value $X(0) \sim N(0,1)$, where $W = \{W(t),\, t \ge 0\}$ is the standard Wiener process, see e.g. [60, Chap. 8, Theorem 7]. It holds also that

$$X \overset{\mathrm{D}}{=} \left\{ e^{-t/2}W\left(e^t\right),\, t \in \mathbb{R} \right\},$$

cf. [60, Chap. 3, p.107].

2. Gaussian Linear Random Function. A *Gaussian linear random function* $X = \{X(t),\ t \in l_2\}$ is defined by $X(t) = \langle N, t \rangle_2$, $t \in l_2$, where $N = \{N_i\}_{i=1}^{\infty}$ is an i.i.d. sequence of $N(0,1)$-distributed random variables, and l_2 is the Hilbert space of sequences $t = \{t_i\}_{i=1}^{\infty}$ such that $\|t\|_2^2 = \sum_{i=1}^{\infty} t_i^2 < \infty$ with scalar product $\langle s, t \rangle_2 = \sum_{i=1}^{\infty} s_i t_i$, $s, t \in l_2$. Since N is not an element of l_2 a.s., the expression $\langle N, t \rangle_2$ is understood formally as the series $\sum_{i=1}^{\infty} N_i t_i$ which converges in the mean square sense, i.e.,

$$\mathbf{E} \left| \sum_{i=n}^{m} N_i t_i \right|^2 = \sum_{i=n}^{m} t_i^2 \to 0, \quad \text{as } n, m \to \infty.$$

It holds that

$$X(t) \sim N(0, \|t\|_2^2), \quad X(t) - X(s) = X(t-s), \quad \mathbf{E}\,(X(s)X(t)) = \langle s, t \rangle_2, \quad s, t \in l_2.$$

The *variogram* $\gamma(h) = 1/2\mathbf{E}\left((X(t+h) - X(t))^2 \right)$ can be computed as

$$\gamma(h) = \frac{1}{2}\mathbf{E}\left(X^2(h) \right) = \frac{\|h\|_2^2}{2}, \quad h \in l_2,$$

see Sect. 11.2.2.3 for more details on variograms. Here we have $\gamma(h) \to \infty$ as $\|h\|_2 \to \infty$. Transferring the notions of stationarity from the index space $\mathbb{R}^d$ to l_2, it is clear that X is intrinsic stationary of order two but not wide sense stationary. Confer [177] for the general theory of Gaussian random functions on Hilbert index spaces. A random field $Y = \{Y(s), s \in \mathbb{R}^d\}$ on $\mathbb{R}^d$ can be derived from X by setting

$$Y(s) = X((s_1, \ldots, s_d, 0, 0, \ldots)^\top), \quad s = (s_1, \ldots, s_d)^\top \in \mathbb{R}^d,$$

i.e., consider sequences in l_2 for which only the first d components are not equal to zero. In this case all sums considered above are finite.

3. Fractional Brownian Fields. A *fractional Brownian field* $X = \{X(t),\ t \in \mathbb{R}^d\}$ is a centered Gaussian field with covariance function

$$\mathbf{E}(X(s)X(t)) = \frac{1}{2}\left(\|s\|^{2H} + \|t\|^{2H} - \|s-t\|^{2H} \right), \quad s, t \in \mathbb{R}^d$$

for some $H \in (0,1]$ where $\|\cdot\|$ is the Euclidean norm in $\mathbb{R}^d$ (see Sect. 11.2.2.1 for more details on covariance functions). The parameter H (often called *Hurst index*) is responsible for the regularity of the paths of X. The larger H is, the smoother are the paths. For $d = 1$, X is called the *fractional Brownian motion*, including the *two–sided* Wiener process (defined on the whole real line $\mathbb{R}$) where $H = 1/2$. In the case $d > 1$, X is called the *Brownian Lévy field* if $H = 1/2$ (see, e.g., [252, Sect. 2]). It is easy to check that X is intrinsically stationary of order two and isotropic. Its variogram $\gamma(h) = 1/2 \cdot \|h\|^{2H}$ is clearly motion invariant. However, this field is not wide-sense stationary as its variance is not constant.

Exercise 11.3. Show that a fractional Brownian field X

1. has stationary increments, which are positively correlated for $H \in (1/2, 1)$ and negatively correlated for $H \in (0, 1/2)$,
2. is H–self-similar, i.e., $X(\lambda t) \stackrel{D}{=} |\lambda|^H X(t)$ for all $\lambda \in \mathbb{R}$ and $t \in \mathbb{R}^d$,
3. has a version with a.s. Hölder-continuous paths of any order $\beta \in (0, H)$,
4. has nowhere differentiable paths for any $H \in (0, 1)$,
5. is a linear process for $d = H = 1$, i.e., $X(t) \stackrel{D}{=} t X_0$, for all $t \in \mathbb{R}$ and some random variable $X_0 \sim N(0, 1)$.

11.2.1.2 Examples: Non-Gaussian Random Fields

We now discuss several examples on non-Gaussian random fields.

1. Lévy Processes with Finite Second Moments. Let $X = \{X(t),\, t \geq 0\}$ be a *Lévy process* with finite second moments. It is usually defined via the *Lévy–Khinchin triplet* coding its jump structure, see e.g. [338] and Chapter 13. It is clear that X is intrinsic stationary of order two, but not wide-sense stationary. For these processes one can calculate the variance of its increments and the variogram. For example, we have

$$\gamma(h) = 1/2 \cdot \mathbf{E}\left((X(t+h) - X(t))^2\right) = \lambda h/2, \quad h, t \geq 0$$

for the stationary Poisson point process with intensity $\lambda > 0$.

2. Poisson Shot-Noise Fields. A *Poisson shot-noise field* $X = \{X(t),\, t \in \mathbb{R}^d\}$ is defined by

$$X(t) = \sum_{X_i \in \Phi} f(t - X_i) = \int_{\mathbb{R}^d} f(t - x)\Phi(dx), \quad t \in \mathbb{R}^d,$$

where Φ is a stationary Poisson point process on $\mathbb{R}^d$ with intensity λ, $f \in L^1(\mathbb{R}^d)$. It follows from [373, Exercise 9.10] that X is strictly stationary. Furthermore, it can be shown that

$$\mathbf{E}X(t) = \lambda \int_{\mathbb{R}^d} f(x)\,dx, \quad t \in \mathbb{R}^d,$$

and, if additionally $f \in L^2(\mathbb{R}^d)$, then

$$\mathbf{cov}\,(X(s), X(t)) = \lambda \int_{\mathbb{R}^d} f(t - s + x)f(x)\,dx, \quad s, t \in \mathbb{R}^d$$

i.e., the Poisson shot-noise field is also wide-sense stationary (cf. [373, Exercise 9.29]). If f is rotation invariant then X is isotropic of order two. See Fig. 11.1(a) for a typical realization of X.

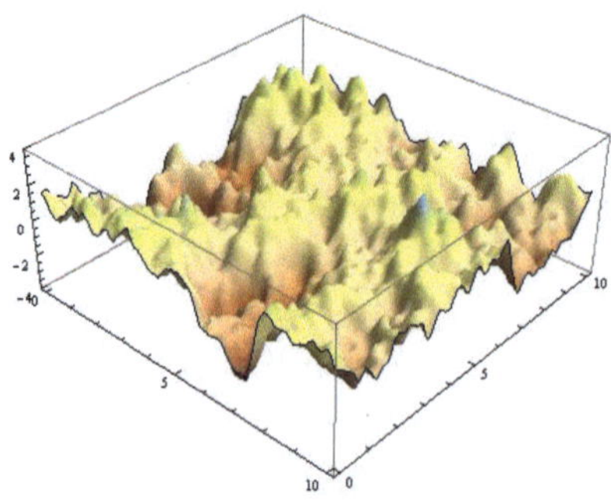

(a) Gaussian random field with Whittle-Matérn–type covariance function (see Sect. 11.2.2.1, Example 6), $a = 2$, $b = v = 1$

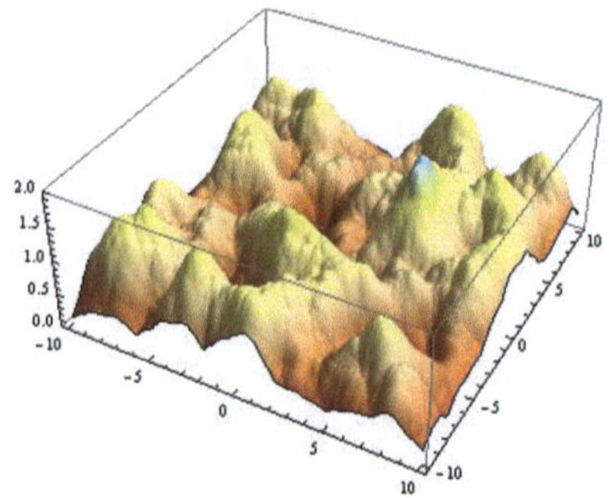

(b) Poisson shot-noise field with $\lambda = 1$ and $f(x) = \frac{1}{2\pi}\left(1 - \frac{1}{4}\|x\|^2\right)\mathbf{1}(\|x\| \leq 2)$

Fig. 11.1 Simulated realizations of (strictly and wide-sense) motion invariant random fields.

3. Boolean Random Functions. Let $\{Z_t(x),\ x \in \mathbb{R}^d\}_{t\in\mathbb{R}}$ be a family of independent lower semi-continuous random functions with subgraphs having almost surely compact sections (see Chap. 5) and $\Pi = \{(X_i, T_i)\}_{i=1}^{\infty}$ be a Poisson point process in $\mathbb{R}^d \times \mathbb{R}$ with intensity measure $v_d \otimes \theta$, where v_d denotes the Lebesgue measure on $\mathbb{R}^d$ and θ is a σ-finite measure on $\mathbb{R}$. The random function $Z = \{Z(x),\ x \in \mathbb{R}^d\}$ with

$$Z(x) = \sup_{(X_k, T_k)\in\Pi} Z_{T_k}(x - X_k),\quad x \in \mathbb{R}^d$$

is called a *Boolean random function*. The functions Z_t are referred to as *primary* functions.

11.2.2 Elements of Correlation Theory for Square Integrable Random Fields

Recall the following basic concepts.

Definition 11.3. A symmetric function $f : \mathbb{R}^d \times \mathbb{R}^d \to \mathbb{R}$ is called *positive semi–definite* if for any $n \in \mathbb{N}$, $w_1, \ldots, w_n \in \mathbb{C}$ and any $t_1, \ldots, t_n \in \mathbb{R}^d$ it holds that

$$\sum_{i,j=1}^{n} w_i \overline{w}_j f(t_i, t_j) \geq 0,$$

where $\overline{w}_j$ denotes the complex conjugate of w_j.

Definition 11.4. A symmetric function $f : \mathbb{R}^d \times \mathbb{R}^d \to \mathbb{R}$ is called *positive definite* if for any $n \in \mathbb{N}$, $w_1, \ldots, w_n \in \mathbb{C}$ such that $(w_1, \ldots, w_n)^\top \neq o \in \mathbb{C}^n$ and any $t_1, \ldots, t_n \in \mathbb{R}^d$ it holds that

$$\sum_{i,j=1}^{n} w_i \overline{w}_j f(t_i, t_j) > 0.$$

Definition 11.5. A symmetric function $f : \mathbb{R}^d \times \mathbb{R}^d \to \mathbb{R}$ is called *conditionally negative semi–definite* if for any $n \in \mathbb{N}$, $w_1, \ldots, w_n \in \mathbb{C}$ such that $\sum_{i=1}^{n} w_i = 0$ and any $t_1, \ldots, t_n \in \mathbb{R}^d$ it holds that

$$\sum_{i,j=1}^{n} w_i \overline{w}_j f(t_i, t_j) \leq 0.$$

Exercise 11.4. Prove that the functions $\cos(ax)$, $a \in \mathbb{R}$ and $e^{-|x|^p}$, $p \in (0, \, 2]$ are positive semi–definite, whereas $e^{-|x|^p}$, $p > 2$, $|\cos x|$ and $a^2 + \cos^2 x$, $a \in \mathbb{R}$ are not.

Exercise 11.5. Find a positive semi-definite function with discrete support.

11.2.2.1 Covariance function

Definition 11.6. For a random field $X = \{X(t), \, t \in \mathbb{R}^d\}$ with $\mathbf{E}X^2(t) < \infty$, $t \in \mathbb{R}^d$, the function $C : \mathbb{R}^d \times \mathbb{R}^d \to \mathbb{R}$ given by

$$C(s,t) = \mathbf{cov}(X(s), X(t)) = \mathbf{E}\left(X(s) - \mathbb{E}X(s)\right)\left(X(t) - \mathbb{E}X(t)\right), \quad s, t \in \mathbb{R}^d$$

is called the *covariance function* of X.

If X is wide-sense stationary (motion invariant), then $C(s,t)$ depends only on $s - t$ ($\|s - t\|$, respectively), $s, t \in \mathbb{R}^d$. For further properties of the covariance function see [373, Sect. 9.4-9.6]. We mention just a few.

1. *Generic property.* A function $f : \mathbb{R}^d \times \mathbb{R}^d \to \mathbb{R}$ is a covariance function of some square integrable random field if and only if it is positive semi–definite.

   Exercise 11.6. Prove this fact. *Hint.* Calculate the variance of the linear combination $\sum_{i=1}^{n} x_i X(t_i)$ for arbitrary $n \in \mathbb{N}$, $t_i \in \mathbb{R}^d$, $x_i \in \mathbb{R}$.

2. *Spectral representation.* By the Bochner-Kchinchin theorem (see, e.g., [46] or [373, Theorem 9.6]), any positive semi–definite function $f : \mathbb{R}^d \to \mathbb{R}$ which is continuous at the origin is a Fourier transform of some symmetric finite measure μ_f on $\mathbb{R}^d$. Thus for a wide-sense stationary and mean- square continuous field X we have

$$\mathbf{cov}(X(s), X(t)) = C(s - t) = \int_{\mathbb{R}^d} e^{i\langle x, s-t \rangle} \mu_C(dx).$$

Here $\langle \cdot, \cdot \rangle$ is the Euclidean scalar product in $\mathbb{R}^d$. The measure μ_C is called a *spectral measure* of X. If μ_C is absolutely continuous with respect to the Lebesgue measure, then its density is called a *spectral density*. The above field X itself has the *spectral representation*

$$X(t) = \int_{\mathbb{R}^d} e^{i\langle x,t\rangle} \Lambda(dx), \tag{11.1}$$

where Λ is a complex-valued orthogonal random measure with $\mathbf{E}\Lambda(A) = 0$ and $\mathbf{E}\left(\Lambda(A)\overline{\Lambda(B)}\right) = \mu_C(A \cap B)$ for any Borel sets $A, B \subset \mathbb{R}^d$. The integral in (11.1) is understood in the mean-square sense, i.e., its integral sums converge in $L^2(\Omega, \mathcal{F}, \mathbf{P})$. For more details on the spectral representation of stationary processes see [60, Sect. 7, §9, §10], [71, Sect. 2.3.3] or [254, Sect. 3.2, pp. 20-21], [412, Sect. 4.2, p. 90]. The spectral representation is used e.g. to simulate stationary Gaussian random fields approximating the integral in (11.1) by its finite integral sums with respect to a Gaussian white noise measure Λ.

11.2.2.2 Examples: Parametric Families of Covariance Functions

In this section, we discuss some examples of parametric families of covariance functions.

1. White Noise Model. For any $s, t \in \mathbb{R}^d$, let

$$C(s,t) = \begin{cases} \sigma^2, & \text{if } s = t, \\ 0, & \text{if } s \neq t . \end{cases}$$

This is the covariance function of a random field $X = \{X(t),\ t \in \mathbb{R}^d\}$ consisting of independent random variables $X(t)$, with variance $\sigma^2 > 0$, for any fixed $d \geq 1$.

2. Normal Scale Mixture. For any $s, t \in \mathbb{R}^d$, let

$$C(s,t) = \int_0^\infty e^{-x\|s-t\|^2} \mu(dx),$$

for some finite measure μ on $[0, \infty)$. This is the covariance function of a motion invariant random field for any $d \geq 1$ (see [349]).

3. Bessel Family. For some constants $a, b > 0$ and any $s, t \in \mathbb{R}^d$, $d \geq 1$, let

$$C(s,t) = b(a\|s-t\|)^{-\nu} J_\nu(a\|s-t\|),$$

where $\nu = \frac{d-2}{2}$ and

$$J_\nu(r) = \sum_{j=0}^\infty \frac{(-1)^j}{j!\,\Gamma(\nu+j+1)} \left(\frac{r}{2}\right)^{\nu+2j}, \qquad r \in \mathbb{R},$$

which is the Bessel function of the 1st kind of order ν (cf. [259]). The positive semi–definiteness of C is proven in [424, p. 367]. The spectral density of C is given by

$$f(h) = \frac{b(a^2 - h^2)^{v-\frac{d}{2}}}{2^v \pi^{\frac{d}{2}} a^{2v} \Gamma(v+1-\frac{d}{2})} \mathbf{1}_{[0,a]}(h).$$

The special case of $d = 3$, i.e., $v = \frac{1}{2}$, yields the so-called *hole-effect model*, where

$$C(s,t) = b\frac{\sin(a\|s-t\|)}{a\|s-t\|}, \quad s,t \in \mathbb{R}^d. \tag{11.2}$$

The function C in (11.2) is a valid covariance function if and only if $d \leq 3$.

4. Cauchy Family. For some constants $a, b, v > 0$ and any $s, t \in \mathbb{R}^d$, let

$$C(s,t) = \frac{b}{(1+(a\|s-t\|)^2)^v}.$$

Up to scaling, this function is positive semi-definite as a normal scale mixture with $\mu(dx) = cx^{v-1}e^{-x}dx$ for some constant $c > 0$.

5. Stable Family. For some $v \in (0,2]$ and any $s, t \in \mathbb{R}^d$, let

$$C(s,t) = be^{-a\|s-t\|^v}.$$

This function is positive semi-definite for all $d \geq 1$ since it is made by substitution $\theta \mapsto \|s-t\|$ out of the characteristic function of a symmetric v-stable random variable, cf. Definition 11.11. In the special case $v = 2$ the stable family yields a *Gaussian model* with $C(s,t) = be^{-a\|s-t\|^2}$. Its spectral density is equal to $f(h) = \frac{b\sqrt{a}}{2}he^{-\frac{ah^2}{4}}$.

6. Whittle-Matérn Family. For any $s, t \in \mathbb{R}^d$, $s \neq t$, let

$$C(s,t) = W_v(\|s-t\|) = b2^{1-v}(a\|s-t\|)^v K_v(a\|s-t\|),$$

where $v, a, b > 0$, $d \geq 1$ and K_v is the *modified Bessel function of third kind*, also called the *Macdonald function*, which is given by

$$K_v(r) = \frac{\pi}{2\sin(\pi v)}(e^{i\frac{\pi}{2}v}J_{-v}(re^{i\frac{\pi}{2}}) - e^{-i\frac{\pi}{2}v}J_v(re^{-i\frac{\pi}{2}})), \quad r \in \mathbb{R}, \quad v \notin \mathbb{N}.$$

For $v = n \in \mathbb{N}$ the above definiton of K_v is understood in the sense of a limit as $v \to n$, see [259, p. 69]. For $s = t$, we put $C(t,t) = b$. The spectral density of C is given by

$$f(h) = b\frac{2\Gamma(v+\frac{d}{2})}{\Gamma(\frac{d}{2})\Gamma(v)}\frac{(ah)^{d-1}}{(1+(ah)^2)^{v+\frac{d}{2}}}\mathbb{1}_{(0,\infty)}(h).$$

If $v = \frac{2d+1}{2}$, then a random field with covariance function C is d times differentiable in mean-square sense. If $v = \frac{1}{2}$, then the *exponential model* is obtained with

$$C(s,t) = be^{-a\|s-t\|}, \quad s,t \in \mathbb{R}^d,$$

which is an important special case. Note that the covariance function of the exponential model belongs to the stable family for $v = 1$. Fi. 11.1 shows a realization of a centered Gaussian random field X with Whittle-Matérn type covariance function.

7. Spherical Model. For $1 \leq d \leq 3$, some constants $a, b > 0$ and any $s, t \in \mathbb{R}^d$, let

$$C(s,t) = b\left(1 - \frac{3}{2}\frac{\|s-t\|}{a} + \frac{1}{2}\frac{\|s-t\|^3}{a^3}\right)\mathbf{1}(\|s-t\| \leq a). \qquad (11.3)$$

If $d = 3$, then (11.3) yields the volume of $B_{\frac{a}{2}}(o) \cap B_{\frac{a}{2}}(x_0)$, where $B_r(x)$ denotes the closed ball with center $x \in \mathbb{R}^d$ and radius $r > 0$, and $x_o \in \mathbb{R}^3$ is chosen such that, $\|x_0\| = \|s-t\|$. This is exactly the way how (11.3) can be generalized to higher dimensions: For any $d \geq 3$, let

$$C(s,t) = bv_d\left(B_{\frac{a}{2}}(0) \cap B_{\frac{a}{2}}(s-t)\right), \quad s,t \in \mathbb{R}^d,$$

where v_d is the d-dimensional Lebesgue measure. The advantage of spherical models is that they have a compact support.

8. Geometric Anisotropy It is easy to see that all covariance models considered above are motion invariant. An example of a stationary but anisotropic covariance structure can be provided by rotating and stretching the argument of a motion invariant covariance model. Let $C_0(\|h\|)$, $h \in \mathbb{R}^d$ be a covariance function of a motion invariant field where $C_0 : \mathbb{R}^+ \to \mathbb{R}^+$. Then, for a positive definite $(d \times d)$–matrix Q,

$$C(h) = C_0(\sqrt{h^\top Qh}), \quad h \in \mathbb{R}^d$$

is a covariance function of some wide-sense stationary anisotropic random field (see [408, Chap. 9]).

9. Cyclone Model For $d = 3$ and for any $s, t \in \mathbb{R}^3$, let

$$C(s,t) = \frac{2^{3/2}\det(S_s)^{1/4}\det(S_t)^{1/4}}{\sqrt{\det(S_s + S_t)}}W_v\left(\sqrt{(s-t)^\top S_s(S_s+S_t)^{-1}S_t(s-t)}\right), \quad (11.4)$$

where $S_s = I_3 + ss^\top$ and I_3 is the (3×3)–identity matrix and W_v is the Whittle-Matérn model. In [342, Theorem 5, Example 16], it is shown that C is a valid covariance function belonging to a more general class of covariances that mimic cyclones.

Exercise 11.7. Show that C given in (11.4) is the covariance function of an isotropic but not wide-sense stationary random field, i.e., $C(s,t) = C(Rs, Rt)$ for any $R \in SO_3$, but $C(s,t)$ does not only depend on $s-t$, $s,t \in \mathbb{R}^3$.

For more sophisticated covariance models including spatio–temporal effects see, e.g. [342] and references therein.

11.2.2.3 Variogram

Definition 11.7. For a random field $X = \{X(t), t \in \mathbb{R}^d\}$ the function

$$\gamma(t,s) = \frac{1}{2}\mathbf{E}\left((X(t) - X(s))^2\right), \quad s,t \in \mathbb{R}^d$$

is called the *variogram* of X whenever it is finite for any $s,t \in \mathbb{R}^d$.

Note that for square-integrable random fields X, it obviously holds that

$$\gamma(s,t) = \frac{1}{2}\mathbf{var}X(s) + \frac{1}{2}\mathbf{var}X(t) - \mathbf{cov}(X(t), X(s)) + \frac{1}{2}(\mathbf{E}X(s) - \mathbf{E}X(t))^2. \quad (11.5)$$

If the field X is intrinsic stationary of order two (motion invariant) then $\gamma(s,t)$ depends only on the difference $s - t$ ($\|s - t\|$, respectively). With slight abuse of notation, in these cases we write $\gamma(s - t)$ and $\gamma(\|s - t\|)$ for functions $\gamma : \mathbb{R}^d \to \mathbb{R}$ and $\gamma : \mathbb{R}_+ \to \mathbb{R}$, respectively. For a wide-sense stationary random field X with covariance function C formula (11.5) reads

$$\gamma(h) = C(0) - C(h), \quad h \in \mathbb{R}^d. \quad (11.6)$$

Basic Properties of Variograms

Let X be a random field with covariance function C and variogram γ. Then the following properties hold

1. $\gamma(t,t) = 0$ for all $t \in \mathbb{R}^d$.
2. *Symmetry:* $\gamma(t,s) = \gamma(s,t)$ for all $s,t \in \mathbb{R}^d$.
3. *Characterization of variograms*

   (a) A function $\gamma : \mathbb{R}^d \times \mathbb{R}^d \to \mathbb{R}_+$ is a variogram of some random field if γ is conditionally negative semi–definite, see, for example, [142, Theorem 1] or [71, Sect. 2.3.3, p.61].

   Exercise 11.8. Prove that the variogram of any intrinsic stationary random field X is a conditionally negative semi–definite function.
   Hint. Compute $\mathbf{var}(\sum_{i=1}^{n} \lambda_i X(t_i))$ applying (11.5) with $\sum_{i=1}^{n} \lambda_i = 0$.

   (b) A continuous even function $\gamma : \mathbb{R}^d \to \mathbb{R}_+$ with $\gamma(0) = 0$ is a variogram of a wide-sense stationary random field if $e^{-\lambda \gamma(h)}$ is a covariance function for all $\lambda > 0$, cf. [352].

4. *Stability.* If γ_1, γ_2 are variograms, then $\gamma = \gamma_1 + \gamma_2$ is a variogram as well.

Exercise 11.9. Prove this fact. Show in particular that the function γ given by $\gamma(h) = \gamma_1(h_i) + \gamma_2(h_j)$, is a variogram, where $h = (h_1, \ldots, h_d)^\top \in \mathbb{R}^d$ and γ_1, γ_2 are univariate variograms.

5. *Mixture.* Let $\gamma_x : \mathbb{R}^d \to \mathbb{R}_+$ be a variogram of an intrinsic stationary (of order two) random field for any $x \in \mathbb{R}$. Then the function γ given by

$$\gamma(h) = \int_{\mathbb{R}} \gamma_x(h)\mu(dx), \quad h \in \mathbb{R}^d$$

is the variogram of some random field provided that μ is a measure on $\mathbb{R}$ and the above integral exists for any $h \in \mathbb{R}^d$, see [71, Sect. 2.3.2, pp. 60-61].
6. If X is wide-sense stationary and $C(\infty) = \lim_{\|h\| \to \infty} C(h) = 0$, then it follows from (11.6) that there exists the so-called *sill* $\gamma(\infty) = \lim_{\|h\| \to \infty} \gamma(h) = C(0)$.
7. If X is mean-square continuous, then $\gamma(h) \leq c\|h\|^2$, for some constant $c > 0$ and for all $h \in \mathbb{R}^d$ with sufficiently large $\|h\|$, see [422, pp. 397-398].
8. If X is mean-square differentiable, then $\lim_{\|h\| \to \infty} \frac{\gamma(h)}{\|h\|^2} = 0$, see [423, pp. 136-137].
9. Let $\gamma : \mathbb{R} \to \mathbb{R}_+$ be an even twice continuously differentiable function with $\gamma(0) = 0$. Then γ is a variogram if and only if the second derivative $\gamma^{(2)}$ is a covariance function, cf. [142, Theorem 7].

Exercise 11.10. Show that for a variogram γ the function $e^{\lambda \gamma(h)}$ is a variogram for any $\lambda > 0$.

Exercise 11.11. Let the function $\gamma : \mathbb{R}^d \to \mathbb{R}_+$ be bounded and the variogram of some random field X which is intrinsic stationary of order two. Consider the function $C(s,t) = \gamma(s) + \gamma(t) - \gamma(s-t)$, $s,t \in \mathbb{R}^d$. Show that C is the covariance function of a random field Z such that $Z(o) = 0$ a.s.

Parametric Families of Variograms

Most parametric models for variograms of stationary random fields, which are widely used in applications, can be constructed from the corresponding families of covariance functions (such as those described in Sect. 11.2.2.1) by applying the formula (11.6). Most models of variograms inherit their names from the corresponding covariance models (e.g., exponential, spherical one, etc.). One of few exceptions is the variogram $\gamma(h) = a(1 - \mathbf{1}_{\{0\}}(\|h\|))$, $h \in \mathbb{R}^d$ corresponding to the white noise, where the jump size $a > 0$ is called a *nugget effect*.

The stability property of variograms mentioned above can be also used to create different anisotropy effects, for instance, the so-called *purely zonal anisotropy*. To explain this on an example, let $\gamma(h) = a\gamma_1(h_x) + b\gamma_2(h_y) + c\gamma_3(h_z)$, $h = (h_x, h_y, h_z) \in \mathbb{R}^3$, $a,b,c \geq 0$, where γ_i $i = 1,2,3$ are variograms in dimension $d = 1$. Then γ is a variogram in dimension $d = 3$ which allows for different dependence ranges in the directions of the three different axes. An example of a *mixed anisotropy* models is given by

$$\gamma(h) = \gamma_1(\|h\|) + \gamma_2\left(\sqrt{h_x^2 + h_y^2}\right) + \gamma_3(h_z), \quad h = (h_x, h_y, h_z) \in \mathbb{R}^3.$$

This is a mixture of a 3D-isotropic variogram γ_1, a 2D-isotropic (in the xy-plane) variogram γ_2 and a 1D-variogram γ_3. The addition of a linear combination of γ_2 and γ_3 creates anisotropy in the direction of the z-axis. For more details on variograms, see [71, Chap. 2].

11.2.2.4 Statistical Estimation of Covariances and Variograms

There are numerous approaches to estimate covariance functions and variograms. They are well described in the literature and therefore will not be reviewed here. The interested reader can consult e.g. [71, Sect. 2.2] and [373, Sect. 9.8] as well as the references therein. We just present two illustrating examples.

Example 11.1. Consider microscopic steel data (see Fig. 11.2(a)). This data can be interpreted as a realization of an isotropic stationary random field with two possible values: zero for white pixels and one for black pixels of the image. Fig. 11.2(b) shows estimates for the corresponding variogram. For this purpose Mathéron's estimator $\hat{\gamma}$ given by

$$\hat{\gamma}(h) = \frac{1}{2N(h)} \sum_{i,j:t_i - t_j \approx h} (X(t_i) - X(t_j))^2$$

(see [373, p. 325]) was calculated for different directions and $0 \le h \le 0.5$, where $t_i - t_j \approx h$ means that $t_i - t_j$ belongs to a certain neighborhood of h and $N(h)$ denotes the number of such pairs (t_i, t_j), $i, j = 1, \ldots, n$. The directions can be distinguished by the color of their plots. Since the estimates differ not too much from each other, the data can be assumed to be isotropic .

Example 11.2. For $d = 2$, construct an example of a zonally anisotropic variogram, in which the value of the sill depends on the direction of the input vector h. Consider

$$\gamma(h) = \gamma_1(h) + \gamma_2(h) \tag{11.7}$$

where γ_1 is an isotropic variogram given by

$$\gamma_1(h) = 1 - e^{-\|h\|}, \quad h \in \mathbb{R}^2,$$

and γ_2 is a geometric anisotropic variogram given by model

$$\gamma_2(h) = 1 - e^{-\frac{\sqrt{h^T Q^2 h}}{5}}, \quad h \in \mathbb{R}^2,$$

with $Q = \sqrt{\Lambda} \cdot R$. Here R is a rotation matrix with rotation angle $\alpha = 2$ and $\Lambda = \text{diag}(5, 1)$ is a diagonal matrix. Fig. 11.3(a) shows a realization of a Gaussian random field on $[-1, 1]^2$ with variogram γ given by (11.7). Fig. 11.3(b) illustrates the elliptic form of the contour lines of the Matheron estimator of the correspond-

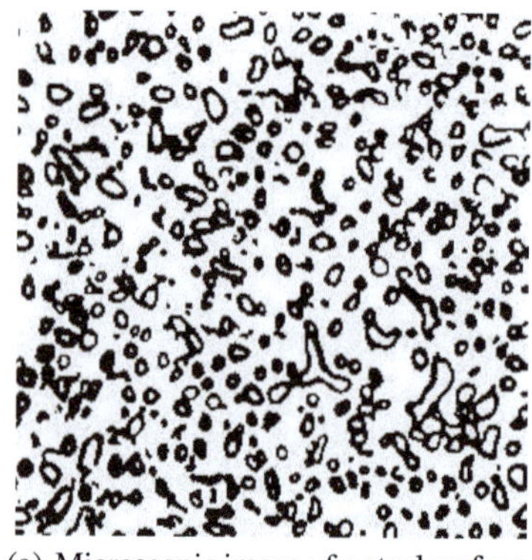

(a) Microscopic image of a steel surface.

(b) Estimates for the x-direction (red), y-direction (green), all directions (black) for values $0 \leq h \leq 0.5$.

Fig. 11.2 Microscopic steel image (left) and its empirical variogram estimated in different directions (right)

ing zonally anisotropic variogram and Fig. 11.3(c) shows the theoretical variogram function γ.

11.2.3 Stable Random Fields

In this section, we review some basic notions of the theories of stable distributions, random measures and fields. A very good reference which covers most of these topics is [337], see also [301], [338, Chap. 3] and [431].

11.2.3.1 Stable Distributions

Let $n \in \mathbb{N}$ be an arbitrary integer. We begin with the definition of stability for random vectors.

Stable Random Vectors

Definition 11.8. A random vector $X = (X_1, \ldots, X_n)^{\mathsf{T}}$ in $\mathbb{R}^n$ is called *stable* if for each $m \geq 2$ there exist $c = c(m) > 0$ and $k = k(m) \in \mathbb{R}^n$ such that

$$X^{(1)} + X^{(2)} + \ldots + X^{(m)} \stackrel{\mathrm{D}}{=} cX + k,$$

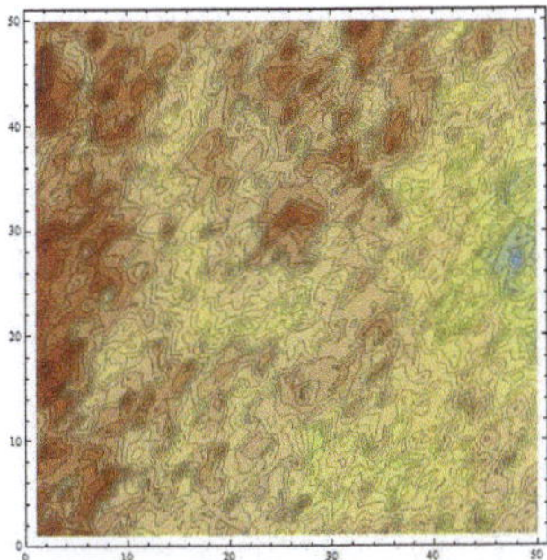

(a) Realization of an anisotropic Gaussian random field with variogram γ given by (11.7), rotation angle $\alpha = 114.59°$ and scaling factors $\lambda_1 = 5$, $\lambda_2 = 1$.

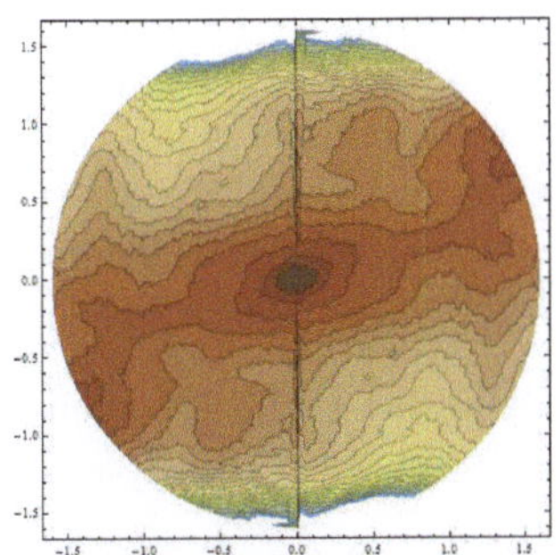

(b) Contour lines of the empirical variogram of the simulation shown in Fig. 11.3(a) for values of $\|h\|$ between 0 and 1.6.

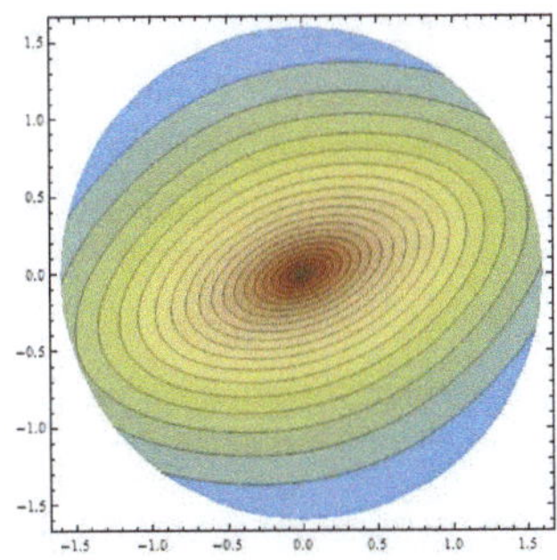

(c) Theoretical variogram function $\gamma = \gamma_1 + \gamma_2$.

Fig. 11.3 Realization of an anisotropic Gaussian random field with the corresponding empirical and theoretical variograms.

where $\{X^{(i)}\}_{i=1}^{m}$ are independent copies of X.

It can be shown that $c = m^{1/\alpha}$ for some $0 < \alpha \leq 2$ which is called the *stability index*, see [337, Theorem 2.1.2]. There is an equivalent definition of stable vectors which is often used in mathematical practice to check stability.

Definition 11.9. Let $\alpha \in (0, 2]$. We say that a random vector $X = (X_1, \ldots, X_n)^{\mathsf{T}}$ in $\mathbb{R}^n$ is *α-stable* if its characteristic function φ_X is given by

$$\varphi_X(\theta) = \begin{cases} e^{-\int_{\mathbb{S}^{n-1}} |\langle \theta, s \rangle|^{\alpha} \left(1 - i\,\mathrm{sign}(\langle \theta, s \rangle) \tan \frac{\pi\alpha}{2}\right) \Gamma(ds) + i\langle \theta, \mu \rangle}, & \text{if } \alpha \neq 1, \\ e^{-\int_{\mathbb{S}^{n-1}} |\langle \theta, s \rangle| \left(1 + i\frac{2}{\pi}\,\mathrm{sign}(\langle \theta, s \rangle) \ln |\langle \theta, s \rangle|\right) \Gamma(ds) + i\langle \theta, \mu \rangle}, & \text{if } \alpha = 1, \end{cases} \tag{11.8}$$

where Γ is a finite measure on the unit sphere $\mathbb{S}^{n-1}$ of $\mathbb{R}^n$ and μ is an arbitrary vector in $\mathbb{R}^n$.

The pair (μ, Γ) gives a unique parametrization of the distribution of α-stable random vectors for $\alpha \in (0, 2)$, and we write $X \sim S_\alpha(\mu, \Gamma)$. This means that there is no other pair (μ', Γ') yielding the same characteristic function φ_X in (11.8). The measure Γ is called the *spectral measure* of X. It contains all information about the dependence between the vector components X_i (see also Exercise 11.15). The vector μ reflects the *shift* with respect to the origin.

Definition 11.10. A random vector $X = (X_1, \ldots, X_n)^\top$ is called *singular* if $\sum_{i=1}^n c_i X_i = 0$ a.s. for some $(c_1, \ldots, c_n)^\top \in \mathbb{R}^n \setminus \{0\}$. Otherwise, it is called *full-dimensional*.

If $\alpha = 2$, then Definition 11.8 yields a *Gaussian* random vector which is equivalently defined via its characteristic function given by

$$\varphi_X(\theta) = \exp\left(i\langle \theta, \mu \rangle - \tfrac{1}{2} \theta^\top \Sigma \theta \right). \tag{11.9}$$

Here $\mu \in \mathbb{R}^n$ is the *mean value* of X and Σ is the symmetric, positive semi–definite $(n \times n)$–*covariance matrix* of X. The matrix Σ has the entries $\sigma_{ij} = \mathbf{E}(X_i - \mu_i)(X_j - \mu_j)$, where X_i and μ_i are the components of vectors X and μ, respectively. It is easy to see that if $\det \Sigma = 0$ then the Gaussian random vector X is singular.

Exercise 11.12. Prove that

1. Definition 11.8 is equivalent to Definition 11.9 for $\alpha \in (0, 2)$, and
2. it is equivalent to the definition of a Gaussian random vector via formula (11.9) for $\alpha = 2$.

Exercise 11.13. Show that for $X \sim S_\alpha(\mu, \Gamma)$ the relation between the drift k in Definition 11.8 and the shift μ in Definition 11.9 is given by $k(m) = \mu(m - m^{1/\alpha})$. *Hint.* Show first that $\sum_{i=1}^m X^{(i)} \sim S_\alpha(m\mu, m\Gamma)$ and $m^{1/\alpha} X + k(m) \sim S_\alpha(m^{1/\alpha}\mu + k(m), m\Gamma)$.

Remark 11.1. For $\alpha = 2$, the characteristic function considered in (11.8) has the form

$$\varphi(\theta) = \exp\left(-\int_{\mathbb{S}^{n-1}} \langle \theta, s \rangle^2 \Gamma(ds) + i\langle \theta, \mu \rangle \right). \tag{11.10}$$

In this case it is easy to find two different finite measures Γ_1 and Γ_2 on $\mathbb{S}^{n-1}$ yielding the same function φ.

Exercise 11.14. Check that the following finite measures Γ_1 and Γ_2 on the unite sphere in $\mathbb{R}^2$

$$\Gamma_1(ds) = \delta_{(\sqrt{2}/2, \sqrt{2}/2)}(ds) + \delta_{(-\sqrt{2}/2, -\sqrt{2}/2)}(ds),$$

$$\Gamma_2(ds) = 2\delta_{(\sqrt{2}/2, \sqrt{2}/2)}(ds)$$

and any shift $\mu \in \mathbb{R}^2$ yield the same expression in (11.8) if $\alpha = 2$, $n = 2$. Here δ_x is the Dirac measure concentrated at $x \in \mathbb{R}^2$. Verify that this expression corresponds to the characteristic function of a Gaussian vector with mean value vector μ and covariance matrix

$$\Sigma = \begin{pmatrix} 2 & 2 \\ 2 & 2 \end{pmatrix}.$$

A random vector X in $\mathbb{R}^n$ is called *symmetric* if $\mathbf{P}(X \in A) = \mathbf{P}(-X \in A)$ for any Borel set $A \subset \mathbb{R}^n$. For symmetric α-stable $(S\alpha S)$ distributions, we use the standard abbreviation $S\alpha S$.

Lemma 11.1 ([337], Theorem 2.4.3). *An α–stable random vector X is symmetric if and only if $\mu = 0$ and its spectral measure Γ is symmetric.*

Exercise 11.15. Let $n = 2$ and $X = (X_1, X_2)^\top$ be an α-stable random vector, $\alpha \in (0,2)$, with spectral measure Γ. Let $\mathrm{supp}(\Gamma)$ be the support of Γ. Show that

1. X_1 is independent of X_2 if and only if $\mathrm{supp}(\Gamma)$ lies within the intersection of the unit sphere with the coordinate axes,
2. $X_1 = c \cdot X_2$ a.s. for some $c \in \mathbb{R}$ (i.e. the vector X is singular) if and only if $\mathrm{supp}(\Gamma)$ is a subset of the unit circle intersected by a line.

Real-valued Stable Random Variables

If $n = 1$ we deal with stable random variables whose distributions are determined by four parameters α, σ, β, and μ.

Definition 11.11. The random variable $X : \Omega \to \mathbb{R}$ is called α-*stable* if its characteristic function φ_X has the form

$$\varphi_X(\theta) = \begin{cases} \exp\left(-\sigma^\alpha |\theta|^\alpha \left(1 - i\beta(\mathrm{sign}(\theta)) \tan \frac{\pi\alpha}{2}\right) + i\mu\theta\right), & \text{if } \alpha \in (0,2], \alpha \neq 1, \\ \exp\left(-\sigma|\theta| \left(1 + i\beta \frac{2}{\pi}(\mathrm{sign}(\theta)) \ln|\theta|\right) + i\mu\theta\right), & \text{if } \alpha = 1, \end{cases}$$

$$(11.11)$$

where we write $X \sim S_\alpha(\sigma, \beta, \mu)$.

In comparison to representation (11.8), two new parameters $\sigma \geq 0$ and $\beta \in [-1,1]$ have been introduced in (11.11) instead of the spectral measure Γ. They are interpreted as scale and skewness parameters, respectively.

Exercise 11.16. Show that the spectral measure Γ of $X \sim S_\alpha(\sigma, \beta, \mu)$ is given by

$$\Gamma(ds) = \frac{\sigma^\alpha}{2}(1 + \beta)\delta_1(ds) + \frac{\sigma^\alpha}{2}(1 - \beta)\delta_{-1}(ds).$$

Hence, it holds that

$$\sigma^\alpha = \Gamma(\{1\}) + \Gamma(\{-1\}), \qquad \beta = \frac{\Gamma(\{1\}) - \Gamma(\{-1\})}{\Gamma(\{1\}) + \Gamma(\{-1\})}.$$

Remark 11.2. Stable distributions are absolutely continuous. Nevertheless, their densities are not known in closed form except for the cases $\alpha = 1/2$, $\alpha = 1$ and $\alpha = 2$. A random variable $X \sim S_\alpha(\sigma, \pm 1, \mu)$ is called *totally skewed*.

Example 11.3. Show that

1. $X \sim S_2(\sigma, 0, \mu)$ is a Gaussian random variable with mean μ and variance $2\sigma^2$.
2. If $\alpha \in [1,2)$ then $X \sim S_\alpha(\sigma, \pm 1, \mu)$ attains values everywhere in $\mathbb{R}$. On the contrary, if $\alpha \in (0,1)$ and $\mu = 0$, then $X \geq 0$, $(X \leq 0)$ a.s. when $\beta = 1$ $(\beta = -1)$, respectively.

Exercise 11.17. Show that the characteristic function of an $S\alpha S$-distributed random variable X is given by $\varphi_X(\theta) = \exp\{-\sigma^\alpha |\theta|^\alpha\}$, i.e., $X \sim S_\alpha(\sigma, 0, 0)$ for some $\sigma > 0$.

Tails and Moments

Non–Gaussian stable distributions are heavy-tailed. In particular, they belong to a subclass of heavy-tailed distributions with especially slow large deviation behavior. For more details on heavy-tailed distributions, see e.g. [122], [261]. For $X \sim S_\alpha(\sigma, \beta, \mu)$ with $\alpha \in (0,2)$ there exists $c > 0$ such that

$$\mathbf{P}(|X| > x) \sim cx^{-\alpha}, \quad \text{as } x \to \infty. \tag{11.12}$$

Here and in what follows we say that $a_x \sim b_x$ if $\lim_{x \to \infty} \frac{a_x}{b_x} = 1$. As a consequence of (11.12), the absolute moments of X behave like

$$\mathbf{E}(|X|^p) = \int_0^\infty \mathbf{P}\{|X| > x^{1/p}\}dx \leq c_1 \int_1^\infty x^{-\alpha/p}dx,$$

for some $c_1 > 0$. They are finite if $p \in (0, \alpha)$ and infinite for any $p \in [\alpha, \infty)$.

Exercise 11.18. Show that the following statements are true.

1. The normal distribution $X \sim N(\mu, \sigma^2)$ is not heavy-tailed (this is equivalent to the statement that the tails are exponentially bounded), i.e.,

$$\mathbf{P}(X < -x) = \mathbf{P}(X > x) \sim \frac{1}{\sqrt{2\pi}\sigma x} e^{-x^2/(2\sigma^2)}, \quad x \to \infty.$$

2. For $X \sim S_\alpha(\sigma, \beta, 0)$, $\alpha \in (0,2)$, $\alpha \neq 1$ it holds that

$$(\mathbf{E}(|X|^p))^{1/p} = c_{\alpha, \beta}(p)\sigma \tag{11.13}$$

for every $p \in (0, \alpha)$. Here $c_{\alpha, \beta}(p) = (\mathbf{E}(|\xi|)^p)^{1/p}$ with $\xi \sim S_\alpha(1, \beta, 0)$. If $\alpha = 1$ then (11.13) holds only for $\beta = 0$.
3. For any α-stable random vector $X = (X_1, X_2)$ the sum $aX_1 + bX_2$, $a, b \in \mathbb{R}$ is again α-stable. In particular, the components X_i of a stable vector $X = (X_1, X_2) \sim$

$S_\alpha(\mu,\Gamma)$ are stable, and it holds that $\sigma_{aX_1+bX_2} = \int_{\mathbb{S}^1} |as_1 + bs_2|^\alpha \Gamma(ds_1,ds_2)$ for any $a,b \in \mathbb{R}$.

Simulation of stable random variables is extensively described in [301].

11.2.3.2 Integration with Respect to Stable Random Measures

Let $(E,\mathcal{E},m)$ be an arbitrary measurable space with σ-finite measure m and $\mathcal{E}_0 = \{A \in \mathcal{E} : m(A) < \infty\}$. Let $\beta : E \to [-1,1]$ be a measurable function.

Definition 11.12. A random function $M = \{M(A),\ A \in \mathcal{E}_0\}$ is called an *independently scattered random measure (random noise)* if

1. for any $n \in \mathbb{N}$ and pairwise disjoint sets $A_1,A_2,\ldots,A_n \in \mathcal{E}_0$ the random variables $M(A_1),\ldots,M(A_n)$ are independent,
2. $M(\bigcup_{j=1}^\infty A_j) = \sum_{j=1}^\infty M(A_j)$ a.s. for arbitrary pairwise disjoint sets $A_1,A_2,\ldots \in \mathcal{E}_0$ with $\bigcup_{j=1}^\infty A_j \in \mathcal{E}_0$.

Definition 11.13. An independently scattered random measure M on $(E,\mathcal{E}_0)$ is called α-*stable* if for each $A \in \mathcal{E}_0$

$$M(A) \sim S_\alpha\left((m(A))^{1/\alpha}, \frac{\int_A \beta(x)m(dx)}{m(A)}, 0\right).$$

The measure m is called the *control measure* of M, and β is the *skewness function* of M.

Our goal is to define the integral $\int_E f(x)M(dx)$ of a measurable function $f : E \to \mathbb{R}$ with respect to an α-stable random measure M. For a simple function $f(x) = \sum_{i=1}^n c_i \mathbf{1}_{A_i}(x)$, where $\{A_i\}_{i=1}^n \subset \mathcal{E}_0$ is a sequence of pairwise disjoint sets, we put

$$\int_E f(x)M(dx) = \sum_{i=1}^n c_i M(A_i).$$

It can be shown that the integral $\int_E f(x)M(dx)$ defined in this way does not depend on the representation of f as a simple function, see [337, Sect. 3.4]. For an arbitrary $f : E \to \mathbb{R}$ such that $\int_E |f(x)|^\alpha m(dx) < \infty$ we consider a pointwise approximation of f by simple functions $f^{(n)}$, where we put

$$\int_E f(x)M(dx) = \mathbf{P}\text{-}\lim_{n\to\infty} \int_E f^{(n)}(x)M(dx).$$

Here $\mathbf{P}$-lim denotes the limit in probability. Note that this definition is independent of the choice of the approximating sequence $\{f^{(n)}\}$, cf. [337, Sect. 3.4] for more details.

Lemma 11.2. *Let $X = \int_E f(x)M(dx)$, where M is an α-stable random measure with control measure m and skewness function β. Then X is an α-stable random variable with zero shift, scale parameter*

$$\sigma_X^\alpha = \int_E |f(x)|^\alpha m(dx), \tag{11.14}$$

and skewness parameter

$$\beta_X = \frac{\int_E f(x)^{<\alpha>} \beta(x)\, m(dx)}{\int_E |f(x)|^\alpha m(dx)},$$

where $a^{<p>} = \mathrm{sgn}\,(a) \cdot |a|^p$ *and* $\mathrm{sgn}\,(a)$ *denotes the sign of a.*

For a proof see [337, Sect.3.4]. Notice that if $\beta(x) = 0$ for all $x \in E$, then the integral X is a $S\alpha S$-distributed random variable.

In case of stable vectors with an integral representation, we have the following criterion of their full–dimensionality and singularity, respectively.

Lemma 11.3. *Consider an n–dimensional α–stable random vector $X = (X_1, \ldots, X_n)^\mathsf{T}$ with $0 < \alpha \leq 2$ and integral representation*

$$X = \left(\int_E f_1(x)M(dx), \ldots, \int_E f_n(x)M(dx) \right)^\mathsf{T}.$$

Then X is singular if and only if $\sum_{i=1}^n c_i f_i(x) = 0$ m–almost everywhere for some vector $(c_1, \ldots, c_n)^\mathsf{T} \in \mathbb{R}^n \setminus \{0\}$.

The proof of Lemma 11.3 immediately follows from Definition 11.10 and the fact that $\sigma_{\sum_{i=1}^n c_i X_i}^\alpha = \int_E |\sum_{i=1}^n c_i f_i(x)|^\alpha m(dx)$, see (11.14).

Remark 11.3. A more universal criterion of singularity for stable random vectors can be given in terms of their spectral measure. If the measure $\Gamma(ds)$ on $\mathbb{S}^{n-1}$ is the spectral measure of an α-stable vector X in $\mathbb{R}^n$ which is concentrated on the intersection of $\mathbb{S}^{n-1}$ with an $(n-1)$–dimensional linear subspace, then the random vector X is singular. Otherwise, X is full–dimensional. For a proof of these statements, see [220].

11.2.3.3 Stable Random Fields with Integral Spectral Representation

Definition 11.14. A random field X is called α-*stable* if its finite-dimensional distributions are α-stable.

Consider a random field $X = \{X(t), t \in \mathbb{R}^d\}$ of the form

$$X(t) = \int_E f_t(x)\, M(dx), \quad t \in \mathbb{R}^d, \tag{11.15}$$

where $f_t : E \to \mathbb{R}$ is a measurable function such that $\int_E |f_t(x)|^\alpha m(dx) < \infty$ and, if $\alpha = 1$, $\int_E |f(x)\beta(x)\ln|f(x)||m(dx) < \infty$ for any $t \in \mathbb{R}^d$. Here M is an α-stable random measure on E with control measure m and skewness function β. Obviously, the marginals of the random field X given in (11.15) are α-stable. If $\beta(x) = 0$ for all

$x \in E$ then the finite-dimensional distributions of X are symmetric α-stable. Thus we call X to be a $S\alpha S$ random field.

A natural question is which stable fields allow for an integral representation (11.15). A necessary and sufficient condition for this is the condition of separability of X in probability, see [337, Theorem 13.2.1].

Definition 11.15. A stable random field $X = \{X(t),\ t \in I\}, I \subseteq \mathbb{R}^d$ is called *separable in probability* if there exists a countable subset $I_0 \subseteq I$ such that for every $t \in I$ and any sequence $\{t_k\}_{k\in\mathbb{N}} \subset I_0$ with $t_k \to t$ as $k \to \infty$ it holds that $X(t) = \mathbf{P}\text{-}\lim_{k\to\infty}X(t_k)$.

In particular, all stochastically continuous α-stable random fields are separable in probability.

11.2.4 Dependence Measures for Stable Random Fields

The relationship between two α-stable random variables cannot be described by using the notion of covariance because of the absence of the second moments if $\alpha < 2$. We consider two different ways of measuring the degree of dependence of two stable random variables.

Covariation

Definition 11.16. Let $X = (X_1, X_2)^\top$ be an α-stable random vector with $\alpha \in (1,2]$ and spectral measure Γ. The *covariation* $[X_1, X_2]_\alpha$ of X_1 on X_2 is given by

$$[X_1, X_2]_\alpha = \int_{\mathbb{S}^1} s_1 s_2^{<\alpha-1>}\Gamma(d(s_1, s_2)).$$

Theorem 11.1 (Properties of covariation). *Let $(X_1, X_2, X_3)^\top$ be an α-stable random vector with $\alpha \in (1,2]$. Then the following properties hold.*

1. *Linearity in the first entry. For any $a, b \in \mathbb{R}$ it holds that*

$$[aX_1 + bX_2, X_3]_\alpha = a[X_1, X_3]_\alpha + b[X_2, X_3]_\alpha.$$

2. *If X_1 and X_2 are independent, then $[X_1, X_2]_\alpha = 0$.*
3. *Gaussian case. For $\alpha = 2$, it holds that $[X_1, X_2]_2 = 1/2 \cdot \mathbf{cov}(X_1, X_2)$.*
4. *Covariation and mixed moments. Let $1 < \alpha < 2$ and Γ be spectral measure of $(X_1, X_2)^\top$ with $X_1 \sim S_\alpha(\sigma_1, \beta_1, 0)$ and $X_2 \sim S_\alpha(\sigma_2, \beta_2, 0)$. Then, for $1 \le p < \alpha$, it holds that*

$$\frac{\mathbf{E}\left(X_1 X_2^{<p-1>}\right)}{\mathbf{E}|X_2|^p} = \frac{[X_1, X_2]_\alpha(1 - c \cdot \beta_2) + c \cdot (X_1, X_2)_\alpha}{\sigma_2^\alpha}, \tag{11.16}$$

where $(X_1, X_2)_\alpha = \int_{\mathbb{S}^1} s_1 |s_2|^{\alpha-1} \Gamma(d(s_1, s_2))$ *and*

$$c = \frac{\tan(\alpha\pi/2)}{1 + \beta_2^2 \tan^2(\alpha\pi/2)} \left[\beta_2 \tan(\alpha\pi/2) - \tan\left(\frac{p}{\alpha} \arctan(\beta_2 \tan(\alpha\pi/2)) \right) \right].$$

Proof. 1. The linearity in the first argument is obvious. However, note that the co-variation is not symmetric, so that there is no linearity in the second argument. 2. To see this, use Exercise 11.15. 3. This assertion (together with the useful relationship $\mathbf{var}X_i = 2 \int_{\mathbb{S}^1} s_i^2 \Gamma(d(s_1, s_2))$, $i = 1, 2$) follows from the comparison of the representations (11.9) and (11.10) of the characteristic function $\varphi(\theta)$ of the Gaussian random vector $(X_1, X_2)^\top$. 4. See [221]. $\qquad\square$

Remark 11.4. If X_2 is symmetric, i.e. $\beta_2 = 0$, then $c = 0$ and formula (11.16) has the following simple form

$$\frac{\mathbf{E}\left(X_1 X_2^{<p-1>} \right)}{\mathbf{E}|X_2|^p} = \frac{[X_1, X_2]_\alpha}{\sigma_2^\alpha},$$

which allows for the estimation of $[X_1, X_2]_\alpha$ via empirical mixed moments of X_1 and X_2.

For a stable random field X with integral representation (11.15), the covariation $[X(t_1), X(t_2)]_\alpha$ can be written in the form

$$[X(t_1), X(t_2)]_\alpha = \int_E f_{t_1}(x)(f_{t_2}(x))^{<\alpha-1>} m(dx). \tag{11.17}$$

Note that the proof of (11.17) given in [337, Proposition 3.5.2] for the $S\alpha S$ case holds true for skewed random fields as well.

Codifference

Drawbacks of the covariation are the lack of symmetry and the impossibility to define it for $\alpha \in (0, 1]$. The following measure of dependence does not have these drawbacks. That is however compensated by a mathematically less convenient form.

Definition 11.17. Let $(X_1, X_2)^\top$ be an α-stable random vector. The *codifference* $\tau(X_1, X_2)$ of X_1 and X_2 is given by

$$\tau(X_1, X_2) = \sigma_{X_1} + \sigma_{X_2} - \sigma_{X_1 - X_2},$$

where σ_Y is the scale parameter of an α-stable random variable Y.

Theorem 11.2 (Properties of Codifference). *Let* $(X_1, X_2)^\top$ *be an α-stable random vector. Then the following properties hold.*

1. *Symmetry.* $\tau(X_1, X_2) = \tau(X_2, X_1)$.

2. *If X_1 and X_2 are independent then $\tau(X_1, X_2) = 0$. The inverse statement holds only for $\alpha \in (0,1)$.*
3. *Gaussian case. For $\alpha = 2$, it holds that $\tau(X_1, X_2) = \mathbf{cov}(X_1, X_2)$.*
4. *Let (X_1, X_2) and (X_1', X_2') be SαS-distributed vectors such that $\sigma_{X_1} = \sigma_{X_2} = \sigma_{X_1'} = \sigma_{X_2'}$. If $\tau(X_1, X_2) \le \tau(X_1', X_2')$ then for any $c > 0$*

$$\mathbf{P}(|X_1 - X_2| > c) \ge \mathbf{P}(|X_1' - X_2'| > c),$$

i.e., the larger the codifference, the greater the degree of dependence.

Proof. 1. The symmetry is obvious. 2. Use Exercise 11.18 to see the first part of the statement. Now let $\tau(X_1, X_2) = 0$. It holds $\sigma_{X_1} + \sigma_{X_2} = \sigma_{X_1 - X_2}$ if and only if

$$\int_{\mathbb{S}^1} |s_1|^\alpha \Gamma(ds) + \int_{\mathbb{S}^1} |s_2|^\alpha \Gamma(ds) = \int_{\mathbb{S}^1} |s_1 - s_2|^\alpha \Gamma(ds).$$

We know however that $|s_1 - s_2|^\alpha = |s_1|^\alpha + |s_2|^\alpha$ if and only if $\alpha < 1$ and $s_1 s_2 = 0$.
3. It holds $\tau(X_1, X_2) = 1/2(\mathbf{var}\, X_1 + \mathbf{var}\, X_2 - \mathbf{var}(X_1 - X_2)) = \mathbf{cov}(X_1, X_2)$. 4. See [337, Property 2.10.6]. $\qquad\square$

11.2.5 Examples of Stable Processes and Fields

1. Stable Lévy Processes. Let $X = \{X(t),\ t \ge 0\}$ be a process defined by $X(t) = M([0,t])$, $t \in \mathbb{R}_+$, where M is an α-stable random measure on $\mathbb{R}_+$ with skewness function β and Lebesgue control measure multiplied by $\sigma > 0$. Then X has representation (11.15) with $f_t(x) = \mathbf{1}(x \in [0,t])$. It obviously holds that $X(0) = 0$ a.s. Moreover, X has independent and stationary increments.

Depending on β the skewness of the process may vary. For instance, for $\alpha < 1$ and $\beta \equiv 1$ we obtain a stable Lévy process with non-decreasing sample paths, the so–called *stable subordinator*. To see this use the one-to-one correspondence between infinitely divisible distributions and Lévy processes, i.e., $X(1) \sim S_\alpha(\sigma, 1, 0)$ corresponds to a Lévy process with the triplet $(0, 0, \frac{\sigma\alpha}{\Gamma(1-\alpha)\cos(\pi\alpha/2)} \frac{dx}{x^{\alpha+1}} \mathbf{1}(x > 0))$, which has positive integrable jumps, see also [338, Examples 21.7 and 24.12].

2. Stable Moving-Average Random Fields. An α-*stable moving-average random field* $X = \{X(t),\ t \in \mathbb{R}^d\}$ is defined by

$$X(t) = \int_{\mathbb{R}^d} f(t - s) M(ds), \quad t \in \mathbb{R}^d,$$

where $f \in L^\alpha(\mathbb{R}^d)$ is called a *kernel function* and M is an α–stable random measure with Lebesgue control measure. It can be easily seen that X is strictly stationary. See Fig. 11.4(a) and (b) for simulated realizations of moving-average random fields in $\mathbb{R}^2$, with the bisquare and the cylindric kernels.

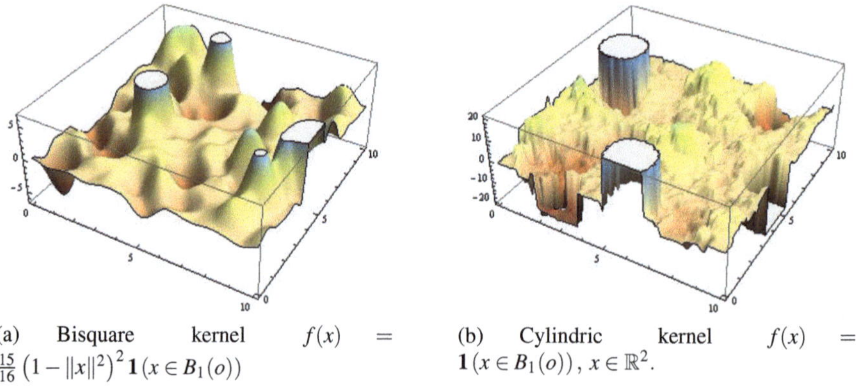

(a) Bisquare kernel $f(x) = \frac{15}{16}\left(1 - \|x\|^2\right)^2 \mathbf{1}\left(x \in B_1(o)\right)$

(b) Cylindric kernel $f(x) = \mathbf{1}\left(x \in B_1(o)\right),\, x \in \mathbb{R}^2.$

Fig. 11.4 Continuous (left) and discontinuous (right) realization of a 0.8-stable moving average random field with $S\alpha S$-distributed random measure M.

For $d = 1$, the *stable Ornstein–Uhlenbeck process* is a stable moving-average process with $X(t) = \int_{-\infty}^{t} e^{-\lambda(t-s)} M(ds)$, $t \in \mathbb{R}$ where M is a $S\alpha S$-distributed random measure on $\mathbb{R}$ with Lebesgue control measure. Note that the stable Ornstein–Uhlenbeck process $X = \{X(t),\, t \in \mathbb{R}\}$ is strictly stationary.

3. Linear Multifractional Stable Motions. A linear multifractional stable motion $X = \{X(t),\, t \in \mathbb{R}\}$ is given by

$$X(t) = \int_{\mathbb{R}} \left((t-x)_+^{H(t)-1/\alpha} - (-x)_+^{H(t)-1/\alpha}\right) M(dx), \quad t \in \mathbb{R},$$

where M is an α-stable random measure with skewness function β and Lebesgue control measure, $\alpha \in (0,2]$, $H : \mathbb{R} \to (0,1)$ is a continuous function which is called a *local scaling exponent*, and $(x)_+ = max\{x,0\}$. It is known that X is a locally self–similar random field, see e.g. [381, 382] for more details. In case $\alpha = 2$ we have a Gaussian process called *multifractional Brownian motion*, cf. [311]. For a constant $H \in (0,1)$, we get the usual *linear fractional stable motion* which has stationary increments and is H–self–similar (see [373, Sect. 9.5]).

Stable Riemann–Liouville Processes Let $X = \{X(t),\, t \geq 0\}$ be given by $X(t) = \int_0^t (t-s)^{H-1/\alpha} M(ds)$, $t \in \mathbb{R}_+$, where M is an α-stable random measure on $\mathbb{R}_+$ and $H > 0$. This is a family of H–self–similar random processes. Note that X has no stationary increments, unless $H = 1/\alpha$. For $\alpha = 2$ we get the Gaussian Riemann–Liouville process, see e.g. [254, Example 3.4].

5. Sub–Gaussian Random Fields Sub-Gaussian random fields are random fields $X = \{X(t),\, t \in \mathbb{R}^d\}$ of the form

$$X \stackrel{\mathrm{D}}{=} \{A^{1/2} G(t),\, t \in \mathbb{R}^d\}, \tag{11.18}$$

where $A \sim S_{\alpha/2}((\cos(\pi\alpha/4))^{2/\alpha},1,0)$ and $G = \{G(t), \, t \in \mathbb{R}^d\}$ is a a zero mean Gaussian random field with a positive definite covariance function which is independent of A. The following lemma (cf. [337, Proposition 3.8.1]) shows that X is α-stable.

Lemma 11.4. *If $A \sim S_{\alpha/2}((\cos(\pi\alpha/4))^{2/\alpha},1,0)$ and $\xi \sim N(0,2\sigma^2)$, where A and ξ are independent, then $X = A^{1/2}\xi \sim S_\alpha(\sigma,0,0)$.*

To prove lemma 11.4, it suffices to determine the characteristic function of X using the conditional expectation provided that A is fixed.

If the Gaussian random field G considered in (11.18) is stationary then the resulting sub–Gaussian field X is strictly stationary as well. However, a strictly stationary sub–Gaussian random field X with a mean-square continuous Gaussian component G is not ergodic since it differs from G by a random scaling. A sufficient condition for the ergodicity of G is that its spectral measure has no atoms, see [390, Theorem A].

11.3 Extrapolation of Stationary Random Fields

Let $X = \{X(t), \, t \in \mathbb{R}^d\}$ be a stationary (in the appropriate sense to be specified later) random field. We are looking for a *linear predictor* $\widehat{X}(t)$ of the unknown random value $X(t)$ of the field X at location $t \in \mathbb{R}^d$ based on the observations $X(t_1),\ldots,X(t_n)$ at locations $t_1,\ldots,t_n$, $n \in \mathbb{N}$, which has the form

$$\widehat{X}(t) = \sum_{i=1}^{n} \lambda_i(t)X(t_i) + \lambda_0(t). \qquad (11.19)$$

The weights $\lambda_0,\ldots,\lambda_n$ are functions of $t,t_1,\ldots,t_n$, which may depend on the distribution of X. For simplicity of notation, we omit all their arguments except for t. These weights $\lambda_0,\ldots,\lambda_n$ have to be determined in a way (which depends on the integrability properties of X) such that the predictor $\widehat{X}(t)$ is in some regard close to $X(t)$.

Definition 11.18. A linear predictor $\widehat{X}(t)$ for $X(t)$ is called

1. *exact* if $\widehat{X}(t) = X(t)$ a.s. whenever $t = t_i$ for any $i \in \{1,\ldots,n\}$. In this case, the predictor $\widehat{X}(\cdot)$ is an extrapolation surface for X with knots $t_1,\ldots,t_n$,
2. *unbiased* if $\mathbf{E}|X(0)| < \infty$ and $\mathbf{E}(\widehat{X}(t) - X(t)) = 0$ for all $t \in \mathbb{R}^d$,
3. *continuous* if the weights λ_i, $i = 0,\ldots,n$ are continuous functions with respect to t, i.e., any realization of $\widehat{X} = \{\widehat{X}(t), \, t \in \mathbb{R}^d\}$ is continuous in $t \in \mathbb{R}^d$.

11.3.1 Kriging Methods for Square Integrable Random Fields

If the field $X = \{X(t),\ t \in \mathbb{R}^d\}$ has finite second moments, then the most popular prediction technique for X in geostatistics is the so–called *kriging*. It is named after D.G. Krige who first applied it (in 1951) to gold mining. Namely, he predicted the size of a gold deposit by collecting the data of gold concentration at some isolated locations. Apart from kriging, there are many other prediction techniques such as inverse distance, spline and nearest neighbor interpolation, and triangulation. For details see [86, Sect. 5.9.2], [410, Chap. 3] and [367]. However, the latter methods ignore the correlation structure contained in the spatial data; see [86, Sect. 3.4.5, p.180; Chap. 5.9], [127], and [242] for their comparison.

The main idea of kriging is to compute the prediction weights λ_i by minimizing the mean-square error between the predictor and the field itself, i.e., to solve the minimization problem

$$\mathbf{E}\left((X(t) - \widehat{X}(t))^2 \right) \to \min_{\lambda_0, \dots \lambda_n} \tag{11.20}$$

under some additional conditions on the λ_i for each fixed $t \in \mathbb{R}^d$.

Depending on the assumptions on X, numerous variants of kriging are avaliable. We mention just few of them and refer the interested reader to the vast literature.

1. *Simple kriging.* For square integrable random fields X with known mean value function $\mathbf{E}X(t) = m(t), t \in \mathbb{R}^d$, see Sect. 11.3.2.
2. *Ordinary kriging.* For second order intrinsic stationary random fields X (with unknown but constant mean), see Sect. 11.3.3.
3. *Kriging with drift.* $\mathbf{E}X(t) = a + b\|t\|$, where the constants $a,\ b \in \mathbb{R}$ are unknown, see [71, Sect. 3.4.6] for details.
4. *Universal kriging.* The unknown mean value function $\mathbf{E}X(t) = m(t) \neq$ const belongs to some parametric family of functions, see [71, 408]. Note that ordinary kriging and kriging with drift are special cases of universal kriging.

11.3.2 Simple Kriging

Let $X = \{X(t),\ t \in \mathbb{R}^d\}$ be a square-integrable random field with known mean value function m. It is easy to see that the minimum of the mean square error

$$\mathbf{E}\left((X(t) - \widehat{X}(t))^2 \right) = \mathbf{var}(X(t) - \widehat{X}(t)) + (\mathbf{E}(X(t) - \widehat{X}(t)))^2$$

is attained exactly when the predictor $\widehat{X}(t)$ is unbiased, i.e. if $\mathbf{E}\widehat{X}(t) = \mathbf{E}X(t)$. This yields that $\lambda_0(t) = m(t) - \sum_{i=1}^n \lambda_i(t)m(t_i)$ and

$$\widehat{X}(t) = \sum_{i=1}^n \lambda_i(t)(X(t_i) - m(t_i)) + m(t).$$

It follows from the above relation that the knowledge of the mean value function m leads to centering the field X (subtracting m) in the prediction.

Taking derivatives of the objective function in (11.20) with respect to λ_i, we obtain

$$\sum_{i=1}^{n} \lambda_i(t)\mathbf{cov}(X(t_i),X(t_j)) = \mathbf{cov}(X(t),X(t_j)), \quad j=1,\ldots,n. \qquad (11.21)$$

The matrix form of this system of equations is

$$\Sigma \cdot \lambda(t) = \sigma(t),$$

where $\Sigma = (\mathbf{cov}(X(t_i),X(t_j)))_{i,j=1}^{n}$ is the covariance matrix of the random vector $(X(t_1),\ldots,X(t_n))^{\top}$, $\lambda(t) = (\lambda_1(t),\ldots,\lambda_n(t))^{\top}$, and

$$\sigma(t) = (\mathbf{cov}(X(t),X(t_1)),\ldots,\mathbf{cov}(X(t),X(t_n)))^{\top}.$$

If Σ is non-degenerate then the solution of (11.21) exists and is unique. The covariance matrix Σ is non-degenerate if the covariance function of X is positive definite and all $t_i,\ i = 1,\ldots,n$ are distinct.

Exercise 11.19. Let the random field $X = \{X(t),\ t \in \mathbb{R}^d\}$ be as above. Show that the random vector $(X(t_1),\ldots,X(t_n))^{\top}$ is singular if and only if $\det \Sigma = 0$. *Hint.* A symmetric matrix is positive definite (positive semi–definite) if and only if its eigenvalues are positive (non–negative).

Finally, we get the following form of the predictor $\widehat{X}(t)$ for $X(t)$:

$$\widehat{X}(t) = \bar{X}^{\top} \Sigma^{-1} \sigma(t), \qquad (11.22)$$

where $\bar{X} = (X(t_1),\ldots,X(t_n))^{\top}$.

Properties of Simple Kriging

1. *Exactness.* To see that $\widehat{X}(t_j) = X(t_j)$ for any j, put $t = t_j$ and check that $\lambda_i(t_j) = \delta_{ij},\ i,j = 1,\ldots,n$ is the solution of (11.21), where $\delta_{ij} = \mathbf{1}(i = j)$ denotes the Kronecker delta.
2. *Continuity and smoothness.* Rewrite (11.22) as $\widehat{X}(t) = b^{\top}\sigma(t)$ with $b = \Sigma^{-1}\bar{X}$ which means that sample path properties of the extrapolation surface such as continuity and smoothness directly depend on the properties of $\sigma(t)$. Thus if the covariance function is continuous and smooth, so is the extrapolation surface.

3. *Shrinkage property.* The mean prediction error $\mathbf{E}\left((\widehat{X}(t) - X(t))^2\right)$ can be determined directly from (11.21). Thus

$$\mathbf{E}((\widehat{X}(t) - X(t))^2) = \mathbf{var}X(t) - \mathbf{var}\widehat{X}(t). \qquad (11.23)$$

Equation (11.23) yields the following shrinkage property

$$\mathbf{var}\widehat{X}(t) \leq \mathbf{var}X(t) \quad \text{for all } t \in \mathbb{R}^d, \qquad (11.24)$$

i.e., the simple kriging predictor is less dispersed than the original random field. In a sense, kriging performs linear averaging (or smoothing) and does not perfectly imitate the trajectory properties of the original random field.

4. *Geometric interpretation.* The predictor $\widehat{X}(t)$ for any fixed t can be seen as a metric projection of $X(t)$ onto the linear subspace $L_n = \mathrm{span}\{X(t_1),\ldots,X(t_n)\}$ of the Hilbert space $L^2(\Omega,\mathcal{F},\mathbf{P})$ with scalar product $\langle X,Y \rangle = \mathbf{E}(XY)$ for $X,Y \in L^2(\Omega,\mathcal{F},\mathbf{P})$. That is,

$$\widehat{X}(t) = \mathrm{Proj}_{L_n} X(t) = \mathrm{argmin}_{\xi \in L_n} \langle X(t) - \xi, X(t) - \xi \rangle. \qquad (11.25)$$

It is known from Hilbert space theory that this projection is unique if the vector $(X(t_1),\ldots,X(t_n))^\top$ is not singular (cf. Definition 11.10).

5. *Orthogonality.* The above projection is also orthogonal, i.e., $\langle \widehat{X}(t) - X(t), \xi \rangle = 0$ for all $\xi \in L_n$. In particular, it holds that

$$\langle \widehat{X}(t) - X(t), X(t_i) \rangle = 0 \quad \text{for all } i = 1,\ldots,n \qquad (11.26)$$

which rewrites as a dependence relation

$$\mathbf{E}(\widehat{X}(t)X(t_i)) = \mathbf{E}(X(t)X(t_i)) \quad \text{for all} \quad i = 1,\ldots,n$$

yielding

$$\mathbf{cov}(\widehat{X}(t) - X(t), \widehat{X}(s)) = 0, \quad s,t \in \mathbb{R}^d.$$

Exercise 11.20. Prove (11.23) via the Pythagorean theorem.

6. *Gaussian case.* Under the assumptions that X is Gaussian and Σ non–singular it is easy to show that

$$\widehat{X}(t) = \mathbf{E}(X(t) \mid X(t_1),\ldots,X(t_n)), \quad t \in \mathbb{R}^d. \qquad (11.27)$$

Exercise 11.21. Prove (11.27) using the uniqueness of the kriging predictor and the following properties the conditional expectation and of the Gaussian multivariate distribution.

1. $\mathbf{E}((\eta - \mathbf{E}(\eta \mid \xi))h(\xi)) = 0$ for arbitrary random variables $\xi, \eta \in L^1(\Omega,\mathcal{F},\mathbf{P})$ and any measurable function $h : \mathbb{R} \to \mathbb{R}$.
2. If $\eta, \xi_1,\ldots,\xi_n$ are jointly Gaussian then there exist real numbers $\{a_i\}_{i=1}^n$ such that $\mathbf{E}(\eta \mid \xi_1,\ldots,\xi_n) = \sum_{i=1}^n a_i \xi_i$.

In the Gaussian case, simple kriging has additional properties.
Conditional unbiasedness. $\mathbf{E}(X(t)|\widehat{X}(t)) = \widehat{X}(t)$ a.s. for any $t \in \mathbb{R}^d$, cf. [71, p. 164]. This property is important in practice for resource assessment problems and selective mining.

Homoscedasticity. The conditional mean-square estimation error does not depend on the data, i.e.,

$$\mathbf{E}\left(\left(\widehat{X}(t)-X(t)\right)^2 \mid X(t_1),\ldots,X(t_n)\right) = \mathbf{E}\left(\widehat{X}(t)-X(t)\right)^2 \text{ a.s. for any } t \in \mathbb{R}^d.$$

11.3.3 Ordinary Kriging

When the mean value function m of a square-integrable random field X is constant but unknown then *ordinary kriging* can be applied. We are looking for a predictor of the form (11.19). For an arbitrary (but fixed) location $t \in \mathbb{R}^d$, the mean-square prediction error is given by

$$\mathbf{E}\left(\left(\widehat{X}(t)-X(t)\right)^2\right) = \mathbf{var}(\widehat{X}(t)-X(t)) + \left(\lambda_0 + \left(\sum_{i=1}^n \lambda_i - 1\right)m\right)^2.$$

Assuming that

$$\lambda_0 = 0, \quad \sum_{i=1}^n \lambda_i = 1, \tag{11.28}$$

we get the smallest possible error together with unbiasedness, i.e. $\mathbf{E}\widehat{X}(t) = \mathbf{E}X(t)$. The ordinary kriging predictor writes then

$$\widehat{X}(t) = \sum_{i=1}^n \lambda_i X(t_i), \quad t \in \mathbb{R}^d.$$

The prediction error can be computed as

$$\mathbf{E}\left(\left(\widehat{X}(t)-X(t)\right)^2\right) = \sum_{i,j=1}^n \lambda_i\lambda_j \mathbf{cov}(X(t_i),X(t_j))$$
$$-2\sum_{i=1}^n \lambda_i \mathbf{cov}(X(t_i),X(t)) + \mathbf{var}X(t).$$

One should minimize this error under the constraint (11.28). Taking partial derivatives of the Lagrange function

$$L(\lambda,\mu) = \mathbf{E}\left((\widehat{X}(t)-X(t))^2\right) + 2\mu\left(\sum_{i=1}^n \lambda_i - 1\right)$$

with respect to $\lambda_i = \lambda_i(t)$, $i=1,\ldots,n$, and $\mu = \mu(t)$ and putting them equal to zero, we obtain the following system of $n+1$ linear equations

$$\sum_{i=1}^n \lambda_i \mathbf{cov}(X(t_i),X(t_j)) + \mu = \mathbf{cov}(X(t_j),X(t)), \quad j=1,\ldots,n,$$
$$\sum_{i=1}^n \lambda_i = 1$$

for each $t \in \mathbb{R}^d$. The solution $(\lambda_1, \ldots, \lambda_n, \mu)^\top$ of this system is unique if and only if the covariance matrix of $(X(t_1), \ldots, X(t_n))^\top$ is non–singular.

The above linear system of equations can be rewritten in terms of the variogram γ. From (11.5) we immediately get the following ordinary kriging system of equations with respect to the weights λ_i, $i = 1, \ldots, n$, and μ:

$$\sum_{i=1}^n \lambda_i \gamma(t_i, t_j) + \mu = \gamma(t_j, t), \quad j = 1, \ldots, n,$$
$$\sum_{i=1}^n \lambda_i = 1.$$

The corresponding mean-square prediction error is given by

$$\sigma_{OK}^2 = \mathbf{E}\left(\left(\widehat{X}(t) - X(t) \right)^2 \right) = \sum_{i=1}^n \lambda_i \gamma(t_i, t) + \mu.$$

Exercise 11.22. Show that $\mu = -(1 - e^\top \Gamma^{-1} \gamma)/e^\top \Gamma^{-1} e$, where e is the unit vector, $\gamma = (\gamma(t_1, t), \ldots, \gamma(t_n, t))^\top$ and $\Gamma = (\gamma(t_i, t_j))_{i,j=1,\ldots,n}$.

The main advantage of this way of posing the problem is that it is solvable even if the variance of $X(t)$ is infinite whereas the variogram is finite, e.g., if X is intrinsic stationary of order two.

Properties of Ordinary Kriging

1. *Exactness.* For $t = t_j$, note that $\lambda_i(t_j) = \delta_{ij}$, $i, j = 1, \ldots, n$, and $\mu(t_j) = 0$ is a solution of the ordinary kriging system.
2. *Orthogonality.* For any real weights a_i, $i = 1, \ldots, n$ with $\sum_{i=1}^n a_i = 1$ it holds that

$$\left\langle \widehat{X}(t) - X(t), \sum_{i=1}^n a_i X(t_i) \right\rangle = 0.$$

3. *Conditional unbiasedness.* The ordinary kriging predictor reduces the conditional bias $\mathbf{E}(X(t) \mid \widehat{X}(t)) - \widehat{X}(t)$. To see this, check the following formula showing that the minimum of the kriging error corresponds to the minimum of the conditional bias error:

$$\mathbf{E}\left(\left(\mathbf{E}\left(X(t) \mid \widehat{X}(t) \right) - \widehat{X}(t) \right)^2 \right) = \mathbf{E}\left((\widehat{X}(t) - X(t))^2 \right) - \mathbf{E}\left(\mathbf{var}(X(t) \mid \widehat{X}(t)) \right),$$

cf. [71, p.185]. To prove this formula, the following *law of total variance* can be used

$$\mathbf{var}Y = \mathbf{var}\left(\mathbf{E}(Y|Z) \right) + \mathbf{E}\left(\mathbf{var}(Y|Z) \right)$$

as well as $\mathbf{E}(Y\mathbf{E}(Z|Y)) = \mathbf{E}(YZ)$ for any $Y, Z \in L^1(\Omega, \mathcal{F}, \mathbf{P})$.

Example 11.4. A simulated realization of a centered stationary isotropic Gaussian random field $X = \{X(t), t \in [0, 10]^2\}$ with Whittle–Matérn–type covariance function $C(s,t) = 2\mathbf{1}(s=t) + 2\mathbf{1}(s \neq t)\|s-t\|K_1(2\|s-t\|)$ exhibiting a nugget effect of height 1 is observed on the grid $\{(3i,2j),\ i,j \in \mathbb{N} \cap [0,3]\}$, see Fig. 11.5(a). The corresponding theoretical variogram together with the Matheron estimator (given in [373, formula (9.67)]) are shown in Fig. 11.6. A Whittle–Matérn–type variogram model with nugget effect σ^2, i.e.,

$$\gamma(s,t) = \mathbf{1}(s \neq t)\left(\sigma^2 + b - b2^{1-\nu}(a\|s-t\|)^\nu K_\nu(a\|s-t\|)\right), \quad s,t \in \mathbb{R}^d,$$

was fitted to the estimated variogram by the ordinary least-squares method yielding the parameter estimates $\widehat{\sigma}^2 = 0.933$, $\widehat{a} = 1.967$, $\widehat{b} = 1.067$. An extrapolation by ordinary kriging with the fitted variogram model γ is shown in Fig. 11.5(b).

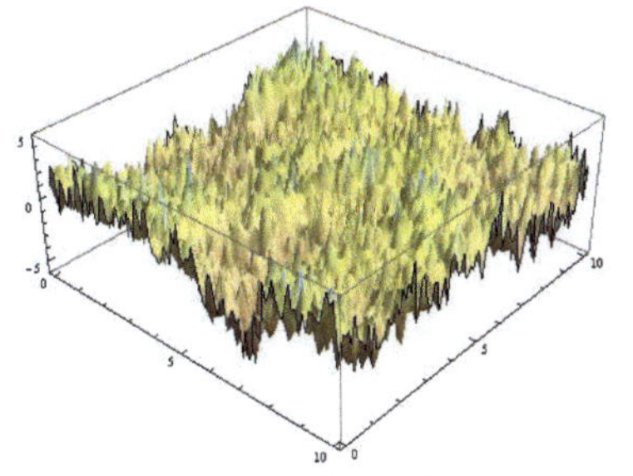

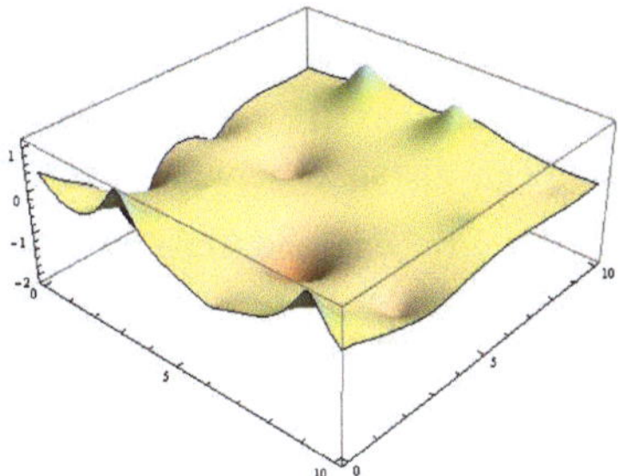

(a) Simulated realization of a stationary Gaussian random field with nugget effect.

(b) Extrapolation by ordinary kriging.

Fig. 11.5 Application of ordinary kriging to simulated data from Example 11.4

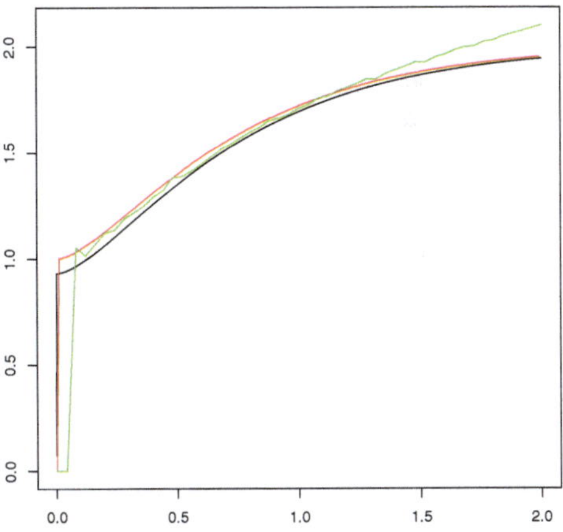

Fig. 11.6 Theoretical variogram (red), estimated (green) and fitted variogram (black) for the realization given in Fig. 11.5(a), see Example 11.4

11.4 Extrapolation of Stable Random Fields

Let $X = \{X(t),\ t \in \mathbb{R}^d\}$ be an α-stable random field admitting the integral representation

$$X(t) = \int_E f_t(x)M(dx), \quad t \in \mathbb{R}^d, \tag{11.29}$$

see (11.15). If $\alpha \in (1,2]$, we assume that the field X is centered. If $\alpha \in (0,1]$, then the mean value of $X(t)$ does not exist. We are looking for a predictor $\widehat{X}(t)$ of $X(t)$ based on the random vector $(X(t_1), \ldots, X(t_n))^\top$ which has the form

$$\widehat{X}(t) = \sum_{i=1}^{n} \lambda_i X(t_i). \tag{11.30}$$

For each $j \in \mathbb{N}$ let $T_j = \{t_{j,1}, \ldots, t_{j,n_j}\}$ be a sequence of locations such that $\mathrm{dist}\,(T_j, \{t\}) \to 0$ as $j \to \infty$, where $\mathrm{dist}\,(A,B) = \inf\{\|x - y\| : x \in A, y \in B\}$ is the Euclidean distance between two arbitrary sets $A, B \subset \mathbb{R}^d$. The predictor $\widehat{X}_j(t) = \sum_{i=1}^{n_j} \lambda_i^{(j)} X(t_{j,i})$ is said to be *weakly consistent* if $\widehat{X}_j(t) \xrightarrow[j \to \infty]{\mathbf{P}} X(t)$ for any $t \in \mathbb{R}^d$. Furthermore, $\widehat{X}_j(t)$ is called *stochastically continuous* if $\widehat{X}_j(s) \xrightarrow[s \to t]{\mathbf{P}} \widehat{X}_j(t)$ for any $j \in \mathbb{N}$ and $t \in \mathbb{R}^d$. Let

$$\|f\|_\alpha = \left(\int_E |f(x)|^\alpha m(dx) \right)^{1/\alpha} \tag{11.31}$$

denote the norm of $f \in L^\alpha(E, \mathcal{E}, m)$, $\alpha \geq 1$.

Theorem 11.3. *Let the α–stable random field X given in (11.29) be stochastically continuous, $\alpha \in (1,2]$. If the predictor $\widehat{X}_j(t)$ defined above exists and if it is unique, exact and stochastically continuous, then $\widehat{X}_j(t)$ is weakly consistent.*

Proof. Fix an arbitrary $t \in \mathbb{R}^d$. By [337, Proposition 3.5.1], to prove weak consistency it is sufficient to show that $\sigma_{\widehat{X}_j(t)-X(t)} \to 0$ as $j \to \infty$. Let $s_j \in T_j$ be the point at which $\mathrm{dist}\,(\{s_j\}, \{t\}) = \mathrm{dist}\,(T_j, \{t\})$ for any $j \in \mathbb{N}$. It is clear that $s_j \to t$ as $j \to \infty$. Since $\widehat{X}_j(t)$ is exact, it holds that $\widehat{X}_j(s_j) = X(s_j)$ for any j. Thus we have

$$\sigma_{\widehat{X}_j(t)-X(t)} = \left\| \sum_{i=1}^{n_j} \lambda_i^{(j)} f_{t_{j,i}} - f_t \right\|_\alpha \leq \|f_{s_j} - f_t\|_\alpha + \left\| \sum_{i=1}^{n_j} \lambda_i^{(j)} f_{t_{j,i}} - f_{s_j} \right\|_\alpha \to 0$$

as $j \to \infty$ by [337, Proposition 3.5.1]), using the stochastic continuity of $X(t)$ and $\widehat{X}_j(t)$ as well as exactness of $\widehat{X}_j(t)$. $\qquad\square$

11.4.1 Least Scale Linear Predictor

For $\alpha \in (0,2]$, consider the following optimization problem

$$\sigma^{\alpha}_{\widehat{X}(t)-X(t)} = \int_E |f_t(x) - \sum_{i=1}^{n} \lambda_i f_{t_i}(x)|^{\alpha} m(dx) \to \min_{\lambda_1,\dots,\lambda_n}. \qquad (11.32)$$

It is clear that the solution of this optimization problem (if it exists and is unique) will be an extrapolation. To see this, put $t = t_j$ and $\lambda_i(t_j) = \delta_{ij}$, $i, j = 1,\dots,n$.

The predictor $\widehat{X}(t)$ based on a solution of (11.32) is called *least scale linear (LSL) predictor*. Note that this approach is similar of the least mean-square error property (11.20) of kriging.

If $\alpha \in (1,2]$, then it is easy to see that any solution of (11.32) is also a solution of the following system of equations

$$\int_E f_{t_j}(x) \left(f_t(x) - \sum_{i=1}^{n} \lambda_i f_{t_i}(x) \right)^{<\alpha-1>} m(dx) = 0, \quad j = 1,\dots,n, \qquad (11.33)$$

or equivalently

$$\left[X(t_j), X(t) - \sum_{i=1}^{n} \lambda_i X(t_i) \right]_{\alpha} = 0, \quad j = 1,\dots,n, \qquad (11.34)$$

where $[\cdot,\cdot]_{\alpha}$ denotes the covariation, see Definition 11.16.

Exercise 11.23. Show that any solution of (11.32) solves also the system of equations (11.33) or (11.34). *Hint.* Use the dominated convergence theorem.

Notice that the expression in (11.32) is nonlinear in $\lambda_1,\dots,\lambda_n$ if $\alpha < 2$ because the covariation is not linear in the second argument (cf. Sect. 11.2.4). Thus, numerical methods have to be applied to solve the optimization problem (11.32).

Properties of the LSL Predictor

Assume that $1 < \alpha \leq 2$. For the case $0 < \alpha \leq 1$, see Sect. 11.4.4.

Theorem 11.4. *The LSL predictor exists. If the random vector $(X(t_1),\dots,X(t_n))^{\top}$ is full–dimensional, then the LSL predictor is unique.*

Proof. We are using the properties of the best approximation in $L^{\alpha}(E,m)$-spaces for $1 < \alpha \leq 2$. Let $L = span\{f_{t_1},\dots,f_{t_n}\}$. This is a finite dimensional space. Put, for simplicity, $f = f_t$ and $E(f) = \inf_{x \in L} \|f - x\|_{\alpha}$. We show that this infimum is attained in L. Consider $\{x_m\}_{m \in \mathbb{N}}$ such that $x_m \in L$ for each $m \in \mathbb{N}$ and $\|x_m - f\|_{\alpha} \to E(f)$ as $m \to \infty$. By the triangle inequality $\|x_m\|_{\alpha} \leq \|f\|_{\alpha} + \|f - x_m\|_{\alpha}$, we get that $\{x_m\}_{m \in \mathbb{N}}$ is a bounded sequence in a finite-dimensional subspace. Thus, there exists a convergent subsequence $\{m_j\}_{j \in \mathbb{N}}$ and $f_0 \in L$ such that $\|x_{m_j} - f_0\|_{\alpha} \to 0$ as $j \to \infty$. Since $\|f - x_{m_j}\|_{\alpha} \to \|f - f_0\|_{\alpha}$ and $\|f - x_{m_j}\|_{\alpha} \to E(f)$ as $j \to \infty$, it holds that $E(f) = \|f - f_0\|_{\alpha}$. Thus, f_0 is the best approximation. To show uniqueness, we use the strict convexity property. If $\alpha > 1$, then the space $L^{\alpha}(E,m)$ is strictly convex (see e.g. [99, p. 59]), i.e. for all $g_1, g_2 \in L^\alpha(E, m)$ such that $\|g_1\|_\alpha = \|g_2\|_\alpha = 1$, $g_1 \neq g_2$ it follows that $\|\beta g_1 + (1 - \beta)g_2\|_\alpha < 1$ for any $\beta \in (0, 1)$. Take $y_j = \sum_{i=1}^{n} \lambda_i^{(j)} f_{t_i} \in L$, $j = 1, 2$ such that $y_1 \neq y_2$ and $\|f - y_1\|_\alpha = \|f - y_2\|_\alpha = E(f)$. Then, by strict convexity we have

$$E(f) \leq \left\| f - \frac{1}{2}(y_1 + y_2) \right\|_\alpha = \left\| \frac{1}{2}(f - y_1) + \frac{1}{2}(f - y_2) \right\|_\alpha < E(f).$$

This leads to a contradiction, and, therefore, $y_1 = y_2 = f_0$. By the full–dimensionality of the random vector X and by Lemma 11.3 one can easily see that the set of weights λ_i in the representation $f_0 = \sum_{i=1}^{n} \lambda_i f_{t_i}$ is unique. $\qquad\qquad\square$

Theorem 11.5 ([221]). *Let the α-stable random field X given in (11.29) be stochastically continuous. If the random vector $(X(t_1), \ldots, X(t_n))^\top$ is full–dimensional, then the LSL predictor is continuous.*

11.4.2 Covariation Orthogonal Predictor

Throughout this section, we assume that $\alpha \in (1, 2]$. The linear predictor considered in (11.30) with weights $\lambda_1, \ldots, \lambda_n$ that are a solution of

$$\left[X(t) - \sum_{i=1}^{n} \lambda_i X(t_i), X(t_j) \right]_\alpha = 0, \quad j = 1, \ldots, n \qquad (11.35)$$

is called a *covariation orthogonal linear (COL) predictor* . If the solution of (11.35) exists and is unique, then it is an exact predictor, since we can put $\lambda_i(t_j) = \delta_{ij}$, $i, j = 1, \ldots, n$. This extrapolation method is similar of the generic orthogonality property of simple kriging, cf. (11.26). It is also symmetric (in a sense) to the LSL predictor, compare (11.34) and (11.35). In contrast to (11.34), the system of equations considered in (11.35) is linear which makes the computation of the weights λ_i easier.

Introduce the *covariation function* $\kappa : \mathbb{R}^d \times \mathbb{R}^d \to \mathbb{R}$ of $X = \{X(t), \ t \in \mathbb{R}^d\}$ by

$$\kappa(s, t) = [X(s), X(t)]_\alpha. \qquad (11.36)$$

Note that this function is not symmetric in its arguments, in distinction to the covariance function, cf. Definition 11.6.

By the additivity of the covariation in the first argument (see Sect. 11.2.4), the system of equations in (11.35) rewrites as

$$\begin{pmatrix} \kappa(t_1, t_1) & \cdots & \kappa(t_n, t_1) \\ \vdots & \ddots & \vdots \\ \kappa(t_1, t_n) & \cdots & \kappa(t_n, t_n) \end{pmatrix} \begin{pmatrix} \lambda_1 \\ \vdots \\ \lambda_n \end{pmatrix} = \begin{pmatrix} \kappa(t, t_1) \\ \vdots \\ \kappa(t, t_n) \end{pmatrix}. \qquad (11.37)$$

If the matrix $K = (\kappa(t_i, t_j))_{i,j=1,\ldots,n}$ is positive definite, the solution of (11.37) exists and is unique.

In the following we show that for moving-average and for sub–Gaussian random fields X sufficient conditions for the positive definiteness of K can be given.

11.4.2.1 The COL Predictor for Moving Averages

Consider a moving-average stable random field $X = \{X(t),\ t \in \mathbb{R}^d\}$ with representation

$$X(t) = \int_{\mathbb{R}^d} f(t-x) M(dx), \quad t \in \mathbb{R}^d,$$

where M is an α–stable random measure with Lebesgue control measure and $f \in L^\alpha(\mathbb{R}^d)$ (see Sect. 11.2.5). By the strict stationarity of X, we get that $[X(h), X(0)]_\alpha = [X(t+h), X(t)]_\alpha$ for all $t,\ h \in \mathbb{R}^d$. With slight abuse of notation, we write $\kappa(s-t) = [X(s-t), X(0)]_\alpha = \kappa(s,t)$, for $s,t \in \mathbb{R}^d$ and the system of equations considered in (11.37) is equivalent to

$$\begin{pmatrix} \kappa(0) & \cdots & \kappa(t_n - t_1) \\ \vdots & \ddots & \vdots \\ \kappa(t_n - t_1) & \cdots & \kappa(0) \end{pmatrix} \begin{pmatrix} \lambda_1 \\ \vdots \\ \lambda_n \end{pmatrix} = \begin{pmatrix} \kappa(t - t_1) \\ \vdots \\ \kappa(t - t_n) \end{pmatrix}. \tag{11.38}$$

The next theorem gives a sufficient condition for the existence and uniqueness of the COL predictor.

Theorem 11.6. *If the kernel $f : \mathbb{R}^d \to \mathbb{R}_+$ is a positive-definite function that is positive on a set of non–zero Lebesgue measure, then κ is positive definite.*

Proof. By formula (11.17), we have

$$\kappa(h) = \int_{\mathbb{R}^d} f(h-x)(f(-x))^{\langle \alpha-1 \rangle}\, dx, \quad h \in \mathbb{R}^d.$$

Thus for any $m \in \mathbb{N}$, $z_1,\ldots,z_m \in \mathbb{R}$, $(z_1,\ldots,z_n)^\top \neq (0,\ldots,0)^\top$ and $s_1,\ldots,s_m \in \mathbb{R}^d$ it holds that

$$\sum_{i,j=1}^m \kappa(s_i - s_j) z_i z_j = \int_{\mathbb{R}^d} \sum_{i,j=1}^m f(s_i - s_j - x) z_i z_j (f(-x))^{\langle \alpha-1 \rangle}\, dx > 0.$$

$\square$

For $d = 1$, an example of a stochastic process $X = \{X(t),\ t \in \mathbb{R}\}$ satisfying the conditions of Theorem 11.6 is the $S\alpha S$ Ornstein–Uhlenbeck process, where for any fixed $\lambda > 0$

$$X(t) = \int_{\mathbb{R}} e^{-\lambda(t-x)} \mathbf{1}(t - x \geq 0) M(dx), \quad t \in \mathbb{R}.$$

Then, by [337, p. 138], we have $\widehat{X}(t) = e^{-\lambda(t-t_n)} X(t_n)$ provided that $t_1 < \ldots < t_n < t$.

Theorem 11.7. *If the covariation function κ is positive-definite and continuous, then the COL predictor is continuous.*

Proof. Since κ is positive-definite and the matrix K is invertible, we have

$$
\begin{pmatrix} \lambda_1(t) \\ \vdots \\ \lambda_n(t) \end{pmatrix} = \begin{pmatrix} \kappa(0) & \cdots & \kappa(t_n - t_1) \\ \vdots & \ddots & \vdots \\ \kappa(t_n - t_1) & \cdots & \kappa(0) \end{pmatrix}^{-1} \begin{pmatrix} \kappa(t - t_1) \\ \vdots \\ \kappa(t - t_n) \end{pmatrix}.
$$

Since κ is continuous, the weights $\lambda_1, \ldots, \lambda_n$ are continuous in t. $\square$

Exercise 11.24. Show that continuous kernel functions with compact support yield a continuous covariation function κ. *Hint.* Use the dominated convergence theorem.

11.4.2.2 The COL Predictor for Gaussian and sub–Gaussian Random Fields

Let $X = \{X(t),\ t \in \mathbb{R}^d\}$ be a sub–Gaussian random field, i.e., $X(t) = A^{1/2}G(t)$, $t \in \mathbb{R}^d$, where $A \sim S_{\alpha/2}((\cos(\pi\alpha/4))^{2/\alpha}, 1, 0)$ and G is a zero-mean stationary Gaussian field independent of A. In [337, Example 2.7.4], it is shown that for sub–Gaussian random fields, the covariation function is given by

$$
\kappa(h) = 2^{-\alpha/2} C(h) C(0)^{(\alpha-2)/2}, \quad h \in \mathbb{R}^d, \tag{11.39}
$$

where $C(\cdot)$ is the covariance function of G.

It is easy to see that in this case, (11.37) coincides with the simple kriging system for G considered in (11.21) for G:

$$
\begin{pmatrix} C(0) & \cdots & C(t_n - t_1) \\ \vdots & \ddots & \vdots \\ C(t_n - t_1) & \cdots & C(0) \end{pmatrix} \begin{pmatrix} \lambda_1 \\ \vdots \\ \lambda_n \end{pmatrix} = \begin{pmatrix} C(t - t_1) \\ \vdots \\ C(t - t_n) \end{pmatrix}. \tag{11.40}
$$

If C is positive-definite, then the corresponding covariance matrix is invertible which ensures the existence and uniqueness of the solution of (11.40).

Theorem 11.8. *If $(X(t_1), \ldots, X(t_n))^\top$ is full–dimensional and the covariance function C of the Gaussian component is continuous, then the COL predictor for sub–Gaussian random fields is continuous.*

The proof is similar to the proof of Theorem 11.7.

Theorem 11.9. *Let $1 < \alpha \leq 2$. For Gaussian and sub–Gaussian random fields, the COL and LSL predictors coincide.*

Proof. Introduce the notation $t_0 = t$. Put $\lambda_0(t_0) = -1$ and

$$
\widehat{X}(t_0) - X(t_0) = A^{1/2} \sum_{i=0}^{n} \lambda_i(t_0) G(t_i).
$$

The characteristic function of random vector $(X(t_0),\dots,X(t_n))^\top$ is given by

$$\mathbf{E}\exp\left(i\sum_{k=0}^{n}\theta_k X(t_k)\right)=\exp\left(-\frac{1}{2}\left|\sum_{i=0}^{n}\sum_{j=0}^{n}\theta_i\theta_j C(t_i-t_j)\right|^{\alpha/2}\right) \tag{11.41}$$

for all $\theta_1,\dots,\theta_n\in\mathbb{R}$, cf. [337, Proposition 2.5.2]. Now it is simple to see that

$$\sigma_{\widehat{X}(t_0)-X(t_0)}=\left(\frac{1}{2}\mathbf{var}\left(\sum_{i=0}^{n}\lambda_i(t_0)G(t_i)\right)\right)^{1/2}=\left(\frac{1}{2}\sum_{i,j=0}^{n}\lambda_i\lambda_j C(t_i-t_j)\right)^{1/2}.$$

Thus, the LSL optimization problem is equivalent to

$$\sum_{i,j=0}^{n}\lambda_i\lambda_j C(t_i-t_j)\to\min_{\lambda_1,\dots,\lambda_n}.$$

Taking derivatives we obtain $\sum_{j=0}^{n}C(t_k-t_j)\lambda_j=0$, $k=1,\dots,n$ which coincides with the COL extrapolation system (11.40). $\qquad\square$

Remark 11.5. It follows from the proof of Theorem 11.9 (which is valid for all $\alpha\in(0,2)$) that the weights of the LSL predictor for sub–Gaussian random fields are a solution of the system (11.40) also in the case $\alpha\in(0,1]$. The statement of Theorem 11.8 holds as well. To summarize, the LSL predictor for stationary sub–Gaussian random fields X exists and is unique and exact for all $\alpha\in(0,2]$ if the covariance function C of the Gaussian component G is positive definite. If C is additionally continuous then this LSL predictor is also continuous.

11.4.3 Maximization of Covariation

In this section, we assume that X is an α–stable random field of the form (11.29) with $\alpha\in(1,2]$. The predictor $\widehat{X}(t)=\sum_{i=1}^{n}\lambda_i(t)X(t_i)$, whose weights $\lambda_1(t),\dots,\lambda_n(t)$ solve the following optimization problem

$$\begin{cases}\left[\widehat{X}(t),X(t)\right]_\alpha=\sum_{i=1}^{n}\lambda_i(t)[X(t_i),X(t)]_\alpha\to\max_{\lambda_1,\dots,\lambda_n},\\[2mm]\sigma_{\widehat{X}(t)}=\sigma_{X(t)}\end{cases} \tag{11.42}$$

for $t\in\mathbb{R}^d$, is called *maximization of covariation linear (MCL) predictor*. The Lagrange function of (11.42) is given by

$$L(\lambda,\gamma)=\sum_{i=1}^{n}\lambda_i[X(t_i),X(t)]_\alpha+\gamma\left(\sigma^{\alpha}_{\sum_{i=1}^{n}\lambda_i X(t_i)}-\sigma^{\alpha}_{X(t)}\right),\ \lambda\in\mathbb{R}^n,\ \gamma\in\mathbb{R}.$$

By taking partial derivatives and putting them equal to zero, we get that

$$\begin{cases} [X(t_j), X(t)]_\alpha + \gamma \cdot \partial \sigma^\alpha_{\sum_{i=1}^n \lambda_i X(t_i)} / \partial \lambda_j = 0, \quad j = 1, \dots, n, \\ \sigma_{\sum_{i=1}^n \lambda_i X(t_i)} = \sigma_{X(t)}. \end{cases} \tag{11.43}$$

Analogously to (11.34) one can show that

$$\frac{\partial \sigma^\alpha_{\sum_{i=1}^n \lambda_i X(t_i)}}{\partial \lambda_j} = \alpha \cdot \left[X(t_j), \sum_{i=1}^n \lambda_i(t) X(t_i) \right]_\alpha.$$

Since $\gamma = -1/\alpha$, $\lambda_i(t_j) = \delta_{ij}$ is obviously a solution of (11.43) for $t = t_j$, $j = 1, \dots, n$, the MCL predictor is exact.

Let us now discuss the properties of the MCL predictor. Note that here no direct analogy with kriging can be drawn. For instance, a counterpart $\sigma_{\widehat{X}(t)} \leq \sigma_{X(t)}$ of the shrinkage property (11.24) is deliberately mutated to the additional condition $\sigma_{\widehat{X}(t)} = \sigma_{X(t)}$. The reason for this is that both conditions lead to the same solutions due to the convexity of the optimization problem (11.42).

Let $\zeta(t) = (\kappa(t_1, t), \dots, \kappa(t_n, t))^\top$, $t \in \mathbb{R}^d$, and let the function $\sigma_0 : \mathbb{R}^d \to \mathbb{R}_+$ be given by $\sigma_0(t) = \sigma_{X(t)} = \kappa(t, t)$. Furthermore, let the function $\Psi : \mathbb{R}^n \to \mathbb{R}_+$ be defined by

$$\Psi(\lambda) = \sigma_{\widehat{X}(t)} = \left\| \sum_{i=1}^n \lambda_i f_{t_i} \right\|_\alpha.$$

Denote the *level set* of Ψ at level $u \in \mathbb{R}$ by $B_u = \{\lambda \in \mathbb{R}^n : \Psi(\lambda) \leq u\}$. The *support set* of any convex set $B \subset \mathbb{R}^n$ at a point $x \in \mathbb{R}^n$ is defined by

$$T(B, x) = \left\{ y \in B : \langle y, x \rangle = \sup_{z \in B} \langle z, x \rangle \right\}.$$

It is well-known that for strictly convex sets B and any non–zero $x \in \mathbb{R}^n$ the support set $T(B, x)$ is a singleton. We denote this single point by $y_{B,x}$.

Theorem 11.10. *Assume that the α-stable random vector $X = (X(t_1), \dots, X(t_n))^\top$ is full–dimensional. Then, the following is true.*

1. *The solution of the optimization problem (11.42) exists for all $t \in \mathbb{R}^d$. If $\kappa(t_i, t) \neq 0$ for some $i = 1, \dots, n$, then the MCL predictor $\widehat{X}(t)$ is unique.*
2. *If κ is a continuous function on $\mathbb{R}^d \times \mathbb{R}^d$ and $\kappa(t_i, t) \neq 0$ for some $i = 1, \dots, n$, then the MCL predictor is continuous in t.*

Proof. For a proof of the existence and uniqueness of an MCL predictor, we refer to [221]. It is also shown in [221] that the vector of MCL weights

$$\lambda(t) = (\lambda_1(t), \dots, \lambda_n(t))^\top$$

is equal to $y_{B_{\sigma_0(t)}, \zeta(t)}$ for any $t \in \mathbb{R}^d$, whereas the set $B_{\sigma_0(t)}$ is strictly convex. We show that $\lambda : \mathbb{R}^d \to \mathbb{R}^n$ is a continuous function. It is easy to see that $B_{\sigma_0(t)} = \frac{1}{\sigma_0(t)} B_1$, because the sets $B_{\sigma_0(t)}$, $t \in \mathbb{R}^d$ are homothetic, i.e. $a B_{\sigma_0(t)} = B_{\sigma_0(t)/a}$, $a > 0$. Thus, by simple geometric considerations, we get that

$$T(B_{\sigma_0(t)}, \zeta(t)) = T\left(\frac{1}{\sigma_0(t)}B_1, \zeta(t)\right) = \frac{1}{\sigma_0(t)}T(B_1, \zeta(t)),$$

and, consequently $\lambda(t) = \frac{1}{\sigma_0(t)}y_{B_1,\zeta(t)}$. Put $B = B_1$ and $x(s) = y_{B,\zeta(s)}$ for any $s \in \mathbb{R}^d$. It remains to show that $\lim_{s \to t} x(s) = x(t)$. This limit exists by the definition of the support set and the continuity of the scalar product. Note that $\zeta(s) \to \zeta(t)$ as $s \to t$ since κ is a continuous function. Moreover, B is a compact, and $x(s) \in B$ for all s. Choose a convergent sequence $s_m \to t$ as $m \to \infty$ such that $x(s_m) \to y$ as $m \to \infty$, where $y \in B$, and show that $y = x(t)$. It is clear that $\langle x(s_m), \zeta(s_m) \rangle \to \langle y, \zeta(t) \rangle$ as $m \to \infty$. Furthermore, for any $x \in B$ it holds that

$$\langle x, \zeta(t) \rangle = \lim_{m \to \infty} \langle x, \zeta(s_m) \rangle \leq \lim_{m \to \infty} \langle x(s_m), \zeta(s_m) \rangle = \langle y, \zeta(t) \rangle.$$

The latter inequality here is obtained from the fact that $\{x(s_m)\} = T(B, \zeta(s_m))$ for any $m \in \mathbb{N}$. Thus $y = y_{B,\zeta(t)}$. $\square$

11.4.4 The Case $\alpha \in (0, 1]$

As mentioned in Sect. 11.2.4, the covariation function is not defined for $\alpha \in (0, 1]$. Moreover, for $\alpha < 1$ the function $\|\cdot\|_\alpha$ defined in (11.31) is not a norm anymore since the triangle inequality fails to hold. The property of strict convexity of $L^\alpha(E, \mathcal{E}, m)$ does not hold as well.

To cope with these drawbacks, one may come up with the idea that the codifference (cf. Definition 11.17) can be used instead of the covariation in the COL and MCL methods. However, this does not seem to make sense in extrapolation. For instance, replacing the covariation by the codifference in the MCL method leads to the optimization problem

$$\begin{cases} \tau(\widehat{X}(t), X(t)) = \sigma_{\widehat{X}(t)} + \sigma_{X(t)} - \sigma_{\widehat{X}(t)-X(t)} \to \max_{\lambda_1,\ldots,\lambda_n}, \\ \sigma_{\widehat{X}(t)} = \sigma_{X(t)}. \end{cases} \tag{11.44}$$

Using the constraint $\sigma_{\widehat{X}(t)} = \sigma_{X(t)}$, the first line in (11.44) rewrites

$$\tau(\widehat{X}(t), X(t)) = 2\sigma_{X(t)} - \sigma_{\widehat{X}(t)-X(t)}.$$

Hence, the optimization problem (11.44) is equivalent to LSL extrapolation, i.e., to minimizing the scale parameter

$$\sigma_{\widehat{X}(t)-X(t)} = \left\| f_t - \sum_{i=1}^n \lambda_i f_{t_i} \right\|_\alpha$$

of $\widehat{X}(t) - X(t)$.

Replacing the covariation by the codifference in the COL approach considered in (11.35), one arrives at the following system of nonlinear equations

$$\tau_{\widehat{X}(t),X(t_i)} = \tau_{X(t),X(t_i)}, \quad i = 1,\ldots,n. \tag{11.45}$$

Note that the solution of (11.45) has to be computed numerically, which can be very time consuming. Furthermore, it is shown in [161] that the solution of (11.45) is not unique. For this reason, we shall not pursue the method given by (11.45).

We also remark that the maximization of $\tau_{\widehat{X}(t),X(t)}$ with respect to the weights $\lambda_1,\ldots,\lambda_n$ does not lead to a unique predictor of the form considered in (11.30). In particular, its existence is not really clear. As an example consider a random field as in (11.29) with kernel function f_t of compact support such that the supports of f_t and $f_{t_1},\ldots,f_{t_n}$ do not overlap. Then it is easy to see that $\tau_{\widehat{X}(t),X(t)} = 0$ holds, which allows allowing for an arbitrary choice of the weights $\lambda_1,\ldots,\lambda_n$.

In the rest of this section, we focus on the properties of the LSL method for α–stable random fields with $\alpha \in (0,1]$. First of all, the fundamental question of existence has to be answered. Here we follow [161] and consider this problem in a more general setting of r–normed vector spaces.

Definition 11.19. Let V be a vector space over a field $\mathbb{K}$. A map $||.||_{(r)} : V \to \mathbb{R}_+$ is called an r–*norm* if there exists $K \geq 1$ and $r > 0$ such that

$$||x||_{(r)} = 0 \ \text{if and only if} \ x = 0,$$
$$||ax||_{(r)} = |a| \cdot ||x||_{(r)} \quad \text{for all } a \in \mathbb{K} \text{ and } x \in V,$$
$$||x+y||_{(r)} \leq K(||x||_{(r)} + ||y||_{(r)}) \quad \text{for all } x,y \in V,$$
$$||x+y||_{(r)}^r \leq ||x||_{(r)}^r + ||y||_{(r)}^r \quad \text{for all } x,y \in V.$$

Now the following existence theorem can be formulated.

Theorem 11.11 ([161]). *Let V be a vector space over $\mathbb{R}$ with the r-norm $||\cdot||_{(r)}$ and let $f_1,\ldots,f_n \in V$ be linearly independent. Then, for any $f_0 \in V$, there exist real numbers $\lambda_1^*,\ldots,\lambda_n^*$ such that*

$$\left\| f_0 - \sum_{i=1}^{n} \lambda_i^* f_i \right\|_{(r)} = \inf_{\lambda_1,\ldots,\lambda_n \in \mathbb{R}} \left\| f_0 - \sum_{i=1}^{n} \lambda_i f_i \right\|_{(r)}.$$

If we put $V = L^\alpha(E,m)$ and note that $||\cdot||_{(\alpha)} = ||\cdot||_\alpha$ defined in (11.31) is an α-norm on $L^\alpha(E,m)$ (even a norm if $\alpha \geq 1$), the existence of the LSL predictor follows immediately from Theorem 11.11. In contrast to the case $\alpha \in (1,2]$ (Theorem 11.4), the uniqueness of the LSL weights $\lambda^* := (\lambda_1^*,\ldots,\lambda_n^*)^\top$ in Theorem 11.11 is not guaranteed. We illustrate this by the following example. Introduce the notation $H_\alpha(\lambda) = \sigma_{\widehat{X}(t)-X(t)}$ for $\lambda = (\lambda_1,\ldots,\lambda_n)^\top \in \mathbb{R}^n$.

Example 11.5. Let $(E,m) = ([0,1], v_1)$ and consider the kernel function $f_t(x) = \mathbf{1}\left(x \in (t+\frac{1}{4}, t+\frac{3}{4})\right)$. Given $t_1 = \frac{1}{4}$, predict the value of the symmetric α–stable process $X(t) = \int_{[0,1]} f_t(x) M(dx)$ at $t = 0$. By elementary computations we obtain

$$H_\alpha^\alpha(\lambda) = \int_{[0,1]} |f_t(x) - \lambda f_{t_1}(x)|^\alpha dx = \frac{1}{4}(1 + |1 - \lambda|^\alpha + |\lambda|^\alpha).$$

It is easy to see that for $0 < \alpha < 1$, H_α has two global minima at $\lambda = 0$ and $\lambda = 1$. If $\alpha = 1$, then the set of all global minimum points equals the interval $[0,1]$. For $\alpha > 1$, the function H_α has a unique global minimum at $\lambda = 0.5$.

The structure of the set of all possible LSL predictors for $\alpha \in (0,1]$ and $n \geq 1$ has been described in papers [215–219]. In order to get an unbiased predictor (provided that the mean value function of X is well defined), the parameter space is often restricted to $\{(\lambda_1,\dots,\lambda_n)^\top \in \mathbb{R}^n : \sum_{i=1}^n \lambda_i = 1\}$. In [161] it has been shown that this restriction does not cause uniqueness of the LSL predictor for $\alpha \in (0,1)$. Alternatively, the following algorithmic approach to choose a unique global minimum in the LSL optimization problem is proposed.

Algorithm 11.1. Let $\{X(t),\ t \in I\}$ be an α–stable random field which admits the integral representation (11.29) with $0 < \alpha < 1$ and $I \subset \mathbb{R}^d$. Let $t_1,\dots,t_n \in I$ be fixed such that functions $f_{t_1},\dots,f_{t_n}$ are linearly independent.

1. Order the points $t_1,\dots,t_n$ so that

$$\|t - t_1\| \leq \|t - t_2\| \leq \dots \leq \|t - t_n\|$$

and if $\|t - t_i\| = \|t - t_{i+1}\|$ for some $i \in \{1,\dots,n-1\}$ then

$$t_i^{(p)} = t_{i+1}^{(p)} \quad \text{for all } p = 1,\dots,k-1 \tag{11.46}$$

$$t_i^{(k)} < t_{i+1}^{(k)} \tag{11.47}$$

for some $k \in \{1,\dots,m\}$, where $t_i^{(p)}$ is the p–th component of t_i.

2. Determine the set A_0 of all critical points

$$A_0 = \{(\lambda_1,\dots,\lambda_n)^\top \in \mathbb{R}^n : H_\alpha(\lambda_1,\dots,\lambda_n) = \inf_{(\mu_1,\dots,\mu_n)\in\mathbb{R}^n} H_\alpha(\mu_1,\dots,\mu_n)\}$$

3. Reduce A_0 step by step to the sets $A_1 \supseteq A_2 \supseteq \cdots \supseteq A_n$ given by

$$A_j = \{(\lambda_1,\dots,\lambda_n)^\top \in A_{j-1} : \lambda_j = \max_{(\mu_1,\dots,\mu_n)\in A_{j-1}} \mu_j\}, \quad j = 1,\dots,n.$$

Clearly, the set A_n consists of just one element.

Definition 11.20. We call $\widehat{X}(t) = \sum_{i=1}^n \lambda_i^* X(t_i)$ the *best LSL predictor* if $(\lambda_1^*,\dots,\lambda_n^*) \in A_n$.

The above construction has a simple intuitive meaning. The points $t_1,\dots,t_n$ are ordered with respect to their distance from t. To get a unique ordering, conditions (11.46) and (11.47) are required. Points with a smaller distance from t are regarded to exert more influence on the value of X at t. Thus their weights should be maximized first.

To show that $A_j \neq \emptyset$, $j = 1,\ldots,n$ we notice that A_0 is nonempty and compact. Therefore, the projection mapping $(x_1,\ldots,x_n)^\top \mapsto x_1$ takes its maximum on A_0. Hence, A_1 is nonempty and compact as well. The sets $A_2,\ldots,A_n$ are not empty by induction.

It can be easily proved that the best LSL predictor is exact. To see this, let $t = t_i$ for some $i \in \{1,\ldots,n\}$ and let $t_1,\ldots,t_n \in \mathbb{R}^d$ be as in Algorithm 11.1. Then, conditions (11.46) and (11.47) imply that $t = t_1$. Trivially $(1,0,\ldots,0)^\top \in A_0$ holds. Due to the linear independence of $f_{t_1},\ldots,f_{t_n}$, it holds that $A_n = A_0 = \{(1,0,\ldots,0)^\top\}$.

For $1 < \alpha \leq 2$, Theorem 11.5 states the continuity of LSL predictor. In contrast, the best LSL predictor is not necessarily continuous for $0 < \alpha \leq 1$ as the next example shows.

Example 11.6. Let $X = \{X(t), \, t \in \mathbb{R}^2\}$ be an α–stable random field which admits the integral representation (11.29) with $0 < \alpha < 1$,

$$f_t(x) = \mathbf{1}\left(x \in (\min\{t^{(1)}, t^{(2)}\}, \max\{t^{(1)}, t^{(2)}\})\right)$$

for $t = (t^{(1)}, t^{(2)})^\top \in \mathbb{R}^2$, $E = \mathbb{R}$ and M being an $S\alpha S$ random measure on $\mathbb{R}$ with Lebesgue control measure. It follows from (11.13), (11.14) and the Markov inequality that that X is stochastically continuous, i.e., it has a.s. no jumps at fixed locations t. For $n = 1$, introduce $t_0 = (\frac{1}{2}, \frac{3}{2})^\top$, $t_1 = (0, 1)^\top$, $t = t_0 + \varepsilon$, where $\varepsilon = (\delta, \delta)^\top \in \mathbb{R}^2$ for some $\delta \in (-\frac{1}{2}, \frac{1}{2})$. Consider the best LSL predictor $\widehat{X}(t)$ of $X(t)$ based on $X(t_1)$. Then, it holds that

$$H_\alpha^\alpha(\lambda) = \int_{\mathbb{R}} |f_{t_0+\varepsilon}(x) - \lambda f_{t_1}(x)|^\alpha dx$$

$$= \left(\frac{1}{2} + \delta\right) \cdot |\lambda|^\alpha + \left(\frac{1}{2} - \delta\right) \cdot |1 - \lambda|^\alpha + \left(\frac{1}{2} + \delta\right).$$

If $\delta > 0$, then H_α has a global minimum at $\lambda = 0$ and, if $\delta < 0$, H_0 has a global minimum at $\lambda = 1$. Thus, $\widehat{X}(t)$ is discontinuous at $t = t_0$.

In addition to the best LSL predictor, it is possible to treat the case $\alpha = 1$ similar to the case $1 < \alpha < 2$. The following approach has been proposed in [161]. For a symmetric 1–stable field $\{X(t) : t \in I\}$ with integral representation

$$X(t) = \int_E f_t(x) M(dx),$$

let $f_t \in L^1(E, \mathcal{E}, m) \cap L^\delta(E, \mathcal{E}, m)$ for some $\delta > 1$. Then we have

$$\int_E \left|f_t(x) - \sum_{i=1}^n \lambda_i f_{t_i}(x)\right|^\gamma m(dx) < \infty$$

for all $\gamma \in [1, \delta]$, $\lambda_1,\ldots,\lambda_n \in \mathbb{R}$ and $t, t_1,\ldots,t_n \in I$. Now fix $t, t_1,\ldots,t_n \in I$ and chose an arbitrary sequence $\{\gamma_k\}_{k \in \mathbb{N}} \subset (1, \delta]$ which converges to 1 as $k \to \infty$. Let $(\lambda_1^{(\gamma_k)},\ldots,\lambda_n^{(\gamma_k)})$ be the unique solution of

$$\int_E |f_t(x) - \sum_{i=1}^n \lambda_i f_{t_i}(x)|^{\gamma_k} m(dx) \to \min_{\lambda_1,\ldots,\lambda_n}. \tag{11.48}$$

Applying the stability theorem in [235, p.225], the convergence

$$\int_E |f_t(x) - \sum_{i=1}^n \lambda_i^{(\gamma_k)} f_{t_i}(x)|^{\gamma_k} m(dx) \to \inf_{\mu_1,\ldots,\mu_n} \int_E |f_t(x) - \sum_{i=1}^n \mu_i f_{t_i}(x)| m(dx) \tag{11.49}$$

follows as $k \to \infty$. Moreover, it can be shown that

$$(\lambda_1^{(\gamma_k)},\ldots,\lambda_n^{(\gamma_k)}) \to (\lambda_1^*,\ldots,\lambda_n^*), \quad \text{as } k \to \infty$$

for some set of weights $(\lambda_1^*,\ldots,\lambda_n^*)$ which is unique if all LSL prediction problems considered in (11.48) with stability indices $\gamma_k > 1$ possess this property. Furthermore the weights $(\lambda_1^*,\ldots,\lambda_n^*)$ do not depend on the choice of the sequence $(\gamma_k)_{k \in \mathbb{N}} \subset (1,\delta]$ such that $\gamma_k \to 1$ as $k \to \infty$.

Definition 11.21. The predictor $\widehat{X}^*(t) = \sum_{i=1}^n \lambda_i^* X(t_i)$, $t \in I$ is called an *index–continuous LSL (ICLSL) predictor* for the symmetric 1–stable random field X.

It is still an open problem to explore the statistical properties of ICLSL predictors.

11.4.5 Numerical Examples

In this section, the LSL, COL and MCL extrapolation methods (as well as maximum likelihood extrapolation and conditional simulation for sub–Gaussian random fields) are applied to simulated data of various α–stable random processes and fields for $\alpha \in (0,2)$.

The random fields are simulated and extrapolated on an equidistant $50 \times 50 -$ grid of points within $I = [0,1]^2$. In Examples 11.7 and 11.8, the simulated field $X = \{X(t),\, t \in [0,1]^2\}$ is observed at the points $t_1,\ldots,t_{16}$ given by their coordinates

$$
\begin{aligned}
t_1 &= (0,0), & t_2 &= (0,0.3), & t_3 &= (0,0.6), & t_4 &= (0,0.9), \\
t_5 &= (0.3,0), & t_6 &= (0.3,0.3), & t_7 &= (0.3,0.6), & t_8 &= (0.3,0.9), \\
t_9 &= (0.6,0), & t_{10} &= (0.6,0.3), & t_{11} &= (0.6,0.6), & t_{12} &= (0.6,0.9), \\
t_{13} &= (0.9,0), & t_{14} &= (0.9,0.3), & t_{15} &= (0.9,0.6), & t_{16} &= (0.9,0.9).
\end{aligned}
$$

Example 11.7 (Sub–Gaussian random fields). Consider a stationary sub–Gaussian random field $X = \{X(t),\, t \in [0,1]^2\}$ described in Example 5 of Section 11.2.5 with $\alpha = 1.2$. The Gaussian part G of this field has a Whittle–Matérn covariance function (cf. Sect. 11.2.2.1, Example 6) with parameters as in Fig. 11.1(a). Fig. 11.7(a) shows a realization of X. Realisations of the corresponding LSL predictor (coinciding with the COL predictor by Theorem 11.9) and MCL predictors are given in Figs.

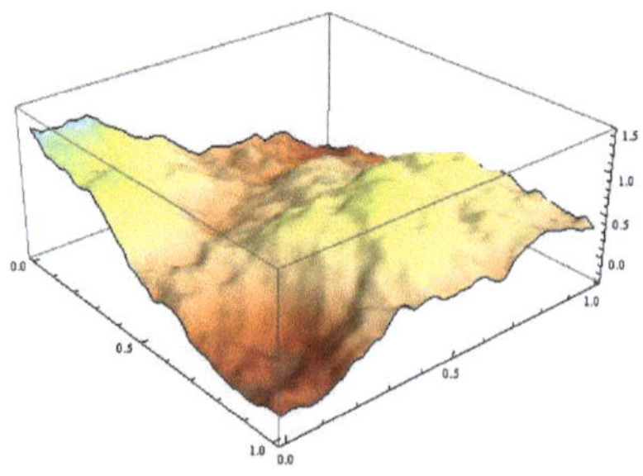

(a) Realization of a sub–Gaussian random field with $\alpha = 1.2$.

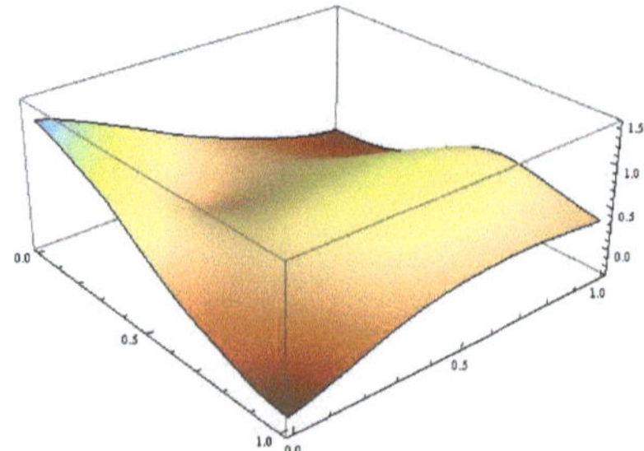

(b) Corresponding LSL (COL) predictor

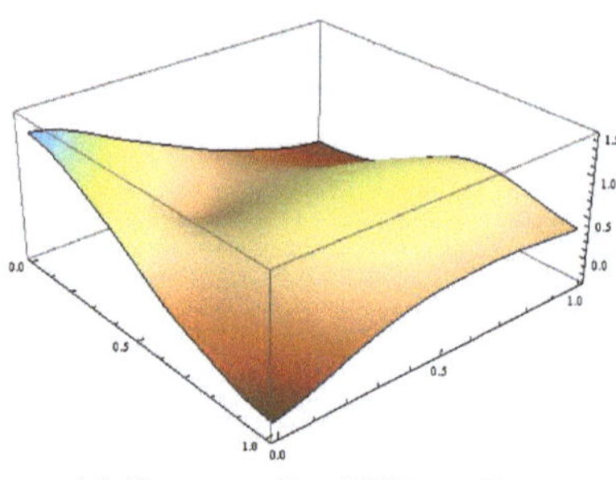

(c) Corresponding MCL predictor

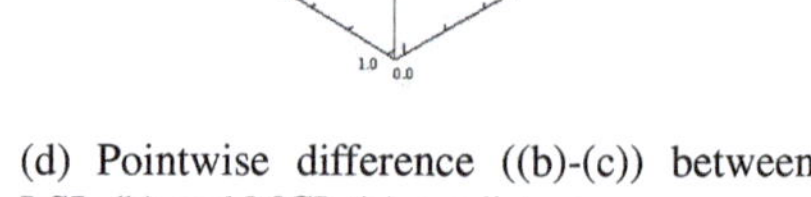

(d) Pointwise difference ((b)-(c)) between LSL (b) and MCL (c) predictors

Fig. 11.7 Realization of a sub–Gaussian random field for $\alpha > 1$ and different predictors

11.7(b) and (c). The realisations of both predictors are smoother than the realization of the field itself. Since the predictions in Figs. 11.7(b) and (c) look quite similar and cannot be distinguished from each other by the naked eye, their difference is given in Fig. 11.7(d). Notice the differences in scale of the z-axes in Fig.11.8 (a),(b),(c) and (d). Fig. 11.8(a) shows a realization of the stationary sub–Gaussian field with $\alpha = 0.8$ and covariance function C of the Gaussian part as above. Note that a maximum likelihood (ML) predictor for sub–Gaussian random fields has been introduced in [221]. It is shown in Theorem 11 of [221] that the LSL, COL and ML methods coincide if $\alpha \in (1, 2)$. However, the proof of this result does not depend on α covering (with regard to Remark 11.5 of the present chapter) the range of all $\alpha \in (0, 2)$. Thus, the LSL and ML predictors coincide for sub–Gaussian random fields with any stability index $\alpha \in (0, 2)$. A possibility of extrapolation of sub–Gaussian random fields X by conditional simulation (CS) of the Gaussian component G of X and subsequent scaling by $\sqrt{A}$ is straightforward; see e.g. [219, p. 112] and [308]. Algorithms for conditional simulation of G are given in [241]. Corresponding extrapolation results for the LSL (ML) and CS methods are given in Figs. 11.8(b) and (c). Note that the ML prediction for this realization of X is much smoother than the CS prediction.

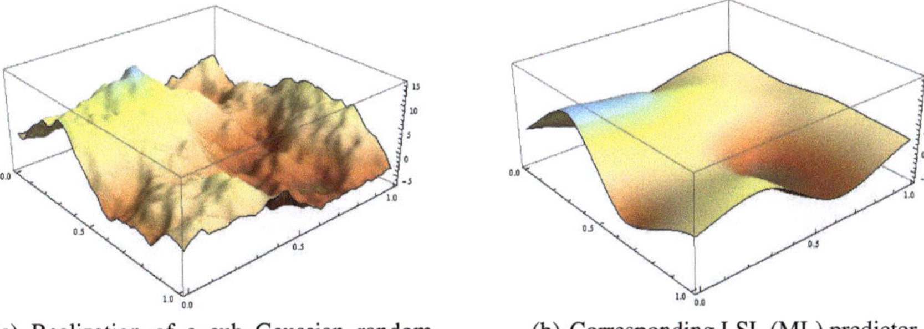

(a) Realization of a sub–Gaussian random field with $\alpha = 0.8$.

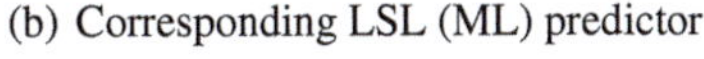

(b) Corresponding LSL (ML) predictor

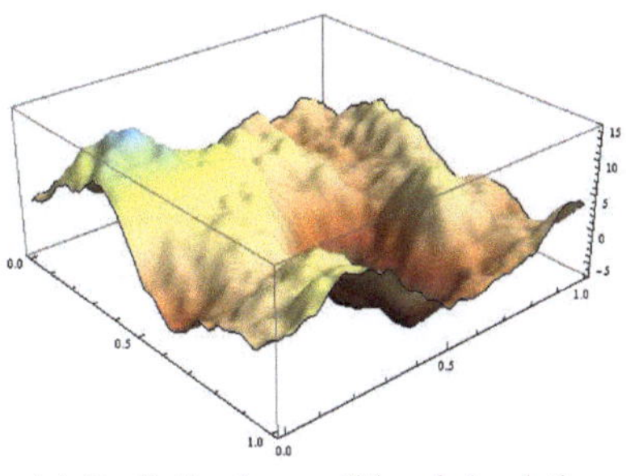

(c) Prediction by conditional simulation

Fig. 11.8 Realization of a sub–Gaussian random field for $\alpha < 1$ and different predictors

Example 11.8 (Skewed stable Lévy motion). Consider the two–dimensional 1.5–stable Lévy motion $X = \{X(t),\ t \in [0,1]^2\}$ defined by

$$X(t) = \int_{[0,1]^2} \mathbf{1}(x_1 \leq t_1, x_2 \leq t_2) M\big(d(x_1,x_2)\big), \quad t = (t_1,t_2)^\top \in [0,1]^2,$$

where M is a non–symmetric centered 1.5–stable random measure with skewness intensity $\beta = 1$. Comparing the realization of X given in Fig. 11.9(a) with the corresponding LSL, COL and MCL predictions (cf. Figs. 11.9(b), (c) and (d)) one can see that the prediction has a smoothing effect.

Example 11.9 (Stable Ornstein–Uhlenbeck process). Let $X = \{X(t),\ t \in [0,10]\}$ be a 1.6-stable Ornstein–Uhlenbeck process with $\lambda = 0.5$ defined in Example 2 of Sect. 11.2.5. Fig. 11.10 shows a trajectory of this process and different interpolators. The process X is observed at positions $t_i = 1, \ldots, 10$ within $[0,10]$. It can be seen that the LSL interpolation is very smooth. In contrast, the COL predictor is piecewise smooth and continuous on the whole interval.

Example 11.10 (Stable moving average). Let $X = \{X(t),\ t \in [0,0.49]^2\}$ be a moving average field (cf. Example 2 of Section 11.2.5) with the kernel function

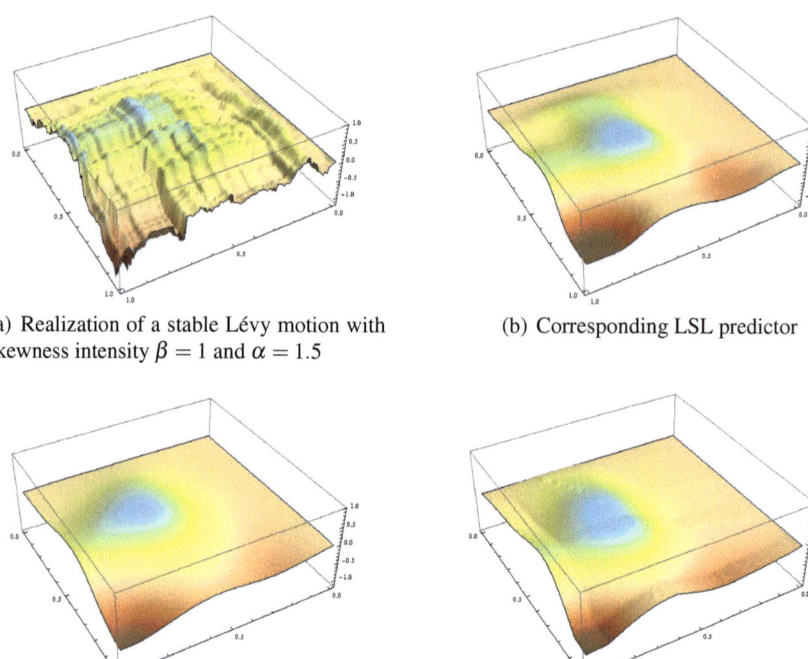

(a) Realization of a stable Lévy motion with skewness intensity $\beta = 1$ and $\alpha = 1.5$

(b) Corresponding LSL predictor

(c) Corresponding COL predictor

(d) Corresponding MCL predictor

Fig. 11.9 Realization of a skewed stable Lévy motion field and different predictors

$$f(x) = 0.5\left(0.04 - \|x\|^2\right)\mathbf{1}\left(\|x\| \le 0.2\right),$$

stability index $\alpha = 0.5$ and skewness intensity $\beta = 0.8$. The random field X is simulated on an equidistant 50×50–grid of points within $[0, 0.49]^2$ using the step function approach proposed in [220] with an accuracy (L^α-error) of $\varepsilon = 0.01$. The field X is observed at the points

$$
\begin{array}{lll}
t_1 = (0,0), & t_2 = (0,0.25), & t_3 = (0,0.49), \\
t_4 = (0.25,0), & t_5 = (0.25,0.25), & t_6 = (0.25,0.49), \\
t_7 = (0.49,0), & t_8 = (0.49,0.25), & t_9 = (0.49,0.49).
\end{array}
$$

To solve the optimization problems for the best LSL prediction (cf. Sect. 11.4.4) numerically, an average of 8 realizations of the simulated annealing algorithm from [233] is used. Figs. 11(a) and 11(b) show a realization of X and its best LSL predictor. The numerical optimization procedure is quite time consuming with 136 minutes of computation time (Pentium Dual Core E5400, 2.70 GHz, 8 GB RAM) per extrapolation.

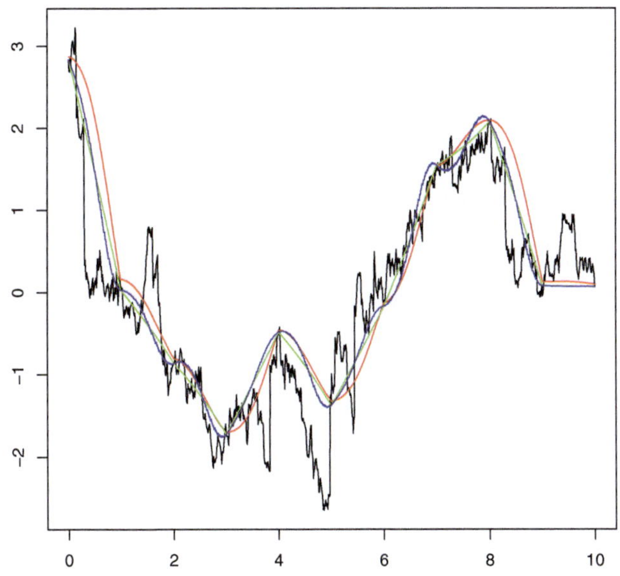

Fig. 11.10 Trajectory (black) of the stable Ornstein–Uhlenbeck process together with LSL (red), COL (green) and MCL (blue) predictors, $\alpha = 1.6$

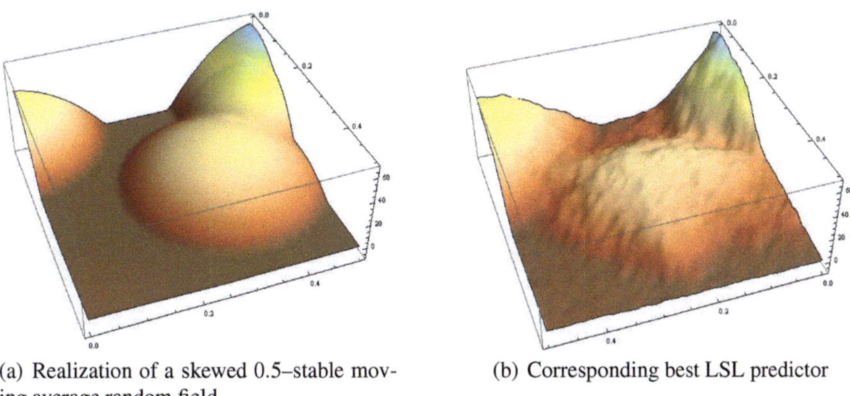

(a) Realization of a skewed 0.5–stable moving average random field

(b) Corresponding best LSL predictor

Fig. 11.11 Realization of a skewed moving average field with $\alpha = 0.5$ and its best LSL predictor

11.5 Open problems

In contrast to kriging methods, there is no common methodology of measuring prediction errors in the stable case. We propose the following measures

$$\sup_{t\in\mathbb{R}^d}\left(\mathbf{E}|X(t)-\widehat{X}(t)|^p\right)^{1/p}=c_\alpha(p)\sup_{t\in\mathbb{R}^d}\|f_t-\sum_{i=1}^n\lambda_if_{t_i}\|_\alpha,\qquad(11.50)$$

where $1<p<\alpha$ and $c_\alpha(p)>0$ is the constant from (11.13), or

$$\mathbf{P}\left(\sup_{t\in\mathbb{R}^d}|X(t)-\widehat{X}(t)|>\varepsilon\right),\quad\varepsilon>0.\qquad(11.51)$$

It is an open problem to find lower and upper bounds for these errors as well as minimax bounds where the infimum over a subclass of stable random fields X is additionally considered in (11.50) and (11.51). Alternatively, one can be interested in the asymptotic behavior of $\mathbf{P}\left(\sup_{t\in\mathbb{R}^d}|X(t)-\widehat{X}(t)|<\varepsilon\right)$ as $\varepsilon\to0$ which is related to small deviation problems.

Acknowledgements This research was partially supported by the DFG – RFBR grant 09–01–91331. The second author was also supported by the Chebyshev Laboratory (Department of Mathematics and Mechanics, St.-Petersburg State University) within RF government grant 11.G34.31.0026. The research was also supported by JSC "Gazprom Neft". The idea of the proof of Theorem 11.31 is due to Adrian Zimmer. Furthermore, the authors are grateful to Pavel Zatitskiy and Dmitry Stolyarov for the help with the proof simplification of Theorem 11.10.

Chapter 12
Spatial Process Simulation

Dirk P. Kroese and Zdravko I. Botev

Abstract The simulation of random spatial data on a computer is an important tool for understanding the behavior of spatial processes. In this chapter we describe how to simulate realizations from the main types of spatial processes, including Gaussian and Markov random fields, point processes, spatial Wiener processes, and Lévy fields. Concrete MATLAB code is provided.

12.1 Introduction

The collection and analysis of spatially arranged measurements and patterns is of interest to many scientific and engineering disciplines, including the earth sciences, materials design, urban planning, and astronomy. Examples of spatial data are geo-statistical measurements, such as groundwater contaminant concentrations, temperature reports from different cities, maps of the locations of meteorite impacts or geological faults, and satellite images or demographic maps.

From a mathematical point of view, a spatial process is a collection of random variables $\{X_t, t \in \mathcal{T}\}$ where the *index set* $\mathcal{T}$ is some subset of the d-dimensional Euclidean space $\mathbb{R}^d$. Then, X_t can be a random quantity associated with a spatial position t rather than time. The set of possible values of X_t is called the *state space* of the spatial process. Thus, spatial processes can be classified into four types, based on whether the index set and state space are continuous or discrete. An example of a spatial process with a discrete index set and a discrete state space is the Ising model in statistical mechanics, where sites arranged on a grid are assigned either a positive

Dirk P. Kroese
School of Mathematics and Physics, The University of Queensland, Brisbane 4072, Australia e-mail: kroese@maths.uq.edu.au

Zdravko I. Botev
School of Mathematics and Statistics, The University of New South Wales, Sydney 2052, Australia
e-mail: botev@unsw.edu.au

© Springer International Publishing Switzerland 2015
V. Schmidt (ed.), *Stochastic Geometry, Spatial Statistics and Random Fields*,
Lecture Notes in Mathematics 2120, DOI 10.1007/978-3-319-10064-7_12

or negative "spin"; see, for example, [387]. Image analysis, where a discrete number of pixels is assigned a continuous gray scale, provides an example of a process with a discrete index set and a continuous state space. A random configuration of points in $\mathbb{R}^d$ can be viewed as an example of a spatial process with a continuous index set and discrete state space $\{0,1\}$. If, in addition, continuous measurements are recorded at these points (e.g., rainfall data), one obtains a process in which both the index set and the state space are continuous.

Spatial processes can also be classified according to their distributional properties. For example, if any choice of random variables in a spatial process has jointly a multivariate normal distribution, the process is said to be *Gaussian*. Another important property is the *Markov* property, which deals with the local conditional independence of the random variables in the spatial process. A prominent class of spatial processes is that of the *point processes*, which are characterized by the random positions of points in space. The most important example is the *Poisson process*, whose foremost property is that the (random) numbers of points in nonoverlapping sets are independent of each other. *Lévy fields* are spatial processes that generalize this independence property. Throughout this chapter we provide computer implementation of the algorithms in MATLAB. Because of its simple syntax, excellent debugging tools, and extensive toolboxes, MATLAB is the de facto choice for numerical analysis and scientific computing. Moreover, it has one of the fastest implementations of the FFT, which is used in this chapter.

The rest of this chapter is organized as follows. We discuss in Sect. 12.2 the simulation of spatial processes that are both Gaussian and Markov. In Sect. 12.3 we consider various classes of spatial point processes, including the Poisson, compound Poisson, cluster, and Cox processes, and explain how to simulate these. Sect. 12.4 looks at ways of simulating spatial processes based on the Wiener process. Finally, Sect. 12.5 deals with the simulation of Lévy processes and fields.

12.2 Gaussian Markov Random Fields

A spatial stochastic process on $\mathbb{R}^2$ or $\mathbb{R}^3$ is often called a *random field*. Fig. 12.1 depicts realizations of three different types of random fields that are characterized by *Gaussian* and *Markovian* properties, which are discussed below.

12.2.1 Gaussian Property

A stochastic process $\{\tilde{X}_t, t \in \mathcal{T}\}$ is said to be *Gaussian* if all its finite-dimensional distributions are Gaussian (normal). That is, if for any choice of n and $t_1, \ldots, t_n \in \mathcal{T}$, we have

$$X \stackrel{\text{def}}{=} (X_1, \ldots, X_n)^\mathsf{T} \stackrel{\text{def}}{=} (\tilde{X}_{t_1}, \ldots, \tilde{X}_{t_n})^\mathsf{T} \sim N(\mu, \Sigma) \tag{12.1}$$

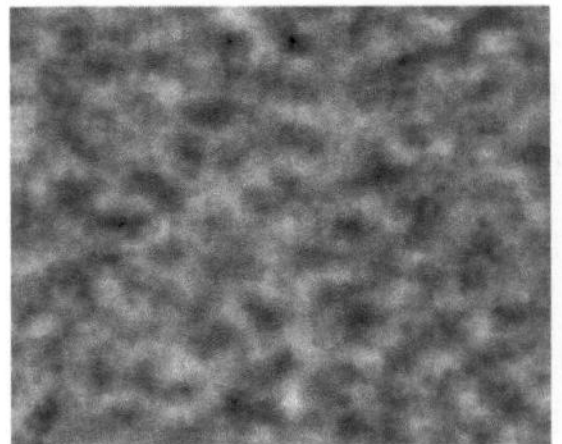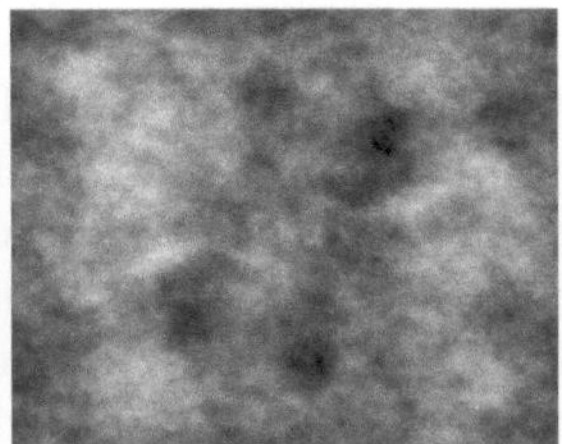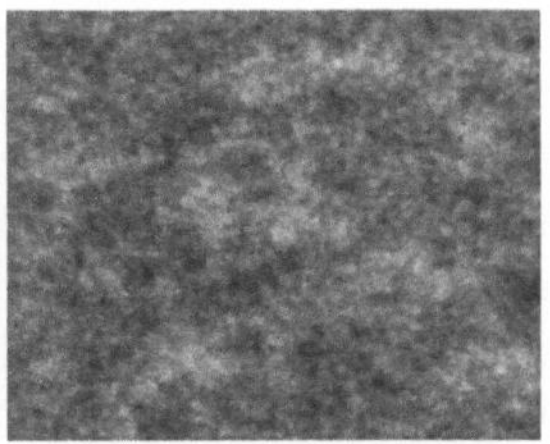

Fig. 12.1 Illustrations of zero-mean Gaussian random fields. Left: Moving average spatial process. Middle: Stationary Gaussian random field on the torus. Right: Gaussian Markov random field

for some *expectation vector* μ and *covariance matrix* Σ. Hence, any linear combination $\sum_{i=1}^{n} b_i \tilde{X}_{t_i}$ has a normal distribution. A Gaussian process is determined completely by its *expectation function* $\tilde{\mu}_t = \mathbf{E}\tilde{X}_t$ and *covariance function* $\tilde{\Sigma}_{s,t} = \mathbf{cov}(\tilde{X}_s, \tilde{X}_t)$. To generate a Gaussian process with expectation function $\tilde{\mu}_t$ and covariance function $\tilde{\Sigma}_{s,t}$ at positions $t_1, \ldots, t_n$, one can sample the multivariate normal random vector in (12.1) via the following algorithm.

Algorithm 12.1 (Gaussian process generator).
1. Construct the mean vector $\mu = (\mu_1, \ldots, \mu_n)^{\mathsf{T}}$ and covariance matrix $\Sigma = (\Sigma_{ij})$ by setting $\mu_i = \tilde{\mu}_{t_i}, i = 1, \ldots, n$ and $\Sigma_{ij} = \tilde{\Sigma}_{t_i, t_j}, i, j = 1, \ldots, n$.
2. Find a square root A of Σ, so that $\Sigma = AA^{\mathsf{T}}$.
3. Generate independent random variables $Z_1, \ldots, Z_n \sim N(0, 1)$. Let $Z = (Z_1, \ldots, Z_n)^{\mathsf{T}}$.
4. Output $X = \mu + AZ$.

Using Cholesky's square-root method, it is always possible to find a real-valued matrix A such that $\Sigma = AA^{\mathsf{T}}$. Sometimes it is easier to work with a decomposition of the form $\Sigma = BB^*$, where $B = B_1 + iB_2$ is a complex matrix with conjugate transpose $B^* = B_1^{\mathsf{T}} - iB_2^{\mathsf{T}}$. Let $Z = Z_1 + iZ_2$, where Z_1 and Z_2 are independent standard normal random vectors, as in Step 3 above. Then, the random vector $X = \Re(BZ) = B_1 Z_1 - B_2 Z_2$ has covariance matrix Σ.

A Gaussian vector $X \sim N(\mu, \Sigma)$ with invertible covariance matrix Σ can also be simulated using its *precision matrix* $\Lambda = \Sigma^{-1}$. Let $Z = DY$, where $Z \sim_{\mathrm{iid}} N(0, 1)$ and DD^{T} is the Cholesky factorization of Λ. Then Y is a zero-mean multivariate normal vector with covariance matrix

$$\mathbf{E}(YY^{\mathsf{T}}) = D^{-1}\mathbf{E}(ZZ^{\mathsf{T}})(D^{-1})^{\mathsf{T}} = (D^{\mathsf{T}}D)^{-1} = (\Lambda^{-1})^{\mathsf{T}} = \Sigma.$$

The following algorithm describes how a Gaussian process can be generated using a precision matrix.

Algorithm 12.2 (Gaussian process generator using a precision matrix).
1. Derive the Cholesky decomposition $\Lambda = DD^{\mathsf{T}}$ of the precision matrix.
2. Draw independent random variables $Z_1, \ldots, Z_n \sim_{\mathrm{iid}} N(0, 1)$. Let $Z = (Z_1, \ldots, Z_n)^{\mathsf{T}}$.
3. Solve Y from $Z = DY$, using forward substitution.
4. Output $X = \mu + Y$.

The Cholesky decomposition of a general $n \times n$ covariance or precision matrix takes $\mathcal{O}(n^3)$ floating point operations. The simulation of high-dimensional Gaussian vectors becomes very time-consuming for large n, unless some extra structure is introduced. In some cases the Cholesky decomposition can be altogether avoided by utilizing the fact that any Gaussian vector can be written as an *affine transformation* $X = \mu + AZ$ of a "white noise" vector $Z \sim N(o,I)$; as in Step 4 of Algorithm 12.1, where o is the n-dimensional zero vector and I the n-dimensional identity matrix. An example where such a transformation can be carried out directly is the following *moving average* Gaussian process $X = \{X_t, t \in \mathcal{T}\}$, where $\mathcal{T}$ is a two-dimensional grid of equally-spaced points. Here each X_t is equal to the average of all white noise terms Z_s with s lying in a disc of radius r around t. That is,

$$ X_t = \frac{1}{N_r} \sum_{s:\,\|t-s\| \leq r} Z_s \,, $$

where N_r is the number of grid points in the disc. Such spatial processes have been used to describe rough energy surfaces for charge transport [51, 223]. The following MATLAB program produces a realization of this process on a 200×200 grid, using a radius $r = 6$. A typical outcome is depicted in the left pane of Fig. 12.1.

```
n = 300;
r = 6; % radius (maximal 49)
noise =  randn(n);
[x,y]=meshgrid(-r:r,-r:r);
mask=((x.^2+y.^2)<=r^2);   %(2*r+1)x(2*r+1) bit mask
x = zeros(n,n);
nmin = 50; nmax = 250;
for i=nmin:nmax
    for j=nmin:nmax
        A = noise((i-r):(i+r), (j-r):(j+r));
        x(i,j) = sum(sum(A.*mask));
    end
end
Nr = sum(sum(mask)); x = x(nmin:nmax, nmin:nmax)/Nr;
imagesc(x); colormap(gray)
```

12.2.2 Generating Stationary Processes via Circulant Embedding

Another approach to efficiently generate Gaussian spatial processes is to exploit the structural properties of stationary Gaussian processes. A Gaussian process $\{\tilde{X}_t, t \in \mathbb{R}^d\}$ is said to be *stationary* if the expectation function, $\mathbf{E}\tilde{X}_t$, is constant and the covariance function, $\mathrm{cov}(\tilde{X}_s, \tilde{X}_t)$, is invariant under translations; that is $\mathbf{cov}(\tilde{X}_{s+u}, \tilde{X}_{t+u}) = \mathbf{cov}(\tilde{X}_s, \tilde{X}_t)$.

An illustrative example is the simulation of a stationary Gaussian process on the unit *torus*; that is, the unit square $[0,1] \times [0,1]$ in which points on opposite sides are identified with each other. In particular, we wish to generate a zero-mean Gaussian random process $\{\tilde{X}_t\}$ on each of the grid points $\{(i,j)/n, i = 0, \ldots, n-1, j = 0, \ldots, n-1\}$ corresponding to a covariance function of the form

$$\mathbf{cov}(\tilde{X}_s, \tilde{X}_t) = \exp\{-c\,\|s-t\|_T^\alpha\}, \tag{12.2}$$

where $\|s-t\|_T = \|(s_1 - t_1, s_2 - t_2)\|_T \stackrel{\text{def}}{=} \sqrt{\sum_{k=1}^{2}(\min\{|s_k - t_k|, 1 - |s_k - t_k|\})^2}$ is the Euclidean distance between $s = (s_1, s_2)$ and $t = (t_1, t_2)$ on the torus. Notice that this renders the process not only stationary but also *isotropic* (that is, the distribution remains the same under rotations).

We can arrange the grid points in the order

$$(0,0), (0, 1/n), \ldots, (0, 1 - 1/n), (1/n, 0), \ldots, (1 - 1/n, 1 - 1/n).$$

The values of the Gaussian process can be accordingly gathered in an $n^2 \times 1$ vector X or, alternatively an $n \times n$ matrix $\underline{X}$. Let Σ be the $n^2 \times n^2$ covariance matrix of X. The key to efficient simulation of X is that Σ is a symmetric *block-circulant matrix with circulant blocks*. That is, Σ has the form

$$\Sigma = \begin{pmatrix} C_1 & C_2 & C_3 & \ldots & C_n \\ C_n & C_1 & C_2 & \ldots & C_{n-1} \\ & & \ldots & & \\ C_2 & C_3 & \ldots & C_n & C_1 \end{pmatrix},$$

where each C_i is a circulant matrix $\mathrm{circ}(c_{i1}, \ldots, c_{in})$. The matrix Σ is thus completely specified by its first row, which we gather in an $n \times n$ matrix G. The eigenstructure of block-circulant matrices with circulant blocks is well known; see, for example, [30]. Specifically, Σ is of the form

$$\Sigma = P^* \mathrm{diag}(\gamma) P, \tag{12.3}$$

where P^* denotes the complex conjugate transpose of P, and P is the Kronecker product of two *discrete Fourier transform* matrices; that is, $P = F \otimes F$, where $F_{jk} = \exp(-2\pi i jk/n)/\sqrt{n}, j, k = 0, 1, \ldots, n-1$. The vector of eigenvalues $\gamma = (\gamma_1, \ldots, \gamma_{n^2})^\mathsf{T}$ ordered as an $n \times n$ matrix Γ satisfies $\Gamma = nF^*GF$. Since Σ is a covariance matrix, the component-wise square root $\sqrt{\Gamma}$ is well-defined and real-valued. The matrix $B = P^* \mathrm{diag}(\sqrt{\gamma})$ is a complex square root of Σ, so that X can be generated by drawing $Z = Z_1 + iZ_2$, where Z_1 and Z_2 are independent standard normal random vectors, and returning the real part of BZ. It will be convenient to gather Z into an $n \times n$ matrix $\underline{Z}$.

The evaluation of both Γ and $\underline{X}$ can be done efficiently by using the (appropriately scaled) *two-dimensional fast Fourier transform* (FFT2). In particular, Γ is the FFT2 of the matrix G and $\underline{X}$ is the real part of the FFT2 of the matrix $\sqrt{\Gamma} \odot \underline{Z}$, where $\odot$ denotes component-wise multiplication. The following MATLAB program generates the outcome of a stationary Gaussian random field on a 256×256 grid, for a covariance function of the form (12.2), with $c = 8$ and $\alpha = 1$. A realization is shown in the middle pane of Fig. 12.1.

```
n = 2^8;
t1 = [0:1/n:1-1/n]; t2 = t1;
for i=1:n    % first row of cov. matrix, arranged in a matrix
   for j=1:n
        G(i,j)=exp(-8*sqrt(min(abs(t1(1)-t1(i))), ...
        1-abs(t1(1)-t1(i)))^2 + min(abs(t2(1)-t2(j)), ...
        1-abs(t2(1)-t2(j)))^2));
   end;
end;
Gamma = fft2(G);    % the eigenvalue matrix n*fft2(G/n)
Z = randn(n,n) + sqrt(-1)*randn(n,n);
X = real(fft2(sqrt(Gamma).*Z/n));
imagesc(X); colormap(gray)
```

The simulation of general stationary Gaussian processes in $\mathbb{R}^d$ with covariance function

$$\mathbf{cov}(\tilde{X}_s, \tilde{X}_t) = \rho(s-t)$$

is discussed in [100] and [65, 418]. Recall that if $\rho(s-t) = \rho(\|s-t\|)$, then the random field is not only stationary, but isotropic.

In [100], the method of *circulant embedding* is proposed, which allows the efficient simulation of a stationary Gaussian field via the FFT. The idea is to embed the covariance matrix into a block circulant matrix with each block being circulant itself (as in the last example), and then construct the matrix square root of the block circulant matrix using FFT techniques. The FFT permits the fast simulation of the Gaussian field with this block circulant covariance matrix. Finally, the marginal distributions of appropriate sub-blocks of this Gaussian field have the desired covariance structure.

Here we consider the two-dimensional case. The aim is to generate a zero-mean stationary Gaussian field over the $n \times m$ rectangular grid

$$\mathcal{G} = \{(i\Delta_x, j\Delta_y),\ i = 0,\ldots,n-1,\ j = 0,\ldots,m-1\},$$

where Δ_x and Δ_y denote the corresponding horizontal and vertical spacing along the grid. The algorithm can broadly be described as follows.

Step 1 (Building and storing the covariance matrix). The grid points can be arranged into a column vector of size mn to yield the $mn \times mn$ covariance matrix $\Omega_{i,j} = \rho(s_i - s_j)$, $i,j = 1,\ldots,mn$, where

$$s_k \overset{\text{def}}{=} \left((k-1) \mod m,\ \left\lfloor \frac{k}{m} \right\rfloor \right),\quad k = 1,\ldots,mn\,.$$

The matrix Ω has symmetric block-Toeplitz structure, where each block is a Toeplitz matrix (not necessarily symmetric). For example, the left panel of Fig. 12.2

shows the block-Toeplitz covariance matrix with $m = n = 3$. Each 3×3 block is itself a Toeplitz matrix with entry values coded in color. For instance, we have $\Omega_{2,4} = \Omega_{3,5} = c$. The matrix Ω is thus uniquely characterized by its first block row $(R_1, \ldots, R_n)$, where each of the n blocks is an $m \times m$ Toeplitz matrix. The k-th $m \times m$ Toeplitz matrix consists of the sub-block of Ω with entries

$$\Omega_{i,j}, \quad i = 1, \ldots, m, \; j = (km+1), \ldots, (k+1)m \, .$$

Notice that each block R_k is itself characterized by its first row and column. (Each Toeplitz block R_k will be characterized by the first row only provided the covariance function has the form $\rho(s-t) = \rho(\|s\|, \|t\|)$, in which case each R_k is symmetric.) Thus, in general, the covariance matrix can be completely characterized by the entries of a pair of $m \times n$ and $n \times m$ matrices storing the first columns and rows of all n Toeplitz blocks $(R_1, \ldots, R_n)$. In typical applications, the computation of these two matrices is the most time-consuming step.

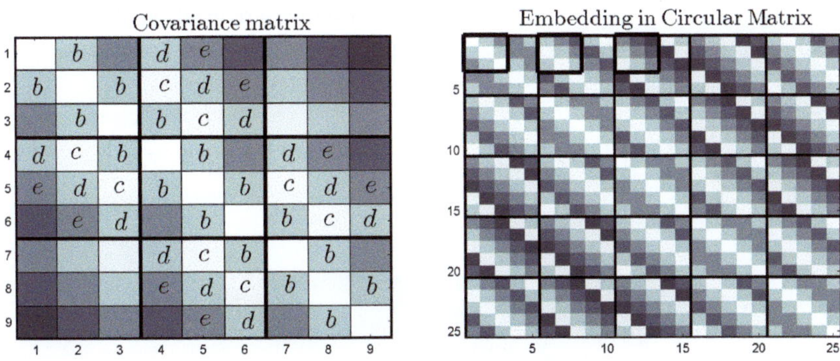

Fig. 12.2 Left panel: the symmetric block-Toeplitz covariance matrix Ω with $n = m = 3$. Right panel: block-circulant matrix Σ of size $((2n-1) \times (2m-1))^2 = 25 \times 25$. The first three block rows of Ω are embedded in the upper left corners of the first three block columns of the circulant matrix Σ

Step 2 (Embedding in block circulant matrix). Each Toeplitz matrix R_k is embedded in the upper left corner of an $2m+1$ circulant matrix C_k. For example, in the right panel of Fig. 12.2 the embedded blocks R_1, R_2, R_3 are shown within bold rectangles. Finally, let Σ be the $(2n-1)(2m-1) \times (2n-1)(2m-1)$ block circulant matrix with first block row given by $(C_1, \ldots, C_n, C_n^\mathsf{T}, \ldots, C_2^\mathsf{T})$. This gives the *minimal embedding* in the sense that there is no block circulant matrix of smaller size that embeds Ω.

Step 3 (Computing the square root of the block circulant matrix). After the embedding we are essentially generating a Gaussian process on a torus with covariance matrix Σ as in the last example. The block circulant matrix Σ can be diagonalized as in (12.3) to yield $\Sigma = P^*\mathrm{diag}(\gamma)P$, where P is the $(2n-1)(2m-1) \times (2n-1)(2m-1)$ two-dimensional discrete Fourier transform matrix. The vector of eigenvalues γ is of length $(2n-1)(2m-1)$ and is arranged in a $(2m-1) \times (2n-1)$ matrix Γ so that the first column of Γ consists of the first $2m-1$ entries of γ, the second column of Γ consists of the next $2m-1$ entries and so on. It follows that if G is an $(2m-1) \times (2n-1)$ matrix storing the entries of the first block row of Σ, then Γ is the FFT2 of G. Assuming that $\gamma > 0$, we obtain the square root factor $B = P^* \mathrm{diag}(\sqrt{\gamma})$, so that $\Sigma = B^* B$.

Step 4 (Extracting the appropriate sub-block). Next, we compute the FFT2 of the array $\sqrt{\Gamma} \odot Z$, where the square root is applied component-wise to Γ and Z is an $(2m-1) \times (2n-1)$ complex Gaussian matrix with entries $\underline{Z}_{j,k} = U_{j,k} + iV_{j,k}$, $U_{j,k}, V_{j,k} \sim N(0,1)$ for all j and k. Finally, the first $m \times n$ sub-blocks of the real and imaginary parts of $\mathrm{FFT2}(\sqrt{\Gamma} \odot \underline{Z})$ represent two independent realization of a stationary Gaussian field with covariance Σ on the grid $\mathcal{G}$. If more realizations are required, we store the values $\sqrt{\Gamma}$ and we repeat Step 4 only. The complexity of the circulant embedding inherits the complexity of the FFT approach, which is of order $\mathcal{O}(mn\log(m+n))$, and compares very favorably with the standard Cholesky decomposition method of order $\mathcal{O}(m^3 n^3)$.

As a numerical example (see [100]), Fig. 12.3 shows a realization of a stationary nonisotropic Gaussian field with $m = 512, n = 384$, $\Delta_x = \Delta_y = 1$, and covariance function

$$\rho(s-t) = \rho(h) = \left(1 - \frac{h_1^2}{50^2} - \frac{h_1 h_2}{50 \times 15} - \frac{h_2^2}{15^2}\right) \exp\left(-\frac{h_1^2}{50^2} - \frac{h_2^2}{15^2}\right). \qquad (12.4)$$

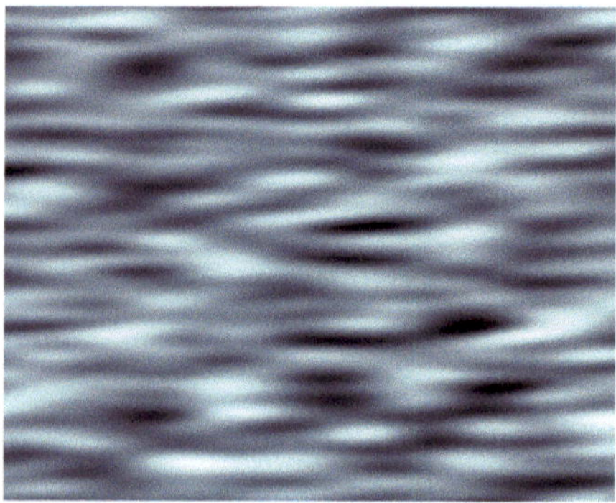

Fig. 12.3 Realization of a stationary nonisotropic Gaussian field with covariance function given by (12.4)

The following MATLAB code implements the procedure described above with co-variance function given by (12.4).

```
n=384; m=512; % size of grid is m*n
% size of covariance matrix is m^2*n^2
tx=[0:n-1]; ty=[0:m-1]; % create grid for field
```

```
rho=@(x,y)((1-x^2/50^2-x*y/(15*50)-y^2/15^2)...
    *exp(-(x^2/50^2+y^2/15^2)));
Rows=zeros(m,n); Cols=Rows;
for i=1:n
    for j=1:m
        Rows(j,i)=rho(tx(i)-tx(1),ty(j)-ty(1)); % rows of blocks
        Cols(j,i)=rho(tx(1)-tx(i),ty(j)-ty(1)); % columns
    end
end
% create the first row of the block circulant matrix with
% circulant blocks and store it as a matrix suitable for fft2
BlkCirc_row=[Rows, Cols(:,end:-1:2);
    Cols(end:-1:2,:), Rows(end:-1:2,end:-1:2)];
% compute eigen-values
lam=real(fft2(BlkCirc_row))/(2*m-1)/(2*n-1);

if abs(min(lam(lam(:)<0)))>10^-15
    error('Could not find positive definite embedding!')
else
    lam(lam(:)<0)=0; lam=sqrt(lam);
end
% generate field with covariance given by block circulant matrix
F=fft2(lam.*complex(randn(2*m-1,2*n-1),randn(2*m-1,2*n-1)));
F=F(1:m,1:n); % extract sub-block with desired covariance
field1=real(F); field2=imag(F); % two independent fields
imagesc(tx,ty,field1), colormap bone
```

Extensions to three and more dimensions are possible, see [100]. For example, in the three-dimensional case the correlation matrix Ω will be symmetric block Toeplitz matrix with each block satisfying the properties of a two-dimensional covariance matrix.

Throughout the discussion so far we have always assumed that the block circulant matrix Σ is a covariance matrix itself. If this is the case, then we say that we have a *nonnegative definite embedding* of Ω. A nonnegative definite embedding ensures that the square root of γ is real. Generally, if the correlation between points on the grid that are sufficiently far apart is zero, then a non-negative embedding will exist, see [100]. A method that exploits this observation is Stein's *intrinsic embedding* method proposed in [377] (see also [143]). Stein's method depends on the construction of a compactly supported covariance function that yields to a nonnegative circulant embedding. The idea is to modify the original covariance function so that it decays smoothly to zero. In more detail, suppose we wish to simulate a process with covariance function ρ over the set $\{h : \|h\| \leq 1, h > 0\}$. To achieve this, we simulate a process with covariance function

$$\psi(h) = \begin{cases} c_0 + c_2\|h\|^2 + \rho(h), & \text{if } \|h\| \leq 1, \\ \varphi(h), & \text{if } 1 \leq \|h\| \leq R, \\ 0, & \text{if } \|h\| \geq R, \end{cases} \qquad (12.5)$$

where the constants c_0, c_2, $R \geq 1$ and function φ are selected so that ψ is a continuous (and as many times differentiable as possible), stationary and isotropic covariance function on $\mathbb{R}^2$. The process will have covariance structure of $c_0 + c_2 \|h\|^2 + \rho(h)$ in the disk $\{h : \|h\| \leq 1, h > 0\}$, which can then be easily transformed into a process with covariance function $\rho(h)$. We give an example of this in Sect. 12.4.4, where we generate fractional Brownian surfaces via the intrinsic embedding technique. Generally, the smoother the original covariance function, the harder it is to embed via Stein's method, because a covariance function that is smoother close to the origin has to be even smoother elsewhere.

12.2.3 Markov Property

A *Markov random field* is a spatial stochastic process $\{X_t, t \in \mathcal{T}\}$ that possesses a *Markov property*, in the sense that

$$(X_t \mid X_s, s \in \mathcal{T} \setminus \{t\}) \sim (X_t \mid X_s, s \in \mathcal{N}_t),$$

where $\mathcal{N}_t$ is the set of "neighbors" of t. Thus, for each $t \in \mathcal{T}$ the conditional distribution of X_t given all other values X_s is equal to the conditional distribution of X_t given only the neighboring values.

Assuming that the index set $\mathcal{T}$ is finite, Markov random fields are often defined via an undirected graph $\mathcal{G} = (V, E)$. In such a *graphical model* the vertices of the graph correspond to the indices $t \in \mathcal{T}$ of the random field, and the edges describe the dependencies between the random variables. In particular, there is *no* edge between nodes s and t in the graph if and only if X_s and X_t are conditionally independent, given all other values $\{X_u, u \neq i, j\}$. In a Markov random field described by a graph $\mathcal{G}$, the set of neighbors $\mathcal{N}_t$ of t corresponds to the set of vertices that share an edge with t; that is, those vertices that are *adjacent* to t. An example of a graphical model for a 2-dimensional Markov random field is shown in Fig. 12.4. In this case vertex corner nodes have two neighbors and interior nodes have four neighbors.

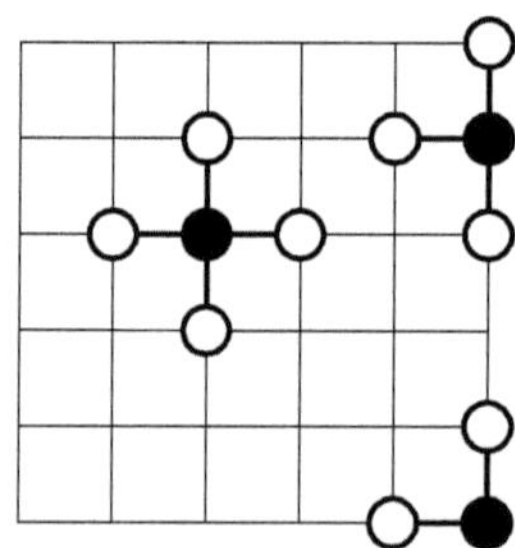

Fig. 12.4 A graphical model for 2D spatial process. Each vertex has at most four neighbors.

Of particular importance are Markov random fields that are also Gaussian. Suppose $\{X_t, t \in \mathcal{T}\}$ is such a *Gaussian Markov random field* (GMRF), with corresponding graphical model $\mathcal{G} = (V, E)$. We may think of $\{X_t, t \in \mathcal{T}\}$ as a column vector of $n = |V|$ elements, or as a spatial arrangement of random variables (pixel values), as in Fig. 12.4. Without loss of generality we assume that $\mathbf{E}X_t = 0$ for each $t \in \mathcal{T}$, and that the index set $\mathcal{T}$ is identified with the vertex set V, whose elements are labeled as $1, 2, \ldots, n$. Because $\{X_t, t \in \mathcal{T}\}$ is Gaussian, the probability density function of $\{X_t, t \in \mathcal{T}\}$ is given by

$$f(x) = (2\pi)^{-n/2} \sqrt{\det(\Lambda)} \, e^{-\frac{1}{2} x^{\mathsf{T}} \Lambda x} \,,$$

where $\Lambda = (\lambda_{ij})$ is the precision matrix. In particular, the joint distribution of two components X_i and X_j is

$$\tilde{f}(x_i, x_j) \propto \exp\left(-\frac{1}{2}(\lambda_{ii} x_i^2 + \lambda_{ij} x_i x_j + \lambda_{jj} x_j^2) \right) \,.$$

This shows that X_i and X_j are conditionally independent given $\{X_k, k \neq i, j\}$, if and only if $\lambda_{ij} = 0$. Consequently, (i, j) is an edge in the graphical model if and only if $\lambda_{ij} \neq 0$. In typical applications (for example in image analysis) each vertex in the graphical model only has a small number of adjacent vertices, as in Fig. 12.4. In such cases the precision matrix is thus *sparse*, and the Gaussian vector can be generated efficiently using, for example, sparse Cholesky factorization [146].

As an illustrative example, consider the graphical model of Fig. 12.4 on a grid of $n = m^2$ points: $\{1, \ldots, m\} \times \{1, \ldots, m\}$. We can efficiently generate a GMRF on this grid via Algorithm 12.2, provided the Cholesky decomposition can be carried out efficiently. For this we can use for example the *band Cholesky* method [146], which takes $n(p^2 + 3p)$ floating point operations, where p is the bandwidth of the matrix; that is, $p = \max_{i,j}\{|i - j| : \lambda_{ij} = 0\}$. The right pane of Fig. 12.1 shows a realization of the GMRF on a 250×250 grid, using parameters $\lambda_{ii} = 2$ for all $i = 1, \ldots, n$ and $\lambda_{ij} = -0.5$ for all neighboring elements $j \in \mathcal{N}_j$ of $i = 1, \ldots, n$. Further details on such construction methods for GMRFs may be found, for example, in [336].

Exercise 12.1. Write MATLAB code to generate a zero-mean Gaussian Markov random field for a 200×200 grid with a neighborhood structure similar to Fig. 12.4, that is, each internal vertex has four neighbors, the boundary ones have three, and the corner ones have two. Choose $\lambda_{ii} = 1$ and $\lambda_{ij} = -0.25$ for each neighboring element j of i. *Hint.* See the MATLAB code in Sect. 5.1 of [237].

12.3 Point Processes

Point processes on $\mathbb{R}^d$ are spatial processes describing random configurations of d-dimensional points. Spatially distributed point patterns occur frequently in nature and in a wide variety of scientific disciplines, such as spatial epidemiology, materials science, forestry, and geography. The positions of accidents on a highway during a fixed time period and the times of earthquakes in Japan are examples of one-dimensional spatial point processes. Two-dimensional examples include the positions of cities on a map, the positions of farms with Mad Cow Disease in the UK, and the positions of broken connections in a communications or energy network. In three dimensions, we observe the positions of stars in the universe, the positions of mineral deposits underground, or the times and positions of earthquakes in Japan. Spatial processes provide excellent models for many of these point patterns [15, 16, 93, 94, 102]. Spatial processes also have an important role in stochastic modeling of complex microstructures, for example, graphite electrodes used in Lithium-ion batteries [380].

Mathematically, point processes can be described in three ways: 1. as random sets of points, 2. as random-sized vectors of random positions, and 3. as random counting measures. In this section we discuss some of the important classes of point processes and their generalizations, including Poisson processes, marked point processes, and cluster processes.

12.3.1 Poisson Process

Poisson processes are used to model random configurations of points in space and time. Let E be some Borel subset of $\mathbb{R}^d$ and let $\mathcal{E}$ be the collection of Borel sets on E. To any collection of random points $\{X_1,\ldots,X_N\}$ in E corresponds a *random counting measure* $X(A), A \in \mathcal{E}$ defined by

$$X(A) = \sum_{i=1}^{N} \mathbf{1}_{\{X_i \in A\}}, \quad A \in \mathcal{E}, \tag{12.6}$$

which counts the random number of points in A. We may *identify* the random measure X defined in (12.6) with the random set $\{X_i, i = 1,\ldots,N\}$, or with the random vector $(X_1,\ldots,X_N)$. The measure $\mu : \mathcal{E} \to [0,\infty]$ given by $\mu(A) = \mathbf{E}X(A), A \in \mathcal{E}$ is called the *mean measure* of X. In many cases the mean measure μ has a density $\lambda : E \to [0,\infty)$, called the *intensity function*; so that

$$\mu(A) = \mathbf{E}X(A) = \int_A \lambda(x)\,dx.$$

We will assume from now on that such an intensity function exists.

The most important class of point processes which holds the key to the analysis of point pattern data is the Poisson process. A random counting measure X is said to be a *Poisson random measure* with locally finite mean measure μ if the following properties hold:

1. For any bounded set $A \in \mathcal{E}$ the random variable $X(A)$ has a Poisson distribution with mean $\mu(A)$. We write $X(A) \sim \text{Pois}(\mu(A))$.

2. For any disjoint sets $A_1, \ldots, A_N \in \mathcal{E}$, the random variables $X(A_1), \ldots, X(A_N)$ are independent.

The Poisson process is said to be *stationary* if the intensity function is constant. When the intensity function is a constant, we simply refer to it as the intensity. An important corollary of Properties 1 and 2 is the following result. Suppose that $0 < \mu(E) < \infty$. Then,

3. conditional upon $X(E) = N$, the points $X_1, \ldots, X_N$ are independent of each other and have the probability density function (pdf) $g(x) = \lambda(x)/\mu(E)$.

This result is the key to generating a Poisson random measure on $E \subset \mathbb{R}^d$ with finite mean measure $\mu(E) < \infty$.

Algorithm 12.3 (Generating a Poisson random measure).
1. Generate a Poisson random variable $N \sim \mathrm{Pois}(\mu(E))$.
2. Draw $X_1, \ldots, X_N \sim g$, where $g(x) = \lambda(x)/\mu(E)$, and return these as the points of the Poisson random measure.

As a specific example, consider the simulation of a 2-dimensional Poisson process with intensity function $\lambda(x_1, x_2) = 300(x_1^2 + x_2^2)$ on the unit square $E = [0,1]^2$. Since the probability density function $g(x_1, x_2) = \lambda(x_1, x_2)/\mu(E) = 3(x_1^2 + x_2^2)/2$ is bounded by 3, drawing from g can be done simply via the acceptance–rejection method [334].

Exercise 12.2. Write MATLAB code that implements this acceptance-rejection method. That is, in Step 2 of Algorithm 12.3, draw (X_1, X_2) uniformly on E and Z uniformly on $[0,3]$, and accept (X_1, X_2) if $g(X_1, X_2) \leq Z$; otherwise repeat.

An alternative, but equivalent, method is to generate a stationary Poisson process on E, with intensity $\lambda = 600$, and to *thin out* the points by accepting each point (x_1, x_2) with probability $\lambda(x_1, x_2)/\lambda$. The following MATLAB code implements this thinning procedure. A typical realization is given in Fig. 12.5.

```
lambda = @(x) 300*(x(:,1).^2 + x(:,2).^2);
lamstar = 600;
N=poissrnd(lamstar); x = rand(N,2); % homogeneous PP
ind = find(rand(N,1) < lambda(x)/lamstar);
xa = x(ind,:); % thinned PP
plot(xa(:,1),xa(:,2))
```

12.3.2 Marked Point Processes

A natural way to extend the notion of a point process is to associate with each point $X_i \in \mathbb{R}^d$ a (random) *mark* $Y_i \in \mathbb{R}^m$, representing an attribute such as width, velocity, weight etc. The collection $\{(X_i, Y_i)\}$ is called a *marked point process*. In

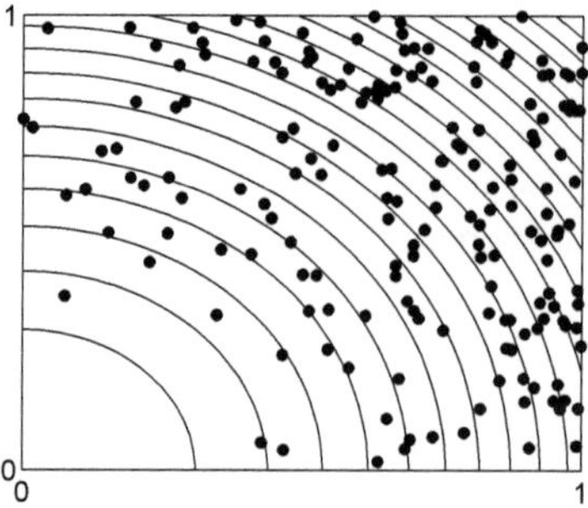

Fig. 12.5 A realization of a non-stationary Poisson process with intensity function $\lambda(x_1, x_2) = 300(x_1^2 + x_2^2)$ (contour lines shown) on the unit square

a marked point process with *independent marking* the marks are assumed to be independent of each other and of the points, and distributed according to a fixed mark distribution. The following gives a useful connection between marked point processes and Poisson processes; see, for example, [93, 94].

Theorem 12.1. *If $\{(X_i, Y_i)\}$ is a Poisson process on $\mathbb{R}^d \times \mathbb{R}^m$ with intensity function $\zeta : \mathbb{R}^d \times \mathbb{R}^m \to [0, \infty)$, and*

$$K(x) = \int \zeta(x, y)\, dy < \infty \quad \text{for all } x \in \mathbb{R}^d, \tag{12.7}$$

then $\{X_i\}$ is a Poisson process on $\mathbb{R}^d$ with intensity function $K : \mathbb{R}^d \to [0, \infty)$ given in (12.7), and $\{(X_i, Y_i)\}$ is a marked Poisson process, where the density function of the marks is given by $\zeta(x, \cdot)/K(x)$ on $\mathbb{R}^m$.

A (spatial) marked Poisson process with independent marking is an important example of a (spatial) *Lévy process*: a stochastic process with independent and stationary increments (discussed in more detail in Sect. 12.5).

The simulation of a marked Poisson process with independent marks is virtually identical to that of an ordinary (that is, non-marked) Poisson process. The only difference is that for each point the mark has to be drawn from the mark distribution. An example of a realization of a marked Poisson process is given in Fig. 12.6. Here the marks are uniformly distributed on [0, 0.1], and the underlying Poisson process is stationary with intensity 100.

12.3.3 Cluster Process

In applications one often observes point patterns that display clustering. An example is the spread of plants from a weed species where the plants are initially introduced

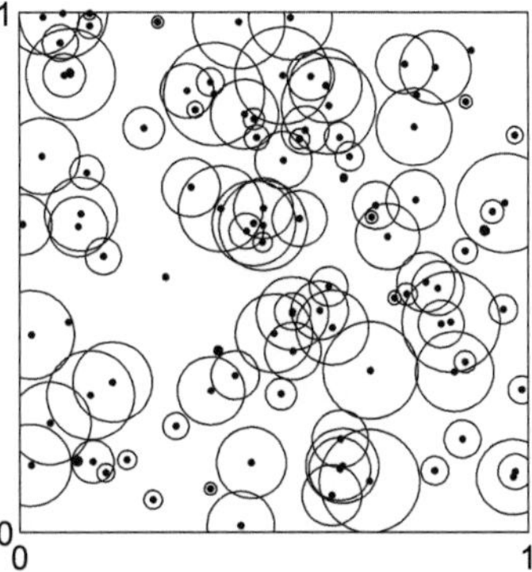

Fig. 12.6 A realization of a marked Poisson process with independent marking on the unit square. The Poisson process has intensity 100. The marks are uniformly distributed on [0,0.1] and correspond to the radii of the circles

by birds at a number of geographically dispersed locations and the plants then spread themselves locally.

Let C be a point process of "centers" and associate with each $c \in C$ a point process X^c, which may include c. The combined set of points $X = \cup_{c \in C} X^c$ constitutes a *cluster process*. If C is a Poisson process, then X is called a *Poisson cluster process*.

As a specific example, consider the following *Hawkes process*. Here, the center process C is a Poisson process on $\mathbb{R}^d$ (or a subset thereof) with some intensity function $\lambda(\cdot)$. The clusters are generated as follows. For each $c \in C$, independently generate "first-generation offspring" according to a Poisson process with intensity function $\rho(x - c)$, where $\rho(\cdot)$ is a positive function on $\mathbb{R}^d$ with integral less than unity. Then, for each first-generation offspring $c^{(1)}$ generate a Poisson process with intensity function $\rho(x - c^{(1)})$, and so on. The collection of all generated points forms the Hawkes process. The requirement that $\int \rho(y)\,dy < 1$ simply means that the expected number of offspring of each point is less than one.

Fig. 12.7 displays a realization for $\mathbb{R}^2$ with the cluster centers forming a Poisson process on the unit square with intensity $\lambda = 30$. The offspring intensity function is here

$$\rho(x_1, x_2) = \frac{\alpha}{2\pi\sigma^2}\, e^{-\frac{1}{2\sigma^2}(x_1^2 + x_2^2)}, \quad (x_1, x_2) \in \mathbb{R}^2, \tag{12.8}$$

with $\alpha = 0.9$ and with $\sigma = 0.02$. This means that the number N of offspring for a single generation from a parent at position (x_1, x_2) has a Poisson distribution with parameter α. And given $N = n$, the offspring locations are independent and identically distributed according to a bivariate normal random vector with independent components with means x_1 and x_2 and both variances σ^2. In Fig. 12.7 the cluster centers are indicated by circles. Note that the process possibly has points outside the displayed box. The MATLAB code is given below.

```
lambda = 30;              %intensity of initial points (centers)
mean_offspring = 0.9;     %mean number of offspring of each point
X = zeros(10^5,2);        %initialise the points
N = poissrnd(lambda);     %number of centers
X(1:N,:) = rand(N,2);     %generate the centers
total_so_far = N;         %total number of points generated
next = 1;
while next < total_so_far
    nextX = X(next,:);    %select next point
    N_offspring = poissrnd(mean_offspring); %number of offspring
    NewX=repmat(nextX,N_offspring,1)+0.02*randn(N_offspring,2);
    X(total_so_far+(1:N_offspring),:) = NewX; %update point list
    total_so_far = total_so_far+N_offspring;
    next = next+1;
end
X=X(1:total_so_far,:); %cut off unused rows
plot(X(:,1),X(:,2),'.')
```

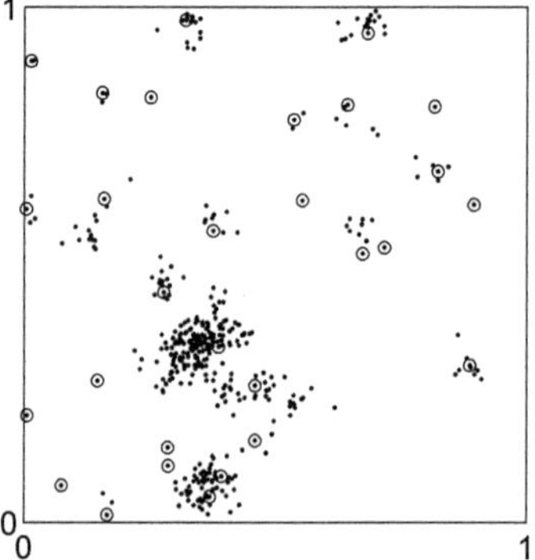

Fig. 12.7 Realization of a two-dimensional Hawkes process with centers (encircled) forming a Poisson process on $[0,1] \times [0,1]$ with intensity $\lambda = 30$. The offspring intensity function is given in (12.8)

12.3.4 Cox Process

Suppose we wish to model a point pattern of trees given that we know the soil quality $h(x_1,x_2)$ at each point (x_1,x_2). We could use a non-stationary Poisson process with an intensity function λ that is an increasing function of h; for example, $\lambda(x_1,x_2) = e^{\alpha+\beta h(x_1,x_2)}$ for some known $\alpha > 0$ and $\beta > 0$. In practice, however, the soil quality itself could be random, and be described via a random field. Consequently, one could

could try instead to model the point pattern as a Poisson process with a random intensity function Λ. Such processes were introduced in [85] as doubly stochastic Poisson processes and are now called Cox processes.

More precisely, we say that X is a *Cox process* driven by the random intensity function Λ if conditional on $\Lambda = \lambda$ the point process X is Poisson with intensity function λ. Simulation of a Cox process on a set $\mathcal{T} \subset \mathbb{R}^d$ is thus a two-step procedure.

Algorithm 12.4 (Simulation of a Cox process).
1. Simulate a realization $\lambda = \{\lambda(x), x \in \mathcal{T}\}$ of the random intensity function Λ.
2. Given $\Lambda = \lambda$, simulate X as an non-stationary Poisson process with intensity function λ.

Fig. 12.8 (a) shows a realization of a Cox process on the unit square $\mathcal{T} = [0,1] \times [0,1]$, with a random intensity function whose realization is given in Fig. 12.8 (b). The random intensity at position (x_1, x_2) is either 3000 or 0, depending on whether the value of a random field on $\mathcal{T}$ is negative or not. The random field that we used is the stationary Gaussian process on the torus described in Sect. 12.2.

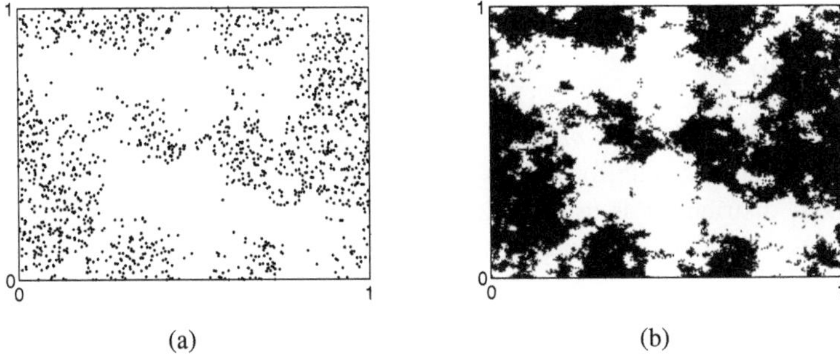

(a)   (b)

Fig. 12.8 Realization of a Cox process: (a) on a square generated by a the random intensity function given in (b). The black points have intensity 3000 and the white points have intensity 0

Given an outcome of the random intensity function, this Cox process (sometimes called random-set Cox or interrupted Poisson process) is constructed by generating a stationary Poisson process with rate 3000 on $\mathcal{T}$ and accepting only those points for which the random intensity function is non-zero. The values of the intensity function are taken to be constant within each square $\{(i+u)/n, (j+v)/n), 0 \le u, v \le 1\}$, $i, j = 0, \ldots, n-1$. The following MATLAB code is to be appended to the code used for the stationary Gaussian process simulation on the torus in Sect. 12.2.

```
Lambda = ones(n,n).*(X < 0); %random intensity function
%imagesc(Lambda); set(gca,'YDir','normal')
Lambda = Lambda(:); %reshape as a column vector
N = poissrnd(3000);
```

```
P = rand(N,2); %generate homogenous PP
Pn = ceil(P*n); %PP scaled by factor n
K = (Pn(:,1)-1)*n + Pn(:,2); % indices of scaled PP
ind = find(Lambda(K)); %indices for which intensity is 1
Cox = P(ind,:);  %realization of the Cox process
plot(Cox(:,1),Cox(:,2),'.');
```

In [299] the following Cox process has been applied to cosmology; see, however, [178], who show the limitations of the model. Used to model the positions of stars in the universe, it now bears the name *Neyman–Scott process*. Suppose C is a stationary Poisson process in $\mathbb{R}^d$ with constant intensity κ. Let the random intensity function Λ be given by

$$\Lambda(x) = \alpha \sum_{c \in C} k(x-c)$$

for some $\alpha > 0$ and some d-dimensional probability density function (pdf) k. Such a Cox process is also a Poisson cluster process. Note that in this case the cluster centers are not part of the Cox process. Given a realization of the cluster center process C, the cluster of points originating from $c \in C$ form a non-stationary Poisson process with intensity function $k(x-c), x \in \mathbb{R}^d$, independently of the other clusters. Drawing such a cluster via Algorithm 12.3 simply means that (1) the number of points N_c in the cluster has a $\text{Pois}(\alpha)$ distribution, and (2) these N_c points are independent and identically distributed according to the density function $k(x-c)$.

A common choice for k is $k(x) \propto \mathbf{1}_{\{\|x\| \leq r\}}$, first proposed in [262]. The resulting process is called a *Matérn process*. Thus, for a Matérn process each point in the cluster with center c is uniformly distributed within a ball of radius r at c. If instead a $N(c, \sigma^2 I)$ distribution is used, where I is the d-dimensional identity matrix, then the process is known as a (modified) *Thomas process* [287].

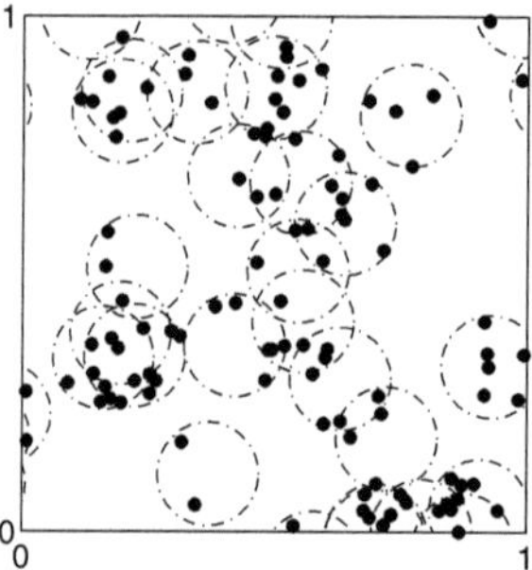

Fig. 12.9 A realization of a Matérn process with parameters $\kappa = 20$, $\alpha = 5$, and $r = 0.1$. The process extends beyond the shown window. The cluster centers (the centers of the circles) are not part of the point pattern

Fig. 12.9 shows a realization of a Matérn process with parameters $\kappa = 20$, $\alpha = 5$ and $r = 0.1$ on $[0, 1] \times [0, 1]$. Note that the cluster centers are assumed to lie on the whole of $\mathbb{R}^2$. To show a genuine outcome of the process within the window $[0,1] \times [0,1]$ it suffices to consider only the points that are generated from centers lying in square $[-r, 1+r] \times [-r, 1+r]$.

The following MATLAB program was used.

```
X = zeros(10^5,2);   %initialise the points
kappa = 20; alpha = 5;  r = 0.1; %parameters
meanpts=kappa*(1 + 2*r)^2;
N = poissrnd(meanpts);       %number of cluster centers
C = rand(N,2)*(1+2*r) - r;   ;%draw cluster centers
total_so_far = 0;
for c=1:N
    NC = poissrnd(alpha);   %number of points in cluster
    k = 0;
    while k < NC %draw uniformly in the n-ball via accept-reject
        Y = 2*r*rand(1,2) - r; %candidate point
        if norm(Y) < r
            X(total_so_far+k+1,:) = C(c,:) + Y;
            k = k+1;
        end
    end
    total_so_far = total_so_far + NC;
end
X = X(1:total_so_far,:);    %cut off unused rows
plot(X(:,1),X(:,2),'.')
axis([0, 1,0, 1])
```

A versatile generalization the Neyman–Scott process is the *shot-noise Cox process* [280], where the random intensity function is of the form

$$\Lambda(x) = \sum_{(c_j, \gamma_j) \in Z} \gamma_j \, k(c_j, x) \,,$$

and $\{(c_j, \gamma_j)\}$ are the points of an non-stationary Poisson process Z on $\mathbb{R}^d \times \mathbb{R}_+$ with intensity function ζ, and k is a *kernel function*; that is, $k(c, \cdot)$ is a probability density function (on $\mathbb{R}^d$) for each $c \in \mathbb{R}^d$. By Theorem 12.1, if there exists a function $K(c) : \mathbb{R}^d \to \mathbb{R}_+$ such that

$$K(c) = \int_0^\infty \zeta(c, \gamma) \, \mathrm{d}\gamma < \infty \,,$$

then $C = \{c : (c, \gamma) \in Z\}$ is a Poisson process with intensity function K, and Z is a marked Poisson process with mark density $\zeta(c, \cdot)/K(c)$. This decomposition of the shot-noise Cox process into a marked Poisson process suggests a method for simulation.

Algorithm 12.5 (Simulation of a shot-noise Cox process).
1. Draw the points C of an non-stationary Poisson process with intensity function $K(c)$.

2. For each $c_j \in C$, draw the corresponding mark γ_j from the density $\zeta(c_j,\cdot)/K(c_j)$.
3. For each $c_j \in C$, draw $N_j \sim \text{Pois}(\gamma_j)$.
4. Draw for each $c_j \in C$, N_j points from the kernel $k(c_j,x)$. The collection of all points drawn at this step constitutes a realization from the shot-noise Cox process.

A special case of the shot-noise Cox process, and one that appears frequently in the literature, is the *shot-noise Gamma Cox process*. Here the intensity function ζ is of the form

$$\zeta(c,\gamma) = \beta\gamma^{\alpha-1}\exp(-\lambda\gamma)/\Gamma(1+\alpha),$$

where $\beta > 0, \alpha > 0, \lambda > 0$, and

$$K(c) = \int_0^\infty \zeta(c,\gamma)\,d\gamma = \beta\lambda^{-\alpha}/\alpha < \infty.$$

Hence, Z is a marked Poisson process, where the intensity function of the centers is $K(c) = \beta\lambda^{-\alpha}/\alpha$ and the marks are independently and identically $Gamma(\alpha,\lambda)$ distributed; see [53] and [287] for more information.

12.3.5 Point Process Densities

Let X be a point process with mean measure μ on a bounded region $E \subset \mathbb{R}^d$. We assume that the expected total number of points $\mu(E)$ is finite. We may view X as a random object taking values in the space $\mathscr{X} = \cup_{n=0}^\infty(\{n\} \times E^n)$, where E^n is the Cartesian product of n copies of E. Note that in this representation the coordinates of X are ordered: $X = (N,(X_1,\ldots,X_N))$. We identify each vector (N,X) with X. If, for every set $\{n\} \times A$ with A a measurable set in E^n we can write

$$\mathbf{P}(X \in \{n\} \times A) = \int_A f(x_1,\ldots,x_n)\,dx_1\ldots dx_n,$$

then $f(x)$ is the probability density function or simply *density* of X on $\mathscr{X}$ (with respect to the Lebesgue measure on $\mathscr{X}$). Using the identification $(n,x) = x$, we can view $f(x)$ as the joint density of the random variable N and the N-dimensional vector X, where each component of X takes values in E. Using a Bayesian notation convention where all probability density functions and conditional probability density functions are indicated by the same symbol f, we have $f(x) = f(n,x) = f(n)f(x|n)$, where $f(n)$ is the (discrete) probability density function of the random number of points N, and $f(x|n)$ is the joint probability density function (pdf) of $X_1,\ldots,X_n$ given $N = n$. As an example, for the Poisson process on E with intensity function $\lambda(x)$ and mean measure μ we have, in correspondence to Algorithm 12.3,

$$f(x) = f(n)f(x|n) = \frac{e^{-\mu(E)}\{\mu(E)\}^n}{n!}\prod_{i=1}^n \frac{\lambda(x_i)}{\mu(E)} = \frac{e^{-\mu(E)}}{n(x)!}\prod_{i=1}^{n(x)}\lambda(x_i), \qquad (12.9)$$

where $n(x)$ is the number of components in x. Conversely, the expression for the pdf in (12.9) shows immediately how X can be generated; that is, via Algorithm 12.3.

A general recipe for generating a point process X is thus:

1. draw N from the discrete pdf $f(n)$;
2. given $N = n$, draw $(X_1,\ldots,X_n)$ from the conditional pdf $f(x\,|\,n)$.

Unfortunately, the pdf $f(n)$ and $f(x\,|\,n)$ may not be available explicitly. Sometimes $f(x)$ is known up to an unknown normalization constant. In such cases one case use *Markov Chain Monte Carlo* (MCMC) to simulate from $f(x)$. The basic idea of MCMC is to run a Markov chain long enough so that its limiting distribution is close to the target distribution. The most well-known MCMC algorithm is the following; see, for example, [334].

Algorithm 12.6. (Metropolis–Hastings algorithm)   Given a *transition density* $q(y\,|\,x)$, and starting from an initial state X_0, repeat the following steps for $t = 1, 2, \ldots$:
1. Generate a candidate $Y \sim q(y\,|\,X_t)$.
2. Generate $U \sim \mathrm{Unif}(0, 1)$ and set

$$X_{t+1} = \begin{cases} Y, & \text{if } U \leq \alpha(X_t, Y), \\ X_t, & \text{otherwise,} \end{cases} \tag{12.10}$$

where $\alpha(x, y)$ is the *acceptance probability*, given by:

$$\alpha(x, y) = \min\left\{ \frac{f(y)\,q(x\,|\,y)}{f(x)\,q(y\,|\,x)}, 1 \right\}. \tag{12.11}$$

This produces a sequence $X_1, X_2, \ldots$ of *dependent* random vectors, with X_t approximately distributed according to $f(x)$, for large t. Since Algorithm 12.6 is of the acceptance–rejection type, its efficiency depends on the acceptance probability $\alpha(x, y)$. Ideally, one would like the proposal transition density $q(y\,|\,x)$ to reproduce the desired pdf $f(y)$ as faithfully as possible. For a *random walk sampler* the proposal state Y, for a given current state x, is given by $Y = x + Z$, where Z is typically generated from some spherically symmetrical distribution. In that case the proposal transition density pdf is symmetric; that is $q(y\,|\,x) = q(x\,|\,y)$. It follows that the acceptance probability is:

$$\alpha(x, y) = \min\left\{ \frac{f(y)}{f(x)}, 1 \right\}. \tag{12.12}$$

As a specific example, suppose we wish to generate a *Strauss process* [222, 384]. This is a point process with density of the form

$$f(x) \propto \beta^{n(x)} \gamma^{s(x)} ,$$

where $\beta, \gamma \geq 0$ and $s(x)$ is the number of pairs of points where the two points are within distance r of each other. As before, $n(x)$ denotes the number of points. The process exists (that is, the normalization constant is finite) if $\gamma \leq 1$; otherwise, it does not exist in general [287].

We first consider simulating from $f(x|n)$ for a fixed n. Thus, $f(x|n) \propto \gamma^{s(x)}$, where $x = (x_1, \ldots, x_n)$. The following MATLAB program implements a Metropolis–Hastings algorithm for simulating a (conditional) Strauss process with $n = 200$ points on the unit square $[0,1] \times [0,1]$, using the parameter values $\gamma = 0.1$ and $r = 0.2$. Given a current state $x = (x_1, \ldots, x_n)$, the proposal state $Y = (Y_1, \ldots, Y_n)$ is identical to x except for the J-th component, where J is a uniformly drawn index from the set $\{1, \ldots, n\}$. Specifically, $Y_J = x_J + Z$, where $Z \sim N(0, (0.1)^2)$. The proposal Y for this random walk sampler is accepted with probability $\alpha(x, Y) = \min\{\gamma^{s(Y)}/\gamma^{s(x)}, 1\}$. The function $s(x)$ is implemented below as `numpairs.m`.

```matlab
gam = 0.1;
r = 0.2;
n = 200;
x = rand(n,2); %initial pp
K = 10000;
np= zeros(K,1);
for i=1:K
    J = ceil(n*rand);
    y = x;
    y(J,:) = y(J,:) + 0.1*randn(1,2); %proposal
    if (max(max(y)) > 1 || min(min(y)) <0)
        alpha =0; %don't accept a point outside the region
    elseif (numpairs(y,r) < numpairs(x,r))
        alpha =1;
    else
        alpha = gam^numpairs(y,r)/gam^numpairs(x,r);
    end
    R = (rand < alpha);
    x = R*y + (1-R)*x; %new x-value
    np(i) = numpairs(x,r);
    plot(x(:,1),x(:,2),'.');
    axis([0,1,0,1])
    refresh; pause(0.0001);
end
```

```matlab
function s = numpairs(x,r)
n = size(x,1);
D = zeros(n,n);
for i = 1:n
    D(i,:) = sqrt(sum((x(i*ones(n,1),:) - x).^2,2));
end
D = D + eye(n);
s = numel(find((D < r)))/2;
end
```

A typical realization of the conditional Strauss process is given in the left pane of Fig. 12.10. We see that the $n = 200$ points are clustered together in groups. This pattern is quite different from a typical realization of the unconditional Strauss process, depicted in the right pane of Fig. 12.10. Not only are there typically far fewer points, but also these points tend to "repel" each other, so that the number of pairs within a distance of r of each other is small. The radius of each circle in the figure is $r/2 = 0.1$. We see that in this case $s(x) = 3$, because 3 circle pairs overlap.

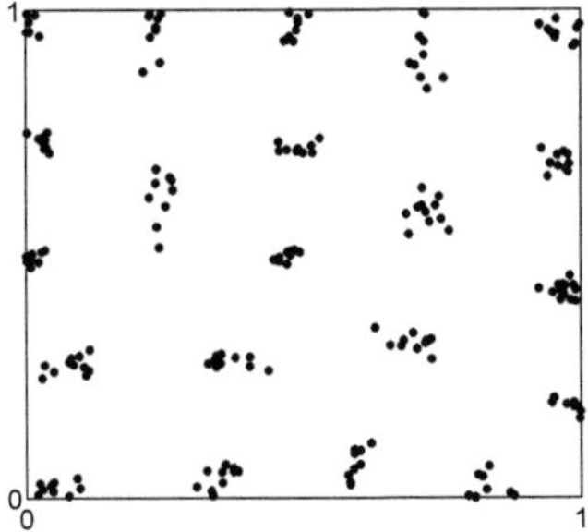
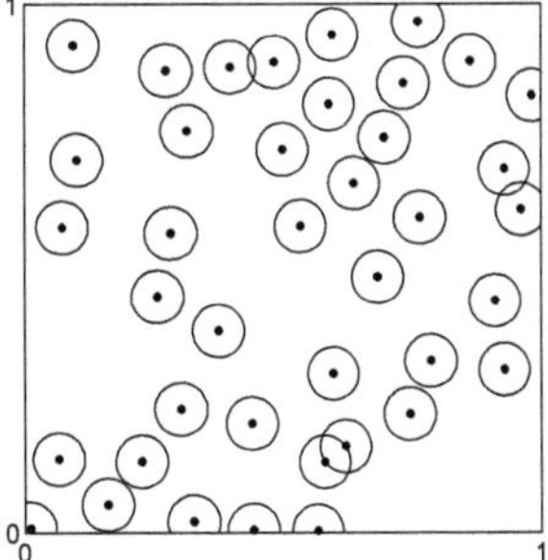

Fig. 12.10 Left: Conditional Strauss process with $n = 200$ points and parameters $\gamma = 0.1$ and $r = 0.2$. Right: Strauss process with $\beta = 100$, $\gamma = 0.1$, and $r = 0.2$

To generate the (unconditional) Strauss process using the Metropolis–Hastings sampler, the sampler needs to be able to "jump" between different dimensions n. The *reversible jump sampler* [150, 334] is an extension of the Metropolis–Hastings algorithm designed for this purpose. Instead of one transition density $q(y\,|\,\mathbf{x})$, it requires a transition density for n, say $q(n\,|\,m)$, and for each n a transition density $q(y\,|\,x,n)$ to jump from x to an n-dimensional vector y.

Given a current state X_t of dimension m, Step 1 of Algorithm 12.6 is replaced with

1a. Generate a candidate dimension $n \sim q(n\,|\,m)$.
1b. Generate an n-dimensional vector $Y \sim q(y\,|\,X_t,n)$.

And in Step 2 the acceptance ratio in (12.11) is replaced with

$$\alpha(x,y) = \min\left\{ \frac{f(n,y)\,q(m\,|\,n)\,q(x\,|\,y,m)}{f(m,x)\,q(n\,|\,m)\,q(y\,|\,x,n)}, 1 \right\}. \qquad (12.13)$$

The MATLAB program below implements a simple version of the reversible jump sampler, suggested in [136]. From a current state x of dimension m, a candidate dimension is chosen to be either $m+1$ or $m-1$, with equal probability. Thus, $q(m+1\,|\,m) = q(m-1\,|\,m) = 1/2, m = 1,2,\ldots$. The first transition corresponds to the birth of a new point; the second to the death of a point. On the occasion of a birth, a candidate Y is generated by simply appending a uniform point on $[0,1]^2$ to x. The corresponding transition density is therefore $q(y\,|\,x,m) = 1$. On the occasion of a death, the candidate Y is obtained from x by removing one of the points of x at random and uniformly. The transition density is thus $q(y\,|\,x,m) = 1/n(x)$. This gives the acceptance ratios:

1. Birth:

$$\alpha(x,y) = \min\left\{\frac{\beta^{n(y)}\gamma^{s(y)}\frac{1}{n(y)}}{\beta^{n(x)}\gamma^{s(x)}1}, 1\right\} = \min\{(\beta\gamma^{s(y)-s(x)})/n(y), 1\}.$$

2. Death:

$$\alpha(x,y) = \min\left\{\frac{\beta^{n(y)}\gamma^{s(y)}1}{\beta^{n(x)}\gamma^{s(x)}\frac{1}{n(x)}}, 1\right\} = \min\{(\gamma^{s(y)-s(x)}n(x))/\beta, 1\}.$$

```
r = 0.1; gam = 0.2; beta = 100; %parameters
n = 200; x = rand(n,2); %initial pp
K = 1000;
for i=1:K
    n = size(x,1);
    B = (rand < 0.5);
    if B %birth
        xnew = rand(1,2);
        y = [x;xnew];
        n = n+1;
    else %death
        y = setdiff(x,x(ceil(n*rand),:),'rows');
    end
    if (max(max(y)) > 1 || min(min(y)) <0)
        alpha =0; %don't accept a point outside the region
    elseif (numpairs(y,r) < numpairs(x,r))
        alpha =1;
    elseif B %birth
        alpha = beta*gam^(numpairs(y,r) - numpairs(x,r))/n;
    else  %death
        alpha = n*gam^(numpairs(y,r) - numpairs(x,r))/beta;
    end
    if (rand < alpha)
        x = y;
    end
    plot(x(:,1),x(:,2),'.');
    axis([0,1,0,1]); refresh; pause(0.0001);
```

For more on Strauss processes, see [401].

12.4 Wiener Surfaces

Brownian motion is one of the simplest continuous-time stochastic processes, and
as such has found myriad applications in the physical sciences [247]. As a first step
toward constructing Brownian motion we introduce the one-dimensional Wiener

process $\{W_t, t \in \mathbb{R}_+\}$, which can be viewed as a spatial process with a continuous index set on $\mathbb{R}_+$ and with a continuous state space $\mathbb{R}$.

12.4.1 Wiener Process

A one-dimensional Wiener process is a stochastic process $\{W_t, t \geq 0\}$ characterized by the following properties:

1. the increments of W_t are stationary and normally distributed, that is, $W_t - W_s \sim N(0, t-s)$ for all $t \geq s \geq 0$;
2. W has independent increments, that is, for any $t_1 < t_2 \leq t_3 < t_4$, the increments $W_{t_4} - W_{t_3}$ and $W_{t_2} - W_{t_1}$ are independent random variables (in other words, $W_t - W_s$, $t > s$ is independent of the past history of $\{W_u, 0 \leq u \leq s\}$);
3. continuity of paths, that is, $\{W_t\}$ has continuous paths with $W_0 = 0$.

The simplest algorithm for generating the process uses the Markovian (independent increments) and Gaussian properties of the Wiener process. Let $0 = t_0 < t_1 < t_2 < \cdots < t_n$ be the set of distinct times for which simulation of the process is desired. Then, the Wiener process is generated at times $t_1, \ldots, t_n$ via

$$W_{t_k} = \sum_{i=1}^{k} \sqrt{t_k - t_{k-1}}\, Z_i, \quad k = 1, \ldots, n\,,$$

where $Z_1, \ldots, Z_n \overset{\text{iid}}{\sim} N(0,1)$. To obtain a continuous path approximation to the path of the Wiener process, one could use linear interpolation on the points $W_{t_1}, \ldots, W_{t_n}$. A realization of a Wiener process is given in the middle panel of Fig. 12.11. Given the Wiener process W_t, we can now define the d-dimensional *Brownian motion process* via

$$\tilde{X}_t = \mu t + \Sigma^{1/2} W_t, \quad W_t = (W_t^{(1)}, \ldots, W_t^{(d)})^\mathsf{T}, \quad t \geq 0\,, \tag{12.14}$$

where $W_t^{(1)}, \ldots, W_t^{(d)}$ are independent Wiener processes and Σ is a $d \times d$ covariance matrix. The parameter $\mu \in \mathbb{R}^d$ is called the *drift* parameter and Σ is called the *diffusion matrix*.

Exercise 12.3. Generate and plot a realization of a three dimensional Wiener process at times $0, 1/n, 2/n, \ldots, 1$ for $n = 10^4$.

One approach to generalizing the Wiener process conceptually or to higher spatial dimensions is to use its characterization as a zero-mean Gaussian process (see Sect. 12.2.1) with continuous sample paths and covariance function $\mathbf{cov}(W_t, W_s) = \min\{t, s\}$ for $t, s > 0$. Since $\frac{1}{2}(|t| + |s| - |t-s|) = \min\{t, s\}$, we can consider the covariance function $\frac{1}{2}(|t| + |s| - |t-s|)$ as a basis for generalization. The first generalization is obtained by considering a continuous zero-mean Gaussian process with covariance $\rho(t, s) = \frac{1}{2}(|t|^\alpha + |s|^\alpha - |t-s|^\alpha)$, where α is a parameter such that

$\alpha = 1$ yields the Wiener process. This generalization gives rise to fractional Brownian motion discussed in the next section.

12.4.2 Fractional Brownian Motion

A continuous zero-mean Gaussian process $\{W_t, t \geq 0\}$ with covariance function

$$\mathbf{cov}(W_t, W_s) = \frac{1}{2}\left(|t|^\alpha + |s|^\alpha - |t-s|^\alpha\right), \quad t, s \geq 0 \tag{12.15}$$

is called *fractional Brownian motion* (fBm) with roughness parameter $\alpha \in (0,2)$. The process is frequently parameterized with respect to $H = \alpha/2$, in which case $H \in (0,1)$ is called the *Hurst* or *self-similarity* parameter. The notion of self-similarity arises, because fBm satisfies the property that the rescaled process $\{c^{-H}W_{ct}, t \geq 0\}$ has the same distribution as $\{W_t, t \geq 0\}$ for all $c > 0$.

Simulation of fBm on the uniformly spaced grid $0 = t_0 < t_1 < t_2 < \cdots < t_n = 1$ can be achieved by first generating the increment process $\{X_1, X_2, \ldots, X_n\}$, where $X_i = W_i - W_{i-1}$, and then delivering the cumulative sum

$$W_{t_i} = c^H \sum_{k=1}^{i} X_k, \quad i = 1, \ldots, n, \quad c = 1/n\,.$$

The increment process $\{X_1, X_2, \ldots, X_n\}$ is called *fractional Gaussian noise* and can be characterized as a discrete zero-mean stationary Gaussian process with covariance

$$\mathbf{cov}(X_i, X_{i+k}) = \frac{1}{2}\left(|k+1|^\alpha - 2|k|^\alpha + |k-1|^\alpha\right), \quad k = 0, 1, 2, \ldots$$

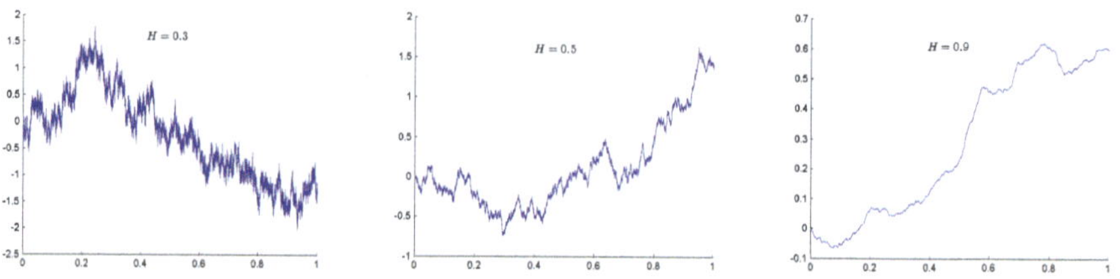

Fig. 12.11 Fractional Brownian motion for different values of the Hurst parameter H. From left to right we have $H = 0.3, 0.5, 0.9$

Since the fractional Gaussian noise is stationary, we can generate it efficiently using the circulant embedding approach in Sect. 12.2.2. First, we compute the first row $(r_1, \ldots, r_{n+1})$ of the symmetric Toeplitz $(n+1) \times (n+1)$ covariance matrix Ω with elements $\Omega_{i+1, j+1} = \mathbf{cov}(X_i, X_j)$, $i, j = 0, \ldots, n$. Second, we build the first row of the $2n \times 2n$ circulant matrix Σ, which embeds Ω in the upper left $(n+1) \times (n+1)$ corner. Thus, the first row of Σ is given by $r = (r_1, \ldots, r_{n+1}, r_n, r_{n-1}, \ldots, r_2)$. We now seek a factorization of the form (12.3). Here, λ is the one-dimensional FFT of r defined as the linear transformation $\lambda = Fr$ with $F_{j,k} = \exp(-2\pi i jk/(2n))/\sqrt{2n}$, $j, k = 0, 1, \ldots, 2n - 1$. Finally, the real and imaginary parts of the first $n + 1$ components of $F^* \mathrm{diag}(\sqrt{\lambda})Z$, where Z is a $2n \times 1$ complex valued Gaussian vector, yield two independent realizations of fractional Brownian noise. Fig. 12.11 shows how the smoothness of fBm depends on the Hurst parameter. Note that $H = 0.5$ corresponds to the Wiener process.

The following MATLAB code implements the circulant embedding method for fBm.

```
n=2^15; % grid points
H = 0.9; %Hurst parameter
r=nan(n+1,1); r(1) = 1;
for k=1:n
    r(k+1) = 0.5*((k+1)^(2*H) - 2*k^(2*H) + (k-1)^(2*H));
end
r=[r; r(end-1:-1:2)]; % first rwo of circulant matrix
lambda=real(fft(r))/(2*n); % eigenvalues
W=fft(sqrt(lambda).*complex(randn(2*n,1),randn(2*n,1)));
W = n^(-H)*cumsum(real(W(1:n+1))); % rescale
plot((0:n)/n,W);
```

12.4.3 Fractional Wiener Sheet in $\mathbb{R}^2$

A simple spatial generalization of the fractional Brownian motion is the fractional Wiener sheet in two dimensions. The *fractional Wiener sheet* process on the unit square is the continuous zero-mean Gaussian process $\{W_t, t \in [0,1]^2\}$ with covariance function

$$\mathbf{cov}(W_t, W_s) = \frac{1}{4}(|s_1|^\alpha + |t_1|^\alpha - |s_1 - t_1|^\alpha)(|s_2|^\alpha + |t_2|^\alpha - |s_2 - t_2|^\alpha), \quad (12.16)$$

where $t = (t_1, t_2)$ and $s = (s_1, s_2)$. Note that (12.16) is simply a product form extension of (12.15).

As in the one-dimensional case of fBm, we can consider the *two-dimensional fractional Gaussian noise* process $\{X_{i,j}, i, j = 1, \ldots, n\}$, which can be used to construct a fractional Wiener sheet on a uniformly spaced square grid via the cumulative sum

$$W_{t_i, t_j} = n^{-2H} \sum_{k=1}^{i} \sum_{l=1}^{j} X_{k,l}, \quad i, j = 1, \ldots, n.$$

Note that this process is self-similar in the sense that $(mn)^{-H} \sum_{k=1}^{m} \sum_{l=1}^{n} X_{k,l}$ has the same distribution as $X_{1,1}$ for all m, n, and H, see [260].

Generating a the two-dimensional fractional Gaussian noise process requires that we generate a zero-mean stationary Gaussian process with covariance [324]

$$\mathbf{cov}(X_{i,j}, X_{i+k,j+l}) = \frac{|k+1|^\alpha - 2|k|^\alpha + |k-1|^\alpha}{2} \times \frac{|l+1|^\alpha - 2|l|^\alpha + |l-1|^\alpha}{2}$$

for $k, l = 0, 1, \ldots, n$. We can thus proceed to generate this process using the circulant embedding method in Sect. 12.2.2. The simulation of the Wiener sheet for $H = 0.5$ is particularly easy since then all of the $X_{i,j}$ are independent standard normally distributed. Fig. 12.12 shows realizations of fractional Wiener sheets for $H = \frac{1}{2}\alpha \in \{0.2, 0.5, 0.8\}$ with $n = 2^9$.

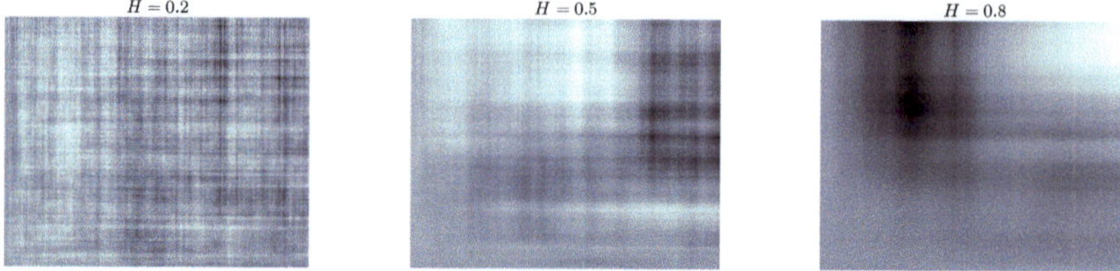

Fig. 12.12 Fractional Wiener sheets with different Hurst parameter H

Exercise 12.4. Show that for the special case of $\alpha = 1$ (that is, $H = 1/2$), we can write the covariance function (12.16) as $\mathbf{cov}(W_t, W_s) = \min\{t_1, s_1\}\min\{t_2, s_2\}$.

Two objects which are closely related to W_t are the *Wiener pillow* and the *Wiener bridge*, which are zero-mean Gaussian processes on $[0,1]^d$ with covariance functions $\mathbf{cov}(W_t, W_s) = \prod_{i=1}^d (\min(t_i, s_i) - t_i s_i)$ and $\prod_{i=1}^d \min(t_i, s_i) - \prod_{i=1}^d t_i s_i$, respectively.

12.4.4 Fractional Brownian Field

Fractional Brownian surface or field in two dimensions can be defined as the zero-mean Gaussian process $\{\tilde{X}_t, t \in \mathbb{R}^2\}$ with nonstationary covariance function

$$\mathbf{cov}(\tilde{X}_s, \tilde{X}_t) = \tilde{\rho}(s,t) = \|s\|^\alpha + \|t\|^\alpha - \|s-t\|^\alpha. \tag{12.17}$$

The parameter $H = \frac{\alpha}{2} \in (0,1)$ is the Hurst parameter controlling the roughness of the random field or surface. Contrast this isotropic generalization of the covariance (12.15) with the product form extension of the covariance in (12.16).

Simulation of $\tilde{X}_t$ over a unit (quarter) disk in the first quadrant involves the following steps. First, we use Dietrich and Newsam's method (Sect. 12.2.2) to generate a stationary Gaussian field $\check{X}_t$ with covariance function over the quarter disk

$\{h : \|h\| \leq 1, h > 0\}$

$$\breve{\rho}(s,t) = c_0 + c_2\|s-t\|^2 - \|s-t\|^\alpha$$

for some constants $c_0, c_2 \geq 0$ whose selection will be discussed later. Once we have generated $\breve{X}_t$, the process $\tilde{X}_t$ is obtained via the adjustment:

$$\tilde{X}_t = \breve{X}_t - \breve{X}_0 + \sqrt{2c_2}\, t^\mathsf{T} Z, \quad Z = (Z_1, Z_2)^\mathsf{T}, \quad Z_1, Z_2 \overset{\text{iid}}{\sim} N(0,1)\,.$$

This adjustment ensures that the covariance structure of $\tilde{X}_t$ over the disk $\{h : \|h\| \leq 1, h > 0\}$ is given by (12.17).

Exercise 12.5. Show that that the covariance structure of $\tilde{X}_t$ is indeed given by $\tilde{\rho}(s,t)$, i.e., prove that for any $s,t \in \mathbb{R}^2$

$$\mathbf{cov}(\tilde{X}_s, \tilde{X}_t) = \tilde{\rho}(s,t)\,.$$

Hint. Verify that

$$\begin{aligned}
\mathbf{cov}(\tilde{X}_s, \tilde{X}_t) &= \mathbf{cov}(\breve{X}_s - \breve{X}_0 + \sqrt{2c_2}\, s^\mathsf{T} Z, \breve{X}_t - \breve{X}_0 + \sqrt{2c_2}\, t^\mathsf{T} Z) \\
&= \breve{\rho}(s,t) - \breve{\rho}(s,0) - \breve{\rho}(0,t) + \breve{\rho}(0,0) + 2c_2\,\mathbf{cov}(s^\mathsf{T} Z, t^\mathsf{T} Z) \\
&= \breve{\rho}(s,t) - (c_0 + c_2\|t\|^2 - \|s\|^\alpha) - (c_0 + c_2\|t\|^2 - \|t\|^\alpha) + c_0 + 2c_2\, s^\mathsf{T} t \\
&= \tilde{\rho}(s,t) + \underbrace{c_2\|s-t\|^2 - c_2\|t\|^2 - c_2\|s\|^2 + 2c_2\, s^\mathsf{T} t}_{=0}\,.
\end{aligned}$$

It now remains to explain how we generate the process $\breve{X}_t$. The idea is to generate the process on $[0,R]^2$, $R \geq 1$ via the intrinsic embedding of Stein (see (12.5)) using the covariance function:

$$\psi(h) = \begin{cases} c_0 + c_2\|h\|^2 - \|h\|^\alpha, & \text{if } \|h\| \leq 1, \\ \frac{\beta(R-\|h\|)^3}{\|h\|}, & \text{if } 1 \leq \|h\| \leq R,, \\ 0, & \text{if } \|h\| \geq R, \end{cases} \qquad (12.18)$$

where depending on the value of α, the constants $R \geq 1, \beta \geq 0, c_2 > 0, c_0 \geq 0$ are defined in Table 12.1.

Note that for $\alpha > 1.5$, the parameters needed for a nonnegative embedding are more complex, because a covariance function that is smoother close to the origin has to be even smoother elsewhere. In particular, for $\alpha > 1.5$ the choice of constants ensures that ψ is twice continuously differentiable as a function of $\|h\|$. Notice that while we generate the process $\tilde{X}_t$ over the square grid $[0,R]^2$, we are only interested in $\tilde{X}_t$ restricted inside the quarter disk with covariance $\breve{\rho}(s,t)$. Thus, in order to reduce the computational effort, we would like to have $R \geq 1$ as close as possible to 1. While the optimal choice $R = 1$ guarantees a nonnegative embedding for all $\alpha \leq 1.5$, in general we need $R > 1$ to ensure the existence of a minimal embedding for $\alpha > 1.5$. The choice $R = 2$ given in Table 12.1 is the most conservative one that

Table 12.1 Parameter values needed to ensure that (12.18) allows for a nonnegative circulant embedding.

	$0 < \alpha \leq 1.5$	$1.5 < \alpha < 2$
R	1	2
β	0	$\frac{\alpha(2-\alpha)}{3R(R^2-1)}$
c_2	$\frac{1}{2}\alpha$	$\frac{\alpha-\beta(R-1)^2(R+2)}{2}$
c_0	$1-c_2$	$\beta(R-1)^3+1-c_2$

guarantees a nonnegative circulant embedding for $\alpha > 1.5$. Smaller values of R that admit a nonnegative circulant embedding can be determined numerically [377, Table 1]. As a numerical example consider generating a fractional Brownian surface with $m = n = 1000$ and for Hurst parameter $H \in (0.2, 0.5, 0.8)$. Fig. 12.13 shows the effect of the parameter on the smoothness of the surface, with larger values providing a smoother surface. We used the following MATLAB code for the generation

Fig. 12.13 Fractional Brownian fields with different roughness parameter $\alpha = 2H$

of the surfaces.

```
H=0.8;          % Hurst parameter
R=2;            % [0,R]^2 grid, may have to extract only  [0,R/2]^2
n=1000; m=n; % size of grid is m*n; covariance matrix is m^2*n^2
tx=[1:n]/n*R; ty=[1:m]/m*R; % create grid for field
Rows=zeros(m,n);
for i=1:n
    for j=1:m % rows of blocks of cov matrix
        Rows(j,i)=rho([tx(i),ty(j)],[tx(1),ty(1)],R,2*H);
    end
end
BlkCirc_row=[Rows, Rows(:,end-1:-1:2);
    Rows(end-1:-1:2,:), Rows(end-1:-1:2,end-1:-1:2)];
% compute eigen-values
lam=real(fft2(BlkCirc_row))/(4*(m-1)*(n-1));
lam=sqrt(lam);
% generate field with covariance given by block circulant matrix
Z=complex(randn(2*(m-1),2*(n-1)),randn(2*(m-1),2*(n-1)));
```

```
F=fft2(lam.*Z);
F=F(1:m,1:n); % extract sub-block with desired covariance
[out,c0,c2]=rho([0,0],[0,0],R,2*H);
field1=real(F); field2=imag(F); % two independent fields
field1=field1-field1(1,1); % set field zero at origin
% make correction for embedding with a term c2*r^2
field1=field1 + kron(ty'*randn,tx*randn)*sqrt(2*c2);
[X,Y]=meshgrid(tx,ty);
field1((X.^2+Y.^2)>1)=nan;
surf(tx(1:n/2),ty(1:m/2),field1(1:n/2,1:m/2),'EdgeColor','none')
colormap bone
```

The code uses the function `rho.m`, which implements the embedding (12.18).

```
function [out,c0,c2]=rho(x,y,R,alpha)
% embedding of covariance function on a [0,R]^2 grid
if alpha<=1.5 % alpha=2*H, where H is the Hurst parameter
    beta=0;c2=alpha/2;c0=1-alpha/2;
else % parameters ensure piecewise function twice differentiable
    beta=alpha*(2-alpha)/(3*R*(R^2-1));
    c2=(alpha-beta*(R-1)^2*(R+2))/2;
    c0=beta*(R-1)^3+1-c2;
end
% create continuous isotropic function
r=sqrt((x(1)-y(1))^2+(x(2)-y(2))^2);
if r<=1
    out=c0-r^alpha+c2*r^2;
elseif r<=R
    out=beta*(R-r)^3/r;
else
    out=0;
end
```

12.5 Spatial Levy Processes

Recall from Sect. 12.4.1 that the Brownian motion process (12.14) can be characterized as a continuous sample path process with stationary and independent Gaussian
increments. The Lévy process is one of the simplest generalizations of the Brownian
motion process, in cases where either the assumption of normality of the increments,
or the continuity of the sample path is not suitable for modeling purposes.

12.5.1 Lévy Process

A d-dimensional *Lévy process* $\{X_t, t \in \mathbb{R}_+\}$ with $X_0 = 0$ is a stochastic process with a continuous index set on $\mathbb{R}_+$ and with a continuous state space $\mathbb{R}^d$ defined by the following properties:

1. the increments of $\{X_t\}$ are stationary, that is, $(X_{t+s} - X_t)$ has the same distribution as X_s for all $t, s \geq 0$;
2. the increments of $\{X_t\}$ are independent, that is, $X_{t_i} - X_{t_{i-1}}$, $i = 1, 2, \dots$ are independent for any $0 \leq t_0 < t_1 < t_2 < \cdots$; and
3. for any $\varepsilon > 0$, we have $\mathbf{P}(\|X_{t+s} - X_t\| \geq \varepsilon) = 0$ as $s \downarrow 0$.

From the definition it is clear that Brownian motion (12.14) is an example of a Lévy process, with normally distributed increments. Brownian motion is the only Lévy process with continuous sample paths. Other basic examples of Lévy processes include the Poisson process $\{N_t, t \geq 0\}$ with intensity $\lambda > 0$, where $N_t \sim \mathrm{Pois}(\lambda t)$ for each t, and the *compound Poisson process* defined via

$$J_t = \sum_{k=1}^{N_t} \delta X_k, \quad N_t \sim \mathrm{Pois}(\lambda t), \tag{12.19}$$

where $\delta X_1, \delta X_2, \dots$ are independent and identically distributed random variables, independent of $\{N_t, t \geq 0\}$. We can more generally express J_t as $J_t = \int_0^t \int_{\mathbb{R}^d} x N(\mathrm{d}s, \mathrm{d}x)$, where $N(\mathrm{d}s, \mathrm{d}x)$ is a Poisson random counting measure on $\mathbb{R}_+ \times \mathbb{R}^d$ (see Sect. 12.3.1) with mean measure $\mathbf{E}N([0,t] \times A)$, $A \in \mathcal{E}$, equal to the expected number of jumps of size A in the interval $[0,t]$.

A crucial property of Lévy processes is infinite divisibility. In particular, if we define $Y_j^{(n)} = X_{jt/n} - X_{(j-1)t/n}$, then using the stationarity and independence properties of the Lévy process, we obtain that for each $n \geq 2$ the $\{Y_1^{(n)}\}$ are independent and identically distributed random variables with the same distribution as $X_{t/n}$. Thus, for a fixed t we can write

$$X_t \sim Y_1^{(n)} + \cdots + Y_n^{(n)}, \quad \text{for any } n \geq 2, \tag{12.20}$$

and hence by definition the random vector X_t is *infinitely divisible* (for a fixed t). The Lévy–Khintchine theorem [338] gives the most general form of the characteristic function of an infinitely divisible random variable. Specifically, the logarithm of the characteristic function of X_t (the so-called *characteristic exponent*) is of the form

$$\log \mathbf{E}\mathrm{e}^{\mathrm{i}s^\mathsf{T} X_t} = \mathrm{i}\,t\,s^\mathsf{T}\mu - \frac{1}{2}t\,s^\mathsf{T}\Sigma s + t \int_{\mathbb{R}^d} \left(\mathrm{e}^{\mathrm{i}s^\mathsf{T}x} - 1 - \mathrm{i}s^\mathsf{T}x\,\mathbf{1}_{\{\|x\|\leq 1\}} \right) \nu(\mathrm{d}x), \tag{12.21}$$

for some $\mu \in \mathbb{R}^d$, covariance matrix Σ and measure ν such that $\nu(\{0\}) = 0$,

$$\int_{\|x\|>1} \nu(\mathrm{d}x) < \infty \ \text{ and } \ \int_{\|x\|\leq 1} \|x\|^2 \nu(\mathrm{d}x) < \infty \ \Leftrightarrow \ \int_{\mathbb{R}^d} \min\{1, \|x\|^2\}\nu(\mathrm{d}x) < \infty. \tag{12.22}$$

The triplet (μ, Σ, ν) is referred to as the *characteristic triplet* defining the Lévy process. The measure ν is referred to as the *Lévy measure*. Note that for a general ν satisfying (12.22), the integral $\int_{\mathbb{R}^d} \left(e^{is^\top x} - 1\right) \nu(dx)$ in (12.21) does not converge separately. In this sense $is^\top x \mathbf{1}_{\{\|x\| \leq 1\}}$ in (12.21) serves the purpose of enforcing convergence under the very general integrability condition (12.22). However, if in addition to (12.22) the measure ν satisfies $\int_{\mathbb{R}^d} \min\{1, \|x\|\} \nu(dx) < \infty$ and $\nu(\mathbb{R}^d) < \infty$, then the integral in (12.21) can be separated as $t \int_{\mathbb{R}^d} (e^{is^\top x} - 1) \nu(dx) - it\, s^\top \int_{\|x\| \leq 1} x\, \nu(dx)$, and the characteristic exponent simplifies to

$$
\underbrace{it\, s^\top \mu^* - \frac{1}{2} t\, s^\top \Sigma\, s}_{\text{Brownian motion term}} + \underbrace{t \int_{\mathbb{R}^d} \left(e^{is^\top x} - 1\right) \nu(dx)}_{\text{Poisson process term}},
$$

where $\mu^* = \mu - \int_{\|x\| \leq 1} x\, \nu(dx)$. We can now recognize this characteristic exponent as the one corresponding to the process $\{X_t\}$ defined by $X_t = t\mu^* + \Sigma^{1/2} W_t + J_t$, where $t\mu^* + \Sigma^{1/2} W_t$ defines a Brownian motion (see (12.14)) and $\{J_t\}$ is a compound Poisson process (12.19) with jump size distribution $\delta X_1 \sim \nu(dx)/\lambda$. Thus, $\nu(dx)$ can be interpreted as the intensity function of the jump sizes in this particular Lévy process. In a similar way, it can be shown that the most general Lévy process $\{X_t\}$ with characteristic triplet (μ, Σ, ν) and integrability condition (12.22) can be represented as the limit in probability of a process $\{X_t^{(\varepsilon)}\}$ as $\varepsilon \downarrow 0$, where $X_t^{(\varepsilon)}$ has the *Lévy–Itô decomposition*

$$
X_t^{(\varepsilon)} = t\mu + \Sigma^{1/2} W_t + J_t + \left(J_t^{(\varepsilon)} - t \int_{\varepsilon < \|x\| \leq 1} x\, \nu(dx) \right) \tag{12.23}
$$

with the following *independent* components:

1. $\{t\mu + \Sigma^{1/2} W_t\}$ is the Brownian motion (12.14), which corresponds to the $it\, s^\top \mu - \frac{1}{2} t\, s^\top \Sigma\, s$ part of (12.21).
2. $\{J_t\}$ is a compound Poisson process of the form (12.19) with $\lambda = \int_{\|x\| > 1} \nu(dx)$ and increment distribution $\delta X_1 \sim \nu(dx)/\lambda$ over $\|x\| > 1$, which corresponds to the $\int_0^t \int_{\|x\| > 1} (e^{is^\top x} - 1) \nu(dx)\, dt$ part in the characteristic exponent (12.21).
3. $\{J_t^{(\varepsilon)}\}$ is a compound Poisson process with $\lambda = \nu(\varepsilon < \|x\| \leq 1)$ and increment distribution $\nu(dx)/\lambda$ over $\varepsilon < \|x\| \leq 1$, so that the *compensated* compound Poisson process $\{J_t^{(\varepsilon)} - t \int_{\varepsilon < \|x\| \leq 1} x\, \nu(dx)\}$ corresponds to the $\int_0^t \int_{\|x\| \leq 1} \left(e^{is^\top x} - 1 - is^\top x\right) \nu(dx)\, dt$ part of (12.21) in the limit $\varepsilon \downarrow 0$.

The Lévy–1tô decomposition immediately suggests an approximate generation method — we generate $\{X_t^{(\varepsilon)}\}$ in (12.23) for a given small ε, where the Brownian motion part is generated via the methods in Sect. 12.4.1 and the compound Poisson process (12.19) in the obvious way. We are thus throwing away the very small jumps of size less than ε. For $d = 1$ it can then be shown [8] that the error process $\{X_t - X_t^{(\varepsilon)}\}$ for this approximation is a Lévy process with characteristic triplet $(0,0,\nu(dx)\mathbf{1}_{\{|x|<\varepsilon\}})$ and variance $\mathbf{var}(X_t - X_t^{(\varepsilon)}) = t\int_{-\varepsilon}^{\varepsilon} x^2\nu(dx)$.

A given approximation $X_t^{(\varepsilon_n)}$ can always be further refined by adding smaller jumps of size $[\varepsilon_{n+1}, \varepsilon_n]$ to obtain:

$$X_t^{(\varepsilon_{n+1})} = X_t^{(\varepsilon_n)} + J_t^{(\varepsilon_{n+1})} - t\int_{\varepsilon_{n+1}<\|x\|\leq\varepsilon_n} x\nu(dx)\,, \quad \varepsilon_n > \varepsilon_{n+1} > 0\,,$$

where the compound Poisson process $\{J_t^{(\varepsilon_{n+1})}\}$ has increment distribution given by $\nu(dx)/\nu(\varepsilon_{n+1} < \|x\| \leq \varepsilon_n)$ for all $\varepsilon_{n+1} < \|x\| \leq \varepsilon_n$. For a more sophisticated method of refining the approximation, see [8].

As an example consider the Lévy process $\{X_t\}$ with characteristic triplet $(\mu,0,\nu)$, where $\nu(dx) = \alpha e^{-x}/x\,dx$ for $\alpha, x > 0$ and $\mu = \int_{|x|\leq 1} x\nu(dx) = \alpha(1 - e^{-1})$. Here, in addition to (12.22), the infinite measure ν satisfies the stronger integrability condition $\int_{\mathbb{R}} \min\{1, |x|\}\nu(dx) < \infty$ and hence we can write $X_t = t\mu - t\int_{|x|\leq 1} x\nu(dx) + \int_0^t \int_{\mathbb{R}} xN(ds, dx) = \int_0^t \int_{\mathbb{R}_+} xN(ds, dx)$, where $\mathbf{E}N(ds, dx) = ds\,\nu(dx)$.

The leftmost panel of Fig. 12.14 shows an outcome of the approximation $X_t^{(\varepsilon_1)}$ in (12.23) over $t \in [0,1]$ for $\varepsilon_1 = 1$ and $\alpha = 10$. The middle and rightmost panels show the refinements $X_t^{(\varepsilon_2)}$ and $X_t^{(\varepsilon_3)}$, respectively, where $(\varepsilon_1, \varepsilon_2, \varepsilon_3) = (1, 0.1, 0.001)$. Note that the refinements add finer and finer jumps to the path.

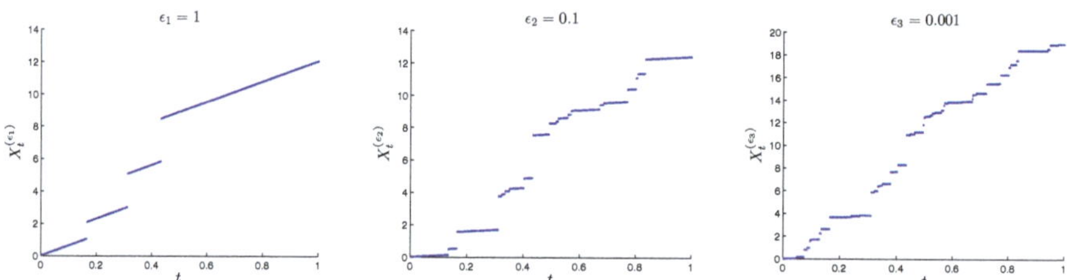

Fig. 12.14 Approximate gamma process realizations obtained by throwing away jumps smaller than ε_i

This process is called a *gamma process* and can be generated more simply using the fact that the increments $\{X_{jt/n} - X_{(j-1)t/n}\}$ have a known gamma distribution [237, p. 212].

Exercise 12.6. Show that the characteristic exponent corresponding to the triplet $(\mu, 0, \nu)$ of the above gamma process is given by

$$\log \mathbf{E}e^{isX_t} = \int_0^\infty \left(e^{isx} - 1\right)\frac{\alpha e^{-x}}{x}\,dx = -\alpha\log(1 - is)\,.$$

Deduce from the characteristic exponent that any increment $X_{t+s} - X_t \sim \Gamma(\alpha s, 1)$. Here $\Gamma(\alpha, \beta)$ denotes the gamma distribution, with probability density function $\beta^\alpha x^{\alpha-1}e^{-\beta x}/\Gamma(\alpha), x \geq 0$.

The gamma process is an example of an increasing Lévy process (that is, $X_t \geq X_s$ almost surely for all $t \geq s$) called a *Lévy subordinator*. A Lévy subordinator $\{X_t, t \geq 0\}$ has characteristic triplet $(\mu, 0, \nu)$ satisfying the positive jump property $\nu((-\infty, 0]) = 0$ and the positive drift property $0 \leq \mu - \int_0^1 x\nu(dx) < \infty$.

12.5.2 Lévy Sheet in $\mathbb{R}^2$

One way of generalizing the Lévy process to the spatial case is to insist on the preservation of the infinite divisibility property. In this generalization, a *spatial Lévy process* or simply a *Lévy sheet* $\{X_t, t \in \mathbb{R}^2\}$ possesses the property that $(X_{t_1}, \ldots, X_{t_n})$ is infinitely divisible for all indices $t_1, \ldots, t_n \in \mathbb{R}^2$ and any integer n (see (12.20)). To construct such an infinitely divisible process, consider stochastic integration with respect to a *random infinitely divisible measure* Λ defined as the stochastic process $\{\Lambda(A), A \in \mathcal{E}\}$ with the following properties:

1. For any set $A \in \mathcal{E}$ the random variable $\Lambda(A)$ has an infinitely divisible distribution with $\Lambda(\emptyset) = 0$ almost surely.
2. For any disjoint sets $A_1, A_2, \ldots \in \mathcal{E}$, the random variables $\Lambda(A_1), \Lambda(A_2), \ldots$ are independent and $\Lambda(\cup_i A_i) = \sum_i \Lambda(A_i)$.

An example of a random infinitely divisible measure is the Poisson random measure (12.6) in Sect. 12.3.1. As a consequence of the independence and infinite divisibility properties, the characteristic exponent of $\Lambda(A)$, $A \in \mathcal{E}$ has the Lévy–Khintchine representation (12.21):

$$\log \mathbf{E}\mathrm{e}^{\mathrm{i}s\Lambda(A)} = \mathrm{i}s\tilde{\mu}(A) - \frac{1}{2}s^2(\tilde{\sigma}(A))^2 + \int_{\mathbb{R}} \left(\mathrm{e}^{\mathrm{i}sx} - 1 - \mathrm{i}sx\mathbf{1}_{\{|x|\leq 1\}} \right) \tilde{\nu}(dx, A) \,,$$

where $\tilde{\mu}$ is an additive set function, $\tilde{\sigma}$ is a measure on the Borel sets $\mathcal{E}$, the measure $\tilde{\nu}(\cdot, A)$ is a Lévy measure for each fixed $A \in \mathcal{E}$ (so that $\tilde{\nu}(\{0\}, A) = 0$ and $\int_{\mathbb{R}} \min\{1, x^2\}\tilde{\nu}(dx, A) < \infty$), and $\tilde{\nu}(dx, \cdot)$ is a Borel measure for each fixed dx. For example, $X_t = \Lambda((0, t])$ defines a one-dimensional Lévy process.

We can then construct a Lévy sheet $\{X_t, t \in \mathbb{R}^2\}$ via the stochastic integral

$$X_t = \int_{\mathbb{R}^d} \kappa_t(x)\Lambda(dx), \quad t \in \mathbb{R}^2 \,, \tag{12.24}$$

where $\kappa_t : \mathbb{R}^d \to \mathbb{R}$ is a Hölder continuous *kernel function* for all $t \in \mathbb{R}^2$, which is integrable with respect to the random infinitely divisible measure Λ. Thus, the Lévy sheet (12.24) is a stochastic integral with a deterministic kernel function as integrand (determining the spatial structure) and a random infinitely divisible measure as integrator [220].

Consider simulating the Lévy sheet (12.24) over $t \in [0, 1]^2$ for $d = 2$. Truncate the region of integration to a bounded domain, say $[0, 1]^2$, so that $\kappa_t(x) = 0$ for each $x \notin [0, 1]^2$ and $t \in [0, 1]^2$. Then, one way of simulating $\{X_t, t \in [0, 1]^2\}$ is to consider the approximation

$$
X_t^{(n)} = \sum_{i=0}^{n-1} \sum_{j=0}^{n-1} \kappa_t(i/n, j/n)\, \Lambda(\triangle_{i,j}), \quad \triangle_{i,j} \equiv \left[\frac{i}{n}, \frac{i+1}{n}\right] \times \left[\frac{j}{n}, \frac{j+1}{n}\right], \quad (12.25)
$$

where all $\Lambda(\triangle_{i,j})$ are independent infinitely divisible random variables with characteristic triplet $(\tilde{\mu}(\triangle_{i,j}), \tilde{\sigma}(\triangle_{i,j}), \tilde{\nu}(\cdot, \triangle_{i,j}))$. Under some technical conditions [220], it can be shown that $X_t^{(n)}$ converges to X_t in probability as $n \uparrow \infty$.

As an example, consider generating (12.25) on the square grid $\{(i/m, j/m), i, j = 0, \ldots, m-1\}$ with the kernel function

$$
\kappa_t(x_1, x_2) = (r^2 - \|x - t\|^2)\, \mathbf{1}_{\{\|x-t\| \le r\}}, \quad t \in [0,1]^2 ,
$$

and $\Lambda(\triangle_{i,j}) \sim \Gamma(\alpha|\triangle_{i,j}|, \beta)$, $|\triangle_{i,j}| = 1/n^2$ for all i, j. The corresponding limiting process $\{X_t\}$ is called a *gamma Lévy sheet*. Fig. 12.15 shows realizations of (12.25) for $m = n = 100$ and $r = 0.05$, $\alpha = \beta \in \{10^2, 10^5\}$, so that we have the scaling $\mathbf{E}\Lambda(\triangle_{i,j}) = \alpha/(\beta n^2) = 1/n^2$.

Fig. 12.15 Gamma Lévy random sheet realizations for different values of the shape parameter α

Note that the sheet exhibits more bumps for smaller values of α than for large values of α. For a method of approximately generating (12.24) using wavelets see [220].

Acknowledgements This work was supported by the Australian Research Council under grant number DP0985177.

Chapter 13
Introduction to Coupling-from-the-Past using R

Wilfrid S. Kendall

Abstract The purpose of this chapter is to exemplify construction of selected coupling-from-the-past algorithms, using simple examples and discussing code which can be run in the statistical scripting language R. The simple examples are: symmetric random walk with two reflecting boundaries, a very basic continuous state-space Markov chain, the Ising model with external field, and random walk with negative drift and a reflecting boundary at the origin. In parallel with this, a discussion is given of the relationship between coupling-from-the-past algorithms on the one hand, and uniform and geometric ergodicity on the other.

13.1 Introduction

Propp and Wilson's coupling-from-the-past algorithm (CFTP) [320, 321] and its generalisations are based on simple but delicately balanced ideas. The purpose of this chapter is to give a careful discussion of exactly how some simple examples of CFTP can be implemented in the popular and flexible statistical language R [323]. Of course R is a scripting language and therefore is not ideal for implementing simulation algorithms; serious work should use compiler-based languages or at least scripting languages with substantial support for numerics (for example, *Python* with the *Numpy* extension, or *Matlab*). However R has several advantages if one wishes to demonstrate precisely what is going on without worrying much about efficiency; in particular R is not only free and open-source but also widely popular within the statistical community. Moreover, by its very nature R allows direct access to statistical procedures, making it easier to check statistical correctness. In addition, the algorithms for CFTP are most clearly communicated if the reader can actually run, test, and vary them within a suitable stable computing environment of this kind. To

Wilfrid S. Kendall
Department of Statistics, University of Warwick, Coventry CV4 7AL, UK, e-mail: `w.s.kendall@warwick.ac.uk`

© Springer International Publishing Switzerland 2015
V. Schmidt (ed.), *Stochastic Geometry, Spatial Statistics and Random Fields*,
Lecture Notes in Mathematics 2120, DOI 10.1007/978-3-319-10064-7_13

facilitate this, the chapter was written using the R report-generation utility software `Sweave` [248], so as to ensure that the R code presented here is identical to the code used to generate the reported results and to test algorithms (though graphics-related code is largely suppressed). Actual R code is presented in boxed displays (sometimes continued over consecutive pages); occasional R output is presented in similar boxes distinguished by their shaded edges.

After this introductory section, the chapter commences (Sect. 13.2) with a discussion of the classic Propp-Wilson version of CFTP, and then demonstrates its use in a (rather inefficient) implementation of CFTP for image analysis using the Ising model, before turning (Sect. 13.3) to the more subtle and generally less understood notion of dominated CFTP. A brief concluding Sect. 13.4 discusses the relationship of CFTP and dominated CFTP to uniform and geometric ergodicity.

Throughout much of the exposition, algorithms are illustrated by short self-contained R scripts, which themselves are discussed in the text. The R scripts (concatenated together into one long R script file) can be found at `go.warwick.ac.uk/wsk/perfect_programs#Ulm-notes`. The exposition focusses on specific examples rather than general principles and theory (which is covered in [227]), though comments on the general theory occur throughout.

13.2 Classic Coupling-from-the-Past

The objective of CFTP is to produce exact draws from the equilibrium distribution of suitable Markov chains X defined on specified state-spaces $\mathcal{X}$. Conventional Markov chain Monte Carlo runs a single realization of a suitable X from the present (time 0) to some distant future (time $T = n$ where n is large), and relies on convergence theorems which assert (under suitable conditions) that the distribution of X_n approximates the equilibrium distribution when n is suitably large. In contrast, CFTP typically seeks to generate a realization of X which starts in the indefinite past (time $T = -n$ for indefinitely large n) and runs until the present (time 0). In the case of classic CFTP, this is achieved by realizing a single simulation of multiple trajectories of X running from all possible starting locations $x \in \mathcal{X}$ and all possible past times $-n$. This is done by constructing a stochastic flow $\{F_{-n,-n+t} : \mathcal{X} \to \mathcal{X} : n,t \geq 0\}$. Here $F_{-n,-n+t} : \mathcal{X} \to \mathcal{X}$ is a random map, and for each $x \in \mathcal{X}$ and each $-n < 0$ the process $\{F_{-n,-n+t}(x) : t = 0,1,\ldots,n\}$ is a realization of the Markov chain X run from time $T = -n$ (with initial state x) through to time 0.

The classic CFTP algorithm succeeds exactly when the stochastic flow is constructed so that the map $F_{-n,0}$ has a one-point random image for all large enough n; we can view this as complete coalescence of the coupled realizations $\{F_{-n,-n+t}(x) : t = 0,1,\ldots,n\}$ for varying $x \in \mathcal{X}$, once n is sufficiently large. Because of the random flow property ($F_{-n-m,0} = F_{-m,0} \circ F_{-n-m,-m}$) it then follows that the random image $\{F_{-n,0}(x) : x \in \mathcal{X}\}$ stabilizes for large enough n, and the point in the stable one-point image is actually an exact draw from the equilibrium distribution of X. See Theorem 3 of [227] for a formal statement and proof.

Practical issues are: firstly, how to construct coalescing stochastic flows; secondly, how to identify a random time such that coalescence has definitely occurred. If the state-space is finite then it is possible in principle to proceed by exhaustive enumeration, but this would typically be unbearably inefficient and computationally infeasible. In this section we will discuss three thematic examples which allow for efficient solutions.

13.2.1 Random Walk CFTP

We begin by illustrating classic CFTP [320] in the very special case of simple symmetric random walk on the integer segment $\{1, 2, \ldots, 10\}$, with (reversible) reflection at both boundaries. This expands on and refines the discussion in Sect. 1.2 of [227]. Here the random flow is implemented as synchronous coupling of random walks begun at different starting points, and the monotonicity of this coupling can be used to detect coalescence in an efficient way. Moreover in this simple case the equilibrium distribution is uniform (this follows directly from a detailed balance calculation) and so the validity of CFTP can be confirmed empirically using a statistical χ^2 test.

This example is completely trivial, but can be viewed as a prototype for a rather less trivial Ising model application, which is discussed below in Sect. 13.2.4.

13.2.1.1 Helper Functions for the Simulations

We begin by describing various *R* functions useful for constructing the random walk CFTP algorithm. First of all, we need a function which generates a sequence or block of innovations for the underlying stochastic flow. Since the flow is composed of synchronously coupled simple symmetric random walks, the innovations can be realized as binary random variables which are equally likely to take the values ± 1. The function `make_block` generates a single block (of prescribed length) of realizations of independent binary random variables of this form, using the *R* simulation function `rbinom` to generate $\text{Ber}(\frac{1}{2})$ random variables (equivalently, $\text{Binom}(1, \frac{1}{2})$) and then transforming them appropriately.

```
make_block <- function(block_length) {
    return(2 * rbinom(block_length, 1, 1/2) - 1)
}
```

The flow for this implementation of random walk CFTP is based on the `update` function. This uses a single ± 1 innovation `innov` as a candidate for the jump of the chosen random walk at a particular instant. If the random walk is prevented by a boundary from using this jump then it simply stays where it is.

```
update <- function(x, innov, lo = 1, hi = 10) {
    return(min(max(x + innov, lo), hi))
}
```

The consequence of this `update` definition is that all random walks are synchronously coupled; they move in parallel except where boundaries prevent them from so doing. In particular the random walk trajectories depend monotonically on their starting positions.

The function `cycle` lies at the heart of the algorithm. It employs a block of ± 1 innovations `innovations` (as supplied by `make_block` above) to construct a synchronously coupled pair of upper and lower random walks using `update`. One of the walks (`upper`) starts from the upper boundary `hi` of the random walk, the other (`lower`) starts from the lower boundary `lo`. Consider all other synchronously coupled simple symmetric walks based on the same stream of innovations started either at the same time as these two, or earlier. By monotonicity of the synchronous coupling, at any given time all these other walks must lie above `lower` and below `upper`. The `cycle` function returns NA (*R*'s "not a number" value) if coalescence has not yet occurred at time 0, and otherwise returns the common coalesced value. Since coalescence is achieved exactly when `upper` and `lower` meet before time 0, it is detected easily.

```
cycle <- function(innovations, lo = 1, hi = 10) {
    lower <- lo
    upper <- hi
    for (i in 1:length(innovations)) {
        lower <- update(lower, innovations[i],
            lo = lo, hi = hi)
        upper <- update(upper, innovations[i],
            lo = lo, hi = hi)
    }
    if (upper != lower)
        return(NA)
    return(upper)
}
```

13.2.1.2 A Typical Run of the Random Walk CFTP Algorithm

We illustrate the CFTP algorithm by carrying out classic random walk CFTP, run statement-by-statement, without regard for efficiency.

First we initialize the seed of the random number generator (to facilitate repeatability of the simulation), and use `make_block` to create an initial vector of ± 1 innovations.

```
set.seed(1)
innovations <- make_block(2)
```

Then we repeatedly apply `cycle`, extending the `innovations` block to the left, until `upper` and `lower` random walks coincide at time zero.

```
while (is.na(cycle(innovations))) {
    innovations <- c(
        make_block(length(innovations)),
        innovations)
}
```

The doubling of the length of the `innovations` block corresponds to a simple binary search for the coalescence time.

Finally we report the result of the CFTP algorithm, which we obtain by re-running `cycle` on the total block of `innovations` which has been accumulated in the previous `while` loop.

```
cycle(innovations)
```

Output is as follows.

```
[1] 4
```

13.2.1.3 Implementing the Algorithm

The random walk CFTP simulation can now be put together succinctly as a single *R* function, using the helper functions defined above.

```
rw_cftp <- function(initial_range, lo = 1,
    hi = 10) {
    innovations <- make_block(initial_range)
    result <- NA
    while (is.na(result)) {
        innovations <- c(
            make_block(length(innovations)),
            innovations)
        result <- cycle(innovations, lo = lo,
            hi = hi)
    }
    return(result)
}
```

13.2.1.4 Statistical Analysis

This R implementation of random walk CFTP is of course rather slow. We are how-ever now able to investigate the algorithm output statistically; we base this on 10000 successive runs of the algorithm

```
data <- sapply(rep(100, 10000), rw_cftp)
tabulate(data)
```

Output is as follows.

```
[1]   959   957 1029 1054 1051 1007   958 1002   987   996
```

Computations took 15 seconds on a notebook computer running Ubuntu 10.10 using an Intel quad processor i7 Core M 620 at 2.67GHz using 2 GB RAM. Fig. 13.1 (a) presents a histogram of these results. A χ^2-test confirms that the output has the correct distribution (uniform on $\{1, 2, \ldots, 10\}$).

```
chisq.test(tabulate(data), p = rep(1/10, 10))
```

Output is as follows.

```
        Chi-squared test for given probabilities

data:   tabulate(data)
X-squared = 11.89, df = 9, p-value = 0.2196
```

13.2.1.5 Graphical Visualization of Random Walk CFTP Algorithm

It is helpful to visualize the algorithm using graphics. Re-writing the algorithm to generate graphical output is now a straightforward exercise; and this can be invalu-able when testing for correctness of implementation. The changes are as follows:

1. The `cycle` part of the algorithm is augmented graphically so that at each cycle it plots the trajectories of upper and lower processes;
2. The main body of the algorithm initializes the graphics, then cycles through the CFTP algorithm using the graphically augmented version of `cycle`.

We can now use the amended algorithm to obtain a graphical demonstration of how the algorithm runs. The result is presented in Fig. 13.1 (b).

13.2.1.6 More Details

The reader is referred to the tutorial [77] (see also [394, 417]) for further discussion of classic CFTP, including illustration of the defective simulation results arising from (a) not re-using randomness, or (b) stopping the CFTP cycle early at the first

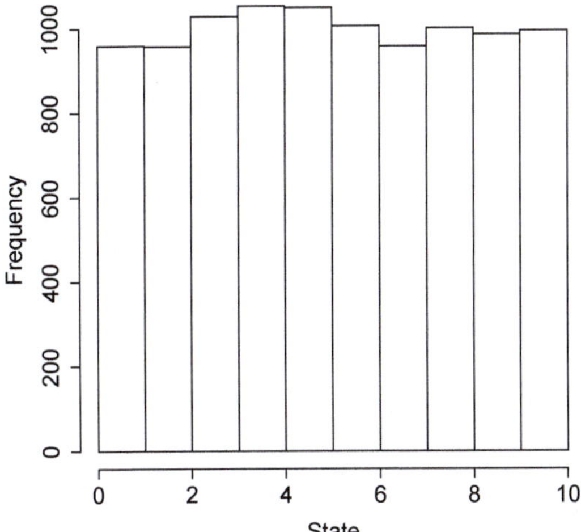

(a) Histogram of results from repeated runs of random walk CFTP. The impression of uniform distribution is confirmed by a χ^2-test.

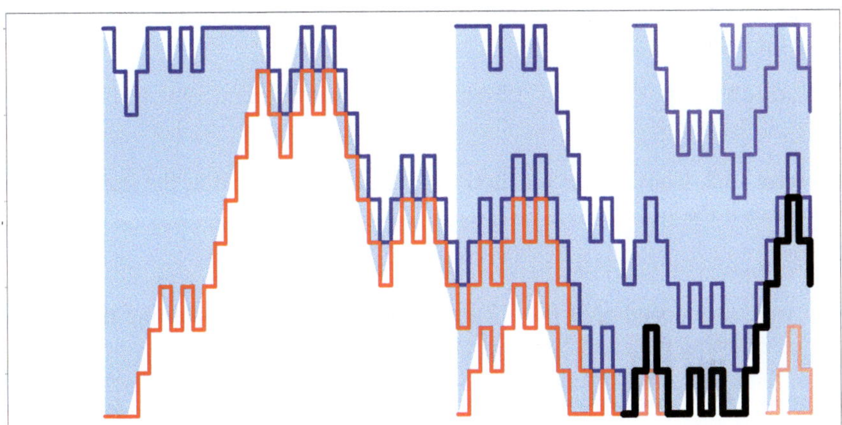

(b) Visualization of a single random walk CFTP run. Initial time range is $\{-1, 0\}$; the algorithm detects coalescence after 8 cycles at which point the time range is $\{-64, -63, \ldots, 0\}$ The thick line indicates the coalesced trajectory. The range between successive upper/lower pairs is indicated by the shaded regions. Note how earlier upper/lower pairs "funnel" between later upper/lower pairs.

Fig. 13.1 Random walk CFTP output: histogram and visualization

point when upper and lower simulations coalesce. See also Exercises 13.1 and 13.3 below.

Note that, despite appearances, the validity of the CFTP algorithm described here actually does *not* depend on the symmetry of the random walk. Indeed the algorithm can be modified to deal with instances in which the target Markov chain is no longer a random walk, nor even a reversible Markov chain; It suffices that the target chain exhibit a stochastic monotonicity sufficient to make sense of being able to couple realizations of the chain starting at all possible initial points so that there is an upper chain and a lower chain, and such that coalescence eventually occurs. For more on the nature of the required monotonicity, see [119, 120]. An alternative general approach to perfect simulation is presented in [117], using a rejection sampling approach, with the advantages of requiring less stringent monotonicity structure and avoiding possible correlation between output and algorithm run-time (compare Exercise 13.2); the relationship between this and classic CFTP has been elucidated in [121].

Exercise 13.1. Consider the function `rw_cftp`. In each cycle of the following loop,

```
while(is.na(result)) {...},
```

the previous block of `innovations` is extended backwards in time. Investigate the statistical consequences of (incorrectly) generating completely new blocks instead, using the following construction.

```
innovations <- make_block(2 * length(innovations))
```

Exercise 13.2. Consider the function `rw_cftp`. What would be the statistical consequences if the

```
while(is.na(result)) {...}
```

loop were abandoned after (say) 2 iterations, in favour of re-starting the function `rw_cftp` using the `initial_range`?

Exercise 13.3. Consider the function `rw_cftp`. What would be the consequences of extending the previous block of `innovations` *forwards* in time rather than back in time? That is to say, generating new blocks using the following construction.

```
innovations <- c(
      innovations,
      make_block(length(innovations)))
```

Exercise 13.4. Modify the function `make_block` so that `rw_cftp` produces a perfect draw from the equilibrium distribution of a specified *asymmetric* random walk on the integer segment $\{1, 2, \ldots, 10\}$. Compute the true equilibrium using detailed balance, and test the modified algorithm statistically.

13.2.2 Read-once CFTP

In this section we consider *read-once CFTP* (RO-CFTP). This idea goes back to
[416], and may be thought of as a variation on classic CFTP which works block-by-
block using successive draws of whole blocks of innovations, and additionally runs
forwards in time rather than backwards. For the sake of clarity we will describe this
in terms of random walk CFTP, although the same principles apply to any instance
of classic CFTP. This is indicated, for example, as part of the treatment of the small-
set CFTP discussed in Sect. 13.2.3.

The implementation uses the same helper functions as are described for random
walk CFTP in Sect. 13.2.1.

13.2.2.1 A Typical Run of RO-CFTP

The idea is to group together fixed sequences of innovations in separate *blocks*,
block length n being chosen to ensure a positive chance of coalescence within the
block.

```
n <- 100
```

Before discussing the details, we work through an instance of read-once CFTP run
statement-by-statement. We begin by repeatedly sampling an initial block of innovations and rejecting the sample until coalescence is achieved. The resulting draw
is therefore of a block of innovations, viewed as running over the time interval
$\{-n, -n+1, \ldots, 0\}$, which is *conditioned* so that the flow map $F_{-n,0}$ has a one-
point image.

```
block <- make_block(n)
while (is.na(cycle(block))) {
    block <- make_block(n)
}
```

This initial (coalescent) block is stored as the first (left-most) contribution to a
whole sequence of `innovations`.

```
innovations <- block
```

Now we generate a sequence of further blocks corresponding to a sequence of flow
functions $\{F_{kn,(k+1)n} : k = 0, 1, \ldots\}$; while no coalescence is achieved these blocks
are successively appended to `innovations`. This iteration is finished once a further coalescent block is obtained, stored separately in `coalescent_block`.

```
block <- make_block(n)
while (is.na(cycle(block))) {
    block <- make_block(n)
```

```
        append(innovations, block)
    }
    coalescent_block <- block
```

Finally we return the final value from the current sequence of `innovations`. A further RO-CFTP run can now start off with the most recent `coalescent_block`, to save the work of repeatedly sampling to generate a suitably conditioned initial block.

```
    cycle(innovations)
```

Output is as follows.

```
[1] 6
```

The validity of this algorithm may be deduced from the following argument. Suppose we carry out classic CFTP, but limit ourselves to detecting coalescence block-by-block and stop when we obtain a coalescent block. We then generate blocks of innovations in succession, $B_{-1}, B_{-2}, \ldots, B_{-(N-1)}, B_{-N}$. By construction these blocks are all non-coalescent except for the final block, B_{-N}. Moreover, conditional on N, the blocks $B_{-1}, B_{-2}, \ldots, B_{-(N-1)}$ are independent draws of blocks conditioned to be non-coalescent, while independently the final block B_{-N} is conditioned to be coalescent. Moreover N has a Geometric distribution with success probability given by the probability of a block being coalescent. For classic CFTP these blocks are arranged thus:

$$B_{-N} \cdot B_{-(N-1)} \cdot \ldots \cdot B_{-2} \cdot B_{-1}, \tag{13.1}$$

and this delivers an exact draw from the equilibrium distribution.

On the other hand the RO-CFTP procedure described above generates an arrangement

$$\widetilde{B}_1 \cdot \widetilde{B}_2 \cdot \ldots \cdot \widetilde{B}_{\widetilde{N}-1} \cdot \widetilde{B}_{\widetilde{N}}, \tag{13.2}$$

where, conditional on $\widetilde{N}$, the *first* block $\widetilde{B}_1$ is conditioned to be coalescent and independently the blocks $\widetilde{B}_2, \widetilde{B}_3, \ldots, \widetilde{B}_{\widetilde{N}}$ are independent draws of blocks conditioned to be non-coalescent. Finally, $\widetilde{N}$ again has a Geometric distribution with success probability given by the probability of a block being coalescent.

Thus the innovation sequences corresponding to the concatenations of (13.1) and (13.2) are statistically identical. It follows that the distributions of the images of the corresponding flow maps are the same, and hence RO-CFTP delivers a draw with the same statistics as a draw from classic CFTP.

This algorithm may be refined to economize on storage: it suffices to establish the coalesced value of the initial coalesced block, and then to keep a record only of the most recent result of updating this coalesced value. Moreover, in a very natural way, RO-CFTP can be made to deliver a sequence of independent exact draws from a sequence of draws of blocks.

13.2.2.2 Packaging the Algorithm

As before, we package the RO-CFTP process as a single *R* function.

```
ro_cftp <- function(block_length) {
    block <- make_block(block_length)
    while (is.na(cycle(block))) {
        block <- make_block(block_length)
    }
    innovations <- block
    block <- make_block(block_length)
    while (is.na(cycle(block))) {
        block <- make_block(block_length)
        append(innovations, block)
    }
    return(cycle(innovations))
}
```

We ignore here the possibility of re-using the final coalescent block, which would economize on the initial rejection-sampling step. Also we do not economize on storage by storing only the most recent value during the iteration loop; this is because the less economical version presented here is easier to augment to produce visualization graphics.

13.2.2.3 Statistical Analysis

As with random walk CFTP, computation is slow but manageable (in [416] it is shown that run-time can be arranged to be comparable to that of classic CFTP). The output appears to be uniformly distributed (Fig. 13.2 (a)).

```
data <- sapply(rep(100, 2000), ro_cftp)
tabulate(data)
```

Output is as follows.

```
[1]  205 212 192 192 212 208 193 204 192 190
```

Uniformity is confirmed by a χ^2-test.

```
chisq.test(tabulate(data), p = rep(1/10, 10))
```

Output is as follows.

```
        Chi-squared test for given probabilities

data:  tabulate(data)
X-squared = 3.67, df = 9, p-value = 0.9318
```

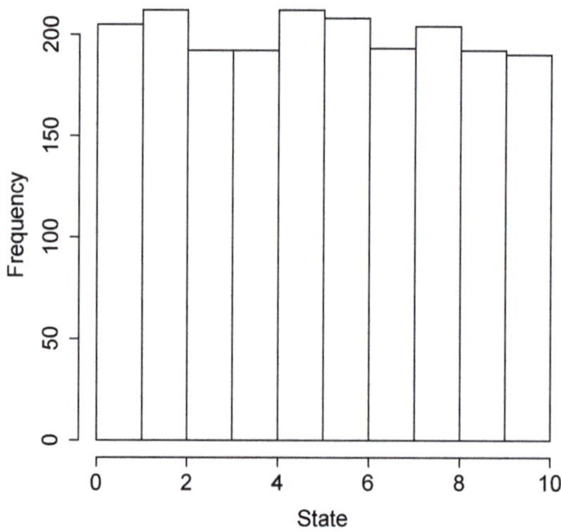

(a) Histogram of results from repeated runs of RO-CFTP. The impression of uniform distribution is confirmed by a χ^2 test.

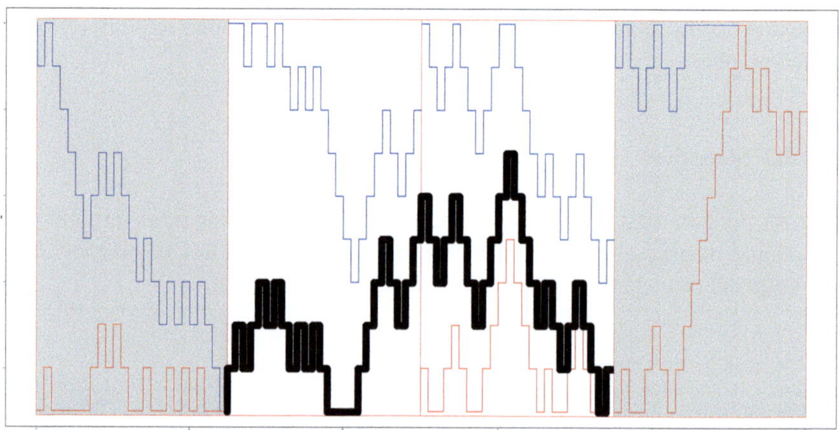

(b) Visualization of an RO-CFTP run producing a single result. The two non-coalescing blocks have white backgrounds. The thick line indicates the coalesced trajectory. This trajectory starts at the end of the first coalesced block, at the coalesced value, and ends at the start of the second coalesced block, there delivering the sampled value.

Fig. 13.2 Illustrations of RO-CFTP output: histogram and visualization

A typical run is visualized graphically in Fig. 13.2 (b)

13.2.2.4 More Details

We have mentioned (Sect. 13.2.2.1) that it is easy to adapt RO-CFTP so as to generate a stream of exact draws for its target distribution. This and other ways of generating streams of exact CFTP draws have been discussed in [294]. Indeed this work led directly to the link [121] between classic CFTP and Fill's version of perfect simulation.

Exercise 13.5. Consider the function `ro_cftp`. Using the modification of the function `make_block` from Exercise 13.4, arrange for `ro_cftp` to produce a perfect draw from the equilibrium distribution of a specified *asymmetric* random walk on the integer segment $\{1, 2, \ldots, 10\}$. Test the modified algorithm statistically.

Exercise 13.6. Modify the function `ro_cftp` to produce a sequence of independent exact draws, with the length of the sequence specified by a second argument of `ro_cftp`, as indicated in Sect. 13.2.2.1.

13.2.3 Small-set (Green-Murdoch) CFTP

So far we have only described CFTP for discrete probability models. But in fact CFTP can be applied to continuous probability models as well. The clue as to how to do this is to be found in the theory of small sets for continuous state-space Markov chains [9, 272, 302].

Definition 13.1 (Small set). Suppose that $X = \{X_0, X_1, X_3, \ldots\}$ is a Markov chain on a state-space $\mathcal{X}$ which is a measurable space. We say $C \subseteq \mathcal{X}$ is *small of lag k* if there is a probability measure ν on $\mathcal{X}$ and $0 < \rho < 1$ such that the following minorization condition is obeyed for all $x \in C$:

$$\mathbf{P}(X_k \in \cdot \mid X_0 = x) \quad \geq \quad \rho \nu(\cdot).$$

(Technically one usually requires X is φ-irreducible and insists that $\varphi(C) > 0$.)

In case $k = 1$ then X can be viewed as having positive probability p of regeneration at every visit to the small set C. If $C = \mathcal{X}$ is the whole state-space then it is shown in [293] how to generate a CFTP algorithm.

13.2.3.1 A Simple Example of Small-Set CFTP

A simple example is given by the Markov chain X on $[0,1]$ with "triangular" transition density: the conditional distribution of X_{n+1} given $X_n = x$ has probability density $p(x,y)$ where

$$p(x,y) \;=\; \begin{cases} 2\,(x/y) & \text{if } 0 < x \le y, \\ 2\,((1-x)/(1-y)) & \text{if } y < x < 1. \end{cases}$$

Thus the density forms a triangle with base the unit segment on the x-axis, and third vertex at $(y,2)$. A visual appreciation of the fact that the whole state-space is small is given by the graph of densities in Fig. 13.3, using the following R function: all possible triangles intersect in a "regeneration triangle".

```
triangle <- function(x, y = 0) {
    if (y == 0)
        return(2 * (1 - x))
    if (y == 1)
        return(x)
    return(((x <= y) * (2 * x/y) + (x > y) * (2 *
        (1 - x)/(1 - y))))
}
```

For the purposes of CFTP we need to be able to generate coupled realizations from all triangle densities. We begin by noting that a simultaneous draw from all densities can be obtained by regarding $p(\cdot,y)$ as obtained from $p(\cdot,0)$ by a horizontal affine shear, while we can draw from $p(\cdot,0)$ by first drawing (x,z) from the rectangle $(0,1) \times (0.2)$ and then folding the rectangle over using the diagonal from (0.2) to $(1,0)$ as in the following R function and illustrated in Fig. 13.4 (a).

```
left_draw <- function(n) {
    x <- runif(n, min = 0, max = 1)
    z <- runif(n, min = 0, max = 2)
    reflect <- (z > (2 - 2 * x))
    return(reflect * cbind(1 - x, 2 - z) + (1 -
        reflect) * cbind(x, z))
}
```

Fig. 13.4 (b) presents the result of shearing the points to produce draws from a different triangular density $p(\cdot,0.75)$.

This coupling construction needs to be modified so that points falling in the "regeneration triangle" are not sheared at all, while points that would fall into or to the right of this triangle after shearing are subject to a further affine shear: see Fig. 13.4 (c) and Fig. 13.4 (d). We compile all this into a single function to deliver the resulting coalescing flow.

```
coalescing_flow <- function(y) {
    draw <- left_draw(1)
    x <- rep(draw[, 1], length(y))
    z <- rep(draw[, 2], length(y))
    regeneration <- (2 * x > z) & (2 * (1 - x) >
        z)
```

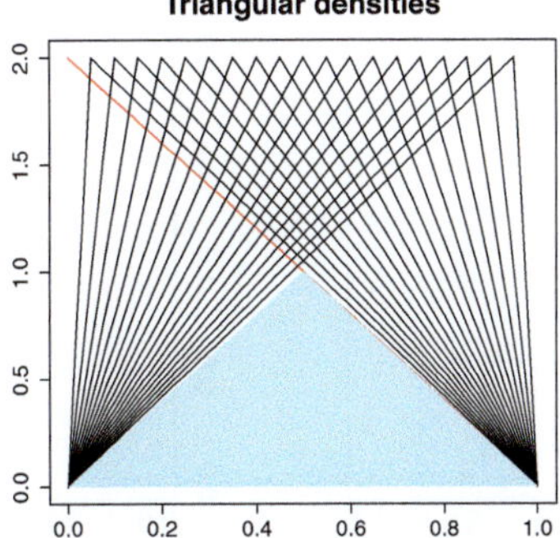

Fig. 13.3 Triangular densities. Note the common area under the densities, indicated by the shaded triangle. This implies that the entire state-space $[0, 1]$ is a small set.

```
    sheared_x <- x + 0.5 * z * y
    needs_second_shear <- (2 * sheared_x > z) &
        (z < 1)
    return(regeneration * x + (1 - regeneration) *
        (sheared_x + needs_second_shear * (1 -
            z))))
}
```

RO-CFTP may be applied to the output to harvest a sequence of exact draws from the target distribution, which is the equilibrium distribution of the Markov chain X. In the following code, we exceptionally include the graphical directives, which produce Fig. 13.5.

```
T1 <- 30
start <- seq(0, 1, 0.2)
state <- start
predecessor <- start[1]
initial <- TRUE
store <- state
t0 <- 1
```

```
plot(0:T1, rep(0, T1 + 1), xlim = c(0, T1),
    ylim = c(0, 1), type = "l",
    main = "Coalescing flow",
    xlab = "Time", ylab = "State")
for (t in 1:T1) {
    if (length(unique(state)) == 1) {
        for (j in 1:dim(store)[2]) lines(t0:t +
            1, store[, j])
```

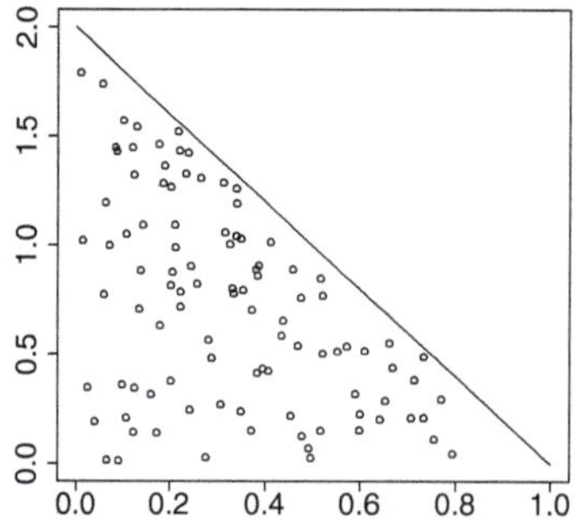

(a) Draws of points lying beneath the left-most triangular density

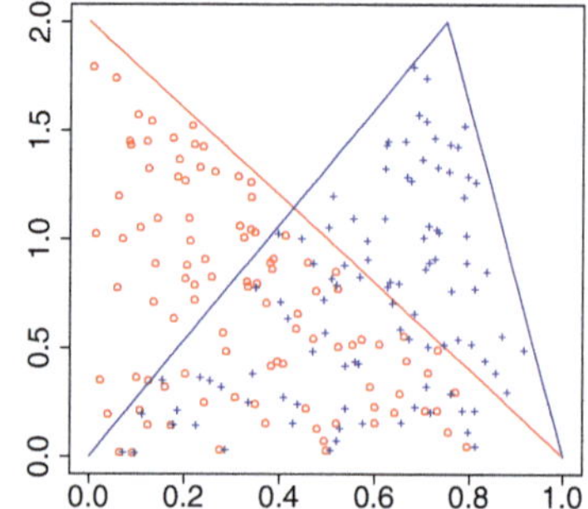

(b) Draws of points as above, and also of their locations after being subjected to a shear so that they are drawn from a different triangular density

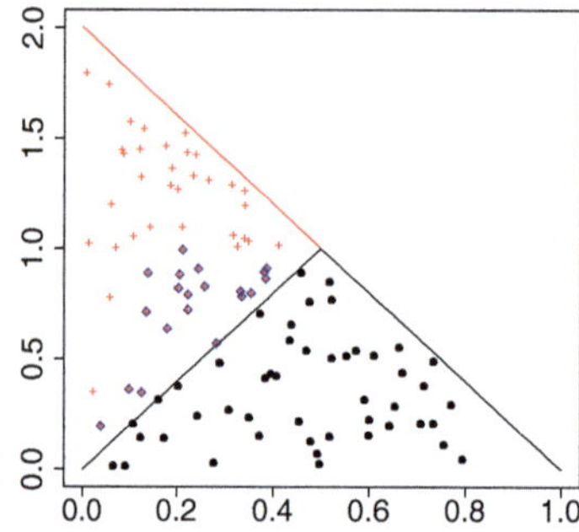

(c) Draws of points lying beneath the left-most triangular density, marking points in "regeneration triangle" and also points which would fall into or to right of "regeneration triangle" after first shearing

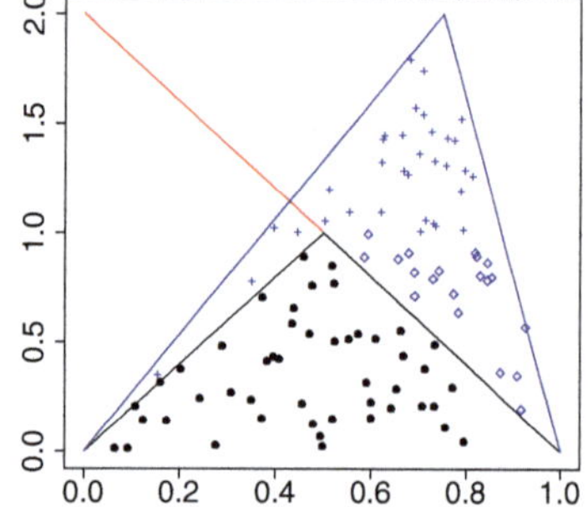

(d) Draws of sheared points, but leaving untouched those points originally in "regeneration triangle" and applying a further shear to points which would fall into or to right of "regeneration triangle" after first shearing

Fig. 13.4 Illustration of triangular densities example

```
if (!initial)
    points(t, predecessor, col = "red",
        pch = 19, cex = 20)
initial <- FALSE
state <- c(unique(state), start)
t0 <- t
store <- state
}
predecessor <- state[1]
```

```
        state <- coalescing_flow(state)
        store <- rbind(store, state)
}
```

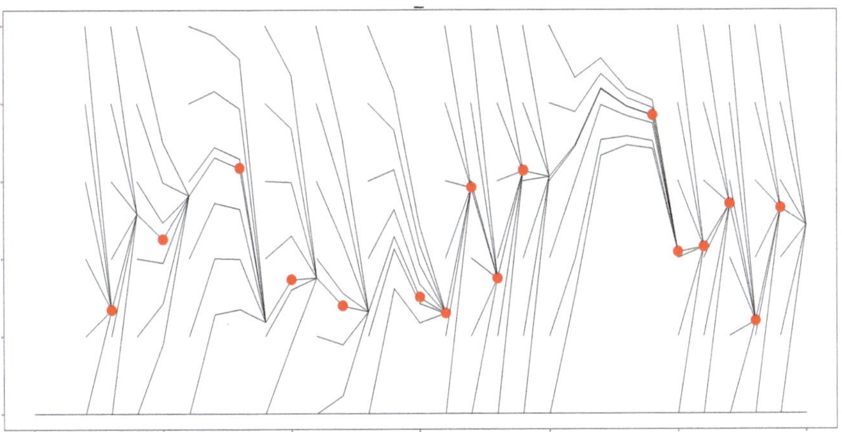

Fig. 13.5 Flow induced by triangular densities example. The circular dots indicate the instants where RO-CFTP permits sampling of exact draws from the equilibrium distribution. These occur at the right-hand ends of trajectories running strictly between adjacent pairs of coalesced blocks.

13.2.3.2 Extensions of Small-set CFTP

In real-world applications either the whole state-space may not form a small set, or the resulting regeneration probability may be too small to be of practical use. In [293] it is shown how to overcome this when the state-space can be partitioned into disjoint small sets of useful regeneration probability. Furthermore, in [10] it is described how the small-set CFTP construction may be adapted to produce a representation of the equilibrium probability distribution (see also [174]). Small-set CFTP provides a strong motivation for the natural question, how prevalent are small sets of small lag? Consider the case of Markov chains with measurable transition densities: in [229] it is shown that there exist examples with no small sets of lag 1 at all, but that all such Markov chains possess so many small sets of lag 2 that, when sub-sampled with periodicity 2, they may be represented in terms of latent discrete-time Markov chains.

Exercise 13.7. Modify the code in this section by removing the graphics directives, and arranging for it to produce a sequence of specified length of perfect draws from the equilibrium distribution of the Markov chain X using RO-CFTP (as applied in Sect. 13.2.3.1). Use R to construct a kernel density estimate of the equilibrium density of X.

13.2.4 Image Analysis and Ising CFTP

One of the earliest instances of perfect simulation concerned the Ising model [320].
Here the major application used Sweeny's algorithm for exact simulation of Fortuin-
Kastelyn-Potts models near the critical temperature; however we will describe the
easier method used in [320] to produce exact draws from high-temperature Ising
models. This method is well-adapted to image analysis problems.

Readers unfamiliar with the Ising model will find a useful and accessible introduction in [232]. Recall that an Ising model is formed by assigning spins $S_i = \pm 1$
to each node i of a fixed graph G (we shall take G to be a finite planar lattice
$\{1,\dots,N\}^2$). Then the Ising model is defined as having the following probability
mass function for assignations of spins:

$$p(S_i : i \in G) \quad = \quad Z(J)^{-1} \exp\left(J \sum_{i \sim j} S_i S_j \right). \tag{13.3}$$

Here $i \sim j$ if i and j are neighbours in the graph G and the sum is taken over all
unordered pairs; we shall take $J > 0$ (the *ferromagnetic* case), so that neighbouring sites with the same spin lead to higher values of the probability mass function.
Finally, $Z(J)$ is the normalizing constant.

For purposes of image analysis it is interesting also to investigate the case when
an "external field" $\{\widetilde{S}_i : i \in G\}$ is applied; given the external field, the probability
mass function is then taken to be proportional to

$$\exp\left(J \sum_{i \sim j} S_i S_j + H \sum_i S_i \widetilde{S}_i \right). \tag{13.4}$$

In the context of binary image analysis the external field represents the observed
noisy image $\{\widetilde{S}_i : i \in G\}$, while $\{S_i : i \in G\}$ is the "true" image. The probability
mass function defined by (13.4) may then be taken to be the Bayesian posterior distribution of the "true" image, based on a Bayesian prior and likelihood determined
by an Ising model connecting $\{\widetilde{S}_i : i \in G\}$, and $\{S_i : i \in G\}$.

We can draw from the Ising model (whether given by (13.3) or by (13.4)) by
viewing it as the equilibrium distribution of the Markov chain given by the single-
site update heat bath algorithm; sites are updated either at random or in systematic
order by re-sampling from their conditional distribution given the configuration at
all other sites. (It follows by detailed balance that the equilibrium will be a draw
from the Ising model.) Thus under (13.3) we update site i as follows:

$$\text{re-sampled } S_i \leftarrow +1, \text{ probability proportional to } \exp\left(+J \sum_{j:j \sim i} S_j \right),$$

$$\text{re-sampled } S_i \leftarrow -1, \text{ probability proportional to } \exp\left(-J \sum_{j:j \sim i} S_j \right).$$

Similarly in the case of (13.4)

$$\text{re-sampled } S_i \leftarrow +1, \text{ probability proportional to } \exp\left(+H\widetilde{S_i}+J\sum_{j:j\sim i} S_j\right),$$

$$\text{re-sampled } S_i \leftarrow -1, \text{ probability proportional to } \exp\left(-H\widetilde{S_i}-J\sum_{j:j\sim i} S_j\right).$$

Since we take $J > 0$, it follows that the probability of the re-sampled S_i being equal to 1 is monotonically increasing in the configuration $\{S_i : i \in G\}$ using the natural partial order. This is the key observation for the purposes of CFTP: as a result we can implement a monotonic coupling of the heat bath process which is a simple generalization of that used for random walk CFTP as described in Sect. 13.2.1.

A further refinement arises from the bipartite structure of the finite lattice $\{1,\ldots,N\}^2$, which allows us to implement a "coding" scheme for the updates. Viewing the lattice as a chessboard, we update all the sites on the black squares, then all the sites on the white squares, and so forth; whatever the detailed order of implementation, the statistical result will be the same. This permits extraction of the heat-bath simulation (in fact, two copies of it; see below) from *simultaneous* updates of all sites. (This refinement is no longer available for more general graphs, such as that arising from the lattice if we choose to make additional connections between diagonally adjacent sites.)

13.2.4.1 Implementation

We begin by constructing a test image, represented as a 256×256 binary matrix. Noisy and clean versions of this test image are illustrated in Fig. 13.6.

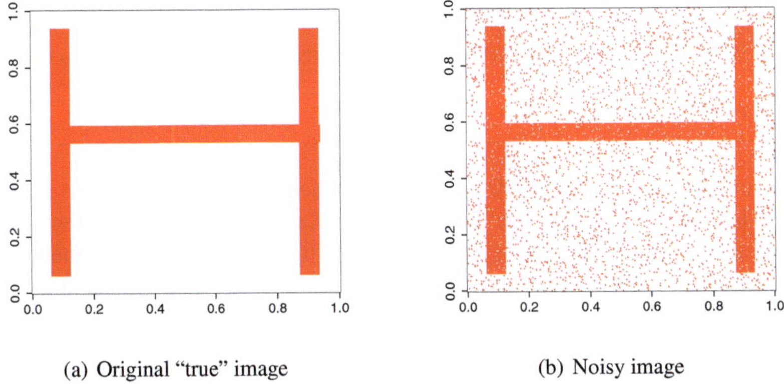

(a) Original "true" image   (b) Noisy image

Fig. 13.6 The test image for Ising CFTP

The heat-bath algorithm is based on a single `update` using an array of innovations. Note that the update is really a parallel update of *two* separate instances of the algorithm: one update confined to the "black squares", one to the "white squares" of the chessboard; this corresponds to an inbuilt period 2 for the heat-bath algorithm for this particular version of the Ising model. The `state` of the algorithm is an $M \times M$ matrix of ± 1, bordered by a single strip of zeros. (This corresponds to free boundary conditions: the construction and algorithm are easily varied to produce other boundary conditions.)

```
update <- function(state, image, innov, J = 0.9,
    H = 1.5) {
    M <- dim(image)[1]
    line <- (1:M) + 1
    work <- cbind(0, rbind(0, state, 0), 0)
    threshold <- 1/(1 + exp(-2 * (J * (work[line,
        line + 1] + work[line, line - 1] +
        work[line + 1, line] +
        work[line - 1, line]) +
        H * image)))
    state[1:M, 1:M] <- -1
    state[innov < threshold] <- +1
    return(state)
}
```

```
M <- dim(im1)[1]
J <- 0.9
H <- 1.5
state <- matrix(data = +1, nrow = M, ncol = M)
line <- (1:M) + 1
for (iter in 1:36) {
    innov <- matrix(data = runif(M * M), nrow = M,
        ncol = M)
    state <- update(state, im1, innov, J = J,
        H = H)
}
```

A sub-sequence of outputs from the parallel heat-bath algorithm is given in Fig. 13.7. Here the initial state was fixed with all spins at $+1$, so the first few updates overcome the discrepancies arising from the fact that most of the noisy image yields spins of -1. By update 24 it appears that equilibrium may nearly be attained. Note that (a) post-processing of the image is required in order to undo the "chess-board" encoding; (b) this post-processing yields *two* draws from equilibrium; (c) the encoding means that each of the two draws entails visiting of each node 12 times in the first 24 updates. In fact the CFTP algorithm will produce exact draws after order of 64 updates or fewer.

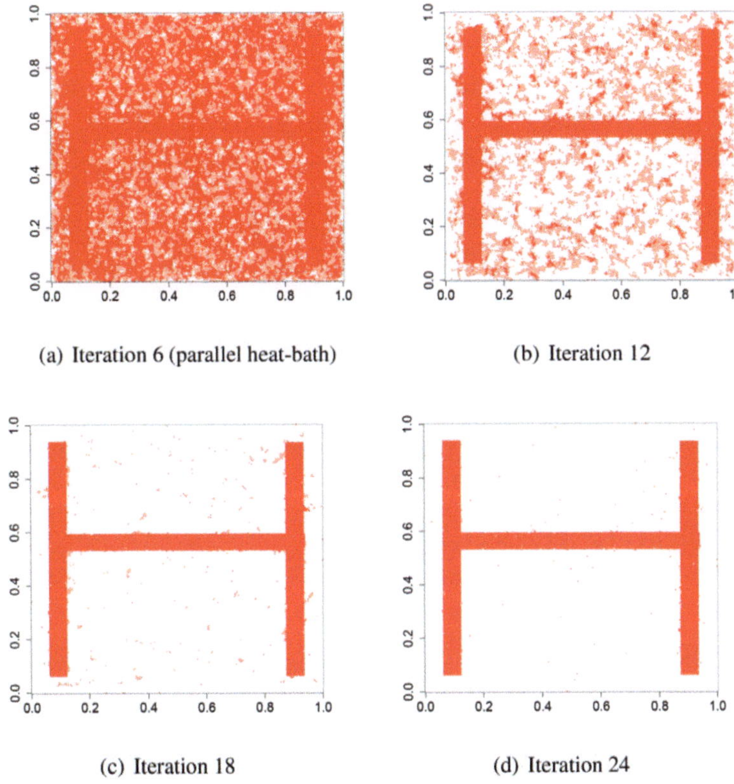

(a) Iteration 6 (parallel heat-bath)   (b) Iteration 12

(c) Iteration 18   (d) Iteration 24

Fig. 13.7 Parallel heat-bath reconstruction ($J = 0.9$ and $H = 1.5$). Post-processing is required.

We now package up a CFTP algorithm, which largely follows the pattern of the original random walk CFTP algorithm discussed above.

```r
make_block <- function(block_length, M) {
    ilist <- list()
    for (iter in 1:block_length) {
        ilist <- c(list(matrix(data = runif(M *
            M), nrow = M, ncol = M)), ilist)
    }
    return(ilist)
}
```

```r
cycle <- function(innovations, image, J = 0.6,
    H = 1.5) {
    M <- dim(image)[1]
    upper <- matrix(data = +1, ncol = M, nrow = M)
```

```r
    lower <- matrix(data = -1, ncol = M, nrow = M)
    for (innov in innovations) {
        upper <- update(upper, im1, innov, J = J,
            H = H)
        lower <- update(lower, im1, innov, J = J,
            H = H)
    }
    if (sum(upper - lower) != 0)
        return(NA)
    else return(upper)
}
```

```r
ising_cftp <- function(image, initial_range,
    J = 0.6, H = 1.5) {
    innovations <- make_block(initial_range,
        dim(image)[1])
    result <- NA
    while (is.na(result)) {
        innovations <- c(
            make_block(length(innovations),
            dim(image)[1]), innovations)
        result <- cycle(innovations, image, J = J,
            H = H)
    }
    return(result)
}
```

We need to post-process the result of the CFTP function `ising_cftp`, since its result interlaces two independent draws from the posterior distribution using the black/white chessboard encoding. Note that the two independent post-processed draws are guaranteed to be taken from the posterior distribution, but are *not* necessarily good results from the point of view of statistical image analysis, which would involve considerations of model adequacy and whether the parameters J and H have been chosen appropriately.

```r
innov <- matrix(data = runif(M * M), nrow = M,
    ncol = M)
result1 <- update(result0, im1, innov, J = J,
    H = H)
chess1 <- outer(1:M, 1:M, function(x, y) ((x +
    y)%%2))
chess0 <- 1 - chess1
image0 <- result0 * chess0 + result1 * chess1
image1 <- result0 * chess1 + result1 * chess0
```

13.2.4.2 More Details

There is room for considerable improvement in this algorithm as far as image analysis is concerned! Certainly it would be helpful to use more general neighbourhood structures, the better to capture boundaries which are non-rectilinear (eg: diagonal of second-nearest neighbours as well as nearest neighbours). However periodicity 2 fails for most more general lattice structures, meaning that the coding technique will also fail. This means that one must use (potentially less efficient) sequential update schemes. Performance is greatly enhanced by using a scripting language with substantial support for numerics: for example *Python* with the *Numpy* extension provides matrix operations which are implemented in a manner that allows one to encode an entire sequential image update in a single command. On the other hand, close inspection of successive updates of the CFTP algorithm (both upper and lower states) makes it apparent that much of the running time is taken up with dealing with small deviations from the equilibrium. In [140] modifications of the CFTP algorithm has been considered which are based on producing draws which are within a set distance from the perfect equilibrium draw, using Wasserstein metric.

We have noted that the major application of CFTP in [320] was to critical Ising models without external field, using Sweeney's algorithm. In [175] it is shown how to use the "bounding chain" technique to produce a CFTP algorithm for the Swendsen-Wang algorithm. Recent results [400] describe bounds for variants on heat-bath dynamics for all temperatures.

Exercise 13.8. Modify the function `update` for suitable M to implement the case of *periodic boundary conditions*: pixels (i_1,i_2) and (j_1,j_2) are neighbours if $i_1 = j_1 \pm 1 \mod M$ and $i_2 = j_2 \pm 1 \mod M$. For which values of M does the "parallel update" chessboard coding scheme still work correctly?

Exercise 13.9. Modify the functions involved in `ising_cftp` to work with rectangular images of size $M_1 \times M_2$ for suitable M_1 and M_2.

13.2.5 Other Remarks

Space precludes treatment of the significant technique of the "multishift sampler", introduced in [417] and developed in [83]. A brief description is given in Sect. 2.4 of [227]; the fundamental observation is that one can draw simultaneously from all $\mathrm{Unif}([x,x+1))$ distributions ($-\infty < x < \infty$) simply by simulating a uniformly randomized integer lattice $X + \{0, \pm 1, \pm 2, \ldots\}$ (for X being $\mathrm{Unif}([0,1))$), and selecting the single point of this random lattice lying in the interval $[x,x+1)$. This is an important tool in CFTP, allowing one to reduce a continuous range of possibilities to a locally finite range.

Very recent work on convergence rates for Gibbs samplers for the n-simplex [371] makes intriguing use of coupling which is *non-co-adapted*; as noted in [371], this can be modified to produce CFTP algorithms (though controlling the range

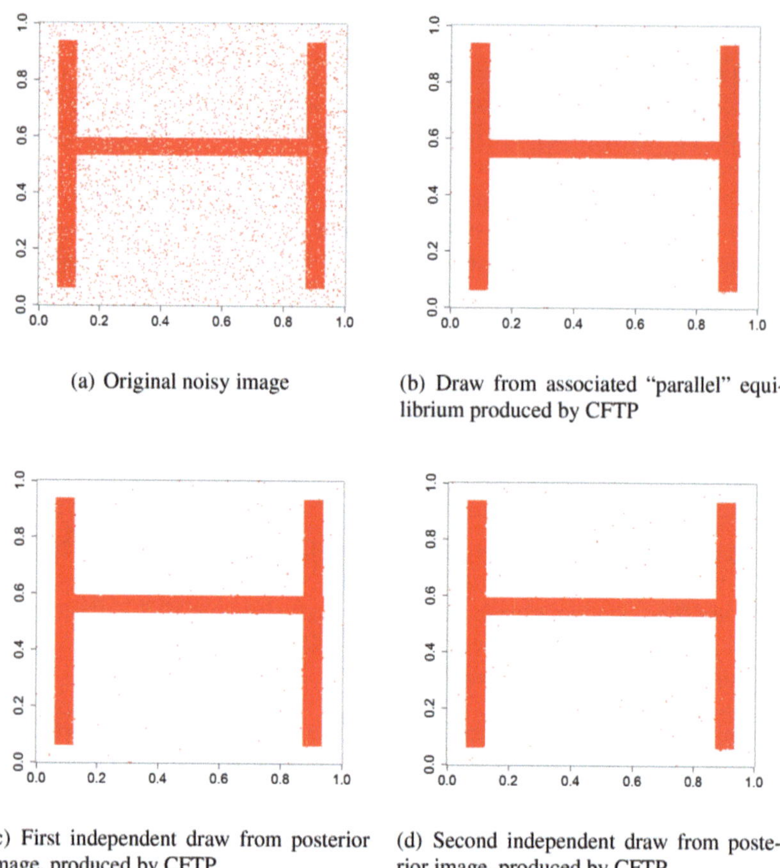

(a) Original noisy image

(b) Draw from associated "parallel" equilibrium produced by CFTP

(c) First independent draw from posterior image, produced by CFTP

(d) Second independent draw from posterior image, produced by CFTP

Fig. 13.8 Results of Ising CFTP using $J = 0.9$ and $H = 1.5$, compared with noisy image

of possibilities then produces a substantial slow factor which probably cannot be avoided).

13.3 Dominated Coupling-from-the-Past

In the first section we introduced classic CFTP by treating three examples (simple symmetric random walk, a Markov chain with triangle densities, the Ising model). The treatment of the first and third of these examples depends heavily on monotonicity, and moreover on there being upper and lower chains. Monotonicity is not evident in the second example, though (as indicated by Green and Murdoch) one may take a set-valued approach to this case, so that suitable monotonicity is seen to follow from the partial ordering imposed by set-inclusion. More generally, the monotonicity requirement can to some extent be avoided by the use of bounding chains [172, 175, 225, 228].

In all three cases the significant constraint is that one is dealing with *uniform ergodicity*. We now state this notion formally, together with the associated notion of geometric ergodicity, formally. First recall the notion of *distance in total variation*

$$\mathrm{dist}_{\mathrm{TV}}(\mu, v) = \sup\{|\mu(A) - v(A)| : \text{measurable } A\},$$

which provides a measure of the amount of agreement between probability measures μ and v.

Definition 13.2 (Uniform ergodicity). The Markov chain X is said to be *uniformly ergodic* with equilibrium distribution π if its distribution converges to the equilibrium distribution π in total variation distance *uniformly in the starting point $X_0 = x$*: for some fixed $C > 0$ and for fixed $\gamma \in (0, 1)$,

$$\sup_{x \in \mathcal{X}} \mathrm{dist}_{\mathrm{TV}}(\mathbf{P}(X_n \in \cdot \mid X_0 = x), \pi) \quad \leq \quad C\gamma^n,$$

where $\mathbf{P}(X_n \in \cdot \mid X_0 = x)$ is the conditional distribution of X_n given $X_0 = x$.

Uniform ergodicity is an easy consequence of the apparently weaker assertion that as $n \to \infty$ so

$$\sup_{x \in \mathcal{X}} \mathrm{dist}_{\mathrm{TV}}(\mathbf{P}(X_t \in \cdot \mid X_0 = x), \pi) \quad \to \quad 0.$$

Definition 13.3 (Geometric ergodicity). The chain X is *geometrically ergodic* with equilibrium distribution π if there is a π-almost surely finite function $V : \mathcal{X} \to [0, \infty]$, and fixed $\gamma \in (0, 1)$, such that

$$\mathrm{dist}_{\mathrm{TV}}\left(P^{(n)}(x, \cdot), \pi\right) \quad \leq \quad V(x)\gamma^n \qquad \text{for all } n, x,$$

where $P^{(n)}(x, \cdot)$ denotes the n-th step transition kernel of X.

It is evident that the very existence of coalescing upper and lower chains implies uniform ergodicity. In [123] it has been shown that the reverse implication also holds at least in principle. Thus the scope of classic CFTP is strictly limited to uniformly ergodic chains. This is a severe limitation; many chains arising in practice are geometrically ergodic but not uniformly ergodic (random walks with negative drift and reflected in the origin, birth-death-immigration processes, queues which do not turn customers away, storage processes with no upper storage limit, ...).

However there are three ways to extend CFTP to extend beyond instances of uniform ergodicity:

1. **Truncation of state-space.** One can approximate the equilibrium distribution *via* constraining the target chain by forbidding any moves which take the chain out of a large but finite subset of state-space (so that the chain does not move

at all instead of carrying out a forbidden move). Under suitable conditions the equilibrium of the constrained chain can be shown to approximate that of the original chain. Particularly in case the target chain is reversible, this holds if the subset is irreducible for the constrained chain, and moreover the constrained equilibrium probabilities are proportional to the original probabilities within the constrained region. However it seems somewhat self-defeating to build an exact simulation of an approximation.

2. **Conversion to uniformly ergodic case.** In [292] it has been noted that one can sometimes convert a geometrically ergodic Markov chain into a uniformly ergodic Markov chain, simply by occasionally substituting in an independence sampler update.

3. **Domination by an amenable process.** It is shown in [225] how (in suitable cases) one can replace the upper bound (and lower bound, if required), by using a stationary random process which can be coupled to the target Markov chain so as to continue to lie above any realization of the target if it does so initially. If one can simulate this stationary process *backwards in time*, and if one can realize the coupling together with some variant on stochastic monotonicity, then this *dominated coupling-from-the-past* (domCFTP) can produce an exact simulation.

In this section we explore the method of domCFTP by describing the algorithm in the special context of a particular and very simple example. We refer to [227, Sect. 3.4, Theorem 31] for a general description of the method and a proof.

13.3.1 Random Walk domCFTP

We demonstrate domCFTP by applying it to the extremely elementary example of a simple random walk with negative drift, reflected in the origin. This has the advantage of being closely related to the random walk discussed in Sect. 13.2.1; therefore the R code will be closely related to the code in that section.

13.3.1.1 Helper Functions for domCFTP Simulation

Here is the `update` function which we will use for dominating and target Markov chains. The ± 1 `innovations` need to be constructed so that the probability of a $+1$ innovation is less than $\frac{1}{2}$, in order for an equilibrium to exist (and in this case it is a straightforward exercise to show that geometric ergodicity applies).

```
update <- function(x, innov) return(max(x +
    innov, 0))
```

Note that the reflection carried out here is the reversible form of reflection also used in Sect. 13.2.1; the recipe can be adapted to other forms of reflection.

The truncation method would amount to replacing this update by an update of the form used in Sect. 13.2.1. Not only does this give only approximate answers, but also coalescence involves waiting for the upper random walk to sink to zero, or the lower random walk to rise to the top level. Clearly good approximations will result in increasing computational demands, unnecessary if compared with the domCFTP method demonstrated below. Conversion to a uniformly ergodic chain is feasible in this simple one-dimensional case, but our purpose here is to demonstrate domCFTP. For dominating process, we choose the same reflected random walk but with a larger value $p^{\mathrm{dom}} \in (p, \frac{1}{2})$ of positive jump. The dominating random walk is reversible, and its equilibrium distribution is easily computed as geometric using detailed balance: setting $\rho^{\mathrm{dom}} = p^{\mathrm{dom}}/(1 - p^{\mathrm{dom}})$,

$$\pi_x^{\mathrm{dom}} = (1 - \rho^{\mathrm{dom}})(\rho^{\mathrm{dom}})^x \quad \text{for } x = 0, 1, 2, \dots . \tag{13.5}$$

We need a function `evolve`, which uses `update` to build up a trajectory from increments using an initial value.

```
evolve <- function(x, dx) {
    result <- c(x)
    for (u in dx) {
        x <- update(x, u)
        result <- append(result, x)
    }
    return(result)
}
```

We can now simulate the dominating process (in statistical equilibrium) by drawing its height at $t = 0$ from the equilibrium distribution and then simulating the same process backwards in time (initially we work back to the initial time specified by the invocation of the algorithm). We base this on generation of innovations, but now these innovations are to be viewed as occurring in reverse time.

```
make_block <- function(block_length, p_dom = 0.4) {
    return(2 * rbinom(block_length, 1, p_dom) - 1)
}
```

13.3.1.2 A Typical Run of the Random Walk domCFTP algorithm

As in Sect. 13.2.1, we build up the algorithm in a sequence of steps. First we determine the trajectory of the dominating process, by using `evolve` to build up the trajectory (in reverse-time) using final value and innovations. The distribution of the dominating process at time 0 is geometric, as given in the detailed balance calculations resulting in (13.5).

```
set.seed(5)
p_dom <- 0.4
p <- 0.25
T <- -9
```

So we work here in this example with $p^{\text{dom}} = 0.4$. We generate reversed dominating innovations `r_d_innovs` for the dominating process *in reversed time* and evolve the dominating `trajectory` accordingly:

```
dom0 <- rgeom(1, 1 - p_dom/(1 - p_dom))
r_d_innovs <- make_block(-T, p_dom = p_dom)
trajectory <- rev(evolve(dom0, r_d_innovs))
```

Next, we need to run the target process forwards in a way which is coupled to the dominating process. Note that the innovations `r_d_innovs` for the dominating process *cannot* be the innovations for the target process; dominating innovations are run backwards in time for the stationary dominating process, so the independence structure is quite different. We construct innovations for the target process by exploiting the reversibility of the dominating process; thus when run forwards in time the dominating process jumps by $+1$ with probability $p^{\text{dom}} = 0.4$. We wish to work with probability $p < p^{\text{dom}}$ for a $+1$ jump; hence -1 jumps of the dominating process remain negative, 0 jumps are converted to -1 jumps, while $+1$ jumps are converted into -1 jumps with probability $1 - p/p^{\text{dom}}$, as performed in the following function.

```
generate_innovations <- function(trajectory,
    old_innovs, p_dom = 0.4, p = 0.25) {
    innovs <- trajectory[2:length(trajectory)] -
        trajectory[1:(length(trajectory) - 1)]
    innovs[innovs == 0] <- -1
    innovs[innovs == 1] <- 1 - 2 *
        rbinom(length(innovs[innovs ==
            1]), 1, 1 - p/p_dom)
    return(c(innovs[1:(length(innovs) -
        length(old_innovs))], old_innovs))
}
```

These innovations can now be used to generate `upper` and `lower` processes. Note the initial value for the `upper` process is given by the trajectory of the dominating process.

```
lower <- evolve(0, innovations)
upper <- evolve(trajectory[1], innovations)
```

Coalescence not yet having occurred, we need to extend the dominating process. We work in reversed time as before.

```
old_upper <- upper
old_lower <- lower
old_T <- T
extend_r_d_innovs <- function(innovs,
    p_dom = 0.4) {
    return(c(innovs, make_block(length(innovs),
        p_dom = p_dom)))
}
trajectory <- rev(evolve(dom0,
    extend_r_d_innovs(r_d_innovs,
    p_dom = p_dom)))
T <- 1 - length(trajectory)
```

Having done this, we may now simulate the new upper and lower processes. Care must be taken to re-use randomness, but this is ensured by the construction of the `generate_innovations` function invoked above.

```
lower <- evolve(0, innovations)
upper <- evolve(trajectory[1], innovations)
```

We now achieve coalescence. Note that the consequence of re-use of randomness is that old upper and lower processes "funnel" between newer upper and lower processes.

13.3.1.3 Implementing the Algorithm

Finally we package the domCFTP algorithm in a function. Note that the output of `dom_cftp(T)` is *not* stable under variations of the start-time parameter T even if the seed of the random number generator is held fixed: this arises because each cycle of this particular domCFTP requires the use of the random number generator to impute innovations for the target process. In order to keep the code simple, this implementation does not synchronize each imputation with the generation of the corresponding entry in `r_d_innovs`, which is what would be required to stabilize the algorithm.

```
dom_cftp <- function(T, p_dom = 0.4, p = 0.25) {
    dom0 <- rgeom(1, 1 - p_dom/(1 - p_dom))
    r_d_innovs <- make_block(-T, p_dom = p_dom)
    trajectory <- rev(evolve(dom0, r_d_innovs))
    innovs <- generate_innovations(trajectory,
        c(), p_dom = p_dom, p = p)
    lower <- evolve(0, innovs)
    upper <- evolve(trajectory[1], innovs)
    while (lower[1 - T] != upper[1 - T]) {
```

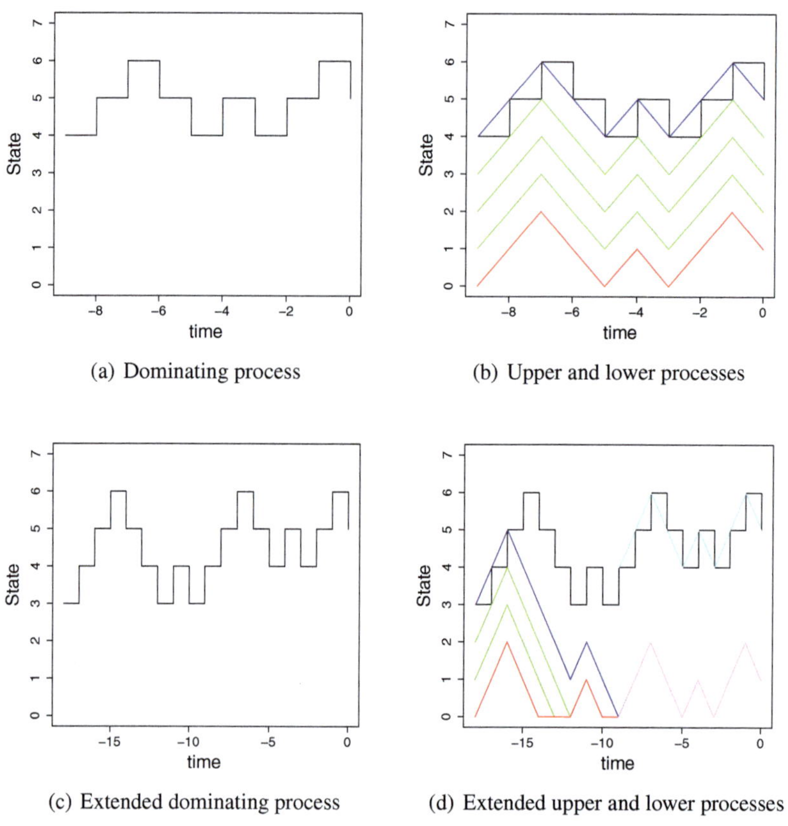

(a) Dominating process

(b) Upper and lower processes

(c) Extended dominating process

(d) Extended upper and lower processes

Fig. 13.9 domCFTP output for a simple random walk example

```
old_innov <- innovs[]
old_lower <- lower[]
old_upper <- upper[]
r_d_innovs <- extend_r_d_innovs(r_d_innovs,
    p_dom = p_dom)
T <- -length(r_d_innovs)
trajectory <- rev(evolve(dom0, r_d_innovs))
innovs <- generate_innovations(trajectory,
    innovs, p_dom = p_dom, p = p)
old_innov_range <- (length(innovs) -
    length(old_innov) + 1):length(innovs)
lower <- evolve(0, innovs)
upper <- evolve(trajectory[1], innovs)
old_range <- (length(lower) + 1 -
```

```
                        length(old_lower)):length(lower)
        }
        return(upper[length(upper)])
    }
```

13.3.1.4 Statistical Analysis

We now test the algorithm. It is worth paying particular attention to whether the
actual implementation maintains the fundamental "funnelling" relationship

```
old_lower ≤ lower ≤ upper ≤ old_upper ≤ trajectory.
```

This can be checked during evolution of the algorithm by using the *R* function
`stopifnot`.

Detailed inspection of the operation of the domCFTP algorithm reveals that in
the following 10000 runs the algorithm reaches back on occasion to time -120 in
order to achieve coalescence.

```
N <- 10000
data <- sapply(rep(-30, N), dom_cftp)
est <- sum(data > 0)/N
sd <- sqrt(est * (1 - est)/N)
print(paste("Empirical estimate of p / (1 - p) (=",
    format(p/(1 - p), digits = 3), "): ",
    format(est, digits = 3), "+/-", format(2 *
        sd, digits = 1), sep = "")) 
```

This yields output giving an empirical estimate for $p/(1-p)$:

```
[1] "Empirical estimate of p / (1 - p) (=0.333): 0.331+/-0.01"
```

The above empirical estimate forms part of a larger analysis, comparing log-
frequencies with their theoretical expectations, which is graphed in Fig. 13.10.

```
rho <- p/(1 - p)
counts <- tabulate(1 + data)
k <- 5
probs <- c(dgeom(0:k, 1 - rho), pgeom(k, 1 -
    rho, lower.tail = FALSE))
counts
```

The above code fragment produces output tabulating the output of the domCFTP
algorithm.

```
[1] 6688 2183  742  268   80   21   13    2    1    1
[11]    0    1
```

We now compute the expected output:

```
N * probs
```

with output as follows.

```
[1] 6666.66667 2222.22222   740.74074    246.91358
[5]    82.30453   27.43484    13.71742
```

Finally we perform a χ^2 test,

```
chisq.test(tabulate(1 + data, nbins = k +
    2), p = probs)
```

with output as follows.

```
        Chi-squared test for given probabilities

data:   tabulate(1 + data, nbins = k + 2)
X-squared = 4.1744, df = 6, p-value = 0.6531
```

The χ^2 test has been carried out on samples of size 10^6, and still reports no significant deviations from the predicted equilibrium distribution.

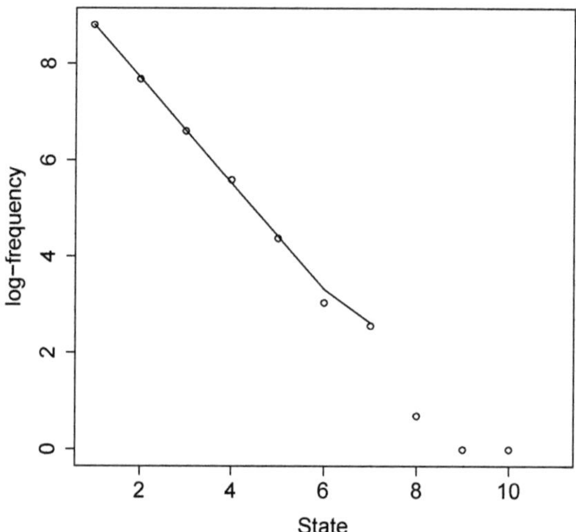

Fig. 13.10 Plot of log-frequencies of domCFTP output for the simple random walk example: the line indicates the theoretical expectation. Note that sampling variation is high at low log-frequencies, so the theoretical line is not plotted there.

While conveniently simple, this example is unrepresentative in a number of ways. Firstly, the dominating process often visits 0, thus forcing coalescence of upper and lower processes at such times. In more general applications of domCFTP such convenient coalescence will be very rare. Speaking technically, one should generally not expect the dominating process in substantial applications to exhibit a practically useful regenerative atom. Secondly, the target process is a random walk (and we are using a rather simple synchronous coupling of upper and lower processes), and this means that upper and lower processes themselves can only coalesce when the upper process hits 0 (in contrast to the continuous-time non-linear birth-death example discussed in [224, 227]). Nevertheless, the simplicity of this example permits clear demonstration of the underlying principles of domCFTP unhindered by technicalities.

13.3.2 Other Remarks

While the random walk domCFTP example is trivial, it serves as a useful prototype for much more complicated and useful examples. There are a variety of instances of more realistic examples of domCFTP. We have already noted the continuous-time non-linear birth-death example discussed in [224, 227; this extends naturally to treatment of spatial point processes (such as area-interaction point processes and Strauss point processes), to be found in [224, 225, 228]; see also the tutorial paper [36]. Note that many spatial point processes are produced by non-attractive spatial birth-and-death processes; in this case a simple crossover trick can be applied to ensure upper and lower processes envelope the perfectly simulated trajectory [225, 228]. A nice application to the (super-stable) M/G/s queue is given in [368]. In [281] the ideas of domCFTP have been applied to produce an intriguing model diagnostic for a spatial point process. Finally, in [118] it is shown how to use domCFTP to produce exact draws from Vervaat's perpetuity.

Exercise 13.10. Modify the function $\texttt{dom\_cftp}$ to work for a target process which evolves as a *modified* random walk X with negative bias: at each time n such that $X_n > 1$ the next step $X_{n+1} - X_n$ is replaced by a double jump -2 with some small positive probability, independently of the past and of the step $X_{n+1} - X_n$.

13.4 Conclusion

As noted in Sect. 13.3, the equivalence of (a) uniform ergodicity and (b) existence (at least in principle) of a classical CFTP algorithm has been shown in [123]. The implication "classical CFTP implies uniform ergodicity" is an immediate calculation, using the consideration that bounds on coalescence time will generate bounds on convergence in total variation, and it is well-known in the field how to improve such bounds to establish geometric decay. The reverse implication follows by noting the fact that uniform ergodicity implies that the whole state-space is a small set of lag k for some $k > 0$. But then the k-sub-sampled chain makes the whole state-space small of lag 1, and then small set CFTP may be applied as in Sect. 13.2.3.

Of course this small set CFTP only works in principle. To make it work in practise, one would need to determine the lag k and the regeneration measure $\rho v(\cdot)$ of the state-space, and one would also need to know how to draw from the k-step transition probability conditioned on regeneration *not* occurring. In practise one would at least need $k = 1$; moreover it would be crucial to consider whether the actual run-time of the algorithm made it a feasible means of achieving draws from the equilibrium. However, whether practical or impractical, the existence of small set CFTP is surely of interest.

It is then natural to ask whether there is a relationship between geometric ergodicity and domCFTP. Even the simple example of Sect. 13.3 provides an example of a geometrically ergodic chain which is not uniformly ergodic, and yet possesses a domCFTP algorithm. However general geometrically ergodic chains do not possess natural monotonicity structure, and this appears to be an obstacle to the existence of domCFTP.

However for Markov chains satisfying the weak property of φ-irreducibility it is actually the case that geometric ergodicity is equivalent to the existence of a geometric Foster-Lyapunov criterion (for φ-irreducibility and this equivalence result we refer to the famous monograph of Meyn and Tweedie [272]). Namely, let us consider the following condition.

Definition 13.4 (Geometric Foster-Lyapunov condition). The Markov chain X satisfies a geometric Foster-Lyapunov condition if there are: a small set C; constants $\alpha \in (0,1)$, $b > 0$; and a scale or drift function $\Lambda : \mathcal{X} \to [1,\infty)$ bounded on C; such that C is a Λ-sub-level set, and for π-almost all $x \in \mathcal{X}$

$$\mathbf{E}\left(\Lambda\left(X_{n+1}\right) \mid X_n = x\right) \quad \leq \quad \alpha\Lambda(x) + b\mathbf{1}_{(x \in C)},$$

where π is the equilibrium measure of X.

The following can then be established [226]: stochastic comparisons using the Markov inequality establish that $\Lambda(X_n)$ can be dominated by D, the scalar multiple of the exponential of the workload of a D/M/1 queue sampled at arrivals. The dominating chain will not itself necessarily be recurrent. However under k-sub-sampling for suitable k and variation of the small set $C = \{x : \Lambda(x) \leq c\}$ the domination can be refined to produce a dominating Markov chain D which *is* recurrent. Although D is not reversible, nonetheless one can use duality to compute its time-reversal under equilibrium. Refining the sub-sampling if necessary, one can then show that a sub-sampling of $\Lambda(X)$ is dominated by a stationary dominating Markov chain D, for which the equilibrium and time-reversed statistics are known, and such that regeneration occurs whenever D sinks below the level c. This suffices to establish a domCFTP algorithm, impractical in the same way that the Foss-Tweedie CFTP is impractical for the case of uniform ergodicity.

Despite its impracticality, this queue-based domCFTP algorithm bears a clear resemblance to the practical domCFTP algorithms discussed in the literature. It suggests the following challenging question: if one can prove that a chain is geometrically ergodic, then shouldn't one try to exhibit a domCFTP algorithm?

It is natural to ask whether geometric ergodicity is equivalent to domCFTP. However the answer to this is negative. It has been shown in [79] that domCFTP can be established for classes of non-geometrically ergodic Markov chains (specifically, polynomially ergodic Markov chains satisfying a technical condition known as "tameness"). This has in turn led to theoretical advances in the theory of polynomially ergodic Markov chains [78].

References

1. Aarts, E., Korst, J.: Simulated Annealing and Boltzmann Machines. A Stochastic Approach to Combinatorial Optimization and Neural Computing. John Wiley & Sons, Chichester (1989)
2. Abramowitz, M., Stegun, I.A.: Handbook of Mathematical Functions with Formulas, Graphs, and Mathematical Tables, *National Bureau of Standards Applied Mathematics Series*, vol. 55. Dover, New York (1964)
3. Adler, R., Taylor, J.: Random Fields and Geometry. Springer, New York (2007)
4. Altendorf, H., Jeulin, D.: Random walk based stochastic modeling of 3d fiber systems. Physical Review E **83**, 041,804 (2011)
5. Andrieu, C., Doucet, A., Holenstein, R.: Particle Markov chain Monte Carlo methods. Journal of the Royal Statistical Society B **72**, 269–342 (2010)
6. Arlot, S., Celisse, A.: A survey of cross-validation procedures for model selection. Statistics Surveys **4**, 40–79 (2010)
7. Arns, C.H., Knackstedt, M.A., Mecke, K.R.: Reconstructing complex materials via effective grain shapes. Physical Review Letters **91**(21), 215,506 (2003)
8. Asmussen, S., Rosiński, J.: Approximations of small jumps of Lévy processes with a view towards simulation. Journal of Applied Probability **38**(2), 482–493 (2001)
9. Athreya, K.B., Ney, P.: A new approach to the limit theory of recurrent Markov chains. Transactions of the American Mathematical Society **245**, 493–501 (1978)
10. Athreya, K.B., Stenflo, O.: Perfect sampling for Doeblin chains. Sankhya **65**(4), 763–777 (2003)
11. Auneau, J., Jensen, E.B.V.: Expressing intrinsic volumes as rotational integrals. Advances in Applied Mathematics **45**, 1–11 (2010)
12. Auneau-Cognacq, J., Rataj, J., Jensen, E.B.V.: Closed form of the rotational Crofton formula. Mathematische Nachrichten **285**, 164–180 (2012)
13. Auneau-Cognacq, J., Ziegel, J., Jensen, E.B.V.: Rotational integral geometry of tensor valuations. Advances in Applied Mathematics **50**, 429–444 (2013)
14. Aurenhammer, F.: A criterion for the affine equivalence of cell complexes in $\mathbb{R}^d$ and convex polyhedra in $\mathbb{R}^{d+1}$. Discrete & Computational Geometry **2**, 49–64 (1987)
15. Baddeley, A.: Spatial point processes and their applications. In: A. Baddeley, I. Barany, R. Schneider, W. Weil (eds.) Stochastic Geometry. *Lecture Notes in Mathematics*, vol. 1892, pp. 1–75. Springer Berlin Heidelberg (2007)
16. Baddeley, A., Gregori, P., Mateu, J., Stoica, R., Stoyan, D. (eds.): Case Studies in Spatial Point Process Modeling, *Lecture Notes in Statistics*, vol. 185. Springer, New York (2006)
17. Baddeley, A., Jensen, E.B.V.: Stereology for Statisticians. No. 103 in Monographs on Statistics and Applied Probability. Chapman & Hall/CRC, Boca Raton (2005)
18. Baddeley, A., Nair, G.: Fast approximation of the intensity of Gibbs point processes. Electronic Journal of Statistics **6**, 1155–1169 (2012)

© Springer International Publishing Switzerland 2015

V. Schmidt (ed.), *Stochastic Geometry, Spatial Statistics and Random Fields*,
Lecture Notes in Mathematics 2120, DOI 10.1007/978-3-319-10064-7

19. Baddeley, A., Turner, R.: Spatstat: an R package for analyzing spatial point patterns. Journal of Statistical Software **12**(6), 1–42 (2005). URL www.jstatsoft.org

20. Baddeley, A.J.: Spatial sampling and censoring. In: W. Kendall, M. van Lieshout, O. Barndorff-Nielsen (eds.) Stochastic Geometry: Likelihood and Computation, pp. 37–78. Chapman & Hall, London (1999)

21. Baddeley, A.J.: Spatial point patterns - models and statistics. In: E. Spodarev (ed.) Stochastic Geometry, Spatial Statistics and Random Fields: Asymptotic Methods, *Lecture Notes in Mathematics*, vol. 2068, pp. 49–116. Springer, Heidelberg (2013)

22. Baddeley, A.J., Møller, J., Waagepetersen, R.: Non- and semiparametric estimation of interaction in inhomogeneous point patterns. Statistica Neerlandica **54**, 329–350 (2000)

23. Balding, D.J., Bishop, M., Cannings, C. (eds.): Handbook of Statistical Genetics, vol. 1,2, 3rd edn. John Wiley & Sons, Chichester (2007)

24. Ballani, F., Daley, D.J., Stoyan, D.: Modelling the microstructure of concrete with spherical grains. Computational Materials Science **35**, 399–407 (2006)

25. Ballani, F., Kabluchko, Z., Schlather, M.: Random marked sets. Advances in Applied Probability **44**, 603–613 (2012)

26. Barbour, A.D.: Stein's method and Poisson process convergence. Journal of Applied Probability **25A**, 175–184 (1988)

27. Barbour, A.D., Brown, T.C.: Stein's method and point process approximation. Stochastic Processes and their Applications **43**(1), 9–31 (1992)

28. Baringhaus, L., Franz, C.: On a new multivariate two-sample test. Journal of Multivariate Analysis **88**(1), 190–206 (2004)

29. Barlow, R.E., Proschan, F.: Mathematical Theory of Reliability. John Wiley & Sons, New York (1965)

30. Barnett, S.: Matrices: Methods and Applications. Clarendon Press, Oxford (1990)

31. Beisbart, C., Barbosa, M.S., Wagner, H., Costa, L.d.F.: Extended morphometric analysis of neuronal cells with Minkowski valuations. European Physical Journal B **52**, 531–546 (2006)

32. Beisbart, C., Dahlke, R., Mecke, K.R., Wagner, H.: Vector- and tensor-valued descriptors for spatial patterns. In: K. Mecke, D. Stoyan (eds.) Morphology of Condensed Matter, *Lecture Notes in Physics*, vol. 600, pp. 249–271. Springer, Berlin (2002)

33. Ben Hough, J., Krishnapur, M., Peres, Y., Virág, B.: Determinantal processes and independence. Probability Surveys **3**, 206–229 (2006)

34. Ben Hough, J., Krishnapur, M., Peres, Y., Virág, B.: Zeros of Gaussian analytic functions and determinantal point processes, *University Lecture Series*, vol. 51. American Mathematical Society (2009)

35. Berry, A.C.: The accuracy of the Gaussian approximation to the sum of independent variates. Transactions of the American Mathematical Society **49**, 122–136 (1941)

36. Berthelsen, K.K., Møller, J.: A primer on perfect simulation for spatial point processes. Bulletin of the Brazilian Mathematical Society **33**, 351–367 (2002)

37. Besag, J.: Spatial interaction and the statistical analysis of lattice systems. Journal of the Royal Statistical Society, Series B **36**(2), 192–236 (1974)

38. Bezrukov, A., Bargiel, M., Stoyan, D.: Statistical analysis of simulated random packings of spheres. Particle & Particle Systems Characterization **19**, 111–118 (2001)

39. Bingham, N.H., Goldie, C.M. (eds.): Probability and Mathematical Genetics. Cambridge University Press, Cambridge (2010)

40. Błaszczyszyn, B., Yogeshwaran, D.: Directionally convex ordering of random measures, shot-noise fields and some applications to wireless networks. Advances in Applied Probability **41**, 623–646 (2009)

41. Blake, A., Kohli, P., Rother, C. (eds.): Markov Random Fields for Vision and Image Processing. MIT Press, Cambridge MA (2011)

42. Błaszczyszyn, B., Yogeshwaran, D.: Connectivity in sub-Poisson networks. In: Proc. of 48 th Annual Allerton Conference. University of Illinois at Urbana-Champaign, IL, USA (2010)

43. Błaszczyszyn, B., Yogeshwaran, D.: Clustering, percolation and directionally convex ordering of point processes. arXiv:1105.4293v1 (2011)

44. Błaszczyszyn, B., Yogeshwaran, D.: Clustering and percolation of point processes. Electronic Journal of Probability **18**(72), 1–20 (2013)
45. Błaszczyszyn, B., Yogeshwaran, D.: On comparison of clustering properties of point processes. Advances in Applied Probability **46**(1), 1–21 (2014)
46. Bochner, S.: Lectures on Fourier Integrals. Princeton University Press, Princeton (1959)
47. Brabec, D., Scherf, U., Dyakonov, V. (eds.): Organic Photovoltaics: Materials, Device Physics, and Manufacturing Technologies. Wiley-VCH, Weinheim (2008)
48. Bradley-Smith, G., Hope, S., Firth, H.V., Hurst, J.A.: Oxford Handbook of Genetics. Oxford University Press, New York (2009)
49. Breiman, L.: Probability. Addison-Wesley, Reading, Massachusetts (1968)
50. Brémaud, P.: Markov Chains: Gibbs Fields, Monte Carlo Simulation, and Queues. Springer, New York (1999)
51. Brereton, T., Kroese, D.P., Stenzel, O., Schmidt, V., Baumeier, G.: Efficient simulation of charge transport in deep-trap media. In: C. Laroque, J. Himmelspach, R. Pasupathy, O. Rose, A.M. Uhrmacher (eds.) Proceedings of the 2012 Winter Simulation Conference. IEEE, Piscataway (2012)
52. Brillinger, D.R.: Modeling spatial trajectories. In: A. Gelfand, P. Diggle, M. Fuentes, P. Guttorp (eds.) Handbook of Spatial Statistics, pp. 463–476. Chapman & Hall/CRC, Boca Raton (2010)
53. Brix, A.: Generalized gamma measures and shot-noise cox processes. Advances in Applied Probability **31**, 929–953 (1999)
54. Bulinski, A., Butkovsky, O., Sadovnichy, V., Shashkin, A., Yaskov, P., Balatskiy, A., Samokhodskaya, L., Tkachuk, V.: Statistical methods of snp data analysis and applications. Open Journal of Statistics **2**, 73–87 (2012)
55. Bulinski, A., Rakitko, A.: Estimation of nonbinary random response. Doklady Mathematics pp. 1–6 (2014)
56. Bulinski, A., Spodarev, E.: Introduction to random fields. In: E. Spodarev (ed.) Stochastic Geometry, Spatial Statistics and Random Fields: Asymptotic Methods., *Lecture Notes in Mathematics*, vol. 2068, pp. 277–335. Springer, Heidelberg (2013)
57. Bulinski, A., Spodarev, E.: Central limit theorems for weakly dependent random fields. In: E. Spodarev (ed.) Stochastic Geometry, Spatial Statistics and Random Fields, Asymptotic Methods, *Lecture Notes in Mathematics*, vol. 2068, pp. 337–383. Springer, Heidelberg (2013)
58. Bulinski, A.V.: To the foundations of the dimesionality reduction method for explanatory variables. Zapiski Nauchnyh Seminarov POMI **408**, 84–101 (2012)
59. Bulinski, A.V.: Central limit theorem related to MDR method. arXiv:1301.6609 (2013)
60. Bulinski, A.V., Shiryaev, A.N.: Theory of Stochastic Processes. Fizmatlit, Moscow (2005). (in Russian)
61. Burton, R., Waymire, E.: Scaling limits for associated random measures. Annals of Applied Probability **13**(4), 1267–1278 (1985)
62. Bush, W.S., Dudek, S.M., Ritchie, M.D.: Parallel multifactor dimensionality reduction: a tool for the large-scale analysis of gene-gene interactions. Bioinformatics **22**, 2173–2174 (2006)
63. Calle, M.L., Urrea, V., Vellalta, G., Malats, N., van Steen, K.: Improving strategies for detecting genetic patterns of disease susceptibility in association studies. Statistics in Medicine **27**, 6532–6546 (2008)
64. Carlsson, G.: Topology and data. Bulletin of the American Mathematical Society (N.S.) **46**(2), 255–308 (2009)
65. Chan, G., Wood, A.T.A.: Simulation of stationary Gaussian vector fields. Statistics and Computing **9**, 265–268 (1999)
66. Chanda, P., Sucheston, L., Zhang, A., Brazeau, D., Freudenheim, J.L., Ambrosone, C., Ramanathan, M.: Ambience: A novel approach and efficient algorithm for identifying informative genetic and environmental associations with complex phenotypes. Genetics **180**, 1191–1210 (2008)
67. Chautru, J.: The use of Boolean random functions in geostatistics. In: M. Armstrong (ed.) Geostatistics, vol. 1, pp. 201–212. Kluwer Academic Publishers, Dordrecht (1989)

68. Chen, L.H.Y., Goldstein, L., Shao, Q.M.: Normal Approximation by Stein's Method. Springer, New York (2011)
69. Chen, L.H.Y., Shao, Q.M.: A non-uniform Berry-Esseen bound via Stein's method. Probability Theory and Related Fields **120**(2), 236–254 (2001)
70. Chen, L.H.Y., Xia, A.: Stein's method, Palm theory and Poisson process approximation. Annals of Probability **32**(3B), 2545–2569 (2004)
71. Chilès, J.P., Delfiner, P.: Geostatistics: Modeling Spatial Uncertainty. John Wiley & Sons, New York (1999)
72. Chiou, J.M., Li, P.L.: Functional clustering and identifying substructures of longitudinal data. Journal of the Royal Statistical Society: Series B. (Statistical Methodology) **69**, 679–699 (2007)
73. Chiu, S.N., Stoyan, D., Kendall, W.S., Mecke, J.: Stochastic Geometry and its Applications. John Wiley & Sons, Chichester (2013)
74. Choquet, G.: Cours d'analyse. Tome II: Topologie. Espaces topologiques et espaces métriques. Fonctions numériques. Espaces vectoriels topologiques. Deuxième édition, revue et corrigée. Masson et Cie, Éditeurs, Paris (1969)
75. Christofides, T.C., Vaggelatou, E.: A connection between supermodular ordering and positive/negative association. Journal of Multivariate Analysis **88**(1), 138–151 (2004)
76. Cole, E.A.B.: Genetic algorithms and simulated annealing. In: E.A.B. Cole (ed.) Mathematical and Numerical Modelling of Heterostructure Semiconductor Devices: From Theory to Programming, pp. 339–376. Springer, London (2009)
77. Connor, S.B.: Perfect Sampling. In: F. Ruggeri, R. Kenett, F. Faltin (eds.) Wiley Encyclopedia of Statistics in Quality and Reliability. John Wiley & Sons, Chichester (2007)
78. Connor, S.B., Fort, G.: State-dependent Foster-Lyapunov criteria for subgeometric convergence of Markov chains. Stochastic Processes and their Applications **119**(12), 4176–4193 (2009)
79. Connor, S.B., Kendall, W.S.: Perfect simulation for a class of positive recurrent markov chains. Annals of Applied Probability **17**(3), 781–808 (2007)
80. Conway, J.B.: A Course in Functional Analysis, 2nd edn. Springer, New York (1990)
81. Cook, R.: Regression Graphics: Ideas for Studying Regressions Through Graphics. John Wiley & Sons, New York (1998)
82. Cook, R.D., Weisberg, S.: Discussion of 'sliced inverse regression for dimension reduction. Journal of the American Statistical Association **86**, 28–33 (1991)
83. Corcoran, J.N., Schneider, U.: Shift and scale coupling methods for perfect simulation. Probability in the Engineering and Informational Sciences **17**(3), 277–303 (2003)
84. Coupier, D., Dereudre, D.: Continuum percolation for quermass model. Electronical Journal of Probability **19** (35), 1–19 (2014)
85. Cox, D.: Some statistical methods connected with series of events. Journal of the Royal Statistical Society, B **17**(24), 129–164 (1955)
86. Cressie, N.: Statistics for Spatial Data. John Wiley & Sons, 2nd edition, New York (1993)
87. Cruz-Orive, L.M.: A new stereological principle for test lines in three-dimensional space. Journal of Microscopy **219**, 18–28 (2005)
88. Cruz-Orive, L.M.: Comparative precision of the pivotal estimators of particle size. Image Analysis & Stereology **27**, 17–22 (2008)
89. Cruz-Orive, L.M.: Flowers and wedges for the stereology of particles. Journal of Microscopy **243**, 86–102 (2011)
90. Cruz-Orive, L.M.: Uniqueness properties of the invariator, leading to simple computations. Image Analysis & Stereology **31**, 89–98 (2012)
91. Cuisenaire, O.: Locally adaptable mathematical morphology using distance transformations. Pattern Recognition **39**(3), 405–416 (2006)
92. Daley, D.J., Last, G.: Descending chains, the lilypond model, and mutual-nearest-neighbour matching. Advances in Applied Probability **37**(3), 604–628 (2005)
93. Daley, D.J., Vere-Jones, D.: An Introduction to the Theory of Point Processes, vol. 1. Springer, New York (2003)

94. Daley, D.J., Vere-Jones, D.: An Introduction to the Theory of Point Processes, vol. 2. Springer, New York (2007)
95. Daniel, C. (ed.): Handbook of Battery Materials. Wiley-VCH, Weinheim (2012)
96. Davies, R.M.: The determination of static and dynamic yield stresses using a steel ball. Proceedings of the Royal Society of London. Series A. Mathematical and Physical Sciences **197**(1050), 416–432 (1949)
97. Davies, S.L., Neath, A.A., Cavanaugh, J.E.: Estimation optimality of corrected AIC and modified C_p in linear regression. International Statistical Review **74**(2), 161–168 (2006)
98. Denis, E.P., Barat, C., Jeulin, D., Ducottet, C.: 3d complex shape characterization by statistical analysis: Application to aluminium alloys. Materials Characterization **59**(3), 338 – 343 (2008)
99. DeVore, R.A., Lorentz, G.G.: Constructive Approximation. Springer, Berlin (1993)
100. Dietrich, C.R., Newsam, G.N.: Fast and exact simulation of stationary Gaussian processes through circulant embedding of the covariance matrix. SIAM Journal on Scientific Computing **18**(4), 1088–1107 (1997)
101. Diggle, P.J.: Spatio-temporal point processes: Methods and Applications. In: B. Finkenstaedt, L. Held, V. Isham (eds.) Statistical Methods for Spatio-temporal Systems, pp. 1–45. Chapman & Hall/CRC, Boca Raton (2007)
102. Diggle, P.J.: Statistical Analysis of Spatial Point Patterns. Oxford University Press, London (2003)
103. Diggle, P.J., Chetwynd, A.G., Häggkvist, R., Morris, S.: Second-order analysis of space-time clustering. Statistical Methods in Medical Research **4**, 124–136 (1995)
104. Doucet, A., de Freitas, N., Gordon, N.: Sequential Monte Carlo Methods in Practice. Springer, New York (2001)
105. Dousse, O., Franceschetti, M., Macris, N., Meester, R., Thiran, P.: Percolation in the signal to interference ratio graph. Journal of Applied Probability **43**(2), 552–562 (2006)
106. Dubhashi, D., Ranjan, D.: Balls and bins: A study in negative dependence. Random Structures & Algorithms **13**, 99–124 (1996)
107. Dudley, R.M.: Real Analysis and Probability. Wadsworth & Brooks/Cole, Pacific Grove, CA (1989)
108. Dvořák, J., Jensen, E.B.V.: On semi-automatic estimation of surface area. Journal of Microscopy **250**, 142–157 (2013)
109. Dvořák, J., Prokešová, M.: Moment estimation methods for stationary spatial Cox processes – a comparison. Kybernetika **48**, 1007–1026 (2012)
110. Edmonds, J., Karp, R.M.: Theoretical improvements in algorithmic efficiency for network flow problems. Journal of the ACM **19**, 248–264 (1972)
111. Edwards, T.L., Torstensen, E., Dudek, S. Martin, E.R., Ritchie, M.D.: A cross-validation procedure for general pedigrees and matched odds ratio fitness metric implemented for the multifactor dimensionality reduction pedigree disequilibrium test mdr-pdt and cross-validation: power studies. Genetic Epidemiology **34**, 194–199 (2010)
112. Ergun, A., Barbieri, R., Eden, U.T., Wilson, M.A., Brown, E.N.: Construction of point process adaptive filter algorithms for neural system using sequential Monte Carlo methods. IEEE Transactions on Biomedical Engineering **54**, 419–428 (2007)
113. Esseen, C.G.: On the Liapounoff limit of error in the theory of probability. Arkiv för matematik, astronomi och fysik **28A**(9), 19 (1942)
114. Esseen, C.G.: A moment inequality with an application to the central limit theorem. Skandinavisk Aktuarietidskrift **39**, 160–170 (1956)
115. Ethier, S.N., Kurtz, T.G.: Markov Processes. John Wiley & Sons, New York (1986)
116. Fan, J., Gijbels, I.: Local Polynomial Modelling and its Applications, 1st. edn. Chapman & Hall, London (1996)
117. Fill, J.A.: An interruptible algorithm for exact sampling via Markov Chains. Annals of Applied Probability **8**, 131–162 (1998)
118. Fill, J.A., Huber, M.L.: Perfect simulation of Vervaat perpetuities. Electronic Journal of Probability **15**(4), 96–109 (2010)

119. Fill, J.A., Machida, M.: Stochastic monotonicity and realizable monotonicity. Annals of Probability **29**(2), 938–978 (2001)
120. Fill, J.A., Machida, M.: Realizable monotonicity and inverse probability transform. In: C.M. Cuadras, J. Fortiana, J. Rodra-Guez-Lallena (eds.) Distributions with Given Marginals and Statistical Modelling, pp. 63–71. Kluwer Academic Publishers, Dordrecht (2002)
121. Fill, J.A., Machida, M., Murdoch, D.J., Rosenthal, J.S.: Extension of Fill's perfect rejection sampling algorithm to general chains. Proceedings of the Ninth International Conference "Random Structures and Algorithms" pp. 290–316 (2000)
122. Foss, S., Korshunov, D.A., Zachary, S.: An Introduction to Heavy-tailed and Subexponential Distributions. Springer, New York (2011)
123. Foss, S.G., Tweedie, R.L.: Perfect simulation and backward coupling. Stochastic Models **14**(1), 187–203 (1998)
124. Fourard, C., Malandain, G., Prohaska, S., Westerhoff, M.: Blockwise processing applied to brain microvascular network study. IEEE Transactions on Medical Imaging **25**, 1319–1328 (2006)
125. François, O., Ancelet, S., Guillot, G.: Bayesian clustering using hidden markov random fields in spatial population genetics. Genetics **174**, 805–816 (2006)
126. Franceschetti, M., Dousse, O., C., T.D.N., P., T.: Closing the gap in the capacity of wireless networks via percolation theory. IEEE Transactions on Information Theory **53**(3), 1009–1018 (2007)
127. Franke, R.: Scattered data interpolation: Tests of some methods. Mathematics of Computation **38**, 181–200 (1982)
128. Frcalová, B., Beneš, V., Klement, D.: Spatio-temporal point process filtering methods with an application. Environmetrics **21**, 240–252 (2010)
129. Fréchet, M.: Les tableaux dont les merges sont données. Annals de l'Univ de Lyon **51**, 53–77 (1951)
130. Fujikoshi, Y., Satoh, K.: Modified AIC and C_p in multivariate linear regression. Biometrika **84**(3), 707–716 (1997)
131. Gaiselmann, G., Kulik, R., Schmidt, V.: Statistical inference for curved fibrous objects in 3D, based on multiple observations of short multivariate autoregression processes. Australian and New Zealand Journal of Statistics, to appear (2014)
132. Gaiselmann, G., Manke, I., Lehnert, W., Schmidt, V.: Extraction of curved fibers from 3D images. Image Analysis and Stereology **32**, 57–63 (2013)
133. Gaiselmann, G., Thiedmann, R., Manke, I., Lehnert, W., Schmidt, V.: Stochastic 3D modeling of fiber-based materials. Computational Materials Science **59**, 75–86 (2012)
134. Gaiselmann, G., Tötzke, C., Froning, D., Manke, I., Lehnert, W., Schmidt, V.: Stochastic 3D modeling of non-woven gdl with wet-proofing agent. International Journal of Hydrogen Energy **38**, 8448–8460 (2013)
135. Gandin, L.S.: Objektivnyj Analiz Meteorologiceskich Polej. Gidrometeoizdat, Leningrad (1963)
136. Geyer, C.J., Møller, J.: Simulation procedures and likelihood inference for spatial point processes. Scandinavian Journal of Statistics **21**(4), pp. 359–373 (1994)
137. Ghosh, S.: Determinantal processes and completeness of random exponentials: the critical case. arXiv:1211.2435 (2012)
138. Ghosh, S., Krishnapur, M., Peres, Y.: Continuum percolation for Gaussian zeroes and Ginibre eigenvalues. arXiv:1211.2514 (2012)
139. Ghrist, R.: Barcodes: the persistent topology of data. Bulletin of the American Mathematical Society (N.S.) **45**(1), 61–75 (2008)
140. Gibbs, A.L.: Convergence in the Wasserstein metric for Markov chain Monte Carlo algorithms with applications to image restoration. Stochastic Models **20**(4), 473–492 (2004)
141. Gilbert, E.N.: Random plane networks. Journal of the Society for Industrial and Applied Mathematics **9**, 533–543 (1961)
142. Gneiting, T., Sasvári, Z., Schlather, M.: Analogies and correspondences between variograms and covariance functions. Advances in Applied Probability **33**, 617–630 (2001)

143. Gneiting, T., Seveikova, H., Percival, D.B., Schlather, M., Jiang, Y.: Fast and exact simulation of large Gaussian lattice systems in $\mathbb{R}^2$: Exploring the limits. Journal of Computational and Graphical Statistics **15**(3), 483–501 (2006)

144. Goldman, A.: The Palm measure and the Voronoi tessellation for the Ginibre process. Annals of Applied Probability **20**(1), 90–128 (2010)

145. Golland, P., Liang, F., Mukherjee, S., Panchenko, D.: Permutation tests for classification. In: P. Aeur, R. Meir (eds.) Learning Theory, *Lecture Notes in Computer Science*, vol. 3559, pp. 501–515, Springer, Berlin (2005)

146. Golub, G.H., Loan, C.F.V.: Matrix Computations. Johns Hopkins University Press, Baltimore (1996)

147. Gordon, R.D.: Values of Mills' ratio of area to bounding ordinate and of the normal probability integral for large values of the argument. Annals of Mathematical Statistics **12**, 364–366 (1941)

148. Götze, F., Tikhomirov, A.N.: Asymptotic distribution of quadratic forms. Annals of Probability **27**(2), 1072–1098 (1999)

149. Grabarnik, P., Myllymäki, M., Stoyan, D.: Correct testing of mark independence for marked point patterns. Ecological Modelling **222**, 3888–3894 (2011)

150. Green, P.J.: Reversible jump Markov chain Monte Carlo computation and Bayesian model determination. Biometrika **82**(4), 711–732 (1995)

151. Grimmett, G.: Probability of Graphs: Random Processes on Graphs and Lattices. Cambridge University Press, Cambridge (2011)

152. Grimmett, G.R.: Percolation. Springer, Heidelberg (1999)

153. Grimmett, G.R., Stirzaker, D.R.: Probability and Random Processes, 3rd edn. Oxford University Press, New York (2001)

154. Gual-Arnau, X., Cruz-Orive, L.M., Nuno-Ballesteros, J.J.: A new rotational integral formula for intrinsic volumes in space forms. Advances in Applied Mathematics **44**, 298–308 (2010)

155. Guan, Y.: A minimum contrast estimation procedure for estimating the second-order parameters of inhomogeneous spatial point processes. Statistics and Its Interface **40**, 603–629 (2008)

156. Guan, Y.: On consistent nonparametric intensity estimation for inhomogeneous spatial point processes. Journal of the American Statistical Association **103**, 1238–1247 (2008)

157. Guan, Y., Wang, H.: Sufficient dimension reduction for spatial point processes directed by gaussian random fields. Journal of the Royal Statistical Society B **72**, 367–387 (2010)

158. Gui, J., Andrew, A.S., Andrews, P., Nelson, H.M., Kelsey, K.T., Karagas, M.R., Moore, J.H.: A robust multifactor dimensionality reduction method for detecting gene-gene interactions with application to the genetic analysis of bladder cancer susceptibility. Annals of Human Genetics **75**, 20–28 (2011)

159. Gundersen, H.J.G.: The nucleator. Journal of Microscopy **151**, 3–21 (1988)

160. Gut, A.: Probability : A Graduate Course, 2nd edn. Springer, New York (2013)

161. Hagel, S.: Extrapolation von stabilen Zufallsfeldern. Master's thesis, Ulm University (2012)

162. Hajek, B.: Cooling schedules for optimal annealing. Mathematics of Operations Research **13**, 311–329 (1988)

163. Hall, P.: Introduction to the Theory of Coverage Processes. John Wiley & Sons, New York (1988)

164. Hansen, L.V., Kiderlen, M., Jensen, E.B.V.: Image-based empirical importance sampling: an efficient way of estimating intensities. Scandinavian Journal of Statistics **38**, 393–408 (2011)

165. Hansen, L.V., Nyengaard, J.R., Andersen, J.B., Jensen, E.B.V.: The semi-automatic nucleator. Journal of Microscopy **242**, 206–215 (2011)

166. Hartnig, C., Jörissen, L., Kerres, J., Lehnert, W., Scholta, J.: Polymer electrolyte membrane fuel cells (PEMFC). In: M. Gasik (ed.) Materials for Fuel Cells, pp. 101–184. Woodhead Publishing, Cambridge (2008)

167. Hastie, T., Tibshirani, R., Friedman, J.: The Elements of Statistical Learning; Data Mining, Inference and Prediction. Springer, New York (2008)

168. He, Z., Zhang, M., Zhan, X., Lu, Q.: Modeling and testing for joint association using a genetic random field model. Biometrics, to appear, arXiv:1302.5493 (2014)

169. Heinrich, L.: Asymptotic methods in statistics of random point processes. In: E. Spodarev (ed.) Stochastic Geometry, Spatial Statistics and Random Fields: Asymptotic Methods, *Lecture Notes in Mathematics*, vol. 2068, pp. 107–136. Springer, Heidelberg (2013)
170. Hellmund, G., Prokešová, M., Vedel Jensen, E.: Lévy based Cox point processes. Advances in Applied Probability **40**, 603–629 (2008)
171. Heuser, H.: Funktionalanalysis. Teubner, Stuttgart (1975)
172. Häggström, O., Nelander, K.: On exact simulation of Markov random fields using coupling from the past. Scandinavian Journal of Statistics. Theory and Applications **26**(3), 395–411 (1999)
173. Hirsch, C., Neuhaeuser, D., Schmidt, V.: Connectivity of random geometric graphs related to minimal spanning forests. Advances in Applied Probability **45**, 20–36 (2013)
174. Hobert, J.P., Robert, C.P.: A mixture representation of pi with applications in Markov chain Monte Carlo. Annals of Applied Probability **14**(3), 1295–1305 (2004)
175. Huber, M.L.: A bounding chain for Swendsen-Wang. Random Structures and Algorithms **22**(1), 43–59 (2003)
176. Hug, D., Schneider, R., Schuster, R.: Integral geometry of tensor valuations. Advances in Applied Mathematics **41**, 482–509 (2008)
177. Ibragimov, I.A., Rozanov, Y.A.: Gaussian Random Processes, *Applications of Mathematics*, vol. 9. Springer, New York-Berlin (1978)
178. Iliev, I.T., Mellema, G., Pen, U.L., Bond, J.R., Shapiro, P.R.: Current models of the observable consequences of cosmic reionization and their detectability. Monthly Notices of the Royal Astronomical Society **384**(3), 863–874 (2008)
179. Illian, J., Penttinen, A., Stoyan, H., Stoyan, D.: Statistical Analysis and Modelling of Spatial Point Patterns. John Wiley & Sons, Chichester (2008)
180. Ilucova, L., Saxl, I., Svoboda, M., Sklenicka, V., Kral, P.: Structure of ECAP aluminium after different number of passes. Image Analysis and Stereology **26**, 27–34 (2007)
181. Itô, K., Nisio, M.: On the oscillation function of Gaussian processes. Mathematica Scandinavica **22**, 209–223 (1968)
182. Jacobi, C.G.J.: De investigando ordine systematis aequationum differentialum vulgarium cujuscunque. Journal für die reine und angewandte Mathematik **64**, 297–320 (1865)
183. James, G., Sugar, C.: Clustering for sparsely sampled functional data. Journal of the American Statistical Association **98**, 397–408 (2003)
184. Jensen, E.B., Gundersen, H.J.G.: Stereological estimation of surface area of arbitrary particles. Acta Stereologica **6**, 25–30 (1987)
185. Jensen, E.B.V.: Local Stereology. World Scientific, Singapore (1998)
186. Jensen, E.B.V., Gundersen, H.J.G.: The rotator. Journal of Microscopy **170**, 35–44 (1993)
187. Jensen, E.B.V., Rataj, J.: A rotational integral formula for intrinsic volumes. Advances in Applied Mathematics **41**, 530–560 (2008)
188. Jensen, E.B.V., Ziegel, J.F.: Local stereology of tensors of convex bodies. Methodology and Computing in Applied Probability **16**, 263–282 (2014)
189. Jeulin, D.: Random image models for microstructure analysis and simulation. Invited Lecture, Tenth Pfefferkorn Conference on Signal and Image Processing in Microscopy and Microanalysis, 16-19 September 1991, Cambridge, UK, Scanning Microscopy Supplement 6, pp. 121–128 (1992)
190. Jeulin, D.: Anisotropic rough surface modelling by random morphological functions. In: Proc. 4th European Symposium of Stereology, vol. 6, pp. 183–189 (1987)
191. Jeulin, D.: Morphological modeling of images by sequential random functions. Signal Processing **16**, 403–431 (1989)
192. Jeulin, D.: Modèles de fonctions aléatoires multivariables. Science de la Terre **30**, 225–256 (1991)
193. Jeulin, D.: Modèles morphologiques de structures aléatoires et de changement d'echelle. Ph.D. thesis, Université of Caen, Caen, France (1991)
194. Jeulin, D.: Multivariate random image models. Acta Stereologica **11**, 59–66 (1992)
195. Jeulin, D.: Morphological modelling of surfaces. Surface Engineering **14**, 199–204 (1998)

196. Jeulin, D.: Random texture models for materials structures. Statistics and Computing **10**, 121–131 (2000)
197. Jeulin, D.: Random structure models for homogenization and fracture statistics. In: D. Jeulin, M. Ostoja-Starzewski (eds.) Mechanics of Random and Multiscale Microstructures, *CISM International Centre for Mechanical Sciences*, vol. 267, pp. 33–91. Springer, Berlin (2001)
198. Jeulin, D.: Random structures in physics. In: M. Bilodeau, F. Meyer, M. Schmitt (eds.) Space, Structure and Randomness, *Lecture Notes in Statistics*, vol. 183, pp. 183–222. Springer, Berlin (2005)
199. Jeulin, D.: Multi scale random models of complex microstructures. In: T. Chandra, N. Wanderka, W. Reimers, M. Ionescu (eds.) Thermec 2009, vol. 638, pp. 81–86. Trans Tech Publications, Berlin (2009)
200. Jeulin, D.: Multi-scale random sets: from morphology to effective properties and to fracture statistics. In: Proc. CMDS12, Journal of Physics: Conf. Ser., vol. 319 (2011)
201. Jeulin, D.: Morphology and effective properties of multi-scale random sets: A review. Comptes rendus de l'Académie des Sciences **240**(4-5), 219–229 (2012)
202. Jeulin, D.: Random tessellations and Boolean random functions. In: C. Hendriks, G. Borgefors, R. Strand (eds.) Mathematical Morphology and Its Applications to Signal and Image Processing, *Lecture Notes in Computer Science*, vol. 7883, pp. 25–36. Springer, Berlin (2013)
203. Jeulin, D., Chardin, H., Grillon, F., Jeandin, M., N'Guyen, F.: Simulation of the shot peening by random processes. Microscopy Microanalysis Microstructures **16**, 299–306 (1997)
204. Jeulin, D., Jeulin, P.: Synthesis of rough surfaces by random morphological models. In: Proc. 3rd European Symposium of Stereology, vol. 3, pp. 239–246 (1981)
205. Jeulin, D., Kurdy, M., Daly, C.: Etude de l'influence de la rugosité sur l'aspect de surface de tôles peintes. Paris School of Mines publication, Paris (1990)
206. Jeulin, D., Laurenge, P.: Probabilistic model of rough surfaces obtained by electro-erosion. In: M.A. Butt, P. Maragos, R.W. Schafer (eds.) Proceedings of the ISMM 96 Conference, pp. 289–296 (1996)
207. Jeulin, D., Laurenge, P.: Simulation of rough surfaces by morphological random functions. Journal of Electronic Imaging **6**, 16–30 (1997)
208. Jeulin, D., Laurenge, P.: Morphological simulation of the roughness transfer on steel sheets. Microscopy Microanalysis Microstructures **7**, 541–547 (1996)
209. Jia, K., Li, X.: Water transport in polymer electrolyte membrane fuel cells. Progress in Energy and Combustion Science **37**, 221–291 (2011)
210. Joag-Dev, K., Proschan, F.: Negative association of random variables with applications. Annals of Statistics **11**(1), 286–295 (1983)
211. Johnson, K.L.: Contact Mechanics. Cambridge University Press, Cambridge (1985)
212. Jónsdóttir, K.Y., Schmiegel, J., Vedel Jensen, E.B.: Lévy-based growth models. Bernoulli **2**, 91–99 (2009)
213. Kadashevich, I., Schneider, H.J., Stoyan, D.: Statistical modeling of the geometrical structure of the system of artificial air pores in autoclaved aerated concrete. Cement and Concrete Research **35**, 1495–1502 (2005)
214. Kahle, M.: Random geometric complexes. Discrete & Computational Geometry **45**(3), 553–573 (2011)
215. Kamuntavichyus, D.: A criterion for the existence of finite-dimensional Chebyshev subspaces in the spaces l_φ. Lithuanian Mathematical Journal **30**(1), 27–35 (1990)
216. Kamuntavichyus, D.: The structure of a metric projection onto a one-dimensional space in l_φ. Lithuanian Mathematical Journal **29**(3), 237–244 (1990)
217. Kamuntavichyus, D.: The absence of nontrivial boundedly compact Chebyshev sets in spaces l_φ. Mathematics of the USSR-Sbornik **69**(2), 431–444 (1991)
218. Kamuntavichyus, D.: Category properties of a set of nonuniqueness points in the space l_φ. Mathematical Notes **48**(3-4), 925–930 (1991)
219. Karcher, W.: On infinitely divisible random fields with an application in insurance. Ph.D. thesis, Ulm University, Ulm (2012)
220. Karcher, W., Scheffler, H.P., Spodarev, E.: Simulation of infinitely divisible random fields. Communications in Statistics - Simulation and Computation **42**, 215–246 (2013)

221. Karcher, W., Shmileva, E., Spodarev, E.: Extrapolation of stable random fields. Journal of Multivariate Analysis **115**, 516–536 (2013)
222. Kelly, F.P., Ripley, B.D.: A note on Strauss's model for clustering. Biometrika **63**(2), 357–360 (1976)
223. Kendall, M., Ord, J.K.: Time Series, 3rd edn. Oxford University Press, Oxford (1990)
224. Kendall, W.S.: On some weighted Boolean models. In: D. Jeulin (ed.) Advances in Theory and Applications of Random Sets, pp. 105–120. World Scientific, Singapore (1997)
225. Kendall, W.S.: Perfect simulation for the area-interaction point process. In: L. Accardi, C.C. Heyde (eds.) Probability Towards 2000, pp. 218–234. Springer, New York (1998)
226. Kendall, W.S.: Geometric ergodicity and perfect simulation. Electronic Communications in Probability **9**(15), 140–151 (2004)
227. Kendall, W.S.: Notes on perfect simulation. In: W.S. Kendall, F. Liang, J.S. Wang (eds.) Markov Chain Monte Carlo: Innovations and Applications, pp. 93–146. World Scientific, Singapore (2005)
228. Kendall, W.S., Møller, J.: Perfect simulation using dominating processes on ordered state spaces, with application to locally stable point processes. Advances in Applied Probability **32**(3), 844–865 (2000)
229. Kendall, W.S., Montana, G.: Small sets and Markov transition densities. Stochastic Processes and their Applications **99**(2), 177–194 (2002)
230. Kiderlen, M.: Introduction into integral geometry and stereology. In: E. Spodarev (ed.) Stochastic Geometry, Spatial Statistics and Random Fields: Asymptotic Methods, *Lecture Notes in Mathematics*, vol. 2068, pp. 21–47. Springer, Berlin (2013)
231. Kiderlen, M., Jensen, E.B.V.: On uniqueness of rotational formulae. Tech. rep., Centre for Stochastic Geometry and Advanced Bioimaging, Department of Mathematics, Aarhus University, Denmark. In preparation (2013)
232. Kindermann, R., Snell, J.L.: Markov Random Fields and their Applications, *Contemporary Mathematics*, vol. 1. American Mathematical Society, Providence, R.I. (1980)
233. Kirkpatrick, S., Gelatt, C.D., Vecchi, M.P.: Optimization by simulated annealing. Science **220**, 671–680 (1983)
234. Kolmogorov, A.: Interpolation und Extrapolation von stationären zufälligen Folgen. Izvestiya Akademii Nauk SSSR **5**, 3–14 (1941)
235. Kosmol, P.: Optimierung und Approximation. Walter de Gruyter, Berlin (1991)
236. Krige, D.G.: A statistical approach to some basic mine valuation problems on the Witwatersrand. Journal of the Chemical, Metallurgical and Mining Society of South Africa **52**(6), 119–139 (1951)
237. Kroese, D.P., Taimre, T., Botev, Z.I.: Handbook of Monte Carlo Methods. John Wiley & Sons, New York (2011)
238. Kuhn, H.W.: The Hungarian method for the assignment problem. Naval Research Logistics Quarterly **2**, 83–97 (1955)
239. Laarhoven, P.J.M., Aarts, E.H.L.: Simulated Annealing: Theory and Applications. Kluwer Academic Publisher, Dortrecht (1987)
240. Laird, N.M., Lange, C.: The Fundamentals of Modern Statistical Genetics. Springer, New York (2011)
241. Lantuéjoul, C.: Geostatistical Simulation: Models and Algorithms. Springer, Berlin (2002)
242. Laslett, G.: Kriging and splines: An empirical comparison of their predictive. performance in some applications. Journal of the American Statistical Association **89**(426), 391–400 (1994)
243. Lautensack, C., Schladitz, K., Särkkä, A.: Modeling the microstructure of sintered copper. In: Proceedings of the 6th International Conference on Stereology, Spatial Statistics and Stochastic Geometry, pp. 1–6. Prague (2006)
244. Lautensack, C., Sych, T.: 3D image analysis of open foams using random tessellations. Image Analysis & Stereology **25**, 87–93 (2006)
245. Lautensack, C., Zuyev, S.: Random Laguerre tessellations. Advances in Applied Probability **40**(3), 630–650 (2008)

246. Lavancier, F., Møller, J., Rubak, E.: Statistical aspects of determinantal point processes Journal of the Royal Statistical Society, to appear, arXiv:1205.4818 (2014)
247. Lawler, G.F.: Introduction to Stochastic Processes. Chapman & Hall/CRC, Boca Raton (2006)
248. Leisch, F.: Sweave: Dynamic generation of statistical reports using literate data analysis. In: W. Hardle, B. Ronz (eds.) Compstat 2002 - Proceedings in Computational Statistics, pp. 575–580. Physika Verlag, Heidelberg (2002)
249. Li, B., Wang, S.: On directional regression for dimension reduction. Journal of the American Statistical Association **102**, 997–1008 (2007)
250. Li, K.: High dimensional data analysis via the SIR/PHD approach Preprint, UCLA (2000)
251. Li, K.C.: Sliced inverse regression for dimension reduction. Journal of the American Statistical Association **86**, 316–327 (1991)
252. Li, M., He, Z., Zhang, M., Zhang, X., Wei, C., Elston, R.C., Lu, Q.: A generalized genetic random field method for the genetic association analysis of sequencing data. Genetic Epidemiology **38**, 242–253 (2014)
253. Liebscher, A., Redenbach, C.: Statistical analysis of the local strut thickness in open cell foams. Image Analysis & Stereology **32**(1), 1–12 (2013)
254. Lifshits, M.: Lectures on Gaussian Processes. Springer, Berlin (2012)
255. Liggett, T.M.: Continuous Time Markov Processes, *Graduate Studies in Mathematics*, vol. 113. American Mathematical Society, Providence, RI (2010)
256. Lin, X.: Variance component testing in generalised linear models with random effects. Biometrika **84**(2), 309–326 (1997)
257. Lou, X.Y., Chen, G.B., Yan, L., Ma, J.Z., Zhu, J., Elston, R.C., Li, M.D.: A generalized combinatorial approach for detecting gene-by-gene and gene-by-environment interactions with application to nicotine dependence. American Journal of Human Genetics **80**, 1125–1137 (2007)
258. Lütkepohl, H.: New Introduction to Multiple Time Series Analysis. Springer, Berlin (2006)
259. Magnus, W., Oberhettinger, F., Soni, R.P.: Formulas and Theorems for the Special Functions of Mathematical Physics, *Die Grundlehren der mathematischen Wissenschaften in Einzeldarstellungen*, vol. 52. Springer, Berlin (1966)
260. Mandelbrot, B.B., van Ness, J.W.: Fractional Brownian motions, fractional noises and applications. SIAM Review **10**(4), 422–437 (1968)
261. Markovich, N.: Nonparametric Analysis of Univariate Heavy-Tailed data: Research and Practice. John Wiley & Sons, Chichester (2007)
262. Matérn, B.: Spatial variation. Meddelanden från Statens Skogsforskningsinstitut **49**, 1–144 (1960)
263. Matheron, G.: Traite de Geostatistique Appliquee, Tome I, *Memoires de Bureau de Recherche Geologiques et Minieres*, vol. 14. Editions Technip, Paris (1962)
264. Matheron, G.: Traite de Geostatistique Appliquee, Tome II: Le Krigeage, *Memoires de Bureau de Recherche Geologiques et Minieres*, vol. 24. Editions Bureau de Recherche Geologiques et Minieres, Paris (1963)
265. Matheron, G.: Eléments Pour une Théorie des Milieux Poreux. Masson, Paris (1967)
266. Matheron, G.: Théorie des Ensembles Aléatoires, Cahiers du Centre de Morphologie Mathématique. Paris School of Mines publication, Paris (1969)
267. Matheron, G.: Random Sets and Integral Geometry. John Wiley & Sons, New York (1975)
268. Mathias, M., Roth, J., Fleming, J., Lehnert, W.: Diffusion media materials and characterisation. In: W. Vielstich, A. Lamm, H. Gasteiger (eds.) Handbook of Fuel Cells, pp. 517–537. John Wiley & Sons, New York (2003)
269. Mecke, J.: Parametric representation of mean values for stationary random mosaics. Mathematische Operationsforschung und Statistik. Series Statistics **3**, 437–442 (1984)
270. Mecke, J., Nagel, W., Weiss, V.: A global construction of homogeneous random planar tessellations that are stable under iteration. Stochastics. An International Journal of Probability and Stochastic Processes **80**, 51–67 (2008)

271. Meester, R., Roy, R.: Continuum Percolation. Cambridge University Press, Cambridge (1996)
272. Meyn, S.P., Tweedie, R.L.: Markov Chains and Stochastic Stability. Springer, New York (1993)
273. Michalski, R.S.: A theory and methodology of inductive learning. Artificial Intelligence **20**, 111–161 (1983)
274. Miyoshi, N., Rolski, T.: Ross-type conjectures on monotonicity of queues. Australian & New Zealand Journal of Statistics **46**(1), 121–131 (2004)
275. Miyoshi, N., Shirai, T.: A cellular network model with Ginibre configurated base stations. Advances in Applied Probability **51**(3), to appear (2014)
276. Mø ller, J., Helisová, K.: Likelihood inference for unions of interacting discs. Scandinavian Journal of Statistics **37**, 365–381 (2010)
277. Molchanov, I.: Foundation of stochastic geometry and theory of random sets. In: E. Spodarev (ed.) Stochastic Geometry, Spatial Statistics and Random Fields: Asymptotic Methods., *Lecture Notes in Mathematics*, vol. 2068, pp. 1–20. Springer, Heidelberg (2013)
278. Møller, J.: Random tessellations in $\mathbb{R}^d$. Advances in Applied Probability **21**, 37–73 (1989)
279. Møller, J.: Random Johnson-Mehl tessellations. Advances in Applied Probability **24**, 814–844 (1992)
280. Møller, J.: Shot noise Cox processes. Advances in Applied Probability **35**, 614–640 (2003)
281. Møller, J., Berthelsen, K.K.: Transforming spatial point processes into Poisson processes using random superposition. Advances in Applied Probability **44**(1), 42–62 (2012)
282. Møller, J., Ghorbani, M.: Aspects of second-order analysis of structured inhomogeneous spatio-temporal point processes. Statistica Neerlandica **66**, 472–491 (2012)
283. Møller, J., Helisová, K.: Power diagrams and interaction process for unions of discs. Advances in Applied Probability **40**, 321–347 (2008)
284. Møller, J., Syversveen, A.R., Waagepetersen, R.P.: Log Gaussian Cox processes. Scandinavian Journal of Statistics **25**(3), 451–482 (1998)
285. Møller, J., Torrisi, G.L.: Generalised shot noise Cox processes. Advances in Applied Probability **37**(1), 48–74 (2005)
286. Møller, J., Waagepetersen, R.: Statistics and Simulations of Spatial Point Processes. World Scientific, Singapore (2004)
287. Møller, J., Waagepetersen, R.P.: Statistical Inference and Simulation for Spatial Point Processes. Chapman & Hall / CRC, Boca Raton (2004)
288. Moore, J.H., Gilbert, J.C., Tsai, C.T., Chiang, F.T., Holden, T., Barney, N., White, B.C.: A flexible computational framework for detecting, characterizing, and interpreting statistical patterns of epistasis in genetic studies of human disease susceptibility. Journal of Theoretical Biology **241**, 252–261 (2006)
289. Müller, A., Stoyan, D.: Comparison Methods for Stochastic Models and Risk. John Wiley & Sons, Chichester (2002)
290. Müller, H.G.: Nonparametric Regression Analysis of Longiutdinal Data, *Lecture Notes in Statistics*, vol. 46. Springer, New York (1988)
291. Müller, H.G., Stadtmüller, U.: Generalized functional linear models. Annals of Statistics **33**, 774–805 (2005)
292. Murdoch, D.J.: Exact sampling for Bayesian inference: unbounded state spaces. In: N. Madras (ed.) Monte Carlo Methods, *Fields Institute Communications*, vol. 26, pp. 111–121. American Mathematical Society, Providence, RI (2000)
293. Murdoch, D.J., Green, P.J.: Exact sampling from a continuous state space. Scandinavian Journal of Statistics **25**, 483–502 (1998)
294. Murdoch, D.J., Rosenthal, J.S.: Efficient use of exact samples. Statistics and Computing **10**, 237–243 (2000)
295. Nagarajan, R., Scutari, M., Lèbre, S.: Bayesian Networks in R with Applications in Systems Biology. Springer, New York (2013)

296. Nagel, W., Weiss, V.: Crack STIT tessellations: characterization of stationary random tessellations stable with respect to iteration. Advances in Applied Probability **37**, 859–883 (2005)
297. Nagel, W., Weiss, V.: STIT tessellations in the plane. Rendiconti del Circolo Matematico di Palermo, Serie II, Suppl. **77**, 441–458 (2006)
298. Nagel, W., Weiss, V.: Mean values for homogeneous STIT tessellations in 3D. Image Analysis & Stereology **27**, 29–37 (2008)
299. Neyman, J., Scott, E.L.: Statistical approach to problems of cosmology. Journal of the Royal Statistical Society. Series B **20**(1), 1–43 (1958)
300. Niu, A., Zhang, S., Sha, Q.: A novel method to detect gene-gene interactions in structured populations: Mdr-sp. Annals of Human Genetics **75**, 742–754 (2011)
301. Nolan, J.: Stable Distributions – Models for Heavy Tailed Data. Birkhauser, Boston (2013)
302. Nummelin, E.: A splitting technique for Harris-recurrent chains. Zeitschrift für Wahrscheinlichkeitstheorie und Verwandte Gebiete **43**(4), 309–318 (1978)
303. Oh, S., Lee, J., Kwon, M.S., Weir, B., Ha, K., Park, T.: A novel method to identify high order gene-gene interactions in genome-wide association studies: Gene-based mdr. BMC Bioinformatics **13**(Suppl 9), S5 (2012)
304. Ohser, J., Mücklich, F.: Statistical Analysis of Microstructures in Materials Science. John Wiley & Sons, Chichester (2000)
305. Ohser, J., Schladitz, K.: 3D Images of Materials Structures – Processing and Analysis. Wiley-VCH, Weinheim (2009)
306. Okabe, A., Boots, B., Sugihara, K., Chiu, S.N.: Spatial Tessellations — Concepts and Applications of Voronoi Diagrams, 2nd edn. John Wiley & Sons, Chichester (2000)
307. Oosterhout, S.D., Wienk, M.M., van Bavel, S.S., Thiedmann, R., Koster, L.J.A., Gilot, J., Loos, J., Schmidt, V., Janssen, R.A.J.: The effect of three-dimensional morphology on the efficiency of hybrid polymer solar cells. Nature Materials **8**, 818–824 (2009)
308. Painter, S.: Numerical method for conditional simulation of Lévy random fields. Journal of Mathematical Geology **30**(2), 163–179 (1998)
309. Papadimitriou, C.H., Steiglitz, K.: Combinatorial Optimization: Algorithms and Complexity. Prentice-Hall Inc., Englewood Cliffs, N.J. (1982)
310. Park, J.: Independent rule in classification of multivariate binary data. Journal of Multivariate Analysis **100**, 2270–2286 (2009)
311. Peltier, R., Lévy-Véhel, J.: Multifractional Brownian motion: definition and preliminary results. Tech. Rep. RR-2645, INRIA, Le Chesnay, France (1995)
312. Pemantle, R.: Towards a theory of negative dependence. Journal of Mathematical Physics **41**(3), 1371–1390 (2000)
313. Pemantle, R., Peres, Y.: Concentration of Lipschitz functionals of determinantal and other strong Rayleigh measures. Combinatorics, Probability and Computing **23**, 140–160 (2013)
314. Penrose, M.D.: Random Geometric Graphs. Oxford University Press, New York (2003)
315. Préteux, F.: Description et interprétation des images par la morphologie mathématique - application à l'imagerie médicale. Ph.D. thesis, Paris VI University, Paris, France. (1987)
316. Préteux, F., Schmitt, M.: Fonctions booléennes et modélisation du spongieux vertébral. 2ème Séminaire International de l'Image Electronique **2**, 476–481 (1986)
317. Préteux, F., Schmitt, M.: Analyse et Synthèse de Fonctions Booléennes : Théorèmes de Caractérisation et Démonstrations. Paris School of Mines publication, Paris (1987)
318. Prokešová, M.: Inhomogeneity in spatial Cox point processes – location dependent thinning is not the only option. Image Analysis & Stereology **29**, 133–141 (2010)
319. Prokešová, M., Dvořák, J.: Statistics for inhomogeneous space-time shot-noise Cox processes. Methodology and Computing in Applied Probability **16**, 433–449 (2014)
320. Propp, J.G., Wilson, D.B.: Exact sampling with coupled Markov chains and applications to statistical mechanics. Random Structures and Algorithms **9**, 223–252 (1996)
321. Propp, J.G., Wilson, D.B.: Coupling from the past: a user's guide. In: D. Aldous, J. Propp (eds.) Microsurveys in discrete probability, *DIMACS Series Discrete Mathematics & Theoretical Computer Science*, vol. 41, pp. 181–192. American Mathematical Society, Providence, RI (1998)

322. Qian, H., Raymond, G.M., Bassingthwaighte, J.B.: On two-dimensional fractional Brownian motion and fractional Brownian random field. Journal of Physics A: Mathematical and General **31**(28), L527–L535 (1998)

323. R Development Core Team: R: A Language and Environment for Statistical Computing. R Foundation for Statistical Computing, Vienna, Austria (2012). URL `http://www.R-project.org/`

324. Ramsay, J., Silverman, B.W.: Functional Data Analysis, 2nd edn. Springer Series in Statistics. Springer, New York (2005)

325. Randle, V., Engler, O.: Introduction to Texture Analysis: Macrotexture, Microtexture and Orientation Mapping. Gordon and Breach, Boca Raton (2000)

326. Rasmusson, A.: Contributions to computational stereology and parallel programming. Ph.D. thesis, Department of Computer Science, Aarhus University, Denmark (2012)

327. Redenbach, C.: Microstructure models for cellular materials. Computational Materials Science **44**(4), 1397–1407 (2009)

328. Redenbach, C., Thäle, C.: On the arrangement of cells in planar STIT and Poisson line tessellations. Methodology and Computing in Applied Probability **15**, 643–654 (2013)

329. Redenbach, C., Thäle, C.: Second-order comparison of three fundamental tessellation models. Statistics **47**, 237–257 (2013)

330. Reitzner, M., Schulte, M.: Central limit theorems for U-statistics of Poisson point processes. Annals of Probability **41**, 3879–3909 (2013)

331. Richardson, T.O.: Spatial and temporal organisation within ant societies. Ph.D. thesis, UWE Bristol, UK (2011)

332. Ritchie, M.D., Hahn, L.W., Roodi, N., Bailey, R.L., Dupont, W.D., Parl, F.F., Moore, J.H.: Multifactor-dimensionality reduction reveals high-order interactions among estrogen-metabolism genes in sporadic breast cancer. American Journal of Human Genetics **69**, 138–147 (2001)

333. Ritchie, M.D., Motsinger, A.A.: Multifactor dimensionality reduction for detecting gene-gene and gene-environment interactions in pharmacogenomics studies. Pharmacogenomics **6**, 823–834 (2005)

334. Robert, C.P., Casella, G.: Monte Carlo Statistical Methods, 2nd edn. Springer, New York (2004)

335. Ruczinski, I., Kooperberg, C., Leblanc, M.: Logic regression. Journal of Computational and Graphical Statistics **12**, 475–511 (2003)

336. Rue, H., Held, L.: Gaussian Markov Random Fields: Theory and Applications. Chapman & Hall, London (2005)

337. Samorodnitsky, G., Taqqu, M.S.: Stable Non-Gaussian Random Processes. Chapman & Hall, Boca Raton (1994)

338. Sato, K.I.: Lévy Processes and Infinitely Divisible Distributions. Cambridge University Press, Cambridge (1999)

339. Schladitz, K., Ohser, J., Nagel, W.: Measurement of intrinsic volumes of sets observed on lattices. In: A. Kuba, L.G. Nyul, K. Palagyi (eds.) 13th International Conference on Discrete Geometry for Computer Imagery, *Lecture Notes in Computer Science*, vol. 4245, pp. 247–258. Springer, Berlin (2006)

340. Schladitz, K., Peters, S., Reinel-Bitzer, D., Wiegmann, A., Ohser, J.: Design of acoustic trim based on geometric modeling and flow simulation for non-woven. Computational Materials Science **38**, 56–66 (2006)

341. Schladitz, K., Redenbach, C., Sych, T., Godehardt, M.: Model based estimation of geometric characteristics of open foams. Methodology and Computing in Applied Probability **14**, 1011–1032 (2012)

342. Schlather, M.: Some covariance models based on normal scale mixtures. Bernoulli **16**(3), 780–797 (2010)

343. Schlather, M., Ribeiro, P.J., Diggle, P.J.: On the second-order characteristics of marked point processes. detecting dependence between marks and locations of marked point processes. Journal of the Royal Statistical Society B **66**(1), 79–93 (2004)

344. Schmitt, M.: Chapter 18. In: J. Serra (ed.) Image Analysis and Mathematical Morphology, vol. 2, Theoretical Advances. Academic Press, New York (1988)
345. Schmitt, M.: A step toward the statistical inference of the Boolean model and the Boolean function. Acta Stereologica **8**, 623–628 (1989)
346. Schneider, R.: Convex bodies: the Brunn-Minkowski theory. Cambridge University Press, Cambridge (1993)
347. Schneider, R., Weil, W.: Stochastic and Integral Geometry. Springer, Berlin (2008)
348. Schoenberg, F.P., Brillinger, D.R., Guttorp, P.M.: Point processes, spatial-temporal. In: A. El-Shaarawi, W. Piegorsch (eds.) Encyclopedia of Environmetrics, vol. 3, pp. 1573–1577. John Wiley & Sons, New York (2002)
349. Schoenberg, I.J.: Metric spaces and complete monotone functions. Annals of Mathematics **39**(4), 811–841 (1938)
350. Schoenberg, I.J.: Metric spaces and positive definite functions. Transactions of the American Mathematical Society **44**(3), 522–536 (1938)
351. Schreiber, T., Thäle, C.: Second-order properties and central limit theory for the vertex process of iteration infinitely divisible and iteration stable random tessellations in the plane. Advances in Applied Probability **42**, 913–935 (2010)
352. Schreiber, T., Yukich, J.E.: Limit theorems for geometric functionals of Gibbs point process. Annales de l'Institut Henri Poincare **49**, 1158–1182 (2013)
353. Schröder-Turk, G.E., Kapfer, S.C., Breidenbach, B., Beisbart, C., Mecke, K.: Tensorial Minkowski functionals and anisotropy measures for planar patterns. Journal of Microscopy **238**, 57–74 (2011)
354. Schröder-Turk, G.E., Mickel, W., Kapfer, S.C., Klatt, M.A., Schaller, F.M., Hoffmann, M.J.F., Kleppmann, N., Armstrong, P., Inayat, A., Hug, D., Reichelsdorfer, M., Peukert, W., Schwieger, W., Mecke, K.: Minkowski tensor shape analysis of cellular, granular and porous structures. Advanced Materials **23**, 2535–2553 (2011)
355. Schuhmacher, D.: Distance estimates for dependent thinnings of point processes with densities. Electronic Journal of Probability **14**, no. 38, 1080–1116 (2009)
356. Schuhmacher, D., Stucki, K.: Gibbs point process approximation: total variation bounds using Stein's method. Annals of Probability **42**(5), 1911–1951 (2014).
357. Schuhmacher, D., Xia, A.: A new metric between distributions of point processes. Advances in Applied Probability **40**(3), 651–672 (2008)
358. Schulte, M., Thaele, C.: The scaling limit of Poisson-driven order statistics with applications in geometric probability. Stochastic Processes and their Applications **122**, 4096–4120 (2012)
359. Schwender, H., Ruczinski, I.: Logic regression and its extensions. Advances in Genetics **72**, 25–45 (2010)
360. Schwender, H., Ruczinski, I., Ickstadt, K.: Testing snps and sets of snps for importance in association studies. Biostatistics **12**, 18–32 (2011)
361. Serra, J.: Image Analysis and Mathematical Morphology, vol. 1. Academic Press, London (1982)
362. Serra, J.: Fonctions Aléatoires Booléennes. Paris School of Mines publication (1985)
363. Serra, J. (ed.): Image Analysis and Mathematical Morphology, vol. 2. Academic Press, London (1988)
364. Serra, J.: Boolean random functions. Journal of Microscopy **156**(1), 41–63 (1989)
365. Shaw, P.E., Rueseckas, A., Samuel, I.D.W.: Exciton diffusion measurements in poly(3-hexylthiophene). Advanced Materials **20**, 3516–3520 (2008)
366. Shevtsova, I.G.: Refinement of estimates for the rate of convergence in Lyapunov's theorem. Doklady Akademii Nauk **435**(1), 26–28 (2010)
367. Sibson, R.: A brief description of natural neighbor interpolation. In: V. Barnett (ed.) Interpreting Multivariate Data, pp. 21–36. John Wiley & Sons, Chichester (1981)
368. Sigman, K.: Exact simulation of the stationary distribution of the fifo m/g/c queue. Journal of Applied Probability **48A**, 209–213 (2011)
369. Silverman, B.W.: Density Estimation for Statistics and Data Analysis. Monographs on Statistics and Applied Probability. Chapman & Hall, London (1986)

370. Simonoff, J.S.: Smoothing Methods in Statistics. Springer, New York (1996)
371. Smith, A.M.: A Gibbs sampler on the n-simplex. Annals of Applied Probability **24**, 114–130 (2014)
372. Soille, P.: Morphological Image Analysis. Springer, Berlin, Heidelberg (1999)
373. Spodarev, E. (ed.): Stochastic Geometry, Spatial Statistics and Random Fields. Asymptotic Methods, *Lecture Notes in Mathematics*, vol. 2068. Springer, Heidelberg (2013)
374. Staněk, J., Beneš, V., Šedivý, O.: On random marked sets with a smaller integer dimension. Methodology and Computing in Applied Probability **16**, 397–410 (2014)
375. Stein, C.: A bound for the error in the normal approximation to the distribution of a sum of dependent random variables. In: Proceedings of the Sixth Berkeley Symposium on Mathematical Statistics and Probability (Univ. California, Berkeley, Calif., 1970/1971), pp. 583–602. University of California Press, Berkeley, California (1972)
376. Stein, C., Diaconis, P., Holmes, S., Reinert, G.: Use of exchangeable pairs in the analysis of simulations. In: P. Diaconis, S. Holmes (eds.) Stein's Method: Expository Lectures and Applications, *IMS Lecture Notes Monograph Series*, vol. 46, pp. 1–26. Institue of Mathematical Statistics, Beachwood, OH (2004)
377. Stein, M.L.: Fast and exact simulation of fractional Brownian motion. Journal of Computational and Graphical Statistics **11**(3), 587–599 (2002)
378. Stenzel, O., Hassfeld, H., Thiedmann, R., Koster, L.J.A., Oosterhout, S., van Bavel, S., Wienk, M.M., Loos, J., Janssen, R.A.J., Schmidt, V.: Spatial modeling of the 3D morphology of hybrid polymer-ZnO solar cells, based on electron tomography data. Annals of Applied Statistics **5**(3), 1920–1947 (2011)
379. Stenzel, O., Koster, L., Oosterhout, S., Janssen, R., Schmidt, V.: A new approach to model-based simulation of disordered polymer blend solar cells. Advanced Functional Materials **22**, 1236–1244 (2012)
380. Stenzel, O., Westhoff, D., Manke, I., Kasper, M., Kroese, D.P., Schmidt, V.: Graph-based simulated annealing: a hybrid approach to stochastic modeling of complex microstructures. Modelling and Simulation in Materials Science and Engineering **21**, 055,004 (2013)
381. Stoev, S., Taqqu, M.: Stochastic properties of the linear multifractional stable motion. Advances in Applied Probability **36**(4), 1085–1115 (2004)
382. Stoev, S., Taqqu, M.: Path properties of the linear multifractional stable motion. Fractals **13**(2), 157–178 (2005)
383. Stoyan, D.: Inequalities and bounds for variances of point processes and fibre processes. Statistics: A Journal of Theoretical and Applied Statistics **14**(3), 409–419 (1983)
384. Strauss, D.J.: Analysing binary lattice data with the nearest-neighbor property. Journal of Applied Probability **12**(4), 702–712 (1975)
385. Stucki, K., Schuhmacher, D.: Bounds for the probability generating functional of a Gibbs point process. Advances in Applied Probability **46**(1), 21–34 (2014)
386. Sugihara, K.: Three-dimensional convex hull as a fruitful source of diagrams. Theoretical Computer Science **235**, 325–337 (2000)
387. Swendsen, R.H., Wang, J.S.: Nonuniversal critical dynamics in Monte Carlo simulations. Physical Review Letters **58**(2), 86–88 (1987)
388. Taylor, R.L., Hu, T.C.: Strong laws of large numbers for arrays of row-wise independent random elements. International Journal of Mathematics and Mathematical Sciences **10**, 805–814 (1987)
389. Tek, F.B., Dempster, A.G., Kale, I.: Blood cell segmentation using minimum area watershed and circle Radon transformations. In: C. Ronse, L. Najman, E. Decencire (eds.) Mathematical Morphology: 40 Years On, *Computational Imaging and Vision*, vol. 30, pp. 441–454. Springer, Dordrecht (2005)
390. Tempelman, A.A.: On ergodicity of Gaussian homogeneous random fields on homogeneous spaces. Theory of Probability and Its Applications **18**(1), 173–175 (1973)
391. Thäle, C., Weiss, V.: New mean values for spatial homogeneous random tessellations stable under iteration. Image Analysis & Stereology **29**, 143–157 (2010)

392. Thiedmann, R., Hassfeld, H., Stenzel, O., Koster, L.J.A., Oosterhout, S.D., van Bavel, S.S., Wienk, M.M., Loos, J., Janssen, R.A.J., Schmidt, V.: A multiscale approach to the representation of 3D images, with applications to polymer solar cells. Image Analysis & Stereology **30**, 19–30 (2011)
393. Thomas, M.: A generalization of Poisson's binomial limit for use in ecology. Biometrika **36**, 18–25 (1949)
394. Thönnes, E.: A primer on perfect simulation. In: K. Mecke, D. Stoyan (eds.) Statistical Physics and Spatial Statistics, *Lecture Notes in Physics*, vol. 554, pp. 349–378. Springer, Berlin (2000)
395. Thórisdóttir, O., Kiderlen, M.: Wicksell's problem in local stereology. Advances in Applied Probability **45**, 925–944 (2013)
396. Tomizawa, N.: On some techniques useful for solution of transportation network problems. Networks **1**, 173–194 (1971)
397. Tuddenham, R.D., Snyder, M.M.: Physical growth of california boys and girls from birth to eighteen years. University of California Publications in Child Development **1**, 183–364 (1954)
398. Tyurin, I.S.: Sharpening the upper bounds for constants in Lyapunov's theorem. Uspekhi Matematicheskikh Nauk **65**(3(393)), 201–202 (2010)
399. Tzeng, J.Y., Zhang, D., Chang, S.M., Thomas, D.C., Davidian, M.: Gene-trait similarity regression for multimarker-based association analysis. Biometrics **65**(3), 822–832 (2009)
400. Ullrich, M.: Exact sampling for the Ising model at all temperatures. In: I. Dimov, K.K. Sabelfeld (eds.) Monte Carlo Methods and Applications: Proceedings of the 8th IMACS Seminar on Monte Carlo Methods, *De Gruyter Proceedings in Mathematics*, pp. 223–233. De Gruyter (2013)
401. Van Lieshout, M.N.M.: Markov Point Processes and their Applications. World Scientific, Singapure (2000)
402. Velez, D.R., White, B.C., Motsinger, A.A., Bush, W.S., Ritchie, M.D., Williams, S.M., Moore, J.H.: A balanced accuracy function for epistasis modeling in imbalanced datasets using multifactor dimensionality reduction. Genetic Epidemiology **31**, 306–315 (2007)
403. Visscher, P.M., Brown, M.A., McCarthy, M.I., Yang, J.: Five years of gwas discovery. American Journal of Human Genetics **90**, 7–24 (2012)
404. Voss, F., Gloaguen, C., Schmidt, V.: Random tessellations and cox processes. In: E. Spodarev (ed.) Stochastic Geometry, Spatial Statistics and Random Fields: Asymptotic Methods., *Lecture Notes in Mathematics*, vol. 2068, pp. 151–182. Springer, Heidelberg (2013)
405. Šedivý, O., Staněk, J., Kratochvílová, B., Beneš, V.: Sliced inverse regression and independence in random marked sets with covariates. Advances in Applied Probability **43**(3), 626–644 (2013)
406. VSG – Visualization Sciences Group – Avizo Standard: `http://www.uni-ulm.de/mawi/mawi-stochastik/software.html`
407. Waagepetersen, R.P., Guan, Y.: Two-step estimation for inhomogeneous spatial point processes. Journal of the Royal Statistical Society B **71**, 685–702 (2009)
408. Wackernagel, H.: Multivariate Geostatistics, 2nd edn. Springer, Berlin (2003)
409. Watkins, P., Walker, A., Verschoor, G.: Dynamical Monte Carlo modelling of organic solar cells: the dependence of internal quantum efficiency on morphology. Nano Letters **5**, 1814–1818 (2005)
410. Webster, R., Oliver, M.: Geostatistics for Environmental Scientists. John Wiley & Sons, Chichester (2007)
411. Wei, Z., Li, H.: A hidden spatial-temporal markov random field model for network-based analysis of time courses gene expression data. Annals of Applied Statistics **2**(1), 408–429
412. Wentzell, A.D.: A Course in the Theory of Stochastic Processes McGrow-Hill, New York (1981) (1981)
413. Whitt, W.: Uniform conditional variability ordering of probability distributions. Journal of Applied Probability **22**(3), 619–633 (1985)
414. Wiener, N.: Extrapolation, Interpolation and Smoothing of Stationary Time Series. John Wiley & Sons, New York (1949)

415. Wikipedia: Photon antibunching — wikipedia, the free encyclopedia (2012). URL `http://en.wikipedia.org/w/index.php?title=Photon_antibunching&oldid=514591605`

416. Wilson, D.B.: How to couple from the past using a read-once source of randomness. Random Structures and Algorithms **16**(1), 85–113 (2000)

417. Wilson, D.B.: Layered multishift coupling for use in perfect sampling algorithms (with a primer on CFTP). In: N. Madras (ed.) Monte Carlo Methods, *Fields Institute Communications*, vol. 26, pp. 143–179. American Mathematical Society, Providence RI (2000)

418. Wood, A., Chan, G.: Simulation of stationary Gaussian process in $[0,1]^d$. Journal of Computational and Graphical Statistics **3**, 409–432 (1994)

419. Wu, M.C., Kraft, P., Epstein, M.P., Taylor, D.M., Chanock, S.J., Hunter, D.J., Lin, X.: Powerful snp-set analysis for case-control genome-wide association studies. The American Journal of Human Genetics **86**(6), 929–942 (2010)

420. Wu, M.C., Lee, S., Cai, T., Li, Y., Boehnke, M., Lin, X.: Rare-variant association testing for sequencing data with the sequence kernel association test. The American Journal of Human Genetics **89**(1), 82–93 (2011)

421. Xia, A.: Stein's method and Poisson process approximation. In: A.D. Barbour, H.Y. Louis Chen (eds.) An Introduction to Stein's Method, *Lecture Notes Series, Institute for Mathematical Sciences, National University of Singapore*, vol. 4, pp. 115–181. Singapore University Press, Singapore (2005)

422. Yaglom, A.M.: Correlation Theory of Stationary and Related Random Functions, vol. 1. Springer, New York (1987)

423. Yaglom, A.M.: Correlation Theory of Stationary and Related Random Functions, vol. 2. Springer, New York (1987)

424. Yao, F., Müller, H.G., Wang, J.L.: Functional data analysis for sparse longitudinal data. Journal of the American Statistical Association **100**, 577–590 (2005)

425. Yogeshwaran, D.: Stochastic geometric networks : connectivity and comparison. Ph.D. thesis, Université Pierre et Marie Curie, Paris, France. (2010). URL `http://tel.archives-ouvertes.fr/tel-00541054`

426. Yogeshwaran, D., Adler, R.J.: On the topology of random complexes built over stationary point processes. arXiv:1211.0061 (2012)

427. Yukich, J.: Limit theorems in discrete stochastic geometry. In: E. Spodarev (ed.) Stochastic Geometry, Spatial Statistics and Random Fields : Asymptotic methods, *Lecture Notes in Mathematics*, vol. 2068, pp. 239–276. Springer, Heidelberg (2013)

428. Zähle, M.: Random processes of Hausdorff rectifiable closet sets. Mathematische Nachrichten **108**(1), 49–72 (1982)

429. Zikmundová, M., Staňková Helisová, K., Beneš, V.: Spatio-temporal model for a random set given by a union of interacting discs. Methodology and Computing in Applied Probability **14**, 883–894 (2012)

430. Zikmundová, M., Staňková Helisová, K., Beneš, V.: On the use of particle Markov chain Monte Carlo in stochastic geometry. Methodology and Computing in Applied Probability **16**, 451–463 (2014)

431. Zolotarev, V.M.: One-dimensional Stable Distributions. Translations of Mathematical Monographs, vol 65, American Mathematical Society, Providence RI (1986)

Index

affine transformation, 372
ambit set, 224
anamorphosis transformation, 152
assignment problem, 18
association, 48

band Cholesky method, 379
basis
    orthonormal, 268
    Schauder, 267
Berry–Esseen bound, 9
Bessel family, 328
best LSL predictor, 361
binomial process, 36
    association, 49
Blaschke-Petkantschin formula, 236
block-circulant matrix, 373
Boolean island, 148
Boolean random closed set, 148
Boolean random function, 147, 326
Brownian Lévy field, 324

Cauchy family, 329
Čech complex, 64
cell reconstruction, 87
central intensity subspace, 188
central subspace, 188
centroid, 77
CFTP, 406
    classic CFTP, 406, 407
    coupling-from-the-past algorithm, 405
    crossover trick, 437
    domCFTP, 430
    funnelling relationship, 435
    montonicity, 428
    random walk CFTP, 409
    read-once CFTP, 413

small-set CFTP, 417
stabilized CFTP, 433
truncation, 429
uniform ergodicity, 429
cluster process, 221, 383
    elliptical, 104
    Matérn, 386
    Poisson, 383
coalescence time, 409
codifference, 342
complete coalescence, 406
conditional intensity, 209
    function, 16
conditional mean mark, 175
conditionally negative semi–definite function, 327
contact distribution function, 216
coupling, 14, 25
covariance, 216
covariance function, 106, 258, 327, 371
covariation, 341
    function, 354
covariation orthogonal linear (COL) predictor, 354
Cox process, 38, 102, 385
    dcx super-Poisson, 54
    dcx ordering, 51
    generalised, 37
    log-Gaussian, 39
    Matérn cluster, 37
    shot-noise, 387
    Thomas (modified), 37
    v-weakly super-Poisson, 47
Crofton formula, 235
    rotational, 242
cyclone model, 330

dcx order, 49, 50

© Springer International Publishing Switzerland 2015

V. Schmidt (ed.), *Stochastic Geometry, Spatial Statistics and Random Fields,*
Lecture Notes in Mathematics 2120, DOI 10.1007/978-3-319-10064-7

of Cox processes, 51
of Lévy-driven Cox processes, 51
of perturbed processes, 50
of shot-noise fields, 51
sub-Poisson perturbed lattice, 54
sub-Poisson process, 53
super-Poisson perturbed lattice, 54
super-Poisson process, 53
detailed balance, 407, 431
determinantal point process, 194
determinantal process, 39
dcx, 54
Ginibre, 40
weakly sub-Poisson, 47
diffusion matrix, 393
dilation, 144
dimension reduction, 187
disector, 254
disorientation, 184

edge, 76
edge correction, 88
eigenequation, 273
eigenspace, 273
eigenvalue, 271
eigenvector, 271
elliptic integral, 237
embedding
intrinsic, 377
minimal, 375
nonnegative definite, 377
estimated prediction error, 299
Euler characteristic, 84
Euler method, 182
Euler-Poincaré characteristic, 234
expectation measure, 15
exponential model, 329

face-to-face tessellation, 76
facet, 76
force-biased algorithm, 89
forest, 309
Fourier transform, 373
discrete, 373
fast, 374
fractional Brownian field, 324
fractional Wiener sheet, 395
Frechet-Hoeffding bound, 177
full-dimensional random vector, 336
functional data analysis, 260

gamma process, 402
Gasser-Müller estimator, 262
Gaussian covariance family, 329

Gaussian linear random function, 324
Gaussian random field, 182
generalized linear model, 292
geometric anisotropy, 330
geometric ergodicity, 429
geometrical covariogram, 154
Georgii–Nguyen–Zessin equation, 17
germ-grain model, 52
coverage, 52, 69
k-percolation, 60
percolation, 56
nonstandard critical radii, 58
Gibbs process, 16, 28, 30
graph, 312
edge, 312
vertex, 312
graphical model, 378

h-minima transform, 88
Hadwiger's characterization theorem, 234
Hausdorff measure, 236
Hawkes process, 383
Hilbert space, 272
hole-effect model, 329
Horvitz-Thompson procedure, 88
Hurst index, 324
hypergeometric function, 237
hyperplane tessellation, 79
hypothesis of orthogonality, 198

immigration-death process, 13
spatial, 25
importance sampling, 207
independent marking, 382
independently scattered random measure, 339
index–continuous LSL predictor, 363
infinitesimal generator, 12
innovation, 407
integral of mean curvature, 84
intensity, 75
intensity function, 15, 219
intensity measure, 75, 100
intrinsic stationarity of order two, 322
intrinsic volume, 84, 235
density, 84
invariance in strict sense, 322
invariance in wide sense, 322
Ising model, 422
Bayesian image analysis, 422
coding scheme, 423
critical temperature, 422
external field, 422
ferromagnetic case, 422
Fortuin-Kastelyn-Potts model, 422

free boundary conditions, 424
  image analysis, 422
  single-site update heat bath algorithm, 422
  spin, 422
  Sweeney's algorithm, 422
isotropy, 75, 77, 322

Johnson-Mehl tessellation, 81

K-function (Ripley's), 41
kernel, 261
  biweight, 261
  Epanechnikov, 261
  estimator, 227
  function, 260, 343, 387
  uniform, 261
kriging, 346
Kullback–Leibler divergence, 309

Lévy basis, 38, 220
Lévy-driven Cox point process, 220
Lévy process, 325, 382
Lévy sheet, 403
Laguerre tessellation, 81
lattice process, 33
  Ripley's K-function of, 42
  simple perturbed, 36
  sub-Poisson perturbed, 36, 54
  super-Poisson perturbed, 37, 54
  Voronoi-perturbed, 36
least scale linear (LSL) predictor, 353
least-squares method, 270
Lebesgue measure, 236
level set, 358
linear multifractional stable motion, 344
linear predictor, 345
linear regression, 279
  with functional response, 288
link function, 307
local polynomial, 264
local stereological estimator, 251
location dependent thinning, 219
logic regression, 308
logistic regression, 308

mark correlation function, 175
mark covariance function, 175
mark variogram, 175
mark-weighted K-function, 175
marked fibre process, 186
Markov chain, 312
  aperiodic, 312
  irreducible, 312
Markov chain Monte Carlo, 206, 389, 406

Markov process, 12, 206
Matérn hard-core process, 102
maximization of covariation linear (MCL)
        predictor, 357
maximum-likelihood, 213, 292
mean-value function, 106, 258
Metropolis–Hastings algorithm, 389
Miles-Lantuejoul correction, 88
minimum contrast estimation, 226
minimum spanning tree, 24
Minkowski addition, 84
Minkowski tensor, 241
  rotational average, 243
mixed anisotropy, 332
model fit, 216
model of interacting discs, 212
modulated Matérn hard-core process, 104
moment measure, 43, 50
  factorial, 43
motion invariance, 322
moving average process, 372
multifactor dimensionality reduction (MDR),
        295
  robust MDR method (RMDR), 306
  Gene-MDR, 306
  generalized (GMDR), 305
  MDR method with independent rule, 304

n-way interaction information, 305
Nadaraya-Watson estimator, 262
nearest neighbour distance, 24
negative association, 48
neurophysiological experiment, 184
Neyman-Scott process, 37, 386
  association, 49
  dcx super-Poisson, 54
non-stationarity function, 223
normal scale mixture, 328
nucleator, 248
nugget effect, 332

ordinary kriging, 349
Ornstein-Uhlenbeck process, 13, 323
overgraph, 145

pair correlation function, 42, 221
pairwise interaction process, 16
Pareto model, 156
particle filter, 207
particle marginal Metropolis-Hastings
        algorithm, 208
path, 313
  of the height $h \geq 0$, 313
Pearson residual, 211

percolation, 56
  nonstandard critical radii, 58
  of germ-grain model, 56, *see also*
      germ-grain model
  of SINR model, 63
permanental process, 39
    simultaneously observable, 54
  weakly super-Poisson, 47
perturbation kernel, 36
  convex ordering, 36, 50
  displacement, 36
  replication, 36
    binomial, 36
    geometric, 37
    hyper-geometric, 36
    negative binomial, 37
    Poisson, 36
    Poisson mixture, 37
phenotype-associated information, 305
point process, 15, 74, 99, 382
  2D elliptical Matérn cluster process, 104
  associated, 48, *see also* association
  binomial, 36, *see also* binomial process
  clustering perturbation, 35
  concentration inequality, 44, 46, 47
  Cox, 38, *see also* Cox process
  dcx order, 49
  determinantal, 39, *see also* determinantal
      process
  Gibbs, 39
  Ginibre, 40
  isotropic, 100
  lattice, 33, *see also* lattice process
  marked, 105, 382
  Matérn hard-core process, 102
  modulated Matérn hard-core process,
      104
  negatively associated, 48
  Neyman-Scott, 37, *see also* Neyman-Scott
      process
  permanental, 39, *see also* permanental
      process
  Poisson, 34
  Poisson process, 100
  simple, 15, 99
  stationary, 100
  sub-Poisson, 32, 36
    α-weakly, 44
    concentration inequality, 47
    dcx, 53
    ν-weakly, 46
    weakly, 47, 48
  super-Poisson
    α-weakly, 44

    dcx, 53
    ν-weakly, 46
    weakly, 47, 48
Poisson cluster process
    association, 49
Poisson hyperplane tessellation, 79
Poisson point process, 147
Poisson process, 16, 28, 34, 75, 100
  as perturbed lattice, 36
  intensity, 100
  mixed, 38
  stationary, 100
Poisson random measure, 380
Poisson shot-noise field, 325
Poisson-Voronoi tessellation, 200
polar coordinates, 238
positive definite function, 326
positive semi–definite function, 326
posterior distribution, 206
precision matrix, 371
predictor
  continuous, 345
  exact, 345
  stochastically continuous, 352
  unbiased, 345
  weakly consistent, 352
preflooded watershed transform, 88
primary grain, 148
principal component analysis, 191, 270
probability distribution
  Bernoulli, 407
  Binomial, 407
  Geometric, 414, 431
  triangular, 417
  Uniform, 427
probability metric, 6, 10
  bounded Lipschitz metric, 6
  bounded Wasserstein metric, 6, 23
  Kantorovich metric, *see* Wasserstein metric
  Kolmogorov metric, 6
  total variation metric, 6, 22
  Wasserstein metric, 6, 23
projection process, 223
proposal density, 207
purely zonal anisotropy, 332

R statistical scripting language, 405
r–norm, 360
random closed set (RACS), 144
random counting measure, 99, 380
random field, 174, 322, 370
  Gaussian, 370
  Gaussian Markov, 378
  Markovian, 378

separable in probability, 341
random geometric graph, 64
random marked closed set, 173
random sequential adsorption, 89
random tessellation, 76, 158
random walk sampler, 389
regeneration, 417
regression model, 260
   nonparametric, 260
reversible jump sampler, 391
rotation invariant, 235
rotational average, 235
rotational integral
   geometry, 235
   of intrinsic volume, 234

second order intensity reweighted stationarity,
   219
semi-continuous (upper, lower) random
   function, 145
separability, 222
sequential Monte Carlo, 206
Shannon entropy, 304
shot-noise Cox process, 220
shot-noise field, 51
   dcx order of , 51
   extremal, 51
     ordering, 52
   level-sets, 60
shot-noise Gamma Cox process, 388
shrinkage property, 348
sill, 332
simple inhibition point process, 193
simple kriging, 346
   conditional unbiasedness, 348
   homoscedasticity, 349
simple symmetric random walk, 407
simplicial complex, 64
simulated annealing, 132, 312
singular random vector, 336
   Gaussian, 336
SINR model, 63
   percolation, 63
size map, 91
skeleton, 91
slice means, 191
sliced inverse regression, 188
small set, 417
   lag, 417
   minorization condition, 417
space-time K-function, 219
space-time germ-grain model, 214
space-time intensity function, 226
space-time point process, 218

spatial Lévy process, 403
spatial random network, 129
specific strut length, 85
spectral density, 327
spectral measure, 327
   of a stable vector, 336
spectral representation, 327
spectral theorem, 273
spherical contact distribution function, 101
spherical model, 330
spiking activity, 185
stability index, 335
stable family, 329
stable Lévy process, 343
stable moving-average random field, 343
stable Ornstein–Uhlenbeck process, 344
stable random field, 340
stable random measure, 339
   control measure, 339
   skewness function, 339
stable random variable, 337
   totally skewed, 338
stable random vector, 334
stable Riemann–Liouville process, 344
stable subordinator, 343
state-space model, 206
stationarity, 322
stationary distribution, 12
Stein equation, 5, 10
Stein's lemma, 4
Steiner formula, 84
stereology, 246
STIT tessellation, 79
stochastic flow, 406
stochastic geometry model with covariates,
   199
stochastic process, 105, 257, 371
   Gaussian, 371
   isotropic, 373
   stationary, 372
   trajectory, 105
   weakly stationary, 106
     process mean, 106
     process variance, 106
Strauss process, 389
strict convexity of $L^{\alpha}(E,m)$, 353
sub–Gaussian random field, 344
subgraph, 145
sufficient dimension reduction subspace, 188
sufficient intensity dimension reduction
   subspace, 188
support set, 358
surface area, 84, 234
Sweave, 406

symmetric distribution, 337
synchronous coupling, 408

tessellation, 75
thickness profile, 91
Thomas process, 386
total variation distance, 429
track modelling, 184
tree, 309
two-dimensional Brownian motion, 181
typical k-face, 78

U-statistic, 64
uniform ergodicity, 429

variance function, 106, 258
variogram, 324, 331

vectorial autoregressive (VAR) process, 106
    Gaussian, 107
vertex, 76
Vietoris-Rips complex, 65
void probability, 45, 50
volume, 84
Voronoi model, 158
Voronoi tessellation, 80

watershed transform, 88
Weibull model, 155
weighted random measure, 173
white noise, 328
Whittle-Matérn family, 329
Wiener bridge, 396
Wiener pillow, 396
Wiener process, 274

Additional technical instructions, if necessary, are available on request from lnm@springer.com.

4. Careful preparation of the manuscripts will help keep production time short besides ensuring satisfactory appearance of the finished book in print and online. After acceptance of the manuscript authors will be asked to prepare the final LaTeX source files and also the corresponding dvi-, pdf- or zipped ps-file. The LaTeX source files are essential for producing the full-text online version of the book (see http://www.springerlink.com/openurl.asp?genre=journal&issn=0075-8434 for the existing online volumes of LNM). The actual production of a Lecture Notes volume takes approximately 12 weeks.

5. Authors receive a total of 50 free copies of their volume, but no royalties. They are entitled to a discount of 33.3 % on the price of Springer books purchased for their personal use, if ordering directly from Springer.

6. Commitment to publish is made by letter of intent rather than by signing a formal contract. Springer-Verlag secures the copyright for each volume. Authors are free to reuse material contained in their LNM volumes in later publications: a brief written (or e-mail) request for formal permission is sufficient.

Addresses:
Professor J.-M. Morel, CMLA,
École Normale Supérieure de Cachan,
61 Avenue du Président Wilson, 94235 Cachan Cedex, France
E-mail: morel@cmla.ens-cachan.fr

Professor B. Teissier, Institut Mathématique de Jussieu,
UMR 7586 du CNRS, Équipe "Géométrie et Dynamique",
175 rue du Chevaleret
75013 Paris, France
E-mail: teissier@math.jussieu.fr

For the "Mathematical Biosciences Subseries" of LNM:

Professor P. K. Maini, Center for Mathematical Biology,
Mathematical Institute, 24-29 St Giles,
Oxford OX1 3LP, UK
E-mail: maini@maths.ox.ac.uk

Springer, Mathematics Editorial, Tiergartenstr. 17,
69121 Heidelberg, Germany,
Tel.: +49 (6221) 4876-8259

Fax: +49 (6221) 4876-8259
E-mail: lnm@springer.com

MIX
Papier aus verantwortungsvollen Quellen
Paper from responsible sources
FSC® C105338

If you have any concerns about our products,
you can contact us on
ProductSafety@springernature.com

In case Publisher is established outside the EU,
the EU authorized representative is:
Springer Nature Customer Service Center GmbH
Europaplatz 3, 69115 Heidelberg, Germany

Printed by Libri Plureos GmbH
in Hamburg, Germany